THE
VACCINE
BOOK

THE VACCINE BOOK

Edited by

Barry R. Bloom

Harvard School of Public Health
Boston, Massachusetts

Paul-Henri Lambert

Centre of Vaccinology
Department of Pathology
University of Geneva, Switzerland

ACADEMIC PRESS
An imprint of Elsevier Science

Amsterdam   Boston   London   New York   Oxford   Paris
San Diego   San Francisco   Singapore   Sydney   Tokyo

Cover image: Original artwork by Alagie Samba, photography by Paul-Henri Lambert.

This book is printed on acid-free paper. ⊗

Academic Press
An imprint of Elsevier Science.
525 B Street, Suite 1900, San Diego, California 92101-4495, USA
http://www.academicpress.com

Academic Press
84 Theobalds Road, London WClX 8RR, UK
http://www.academicpress.com

Library of Congress Catalog Card Number: 2002108521

International Standard Book Number: 0-12-107258-4

PRINTED IN THE UNITED STATES OF AMERICA
02   03   04   05   06   07   9   8   7   6   5   4   3   2   1

To Bill and Melinda Gates,
whose vision and generosity gave new meaning
to the concept of public private partnership and
new life to vaccines and immunization for all children.

■ CONTENTS

■1■ GLOBAL BURDEN OF DISEASE

A The Burden of Vaccine-Preventable Diseases
MAUREEN BIRMINGHAM AND CLAUDIA STEIN

B Cost Effectiveness of Immunization: Asking the Right Questions

RUTH E. LEVINE

C Potential and Existing Impact of Vaccines on Disease Epidemiology

ELIZABETH MILLER

2 IMMUNOLOGY

A Basic Immunology of Vaccine Development
ROBERT A. SEDER AND JOHN R. MASCOLA

B Immunological Requirements for Vaccines to Be Used in Early Life
CLAIRE-ANNE SIEGRIST

3   TRIAL DESIGN FOR VACCINES

A Clinical Development of New Vaccines: Phase 1 and 2 Trials
W. RIPLEY BALLOU

B Phase 3 Studies of Vaccines
JOHN D. CLEMENS AND HYE-WON KOO

4 ETHICS AND VACCINES

RUTH MACKLIN AND BRIAN GREENWOOD

5 UNDERSTANDING MICROBIAL PATHOGENESIS AS A BASIS FOR VACCINE DESIGN

A Bacteria

B. BRETT FINLAY

B Disease-Oriented Approach to the Discovery of Novel Vaccines
JEFFREY N. WEISER AND ELAINE I. TUOMANEN

C Immunological Memory and Vaccines against Acute Cytopathic and Noncytopathic Infections
ROLF M. ZINKERNAGEL

D Parasitic Diseases, with an Emphasis on Experimental Cutaneous Leishmaniasis
PASCAL LAUNOIS, HEIKE VOIGT, ALAIN GUMY, ABRAHAM ASEFFA, FABIENNE TACCHINI-COTTIER, MARTIN RÖCKEN, AND JACQUES A. LOUIS

6 Disease States and Vaccines: Selected Cases

A Introduction

STANLEY A. PLOTKIN

E Rotavirus

JOSEPH S. BRESEE, ROGER I. GLASS, UMESH PARASHAR AND JON GENTSCH

I Malaria

STEPHEN L. HOFFMAN AND THOMAS L. RICHIE

J Human Papillomaviruses

LUTZ GISSMANN

8 VACCINE SAFETY: REAL AND PERCEIVED ISSUES

NEAL A. HALSEY

9 INTRODUCTION OF NEW VACCINES IN THE HEALTHCARE SYSTEM

EDWARD KIM MULHOLLAND AND BJARNE BJORVATN

10 FUTURE CHALLENGES FOR VACCINES AND IMMUNIZATION
PAUL-HENRI LAMBERT AND BARRY R. BLOOM

CONTRIBUTORS

Numbers in parentheses indicate the pages on which the authors' contributions begin.

Abraham Aseffa (165) World Health Organization Immunology Research and Training Center, Institute of Biochemistry, University of Lausanne, 1066 Epalinges, Switzerland

W. Ripley Ballou (85) Immunology and Infectious Diseases, Clinical Development, Medimmune, Inc., Gaithersburg, Maryland 20878

Amie Batson (345) Health, Nutrition and Population Unit, Human Development Network, The World Bank, Washington, DC 20433

Jennifer M. Best (197) Department of Infection, Guy's, King's and St. Thomas' School of Medicine, Kings College London, St. Thomas' Campus, London SE1 7EH, United Kingdom

Maureen Birmingham (1) Vaccine Assessment and Monitoring, Department of Vaccines and Biologicals, World Health Organization, 1211 Geneva 27, Switzerland

Bjarne Bjorvatn (391) Centre for International Health and Department of Microbiology and Immunology, University of Bergen, 5020 Bergen, Norway

Barry R. Bloom (411) Harvard School of Public Health, Boston, Massachusetts 02115

Joseph S. Bresee (225) Viral Gastroenteritis Section, Respiratory and Enteric Virus Branch, National Center for Infectious Diseases, Centers for Disease Control and Prevention, Atlanta, Georgia 30333

Elisa I. Choi (245) Harvard Medical School, Beth Israel Deaconess Medical Center, Boston, Massachusetts 02215

John D. Clemens (95) International Vaccine Institute, Seoul National University Campus, Seoul, Korea 151-742

Giuseppe Del Giudice (323) IRIS Research Center, Chiron SpA, 53100 Siena, Italy

Ciro A. de Quadros (189) Pan American Health Organization, Washington, D.C. 20037

B. Brett Finlay (129) Departments of Microbiology and Immunology, Biochemistry, and Molecular Biology and Biotechnology Laboratory, University of British Columbia, Vancouver, British Columbia, Canada V6T 1Z3

Jon Gentsch (225) Viral Gastroenteritis Section, Respiratory and Enteric Virus Branch, National Center for Infectious Diseases, Centers for Disease Control and Prevention, Atlanta, Georgia 30333

Lutz Gissmann (311) Deutsches Krebsforschungszentrum, Im Neuenheimer Feld 242, 69120 Heidelberg, Germany

Roger I. Glass (225) Viral Gastroenteritis Section, Respiratory and Enteric Virus Branch, National Center for Infectious Diseases, Centers for Disease Control and Prevention, Atlanta, Georgia 30333

Sarah Glass (345) Health, Nutrition and Population Unit, Human Development Network, The World Bank, Washington, DC 20433

Brian Greenwood (119) London School of Hygiene and Tropical Medicine, London WCIE 7HT, United Kingdom

Alain Gumy (165) World Health Organization Immunology Research and Training Center, Institute of Biochemistry, University of Lausanne, 1066 Epalinges, Switzerland

Jill G. Hackell (257) Wyeth Vaccines Research, Pearl River, New York 10965

Neal A. Halsey (371) Institute for Vaccine Safety, Johns Hopkins University, Baltimore, Maryland 21205

Stephen L. Hoffman[*] (291) Celera Genomics, Rockville, Maryland 20850

Hye-Won Koo (95) International Vaccine Institute, Seoul National University Campus, Seoul, Korea 151-742

Paul-Henri Lambert (411) Centre of Vaccinology, Department of Pathology, University of Geneva, 1290 Geneva 4, Switzerland

Pascal Launois (165) World Health Organization Immunology Research and Training Center, Institute of Biochemistry, University of Lausanne, 1066 Epalinges, Switzerland

Norman L. Letvin (245) Harvard Medical School, Beth Israel Deaconess Medical Center, Boston, Massachusetts 02215

Ruth E. Levine (23) Center for Global Development, Washington, DC 20036

[*]Current address: Sanaria, Gaithersburg, Maryland 20878

Jacques A. Louis (165) World Health Organization Immunology Research and Training Center, Institute of Biochemistry, University of Lausanne, 1066 Epalinges, Switzerland

Ruth Macklin (119) Department of Epidemiology and Social Medicine, Albert Einstein College of Medicine, Bronx, New York 10461

John R. Mascola (51) Vaccine Research Center, National Institute of Allergy and Infectious Diseases, National Institutes of Health, Bethesda, Maryland 20892

Vega Masignani (333) IRIS Research Center, Chiron SpA, 53100 Siena, Italy

Elizabeth Miller (37) Immunization Division, Public Health Laboratory Service, Communicable Disease Surveillance Centre, London NW9 5EQ, United Kingdom

Edward Kim Mulholland (391) Department of Paediatrics, University of Melbourne, Royal Children's Hospital, Melbourne, Australia

Siobhan O'Shea (197) Department of Infection, Guy's and St. Thomas' Hospital Trust, London SE1 7EH, United Kingdom

Umesh Parashar (225) Viral Gastroenteritis Section, Respiratory and Enteric Virus Branch, National Center for Infectious Diseases, Centers for Disease Control and Prevention, Atlanta, Georgia 30333

Mariagrazia Pizza (333) IRIS Research Center, Chiron SpA, 53100 Siena, Italy

Stanley A. Plotkin (179) Aventis Pasteur, Doylestown, Pennsylvania 18901

Rino Rappuoli (323, 333) IRIS Research Center, Chiron SpA, 53100 Siena, Italy

Thomas L. Richie (291) Naval Medical Research Center, Silver Spring, Maryland 20910

Martin Röcken (165) Department of Dermatology and Allergy, Ludwig-Maximilians University, 80337 Munich, Germany

Stefania Salmaso (211) Reparto Malattie Infettive, Laboratorio di Epidemiologia e Biostatica, Instituto Superiore di Sanità, 00161 Rome, Italy

Robert A. Seder (51) Vaccine Research Center, National Institute of Allergy and Infectious Diseases, National Institutes of Health, Bethesda, Maryland 20892

Erica Seiguer* (345) Global Health Program, Bill & Melinda Gates Foundation, Seattle, Washington 98102

Claire-Anne Siegrist (73) WHO Collaborating Center for Neonatal Vaccinology, Departments of Pediatrics and Pathology, University of Geneva, 1211 Geneva 4, Switzerland

Claudia Stein (1) Epidemiology and Burden of Disease, Global Programme on Evidence for Health Policy, World Health Organization, 1211 Geneva 27, Switzerland

*Current address: Harvard Medical School, Boston, Massachusetts 02115

Fabienne Tacchini-Cottier (165) World Health Organization Immunology Research and Training Center, Institute of Biochemistry, University of Lausanne, 1066 Epalinges, Switzerland

Elaine I. Tuomanen (139) Department of Infectious Diseases and the Children's Infection Defense Center, St. Jude Children's Research Hospital, Memphis, Tennessee 38105

Heike Voigt (165) World Health Organization Immunology Research and Training Center, Institute of Biochemistry, University of Lausanne, 1066 Epalinges, Switzerland

Jeffrey N. Weiser (139) Departments of Microbiology and Pediatrics, University of Pennsylvania School of Medicine, Philadelphia, Pennsylvania 19104

Douglas B. Young (279) Centre for Molecular Microbiology and Infection, Imperial College of Science, Technology and Medicine, London SW7 2AZ, United Kingdom

Rolf M. Zinkernagel (149) Institute of Experimental Immunology, University Hospital, 8091 Zurich, Switzerland

■ PREFACE

Vaccines represent the most cost-effective medical intervention known to prevent death and disease. With the creation of the first vaccine by Edward Jenner in 1796, smallpox was the first human disease eradicated from the face of the earth by a global immunization campaign. Yet in 1974, only 5% of the world's children received the six childhood vaccines recommended by WHO. Since then, through extraordinary international public and private collaborations, the number of children receiving these basic vaccines has risen to over 80%, and in each of the ensuing years, over three million children's lives have been saved. Through creative basic and applied research, several new vaccines have been developed, and many more are in progress.

Over the past century the world has experienced a *demographic transition*, with people living longer, an increase in the aging population in most societies, and more people living in cities. That has been paralleled by an *epidemiological transition*, in which the diseases that have taken their toll on humankind have changed dramatically. Vaccines have contributed in a significant way to this epidemiological transition by reducing the number of children dying before the age of 6, leading to an increased life expectancy, a concomitant increase in chronic and degenerative diseases, and to the remarkable achievement that infectious diseases no longer represent the largest cause of death in the world. Yet there are major barriers to providing children and adults with vaccines in major parts of the developing world. And, as many infectious diseases have been reduced, there is the concern that the public and the scientific community may become complacent and forget the ever present risk of re-emerging infectious diseases. We hope this book will convey some of the power of vaccines and immunization to prevent disease and the challenges yet to be overcome.

The aims of this book are to share some of the knowledge acquired over the past quarter century and to create excitement about the future potential of vaccines to prevent infectious diseases with a wide audience: students, health professionals, and anyone interested in the field of vaccines. We have sought to engage readers who are nonexperts as well as scientists with a specific interest in immunization by presenting a very broad view of vaccines and immunization. We have received the generous support of many contributors who have summarized here the best current knowledge and experience in vaccines around the world. The book is purposely not designed to be comprehensive, but rather to be selective in presenting innovative approaches and problems that we think are interesting and challenging. For example, we consider some of the special barriers and initiatives to create vaccines against HIV/AIDS, malaria, and tuberculosis, which are urgently needed to fight the greatest remaining infectious killers. There are valuable lessons learned from polio and the global efforts to eliminate it as a public health problem. We consider immunization against *Helicobacter pylori* and papillomavirus, which not only reduce major infectious diseases, but represent vaccines to prevent cancer.

Beyond the challenges of these and other specific vaccines, we believe it is important to understand immunization in a broader context. There are chapters that evaluate the impact of current vaccines on world health and make projections of future impacts if new vaccines could be developed against some of the major killers of humankind. We have included a broad overview, beyond the challenges of the laboratory, of issues critical to the success of any vaccine. For example, chapters deal with the somewhat unique economics of the vaccine industry, critical issues of vaccine safety, and concerns about risks and the regulatory environment. Other chapters address how vaccine clinical trials are designed, the infrastructures required to introduce and deliver vaccines effectively, and the special ethical issues posed by vaccines. By striving to make its contents accessible, we hope this book will have its own impact in stimulating readers to contribute in various ways to realizing the potential of vaccines to save millions of lives in the future.

Barry R. Bloom
Paul-Henri Lambert

ACKNOWLEDGMENTS

I thank Robert Harington, Dolores Wright, and Kristi Anderson at Academic Press for their work in developing and producing this volume. Also, our sincere appreciation goes to all the authors who took time from their busy schedules to contribute to this book, which has made it such a unique volume for a diverse group of readers with an interest in global health and vaccines.

1 GLOBAL BURDEN OF DISEASE

PART A. The Burden of Vaccine-Preventable Diseases

MAUREEN BIRMINGHAM* AND CLAUDIA STEIN†

Vaccine Assessment and Monitoring, Department of Vaccines and Biologicals, World Health Organization, 1211 Geneva 27, Switzerland

†Epidemiology and Burden of Disease, Global Programme on Evidence for Health Policy, World Health Organization, 1211 Geneva 27, Switzerland*

I. INTRODUCTION. THE GLOBAL BURDEN OF INFECTIOUS AND PARASITIC DISEASES

Despite major progress in microbiological research, infectious and parasitic diseases continue to be a significant burden of disease worldwide. Traditionally, these diseases were described using health data from routine surveillance systems or epidemiological studies. Such data sources are often incomplete or unreliable, and studies that investigate particular conditions are often found to exaggerate claims on mortality. The latter is largely a reflection of comorbidity, in which several coexisting pathologies contribute to and compete for the cause of death.

The Global Burden of Disease (GBD) study of 1990 addressed these problems and quantified the burden of over 100 major disease groups, including infectious diseases, by using a novel and single metric, the disability-adjusted life year (DALY). DALYs express the years of life lost to premature death (YLL) and the years lived with disability (YLD) for varying degrees of severity, making time itself the common metric for death and disability [1]. One DALY is, therefore, a health gap measure, equaling 1 year of healthy life lost. Moreover, DALYs are internally consistent and disaggregate comorbidity, hence decoupling epidemiological measures from advocacy.

The GBD study estimated that, in 1990, over 31% of DALYs worldwide were due to communicable diseases (Table 1) [1, 2]. The vast majority of those occurred in developing countries, where 35% of DALYs were lost to infectious and parasitic diseases, whereas only a little over 4% of the burden in developed regions could be attributed to these disorders. The past 50 years have

TABLE 1   Global Burden of Disease in 1990 and 2000

| Cause | Percentage distribution of DALYs among specific causes[a] | | | | | |
| | 1990 | | | 2000 | | |
	World	Developed	Developing	World	Developed	Developing
All causes	100.0	100.0	100.0	100.0	100.0	100.0
All infectious and parasitic diseases[b]	31.4	4.3	35.0	29.7	4.0	31.9
Lower respiratory infections	8.2	1.5	9.1	6.4	1.6	6.8
Diarrheal diseases	7.2	0.3	8.1	4.2	0.2	4.6
Tuberculosis	2.8	0.3	3.1	2.4	0.1	2.6
Measles	2.7	0.0	3.0	1.9	0.0	2.0
Malaria	2.3	0.0	2.6	2.7	0.0	3.0
HIV	0.8	0.8	0.8	6.1	0.7	6.6

[a]DALYs, disability-adjusted life years.
[b]Includes respiratory infections.

seen a dramatic decline in the burden of communicable diseases in established market economies, largely as a result of increased income and education and technological advances in sanitation, antimicrobials, and vaccines. It was, therefore, predicted that the burden of infectious diseases may decrease in less developed regions as their incomes rise and access to technological progress improves.

Although progress has been made with respect to some infectious diseases, including respiratory infections, diarrheal diseases, and measles (Table 1), the overall picture has changed little a decade later. Infectious and parasitic diseases still account for nearly 30% of the global burden in the year 2000 (Table 1) and for nearly 32% of the burden in developing countries [2]. Among the top ten leading causes of DALYs worldwide in the year 2000, five were infectious diseases: lower respiratory infections (the number one cause), HIV–AIDS, diarrheal diseases, malaria, and tuberculosis. The situation is particularly bleak in Africa, where 60% of the burden is caused by infectious and parasitic diseases [2]. The surge of the HIV–AIDS epidemic, which is currently causing 6.1% of the total burden and 5.3% of all deaths worldwide, is the major factor responsible for this discouraging score [2]. It largely explains why 26% of all deaths worldwide are still caused by communicable diseases, compared to 27% in 1990.

Whereas the number and rates of death from some communicable diseases, including measles, tuberculosis, and diarrheal diseases, have fallen over the past 10 years, others such as malaria and HIV have increased, particularly in developing countries [2]. This has been accompanied by a worldwide increase in the proportion of DALYs due to noncommunicable diseases from 36% in 1990 to nearly 40% in 2000. Developing countries therefore are experiencing a "double burden" from both communicable and noncommunicable diseases [2]. Given the enormous scope of vaccine technology in combating infectious diseases, efforts to improve access to these resources must therefore remain a major priority in health policy. The analysis of the global burden of infectious disease over the past decade shows that public health interventions against communicable diseases have seen both triumphs and setbacks; it is the purpose of this article to take the reader through past achievements and future opportunities in vaccinology.

II. BRIEF HISTORICAL PERSPECTIVE OF VACCINOLOGY

Vaccination, or the practice of artificially inducing immunity, has been and continues to be one of the most important public health interventions in history. The practice of vaccination can be traced to as early as the seventh century when Indian Buddhists drank snake venom to induce immunity, possibly through a toxoid-like effect [3]. Variolation against smallpox probably began early in the second millenium in central Asia and then spread east to China and west to Turkey [4]. Lady Mary Wortley Montagu introduced variolation into England during 1721 after observing the practice in Constantinople. By using local knowledge regarding the protective effects of cowpox and experimenting further with it, in 1798 Edward Jenner introduced the notion of large-scale, systematic immunization against smallpox by person-to-person inoculation with cowpox virus [4].

The field of vaccinology experienced its next major advance in the 1800s with the concepts of attenuation and virulence. The concept of "passage" of the immunizing agent between animals or humans was recognized as important to maintain its "strength." In 1885, Louis Pasteur developed the first human vaccine against rabies [5]. In 1886, the concept of killed vaccine was developed when Salmon and Smith produced a killed hog cholera vaccine [6, 7]. This led to three new killed vaccines for humans: against cholera and typhoid in 1896 and against plague in 1897 [4]. Thus, by the end of the nineteenth century, two live attenuated viral vaccines (rabies and smallpox) and three killed bacterial vaccines (typhoid, plague, and cholera) were available for humans.

In the early twentieth century, toxoids were developed against diphtheria and tetanus [8–10], and by 1927, bacille Calmette–Guérin (BCG) vaccine was also available [11]. The isolation of the yellow fever virus in 1927 led to the French strain yellow fever vaccine, followed by the 17D strain, which had fewer side effects [4]. By 1936, two killed influenza vaccines were developed, followed by a live one in 1937. Vaccines against rickettsia ensued, in particular against Q-fever and typhus, the latter in heavy demand during World War II [11]. The first trials for pertussis vaccine, conducted in 1923 and 1924 on the Faroe Islands, led to the availability of several whole cell pertussis vaccines of varying efficacy by the 1940s [12, 13].

The advent of tissue culture in the 1940s sparked the next flurry of vaccine development, allowing for large-scale vaccine production [11]. This new technique made it possible for Enders, Robbins, and Weller to grow poliovirus in tissue culture, winning a Nobel prize for their achievements and leading to the licensure of the Salk and Sabin polio vaccines in 1955 and 1960, respectively [14]. During the 1960s, measles and mumps vaccines were licensed as well as the Japanese encephalitis vaccine, followed by the licensure of rubella and varicella vaccines in the 1970s [4]. The availability of many new vaccines created enormous potential to control several lethal diseases.

III. THE SMALLPOX ERADICATION ERA

Smallpox, one of the deadliest scourges known to humans, was prevalent throughout the world by the seventeenth century. The disease had a profound impact on history, wiping out large populations where it was introduced. It decimated Native American populations and contributed to the demise of the Inca and Aztec empires. In Europe alone, approximately 400,000 persons died annually from smallpox by the end of the eighteenth century [15]. The disease was endemic in populations large enough to sustain transmission, with periodic epidemics every 4–7 years [16]. In smaller or more isolated populations, the disease tended to "burn out," but would resurge when the virus was reintroduced among enough susceptibles to sustain transmission. By using the cowpox-derived vaccine, smallpox transmission was greatly reduced in Europe after World War I and virtually stopped in Europe and North America after World War II. However, the disease continued to ravage populations in the developing world due to problems with keeping the vaccine viable up to the point of use in difficult field settings. This problem was resolved with the development of a stable, freeze-dried vaccine.

With such a vaccine available, in 1950 the Pan American Sanitary Organization declared a goal of smallpox eradication from the western hemisphere and achieved it in 1967 (with the exception of Brazil). In 1958, the Soviet Union proposed to the World Health Assembly (WHA) that global smallpox eradication be pursued, and a WHA resolution was passed in 1959. However, little progress was made until 1967, when the WHA intensified the initiative with an infusion of dedicated resources. At this time, the burden of smallpox was estimated to be 10–15 million cases in 31 endemic countries (representing 1.2 billion of the world's population), although only 132,000 cases were reported [15].

The smallpox eradication initiative involved two strategies: (1) vaccination campaigns aimed to reach 80% of the population, and (2) surveillance and containment of cases and outbreaks. The huge operational obstacles that were overcome to achieve the eradication goal cannot be overstated—mostly related to management, supervision, reaching displaced or mobile populations, cultural beliefs, vaccine shortages, and insufficient funds. By 1977, smallpox eradication was achieved, and on May 8, 1980, the world was certified as smallpox-free, representing one of the most significant public health achievements in history [16].

IV. ESTABLISHMENT OF THE EXPANDED PROGRAM ON IMMUNIZATION (EPI)

Although high-quality vaccines were available by the mid-1950s against major diseases causing child mortality, less than 5% of world's children were being protected by these vaccines 20 years later, yet their potential impact was enormous [17]. Prior to immunization, most of the population experienced pertussis and measles by 15 years of age [18, 19]. WHO estimated that 106 million measles cases occurred in the preimmunization era, 5.7 million of which died annually [20]. Polio, although not yet fully recognized as an endemic disease in developing countries, was already occurring as seasonal epidemics in the industrialized world with increasing severity and age, the latter associated with higher mortality. Neonatal tetanus (NT) remained a "silent killer" until community-based mortality surveys later exposed its enormous toll.

By 1973, smallpox was restricted to only five countries in Asia and Africa [16]. Given the impending success of the eradication initiative, in 1974 WHO launched the Expanded Programme on Immunization (EPI) to exploit the momentum and infrastructure established by the smallpox initiative [21]. The principal aim of EPI was maximum disease control with selected antigens through high levels of immunization coverage in the greatest number of countries [21].

From 1974 to 1976, during which smallpox eradication remained the highest priority, the basic principles of EPI were established. One such principle was that immunization services be delivered primarily through routine health services rather than campaigns [21]. The antigens originally included in EPI for infant immunization were bacille Calmette–Guérin (BCG), diphtheria–tetanus–pertussis (DTP), polio using the oral vaccine (OPV), and measles. Also included was tetanus toxoid (TT) for pregnant women to prevent maternal and neonatal tetanus [22]. The choice of antigens was based on their potential public health impact, availability, and cost effectiveness [23].

In 1977, the year of the last naturally occurring smallpox case, a WHA resolution was passed declaring the goal of EPI to immunize at least 80% of the world's children with six antigens by 1990 [24]. Enormous operational challenges were confronted in pursuing this goal. Logistics systems were established to distribute supplies and maintain vaccines in a "cold chain" to preserve their efficacy from the point of manufacture to the point of use. Training with standard materials on immunization management at the senior, middle, and service-delivery levels was provided. Systems for monitoring coverage and disease incidence were established to track progress and impact [21]. Methods for evaluating national immunization programs at all levels were developed, as were simple survey methods to corroborate reported immunization coverage levels [25]. By 1994, nearly 300 program evaluations were conducted, usually in combination with an immunization coverage survey [25].

Although the number of reported cases of EPI target diseases underestimated the true number occurring, the immunization coverage and disease trend data were powerful statements of progress and impact (Figs. 1–3). Despite the impressive momentum of the early EPI era, the introduction of polio and tetanus immunization was delayed in some countries until community-based surveys could demonstrate a significant disease burden [26–29b]. In the late 1970s and early 1980s, more than 40 developing countries, including 18 of the 25 most populous countries, undertook surveys to assess the burden of neonatal tetanus, and more than 100 countries conducted polio lameness surveys. Neonatal tetanus (NT) mortality rates ranged from 1–2 per 1000 live births in Jordan, Sri Lanka, and Tunisia to 67 per 1000 live births in some

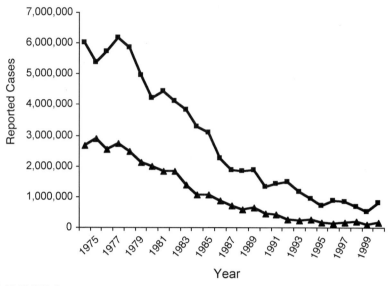

FIGURE 1   Measles (■) and pertussis (▲) cases reported to WHO, 1974–2000.

areas of India and accounted for as much as 50% of all neonatal deaths in the developing world [29a]. On the basis of these studies, in 1987 it was estimated that 800,000 NT deaths occurred globally [29a]. The prevalence of lameness associated with polio ranged from <1 to as high as 25 per 1000 children surveyed [26]. The surveys were critical in prompting many countries to provide routine polio and tetanus immunization services [26, 29a].

In 1984, a high-level meeting was hosted by the Rockefeller Foundation and cosponsored by the United Nations Development Programme (UNDP), United Nations Children's Fund (UNICEF), WHO, and the World Bank to plan how EPI could be accelerated to achieve the 1990 goal of 80% coverage [21].

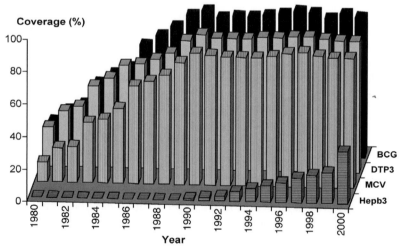

FIGURE 2   Global immunization coverage of selected antigens based on official reports from countries to WHO, 1974–2000.

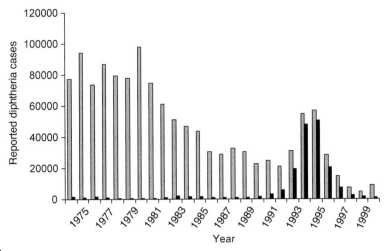

FIGURE 3 Diphtheria cases reported to WHO globally (shaded bars) and from the European region (solid bars), 1974–2000.

One result of this meeting was the establishment of the Task Force on Child Survival to coordinate support for EPI. A substantial increase in immunization activities ensued, culminating in a declaration that the goal of universal child immunization (UCI) had been achieved [30].

By the end of the 1980s, the number of reported cases of disease targeted by EPI had plummeted compared to the preimmunization era (Figs. 1 and 3). The estimated number of cases and deaths prevented by immunization further emphasized the enormous public health impact and cost effectiveness of immunization. For example, by 1988, it was already estimated that 751,000, 317,000, and 194,000 deaths due to measles, pertussis, and tetanus, respectively, were prevented in developing countries [17].

V. ERA OF ACCELERATED DISEASE CONTROL

A. Polio Eradication

In 1985, the Pan American Health Organization (PAHO) declared a goal to eliminate polio from the western hemisphere by 1990 and proposed three strategies: (1) high routine immunization coverage, (2) mass campaigns using oral polio vaccine (OPV), and (3) surveillance to detect and "mop up" any remaining chains of transmission. The initiative and progress toward polio elimination in the western hemisphere stimulated the adoption of a WHA resolution in 1988 of global polio eradication by the year 2000 [31]. During this year, 35,000 cases were notified and 350,000 cases were estimated to be occurring. By the year 2000, 20 endemic countries remained and polio incidence was reduced to an estimated 4000 cases, representing a 99% decrease since 1988 (Fig. 4) [32]. The burden of polio as measured by DALYs fell from 3.4 million in 1990 to 184,000 in 2001 [33, 2]. A massive acceleration of activities is underway to interrupt the remaining chains of transmission [34].

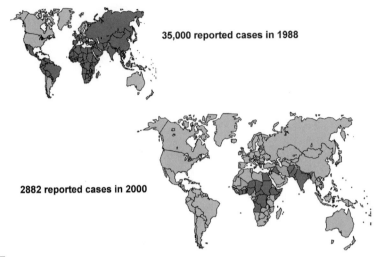

35,000 reported cases in 1988

2882 reported cases in 2000

FIGURE 4   Countries reporting confirmed polio cases, 1988 and 2000.

The initiative has demonstrated how to immunize even the most difficult-to-reach populations, including those affected by conflict [35, 36]. The polio initiative has also established an infrastructure of surveillance, laboratory networks, and trained field staff with improved managerial skills. Much has been learned during the polio initiative about how to design eradication programs that also maximize health system development [37].

In 1989, two additional WHA resolutions were adopted of NT elimination and measles morbidity and mortality reduction by 90% and 95%, respectively, by 1995 [38]. The three goals were also adopted a year later by the World Summit for Children (WSC), where 71 heads of state and other leaders signed the World Declaration on Survival, Protection, and Development of Children and adopted a plan of action to achieve 27 measurable, time-bound goals [39].

B.   Measles Control and Elimination

At the time these goals were set in 1990, an estimated 44,000,000 measles cases and 1,000,000 measles deaths were occurring annually, generating approximately 37 million DALYs [1, 33]. Measles was considered the leading cause of childhood mortality preventable by routine immunization. Approximately 33% and 58% of all measles deaths globally were estimated to occur in Asia and Africa, respectively. Mortality from measles in developing countries is affected by many factors, including malnutrition (especially vitamin A deficiency), age of infection, and intensity of exposure, which increases from primary to secondary cases [40–43].

Measles is also associated with serious complications and can follow a disastrous course in developing countries (particularly in refugee settings or among internally displaced persons), leading to pneumonia, diarrhea (which can exacerbate an already malnourished condition), keratitis, corneal ulcerations, which along with vitamin A deficiency can result in blindness and mediastinal and subcutaneous emphysema, and gangrene of the extremities [18, 44].

In developing countries, measles case fatality ratios (CFRs) generally range from 5 to 15% [18], but higher CFRs in crowded conditions or in outbreak settings particularly among isolated populations have been documented [45–47].

With increasing immunization coverage, measles transmission declines, which is often referred to as the "honeymoon period." This period is followed by periodic outbreaks due to the accumulation of susceptibles in older age groups usually with lower CFRs [48–50]. The duration between outbreaks is positively correlated with increasing immunization coverage levels [51].

Measles coverage in the world has increased dramatically since the launch of EPI in many developing countries, resulting in an 83% decline in reported measles cases from 4,200,000 in 1980 to 700,000 in 1995 [52]. The estimated reductions in morbidity and mortality between the preimmunization era and 1995 were 78% and 88%, respectively, also showing dramatic progress [20]. The number of DALYs from measles fell from 37 million in 1990 to 28 million in 2001 [1, 2]. However, on the basis of these estimates, the WHA and WSC goals for morbidity and mortality reduction were achieved in only 39% and 67% of countries, respectively, by 1995, representing only 33% and 48% of the global population, respectively [20].

In the Americas, following regional polio-free certification in 1994, an ambitious goal of measles elimination was launched [53]. The strategies in the Americas consist of (1) a "catch-up" measles immunization campaign to mop up all susceptibles among all children 1–14 years of age; (2) routine "keep-up" immunization among 1-year-olds; and (3) national "follow-up" campaigns every 4 years among all children 1–4 years of age. The results in the Americas have been dramatic (Fig. 5) and have stimulated much discussion about the feasiblity of global measles eradication in the future [54, 55]. Moreover, many countries are using a combined measles–rubella vaccine in their campaigns and demonstrating the additional impact on rubella and congenital rubella syndrome; an estimated 100,000 cases of the latter occur annually in developing countries alone [56–58]. Despite the progress, an estimated 780,000 measles deaths were still occurring in the world in 2000, the large majority in Asia and sub-Saharan Africa [2].

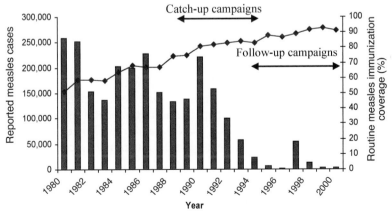

FIGURE 5   Measles immunization coverage (symbols) and reported measles cases (bars) in the Americas, 1980–2000.

A few small studies claimed a potential link between measles immunization and chronic inflammatory bowel disease and autism [59, 60]; however, several other larger scale studies have concluded that the available evidence does not support such a hypothesis [61–66]. Nevertheless, the allegations reduced measles vaccine coverage, particularly in the United Kingdom [67].

In the year 2001, a global strategic plan for measles was launched by WHO, UNICEF, and other key partners to halve measles mortality by the year 2005 [68]. The key strategies include providing the first dose of measles vaccine to successive cohorts of infants, ensuring that all children have a second opportunity for measles immunization, enhancing measles surveillance, and improving measles case management. Promising research is ongoing to develop a mucosal measles vaccine with high efficacy that could be administered during early infancy. Such a vaccine would greatly enhance the operational feasibility of global elimination or eradication.

C. Tetanus

By 1990, NT remained the second leading cause of childhood mortality that was vaccine-preventable, with an estimated 580,000 and 408,000 cases and deaths, respectively [69]. The large majority of all non-neonatal tetanus cases and deaths occurred in children under 15 years of age in Asia and Africa [33]. An estimated 30,000 maternal tetanus deaths occurred postpartum or postabortions, representing 5% of all maternal mortality [70].

The tetanus burden in developed countries by 1990 was quite different, accounting for less than 1000 cases annually and affecting primarily the elderly due to a lack of vaccination or waning immunity. Serosurveys from developed countries since 1977 indicate that 49–66% of elderly persons lack protective immunity against tetanus [71–74]. The low overall incidence with a high proportion of cases among older age groups in developed countries mirrors the population immunity achieved through routine vaccination with tetanus toxoid, and the potential to achieve the same impact in developing countries through better routine immunization services.

NT elimination has been defined as a rate of <1 case/1000 live births in every district of every country [69]. By the end of 1995, the deadline set by the WHA and WSC, 97 developing countries had achieved NT elimination, but a remaining 63 had not and an estimated 450,000 NT deaths were still occurring annually [75]. By the end of 2000, an estimated 200,000 deaths were occurring, representing a 51% reduction in NT deaths since 1990; however, 57 developing countries still had yet to achieve NT elimination. A new 5-year strategic plan for maternal and neonatal tetanus elimination was launched jointly by UNICEF, WHO, and the United Nations Population Fund (UNFPA) to achieve the goal in the remaining 57 countries by 2005 [69]. The principal strategies of the plan are (1) provision of tetanus toxoid (TT) immunization through routine and supplemental immunization, particularly in high-risk populations; (2) strengthening of clean delivery (CD) services; and (3) effective surveillance to detect areas and populations at high risk for NT.

D. Diphtheria

In the preimmunization era, diphtheria, caused by *Corynebacterium diphtheriae*, was a deadly disease, often occurring as periodic epidemics. The year 1613 is referred to as the "year of diphtheria" in Spanish history, and epidemics were known to occur about every 10 years [76]. There are reports of a major diphtheria epidemic in the 1700s in the New England colonies of the Americas, during which 2.5% of the population died and in Kingston, NH, almost every family lost at least one child [77]. In the nineteenth century, a pandemic occurred in both the United States and Europe. Reports during the 1880s in New York City indicated that at least 1% of children under 10 years of age died from diphtheria, and in Boston the diphtheria CFR during the same epidemic ranged from 48 to 52% [76, 78, 79]. The average mortality rate in Europe in the 1880s was about 40 per 100,000 but ranged as high as 100 per 100,000 [80].

In the late 1800s, Emil von Behring produced the first antitoxin against diphtheria and won a Nobel prize for the achievement in 1901 [76]. CFRs declined to about 15% during World War I probably due in large part to therapeutic uses of antitoxin and better supportive care, although incidence rates remained unaffected. In the 1930s, active immunization with diphtheria toxoid began in some countries, and dramatic results were documented [81]. Diphtheria ravaged Europe during World War II with rates exceeding 100 per 100,000 [82]. In 1943 alone, an estimated 1 million cases and 50,000 deaths occurred in Europe, along with exportation of the disease to North America [83–86].

Since the 1950s, more widespread use of diphtheria toxoid resulted in a steady decline in the incidence of the disease, with an average CFR of about 10% in developed countries. Children, who prior to the immunization era were clearly the most vulnerable age group, acquired high levels of immunity resulting in a decreased incidence. However, even before the introduction of routine immunization, diphtheria began shifting to older age groups, probably related to improving socioeconomic conditions that resulted in less exposure to the organism and waning immunity due to a lack of natural boostering [87].

Coverage with a third dose of diphtheria, tetanus, and pertussis vaccine (DTP3) has risen steadily from 24% globally in 1980 to 83% in 1990, when 106,000 cases and 11,000 deaths were estimated to occur [33, 52]. The level of global DTP3 coverage has remained generally stable up to the present and has resulted in a 91% reduction in the number of reported cases from 98,000 in 1980 to 9000 in 2000 [52]. Immunity in many developing countries is still being maintained through the natural boostering effect of skin diphtheria; however, sustained high DTP coverage and socioeconomic changes such as urbanization and better hygiene are changing the epidemiology of diphtheria in parts of the developing world. Outbreaks have occurred among young and older age groups of the serious laryngeal forms that are associated with high CFRs and complication rates [87, 88].

Despite the dramatic progress toward diphtheria control in the world, in the early 1990s diphtheria made a formidable comeback in several countries of the former Soviet Union, resulting in over 157,000 reported cases between 1991 and 1998 (Fig. 3) [52, 82]. The resurgence was due to a decline in immunization coverage associated with the collapse of the public health infrastructure in countries of the former Soviet Union, waning immunity in adults, and socioeconomic hardship resulting in major population movements. A major lesson learned from the epidemic is that high population immunity must be maintained not only among children but also among adolescents and adults.

E. Pertussis

Pertussis, or whooping cough, is an exhausting and sometimes frightening illness that can last for several months, often becoming a "family affair" due to its high degree of communicability. In the preimmunization era, it is likely that 50% of children had pertussis before school age, and 80–100% had it by 20 years of age [12, 19]. The majority of pertussis cases are due to *Bordetella pertussis*, although *Bordetella parapertussis* causes a less severe pertussis-like syndrome in adults. Neonates receive little passive protection from maternal antibodies, and CFRs are highest during infancy. CFRs are usually less than 0.5% in developed countries and range from 0.3 to 15% in developing countries [89]. The long-term complications of pertussis are mainly neurologic, nutritional, and respiratory; the latter are more frequent, particularly among young infants. Bronchopneumonia has been reported in 29% of pertussis cases and is responsible for about half of all pertussis deaths [89].

In the nineteenth century, pertussis caused considerable child mortality and was a public health concern comparable to diphtheria, measles, and scarlet fever [90]. With systematic vaccination, the number of pertussis cases reported to WHO has declined by a dramatic 92% from 2 million cases in 1980 to 150,000 in 2000 [52]. High immunization coverage over several years has changed the epidemiology of the disease in some countries by shifting it into older age groups in whom the disease is milder and the CFR lower.

Although the declining trend of pertussis cases reported globally suggests remarkable impact due to vaccination, WHO estimates that the overall number of cases reported globally represents less than 1% of the real number occurring. Even in developed countries, surveillance could be improved. Only 5–32% of all pertussis cases have been reported in the United States, England, and Wales [91–93]. In Canada, a 9-fold increase in surveillance sensitivity was achieved when an active and enhanced surveillance system was introduced [94].

Problems with diagnosis of pertussis also contribute to underreporting. Classic pertussis in infants and young children is easily diagnosed on clinical grounds due to the characteristic whoop. However, atypical clinical presentations in young infants (particularly among those partially vaccinated) and mild forms of the disease in older children and adults (often due to waning immunity) go undiagnosed and unreported [95]. Laboratory diagnosis is also nonstandardized or nonexistent in many countries, contributing to problems of diagnosis and underreporting [96]. Thus, in many countries, surveillance including laboratory diagnosis must be strengthened to better appreciate the changing epidemiology and burden of pertussis.

Some industrialized countries have achieved good pertussis control but then experienced problems in maintaining it due to physician apathy, anti-immunization lobbies, and litigation regarding alleged vaccine-related injuries.

The result has been a resurgence of the disease and the development of an acellular pertussis vaccine that is associated with fewer side effects [97, 98].

The burden of pertussis remains highly underrated, particularly in the developing world, in terms of its mortality as well as the direct and indirect damage it causes to the health, growth, and resistance of young children [90]. On the basis of the best available data and models, pertussis continues to cause approximately 50 million cases and about 300,000 deaths annually with the large majority of these deaths in developing countries [99]. Modeling of the pertussis disease burden has demonstrated the importance of not only *high* vaccination coverage but *timely* vaccination to reduce severe pertussis among young infants [99].

VI. NEW ERA OF CHALLENGES AND OPPORTUNITIES IN VACCINOLOGY

The 1990s experienced new opportunities and challenges in the field of vaccinology. One challenge was adapting to the health sector reform process occurring in many developing countries. This process often involves decentralization of the management and financing of health services (including immunizations) and greater integration of immunization activities with the health system. Other challenges include sustaining the current coverage achieved by national immunization services despite the many changes occurring in health systems, replacing aging cold chain equipment, strengthening logistics systems and managerial capacity, dealing with competing health priorities, and working amid civil turmoil. To address equity issues by going beyond the current reach of the routine immunization system to immunize chronically underserved populations, new innovative strategies such as the Sustained Outreach Services have been required [100].

Immunization safety issues have become paramount, particularly injection safety assurance to avoid the transmission of blood-borne pathogens. The maintenance of public confidence in immunization has also become challenging. Vaccine-preventable diseases (VPDs) have become rare enough in many countries that any vaccine-associated side effects may be viewed as unacceptable. Moreover, the low incidence of VPDs can result in insufficient public awareness of their dangers and the benefits of immunization. This lack of awareness coupled with physician apathy and efforts by antivaccine lobbies can reduce public confidence in immunizations. The public health consequences of such occurrences can be considerable [67, 87, 97, 101, 102].

By the 1990s, the field of vaccinology was already moving beyond the classic six vaccines originally introduced by EPI. By 2000, more than 100 countries had introduced hepatitis B, mumps, and rubella vaccines into their routine infant immunization schedule, and 69 had introduced *Haemophilus influenzae* type B (Hib) conjugate vaccine [52]. The latter potentially could avert most of the estimated 3 million cases of serious Hib disease and 400,000–700,000 deaths annually [103].

Pneumococcal disease caused by *Streptococcus pneumoniae* can range from a mild upper respiratory infection and otitis media to more invasive diseases such as pneumonia, septicemia, and meningitis [104]. Severe disease primarily affects the elderly in developed countries and infants in developing countries. As with Hib, conjugate pneumococcal vaccines overcome the limitations of polysaccharide vaccines (i.e., low efficacy in infants). One conjugate vaccine has been licensed in the United States. Several clinical vaccine trials are now underway to evaluate efficacy in developing countries using various serotypes in the vaccine. Widespread use of an efficacious pneumococcal vaccine could help to avert an estimated 1 million deaths most of which occur in disadvantaged populations in developing countries [104]. The increasing antimicrobial resistance of *Streptococcus pneumoniae* makes vaccine development and availability even more urgent.

Two licensed typhoid vaccines (one oral and one parenteral) are now available with 50–70% efficacy in children >5 years of age and young adults and are considered far less reactogenic than the old whole cell typhoid vaccine [105]. Such vaccines, if used more widely, would help to avert an estimated 16 million cases and 600,000 deaths globally, 70% of which occur in Asia [105]. The emergence of multidrug-resistant strains of *Salmonella typhi* makes immunization an increasingly attractive and cost-effective option for control. Although hospital-based and passive surveillance data have suggested a peak incidence in the age group 5–12 years, community-based data from India suggest that peak incidence may occur in children under 5 years of age in some endemic areas, which would necessitate a different approach in terms of mechanism of protection, vaccine composition, and formulation [105].

Promising research and development efforts are underway for new or improved vaccines against several agents. These include vaccines against cholera, dengue, enterotoxigenic *Escherichia coli* (ETEC), enterohemorrhagic *E. coli* (EHEC), group A streptococcus, human papillomavirus, Japanese encephalitis, meningococcus A and C, respiratory syncytial virus, rotavirus, and shigella. Given the estimated burden of disease caused by these agents, immunization against them could have a profound public health impact, particularly in the developing world [106–113].

In 2000, the Global Alliance for Vaccines and Immunization (GAVI) was launched, which involves key UN agencies, governments, multi- and bilateral agencies, philanthropic foundations, the research and development community, national technical agencies, and the private sector. GAVI's objectives are to improve access to sustainable immunization services, expand the use of all existing cost-effective vaccines, accelerate the development and introduction of new vaccines, and make immunization coverage an integral part of the design and assessment of health systems and international development efforts. A Global Children's Vaccine Fund is providing considerable new resources to help meet the GAVI objectives [114].

VII. THE QUEST FOR VACCINES AGAINST THE "BIG THREE"

HIV–AIDS, malaria, and tuberculosis claim approximately 5.7 million lives per year worldwide [2]. Although effective treatments exist, disadvantaged populations are often at greatest risk and have the least access to treatment. Control of these three diseases is compounded by increasing resistance to antimalarial and antituberculosis drugs and the difficulty of successfully promoting life style changes required for the prevention of HIV–AIDS. Vaccines therefore are a desirable and potentially cost-effective tool to control these diseases.

A. HIV–AIDS

The issues surrounding a potential HIV vaccine are tremendously complex and have sparked highly publicized scientific, social, political, and ethical debates. HIV–AIDS is preventable through adequate blood and injection safety as well as behavioral change, all of which are likely to remain essential even if HIV vaccines become available.

In the development of this vaccine, daunting scientific hurdles must be overcome. The virus is highly variable and mutates rapidly; therefore, a vaccine protecting against one particular type of the virus may be ineffective against another. There are no human immune responses that protect against infection, and no adequate animal model exists. There are safety concerns about the use of "killed" or attenuated virus methods, in which a defective vaccine could infect the recipient with fatal consequences. It may therefore never be possible to develop a vaccine that prevents infection, and efforts must also focus on therapeutic vaccines that slow the progression of AIDS.

Funding for vaccine development has to compete with the provision of retroviral therapy for millions of people living with HIV–AIDS. In poor countries with high HIV prevalence, the need for and investment in high-tech prevention strategies therefore may seem less pressing. Participation in vaccine trials, which are likely to be carried out under intense media interest, can lead to stigmatization; participants may be identified as belonging to a "high-risk" group, which could discourage trial participation and slow progress. It is unlikely that any HIV vaccine will be 100% effective, and concerns have been expressed about a potential "false sense of security" among recipients who may feel safe from infection and subsequently engage in risky behavior [115]. The successful pursuit of a vaccine against HIV–AIDS may still demand years of work and an open debate about the multifaceted ethical and social issues involved.

B. Malaria

The concept of a vaccine that would prevent infection by priming the body's immune system to destroy the invading organism is challenged when considering malaria. Due to the size and complexity of the parasite, the immune system responds to thousands of antigens during infection. It remains unclear which of these antigens should be targeted for vaccine development. Moreover, the parasite goes through several life stages in the host, thus continuously presenting different molecular characteristics to the immune system.

Malaria parasites cannot be cultured in cell-free systems, and no good animal model for human malaria exists. Vaccine development is further complicated by the various species and strains of malaria that may infect the body simultaneously. Vaccine candidates have been tested that build on different characteristics of the parasite [116]. One candidate provides long-lasting immunity by repeated injections of inactivated (irradiated) parasites, but it is impractical for widespread use. Another approach involves the use of immunoglobulins purified from the blood of people living in malaria-endemic areas. The immunoglobulins prevent the parasite from invading new red cells, thus reducing the impact and clinical severity of the disease.

Despite the challenges, a vaccine against malaria may be feasible, and vaccine development has been accelerated by an infusion of resources from several major donors.

C. Tuberculosis

Worldwide tuberculosis (TB) is the leading cause of death in adults due to a single infectious agent [117]. The rise of the HIV epidemic has increased prevalence rates of TB, which is likely to continue. Although effective chemotherapy is available, a surge in multidrug-resistant TB has renewed calls for an effective TB vaccine.

For decades the BCG vaccine has been used widely with known efficacy against severe, systemic forms of TB during infancy, but varying efficacy against adult TB; the latter ranges from 0 to 80% [118, 119]. BCG immunization therefore has had little impact on the epidemiology of TB, although some studies have demonstrated its protective efficacy against leprosy [120, 121].

Major efforts are underway to develop a new TB vaccine. The complete genome of *Mycobacterium tuberculosis* has been sequenced, which will help us to understand the virulence of the organism. New products (DNA vaccine, attenuated live vaccine) are being tested in animal models and should be available for human testing within the next few years [122].

D. Summary

The use of effective vaccines against the "big three" would be an important step in the elimination of malaria, HIV–AIDS, and TB as global health problems, but their development presents daunting technical and social challenges. The greatest challenge, however, may be ensuring access to these vaccines for the people who need them most and can afford them least.

VIII. CONCLUSION

The field of vaccinology has already had a formidable public health impact throughout history, resulting in the eradication of smallpox and reducing other diseases to levels where they are no longer recognized as public health problems. New challenges present themselves for the twenty-first century. These include immunizing populations beyond the reach of routine health services, sustaining high immunization coverage levels, integrating immunization services more fully into national health systems particularly in the context of health sector reform, ensuring high levels of immunization safety, and maintaining public confidence in immunization. Important opportunities have also presented themselves. Exciting advances in biotechnology are enabling the development of new vaccines with enormous possibilities that were considered unthinkable. The infusion of new resources and the organization of an alliance of dedicated agencies, governments, and industries are other important factors to ensure that immunization services are strengthened in countries where they are weak and that new vaccines are accessible to even the poorest countries where the public health impact could be greatest.

REFERENCES

1. Murray, C. J. L., and Lopez, A. D. (1996). *In* "Global Burden of Disease and Injury Series, Volume I. A Comprehensive Assessment of Mortality and Disability from Diseases, Injuries and Risk Factors in 1990 and Projected to 2020." Cambridge: Harvard University Press.
2. World Health Organization (2001). "The World Health Report 2001. Mental Health: New Understanding, New Hope." Geneva, Switzerland.
3. DeBary, W. T., Ed. (1972). "The Buddhist Tradition in India, China and Japan." New York: Vintage Books.
4. Plotkin, S. L., and Plotkin, S. A. (1999). A short history of vaccination. *In* "Vaccines," 3rd ed., S. A. Plotkin and W. A. Orenstein, Eds., pp. 1–12. Philadelphia: Saunders.
5. Pasteur, L. (1885). Méthode pour prévenir la rage après morsure. *C.R. Acad. Sci. Paris* **101:**765–772.
6. Salmon, D. E. (1886). The theory of immunity from contagious diseases. *Proc. Am. Assoc. Adv. Sci.* **35:**262–266.
7. Salmon, D. E., and Smith, T. (1886). On a new method of producing immunity from contagious diseases. *Am. Vet. Rev.* **10:**63–69.
8. Ramon, G. (1923). Sur le pouvoir floculant et sur les propriétés immunisantes d'une toxine diphtérique rendue anatoxique (anatoxine). *C.R. Acad. Sci. Paris* **177:**1338–1340.
9. Ramon, G., and Zoeller, C. (1926). De la valeur antigénique de l'anatoxine tétanique chez l'homme. *C.R. Acad. Sci. Paris* **182:**245–247.
10. Ramon, G., and Zoeller, C. (1927). L'anatoxine tétanique et l'immunisation active de l'homme vis-à-vis du tétanos. *Ann. Inst. Pasteur Paris* **41:**803–833.
11. Parish, H. J. (1965). "A History of Immunization." London: E & S Livingstone.
12. Madsen, T. (1933). Vaccination against whooping cough. *J. Am. Med. Assoc.* **101:**187–188.
13. Edwards, K. M., Decker, M. D., and Mortimer, E. A (1999). Pertussis vaccine. *In* "Vaccines," 3rd ed., S. A. Plotkin and W. A. Orenstein, Eds., pp. 293–344. Philadelphia: Saunders.
14. Robbins, F. C., and Daniel, T. M. (1997). A history of poliomyelitis. *In* "Polio," T. M. Daniel and F. C. Robbins, Eds., pp. 5–22. Rochester, NY: University of Rochester Press.
15. Henderson, D. A., and Moss, B. (1999). Smallpox and Vaccinia. *In* "Vaccines," 3rd ed., S. A. Plotkin and W. A. Orenstein, Eds., pp. 74–97. Philadelphia: Saunders.
16. Fenner, F., Henderson, D. A., Arita, I., Jezek, Z., and Ladnyi, I. D. (1988). The intensified smallpox eradication programme, 1967–1980. *In* "Smallpox and Its Eradication," pp. 421–538. Geneva, Switzerland: World Health Organization.
17. Henderson, R. H., Keja, K., Galazka, A. M., and Chan, C. A. (1990). Reaping the benefits: Getting vaccines to those who need them. *In* "New Generation Vaccines," G. C. Woodrow and M. M. Levine, Eds., pp. 69–82. New York: Marcel Dekker.
18. Redd, S. C., Markowitz, L. E., and Katz, S. L. (1999). Measles vaccine. *In* "Vaccines," 3rd ed., S. A. Plotkin and W. A. Orenstein, Eds., pp. 222–266. Philadelphia: Saunders.
19. Thomas, M. G. (1989). Epidemiology of pertussis. *Rev. Infect. Dis.* **11:**255–262.
20. World Health Organization (1996). Progress toward the 1995 measles control goals and implications for the future. GPV/SAGE.96/WP.04.
21. Lee, J. W., Aylward, B. R., Hull, H. F., Batson, A., Birmingham, M. E., and Lloyd, J. (1997). Reaping the benefits: Getting vaccines to those who need them. *In* "New Generation Vaccines," G. C. Woodrow and M. M. Levine, Eds., pp. 79–88. New York: Marcel Dekker.
22. World Health Organization (1977). Expanded Programme on Immunization. *Weekly Epidemiol. Record* **7:**74–75.
23. Henderson, R. H. (1984). Providing immunization: The state of the art. *In* "Protecting the World's Children: Vaccine and Immunization within Primary Health Care. Proceedings of the Bellagio Conference," pp. 17–37. New York: The Rockefeller Foundation.

24. World Health Organization (1977). WHA resolution 30.53. *WHO Official Records* **240**(33):140.
25. Henderson, R. H., and Sundaresan, T. (1982). Cluster sampling to assess immunization coverage: A review of experience with a simplified sampling method. *Bull WHO* **60**:253–260.
26. Bernier, R. H. (1984). Some observations on poliomyelitis lameness surveys. *Rev. Infect. Dis.* **6**(Suppl. 2):S371–S375.
27. Stanfield, J. P., and Galazka, A. (1984). Neonatal tetanus in the world today. *Bull. WHO* **62**(4):647–669.
28. Galazka, A., and Stroh, G. (1986). Neonatal tetanus. Guidelines on the community-based survey on neonatal tetanus mortality. WHO/EPI/GEN/86/8.
29a. Galazka, A., Gasse, F., and Henderson, R. H. (1989). Neonatal tetanus in the world and the global Expanded Programme on Immunization. *In* "8th International Conference on Tetanus," G. Nistico, B. Bizzini, B. Bytchenko, and R. Triau, Eds., pp. 470–487. Rome–Milan: Pythagora Press.
29b. LaForce, F. M., Lichnevski, M. S., Keja, J., and Henderson, R. H. (1980). Clinical survey techniques to estimate prevalence and annual incidence of poliomyelitis in developing countries. *Bull. WHO* **58**:609–620.
30. Grant, J. P. (1991). "The State of the World's Children in 1991." Oxford, England: Oxford University Press.
31. World Health Organization (1993). Resolution WHA 41.28. *In* "Handbook of Resolutions and Decisions of the World Health Assembly and the Executive Board (1985–1992)," Vol. III, 3rd ed., pp. 100–101. Geneva, Switzerland: World Health Organization.
32. World Health Organization (2001). Global Polio Eradication Progress 2000. WHO/polio/01.03.
33. Murray, C. J. L., and Lopez, A. D. (1996). *In* "Global Burden of Disease and Injury Series, Volume II. Global Health Statistics. A Compendium of Incidence, Prevalence and Mortality Estimates for over 200 Conditions." Cambridge: Harvard University Press.
34. World Health Organization (2000). Global polio eradication initiative strategic plan 2001–2005. WHO/polio/00.05.
35. Tangermann, R. H., Hull, H. F., Jafari, H., Nkowane, B., Everts, H., and Aylward, R. B. (2000). Eradication of poliomyelitis in countries affected by conflict. *Bull. WHO* **78**(3):330–338.
36. Sutter, R. W., Tangermann, R. H., Aylward, R. B., and Cochi, S. L. (2001). Poliomyelitis eradication: Progress challenges for the end game, and preparation for the posteradication era. *Infect. Dis. Clinics North America* **15**(1):41–64.
37. Melgaard, B., Creese, A., Aylward, B., Olive, J. M., Maher, C., Okwo-Bele, J. M., Lee, J. W. (1998). Disease eradication and health systems development. *Bull. WHO* **76**(Suppl. 2):26–31.
38. World Health Organization (1993). Resolution WHA 42.32. *In* "Handbook of Resolutions and Decisions of the World Health Assembly and the Executive Board (1985–1992)," Vol. III, 3rd ed., p. 102. Geneva, Switzerland: World Health Organization.
39. Grant, J. P. (1990). First call for children. "World Declaration and Plan of Action from the WSC, Convention on the Rights of the Child." New York: UNICEF.
40. Hull, H. (1988). Increased measles mortality in households with multiple cases in the Gambia, 1981. *Rev. Infect. Dis.* **10**:463–467.
41. Markowitz, L. E., Nzilambi, N., Driskell, W. J., Sension, M. G., Rovira, E. Z., Nieburg, P., and Ryder, R. W. (1989). Vitamin A levels and mortality among hospitalised measles patients, Kinshasa, Zaire. *J. Tropical Pediatr.* **35**:109–112.
42. Aaby, P., and Leeuwenburg, J. (1990). Patterns of transmission and severity of measles infection: A reanalysis of data from the Machakos Area, Kenya. *J. Infect. Dis.* **161**:171–174.
43. Sutter, R. W., Markowitz, L. E., Bennetch, J. M., Morris, W., Zell, E. R., and Preblud, S. R. (1991). Measles among the Amish: A comparative study of measles severity in primary and secondary cases in households. *J. Infect. Dis.* **163**:12–16.
44. Markowitz, L. E., and Nieberg, P. (1991). The burden of acute respiratory tract infection due to measles in developing countries and the potential impact of measles vaccine. *Rev. Infect. Dis.* **13**(Suppl. 6):S555–S561.
45. Aaby, P. (1988). Introduction to community studies of severe measles: Comparative test of crowding/exposure hypothesis. *Rev. Infect. Dis.* **10**:451.

46. Narain, J. P., Khare, S., Rana, S. R., and Banerjee, K. B. (1989). Epidemic measles in an isolated unvaccinated population, India. *Int. J. Epidemiol.* **18**:952–958.
47. World Health Organization (1993). High measles case-fatality rates during an outbreak in a rural area (Niger). *Weekly Epidemiol. Record* **68**:142–145.
48. Agocs, M. M., Markowitz, L. E., Straub, I., and Domok, I. (1992). The 1988–1990 measles epidemic in Hungary: Assessment of vaccine failure. *Int. J. Epidemiol.* **21**:1007–1013.
49. Kambarami, R. A., Nathoo, K. J., Nkrumah, F. K., and Pirie, D. J. (1991). Measles epidemic in Harare, Zimbabwe, despite high measles immunization coverage rates. *Bull. WHO* **69**:213–219.
50. Chen, R. T., Weierbach, R., Bisoffi, Z., Cutts, F., Rhodes, P., Ramaroson, S., Htembagara, C., and Bizimana, F. (1994). A "post-honeymoon period" measles outbreak in Muyinga sector, Burundi. *Int. J. Epidmiol.* **23**:185–193.
51. McLean, A. R., and Anderson, R. M. (1988). Measles in developing countries. Part II. The predicted impact of mass vaccination. *Epidmiol. Infect.* **100**:419–442.
52. World Health Organization (2001). Vaccine-Preventable Disease Monitoring System. 2001 Global Summary. WHO/V&B/01.34.
53. de Quadros, C. A., Olive, J. M., Hersh, B. S., Strassburg, M. A., Henderson, D. A., Brandling-Bennett, D., and Alleyne, G. A. O. (1996). Measles elimination in the Americas: Evolving strategies. *J. Am. Med. Assoc.* **275**:224–229.
54. Centers for Disease Control and Prevention (1997). Measles eradication: Recommendations from a meeting cosponsored by the World Health Organization, the Pan American Health Organization, and CDC. *Morb. Mortal. Wkly. Rep.* **46**(RR11):1–20.
55. World Health Organization (2000). Progress toward interrupting indigenous measles transmission in the western hemisphere. *Weekly Epidemiol. Record* **75**(40):354–358.
56. World Health Organization (1997). Expanded Programme on Immunization. Rubella and congenital syndrome in the Americas. *Weekly Epidemiol. Record* **72**(40):301–303.
57. World Health Organization (2001). Accelerated rubella and congenital rubella syndrome programme, Costa Rica. *Weekly Epidemiol. Record* **76**(35):265–270.
58. World Health Organization (2000). Preventing congenital rubella syndrome. *Weekly Epidemiol. Record* **75**(36):290–295.
59. Wakefield, A. J. (1999). MMR vaccination and autism. *Lancet* **354**(9182):949–950.
60. Wakefield, A. J., Murch, S. H., Anthony, A., Linnell, J., Casson, D. M., Malik, M., Berelowitz, M., Dhillon, A. P., Thomson, M. A., Harvey, P., Valentine, A., Davies, S. E., and Walker-Smith, J. A. (1998). Ileal–lymphoid–nodular hyperplasia, non-specific colitis, and pervasive developmental disorder in children. *Lancet* **351**(9103):637–641.
61. World Health Organization (1998). Expanded Programme on Immunization (EPI). Association between measles infection and the occurrence of chronic inflammatory bowel disease. *Weekly Epidemiol. Record* **73**(6):33–39.
62. Peltola, H., Patja, A., Leinikki, P., Valle, M., Davidkin, I., and Paunio, M. (1998). No evidence for measles, mumps, and rubella vaccine-associated inflammatory bowel disease or autism in a 14-year prospective study. *Lancet* **351**(9112):1327–1328.
63. DeStefano, F., and Chen, R. T. (2000). Autism and measles, mumps, and rubella vaccine: No epidemiological evidence for a causal association. *J. Pediatr.* **136**(1):125–126.
64. Halsey, N. A., and Hyman, S. L., Conference Writing Panel. (2001). Measles–mumps–rubella vaccine and autistic spectrum disorder: Report from the New Challenges in Childhood Immunizations Conference convened in Oak Brook, IL, June 12–13, 2000. *Pediatrics* **107**(5):E84.
65. Kastner, J. L., and Gellin, B. G. (2001). Measles–mumps–rubella vaccine and autism: The rise (and fall?) of a hypothesis. *Pediatr. Ann.* **30**(7):408–415.
66. Kaye, J. A., Maria Del Mar, M. M., and Jick, H. (2001). Mumps, measles, and rubella vaccine and the incidence of autism recorded by general practitioners: A time–trend analysis *West. J. Med.* **174**(6):387–390.
67. Anonymous (1999). Fall in MMR vaccine coverage reported as further evidence of vaccine safety is published. *Commun. Dis. Rep. Wkly.* **9**(26):227, 230.
68. World Health Organization (2001). Measles mortality reduction and regional elimination strategic plan, 2001–2005. WHO/V&B/01.13.
69. UNICEF, WHO, UNFPA (2000). Maternal and neonatal tetanus elimination by 2005. "Strategies for Achieving and Maintaining Elimination." New York: UNICEF.

70. Fauveau, V., Mamdani, M., Steinglass, R., and Koblinsky, M. (1993). Maternal tetanus: Magnitude, epidemiology and potential control measures. *Int. J. Gynecol. Obstet.* **40**:3–12.

71. Centers for Disease Control and Prevention (1985). Tetanus—United States, 1982–1984. *Morb. Mort. Wkly. Rep.* **34**(39):602–611.

72. Crossley, K., Irvine, P., Warren, B., Lee, B., and Mead, K. (1979). Tetanus and diphtheria immunity in urban Minnesota adults. *J. Am. Med. Assoc.* **242**:2298–2300.

73. Simonsen, O., Block, A. V., Klaerke, A., Klaerke, M., Kjeldsen, K., and Heron, I. (1987). Immunity against tetanus and response to revaccination in surgical patients more than 50 years of age. *Surg. Gynecol. Obstet.* **164**:329–334.

74. Kjeldsen, K., Simonsen, O., and Heron, I. (1988). Immunity against diphtheria and tetanus in the age group 30–70 yrs. *Scand. J. Infect. Dis.* **20**:177–185.

75. World Health Organization (1996). Eliminating neonatal tetanus. How near, how far? WHO/EPI/GEN/96.01.

76. Mortimer, E. A., and Wharton, M. (1997). Diphtheria toxoid. *In* "Vaccines," 3rd ed., S. A. Plotkin and W. A. Orenstein, Eds., pp. 1–12. Philadelphia: Saunders.

77. English, P. C. (1985). Diphtheria and theories of infectious disease: Centennial appreciation of the critical role of diphtheria in the history of medicine. *Pediatrics* **76**:1–9.

78. Doull, J. A. (1952). The bacteriological era (1876–1920). *In* "The History of American Epidemiology," F. H. Top, Ed., pp. 74–103. St Louis: Mosby Co.

79. Kass, A. M. (1993). Infectious diseases at the Boston City Hospital: The first 60 years. *Clin. Infect. Dis.* **17**:276–282.

80. Kostrzewski, J. (1964). Blonica (diphtheria). *In* "Infectious Diseases in Poland and Their Control in 1919–1962" (in Polish), pp. 96–117. Warsaw, Poland: Panstwowy Zaklad Wydawnictw Lekarskich.

81. McKinnon, N. E. (1942). Diphtheria prevented. *In* "Control of the Common Fevers," R. Cruickshank, Ed., pp. 41–52. London: The Lancet Ltd.

82. Galazka, A. M., Robertson, S. E., and Oblapenko, G. P. (1995). Resurgence of diphtheria. *Eur. J. Epidemiol.* **11**(1):95–105.

83. Stowman, K. (1945). Diphtheria rebounds. *Epidemiol. Inf. Bull.* (United Nations Relief and Rehabilitation Administration) **1**:157–168.

84. Stuart, G. (1945). A note on diphtheria incidence in certain countries. *Br. Med. J.* **2**:613–615.

85. Wheeler, S. M., and Morton, A. R. (1942). Epidemiological observations in the Halifax epidemic. *Am. J. Public Health* **32**:942–956.

86. Fleck, S., Kellam, J. W., and Klippen, A. J. (1944). Diphtheria among German prisoners of war. *Bull. U.S. Army Med. Dept.* **74**:80–89.

87. Galazka, A. M., and Robertson, S. E. (1995). Diphtheria—changing patterns in the developing world and the industrialized world. *Eur. J. Epidemiol.* **11**(1):107–117.

88. World Health Organization (1997). Expanded Programme on Immunization. Diphtheria control. *Weekly Epidemiol. Record* **72**(18):128–130.

89. Galazka, A. (1992). Control of pertussis in the world. *World Health Stat. Q.* **45**:238–247.

90. Wright, P. F. (1991). Pertussis in developing countries. Definition of the problem and prospects for control. *Rev. Infect. Dis.* **13**(Suppl.):S528–S534.

91. Centers for Disease Control and Prevention (1990). Pertussis surveillance—United States, 1986–1988. *Morb. Mortal. Wkly. Rep.* **39**:57–66.

92. Clarkson, J. A., and Fine, P. E. M. (1985). The efficacy of measles and pertussis notification in England and Wales. *Int. J. Epidemiol.* **14**:153–168.

93. Sutter, R. W., and Cochi, S. L. (1992). Pertussis hospitalizations and mortality in the United States, 1985–1988. *J. Am. Med. Assoc.* **267**(3):386–391.

94. Halperin, S. A., Bortolussi, R., Maclean, D., and Chisholm, N. (1989). Persistance of pertussis in an immunized population: Results of the Nova Scotia enhanced pertussis surveillance programme. *J. Pediatr.* **115**:686–693.

95. Cherry, J. P., Brunell, P. A., Golden, G. S., and Karzon, D. T. (1988). Report of the task force on pertussis and pertussis immunization. *Pediatr. Suppl.* **81**(6, Part 2):939–984.

96. Onorato, I. M., and Wassilak, S. G. F. (1987). Laboratory diagnosis of pertussis: The state of art. *Pediatr. Infect. Dis. J.* **6**:145–151.

97. Gangarosa, E. J., Galazka, A. M., Wolfe, C. R., Phillips, L. M., Gangarosa, R. E., Miller, E., and Chen, R. T.(1998). Impact of anti-vaccine movements on pertussis control: The untold story. *Lancet* **351**(9099):356–361.

98. World Health Organization (1999). Pertussis vaccines. WHO position paper. *Weekly Epidemiol. Record* **74**(18):137–143.

99. World Health Organization (2001). Pertussis surveillance. A global meeting, Geneva, 16–18 October 2000. WHO/V&B/01.19.

100. World Health Organization (2000). Sustainable outreach services (SOS). A strategy for reaching the unreached with immunization and other services. WHO/V&B/00.37.

101. Salmon, D. A., Haber, M., Gangarosa, E. J., Phillips, L., Smith, N. J., and Chen, R. T. (1999). Health consequences of religious and philosophical exemptions from laws: Individual and societal risk of measles. *J. Am. Med. Assoc.* **282**(1):47–53.

102. World Health Organization (1997). Expanded Programme on Immunization. Lack of evidence that hepatitis B vaccine causes multiple sclerosis. *Weekly Epidemiol. Record* **72**(21): 149–153.

103. World Health Organization (1998). Global Programme for Vaccines and Immunization (GPV). The WHO position paper on *Haemophilus influenzae* type B conjugate vaccines. *Weekly Epidemiol. Record* **73**(10):64–68.

104. World Health Organization (1999). Pneumococcal vaccines. WHO position paper. *Weekly Epidemiol. Record* **74**(23):177–183.

105. World Health Organization (2000). Typhoid vaccines. WHO position paper. *Weekly Epidemiol. Record* **75**(32):257–264.

106. World Health Organization (1997). Expanded Programme on Immunization. *Weekly Epidemiol. Record* **72**(6):36–40.

107. World Health Organization (1997). Vaccine research and development. New strategies for accelerating *Shigella* vaccine development. *Weekly Epidemiol. Record* **72**(11):73–79.

108. World Health Organization (1999). New frontiers in the development of vaccines against enterotoxigenic (ETEC) and enterohemorrhagic (EHEC) *E. coli. Weekly Epidemiol. Record* **74**(13):98–101.

109. World Health Organization (1999). Group A and C meningococcal vaccines. WHO position paper. *Weekly Epidemiol. Record* **74**(36):297–303.

110. World Health Organization (1999). The current status of development of prophylactic vaccines against human papilloma virus infections. Report of a technical meeting, Geneva, 16–18 February 1999; WHO/V&B/99.04.

111. World Health Organization (2000). Proceedings of the first global vaccine research forum. Montreux, 7–9 June 2000. WHO/V&B/01.21.

112. World Health Organization (2001). Cholera vaccines. WHO position paper. *Weekly Epidemiol. Record* **76**(16):117–124.

113. World Health Organization (2001). Chiangmai declaration on dengue/dengue hemorrhagic fever. Strengthening efforts to control dengue in the new millenium. *Weekly Epidemiol. Record* **76**(4):29–30.

114. Global Alliance for Vaccines and Immunization (1999). Meeting of the Proto Board. Seattle, 12–13 July 1999. GAVI/99.01.

115. Collins, C. (1996). Sustaining support for domestic HIV vaccine research: Social issues over the long haul of human trials. Occasional Paper No. 2, Center for AIDS Prevention Studies (CAPS), Atlanta.

116. Malaria Vaccine Initiative (2001). www.malariavaccine.org.

117. World Health Organization (2001). WHO Report 2001—Global Tuberculosis Control. WHO/CDS/TB/2001.287.

118. Rodrigues, L. C., and Smith, P. G. (1990). Tuberculosis in developing countries and methods for its control. *Trans. R. Soc. Trop. Med. Hyg.* **84**:739–744.

119. World Health Organization (1999). Issues relating to the use of BCG in immunization programmes—A discussion document. WHO/V&B/99.23.

120. Styblo, K., and Meijer, J. (1976). Impact of BCG vaccination programmes in children and young adults on the tuberculosis problem. *Tubercle* **57**(1):17–43.

121. Ponnighaus, J. M., Fine, P. E., Sterne, J. A., Wilson, R. J., Msosa, E., Gruer, P. J., Jenkins, P. A., Lucas, S. B., Liomba, N. G., Bliss, L. (1992). Efficacy of BCG vaccine against leprosy and tuberculosis in northern Malawi. *Lancet* **339**:636–639.

122. Orme, I. M. (1997). Progress in the development of new vaccines against tuberculosis. *Int. J. Tuberc. Lung Dis.* **1**(2):95–100.

1 GLOBAL BURDEN OF DISEASE

PART B. Cost Effectiveness of Immunization: Asking the Right Questions

RUTH E. LEVINE
Center for Global Development, Washington, DC 20036

I. INTRODUCTION

On any list of high-priority health interventions, immunization appears toward the top. It is widely accepted that traditional childhood vaccines, including measles, polio, DTP, and other basic antigens, are among the very best uses for scarce healthcare dollars because they are now inexpensive to produce and deliver, safe, and highly effective at preventing common—but potentially serious—childhood ailments that would be costly to treat. Common sense alone suggests that a preventive health measure that yields many benefits and costs little is bound to be a good buy.

But whereas the notion that basic childhood vaccines are a good use of health resources is held almost universally—and has been demonstrated in many parts of the world with solid economic evaluations—important work remains to better understand whether and when it makes economic sense to introduce new (or new and more costly formulations of traditional) vaccines, to scale up programs to achieve higher coverage levels, or to engage in campaigns to eradicate major diseases.

Cost-effectiveness analysis, a standard part of the economist's armamentarium, often is used to inform—or at least rationalize—decisions about the use of healthcare resources, including those related to vaccines and immunization strategies. In the starkest terms, cost-effectiveness analysis permits us to determine whether it is cheaper to save a life with one type of health intervention than with another. The shift of resources toward the most highly cost effective interventions has the potential to obtain the maximum amount of "good" (lives saved) from a fixed budget.

23

In this part of Chapter 1, we look carefully at the concept of cost effectiveness as an important tool for resource allocation decisions in the health sector, with specific attention to current issues in the field of vaccines and immunization. We first define cost-effectiveness analysis and compare it to other economic evaluation methods. We then summarize a portion of the vast number of estimates of the cost effectiveness of immunization to provide a sense of what is known and, importantly, what is not yet known. This exercise permits us to illustrate how variations in health systems and epidemiological and demographic patterns are reflected in estimates of cost effectiveness of immunization. Next we examine more carefully what the cost-effectiveness estimates mean and how they can be (and have been) used for policymaking and programmatic decision making. We then turn to three (out of many) questions that call for the attention of economists: (1) How can cost-effectiveness analysis be used to determine the right strategy to increase coverage? (2) What does cost-effectiveness analysis tell us about whether and how to introduce newer vaccines, such as hepatitis B (HepB) and *Haemophilus influenzae* type b (Hib)? (3) Can decisions about whether to fund eradication campaigns be based on cost-effectiveness analysis? Finally, we attempt to chart part of the research and policy path into the future.

II. COST-EFFECTIVENESS ANALYSIS DEFINED

Cost-effectiveness analysis (CEA), one of several methods for the economic evaluation of health interventions, compares two or more health interventions that have a common objective and are competing for common resources. In essence, it is a measure of efficiency in the production of health. For each of the competing interventions, CEA relates the net cost—for example, the cost of delivering a standard set of antigens through routine infant immunization in fixed-site clinics minus the treatment and other costs *not incurred* because of the beneficial effects of the immunization—to a desired health outcome, such as the reduction of illness or death from the vaccine-preventable diseases. A ratio is calculated for each of the interventions: the numerator is the cost, expressed in money terms (dollars), and the denominator is the measurable health outcome. The health outcome typically is expressed in terms of the gain in years of life, although there are many variants of the effectiveness measure, including gains in disability-adjusted life years (DALYs), which seeks to capture both morbidity and mortality in a single metric. Thus, the cost effectiveness of a given intervention is expressed as dollars per health benefit, for instance, $245 per life year gained. The higher the dollar figure, the lower the cost effectiveness. Thus, when the cost-effectiveness ratios of two or more health interventions are compared, a somewhat naïve interpretation of economic logic implies that the more cost-effective intervention—the one with the smaller cost per life year gained—should be funded, whereas the less cost-effective intervention should not.

It is useful to distinguish between full cost effectiveness and incremental, or marginal, cost effectiveness. Full cost-effectiveness analysis requires that the total net costs and benefits are taken into consideration, as if no expenditures on a given intervention were being made and no health benefits were being generated.

In the real world, however, the decision being made often is whether a given health program should be expanded or be adapted in one direction or another. In this case, an incremental cost-effectiveness analysis could be used to estimate the additional costs required to generate an additional unit of health benefit.

In the past decade or so, CEA has been warmly embraced by health economists, who see it as a tool to promote more economically rational decision making about the use of resources, public health professionals, who find most CEA results intuitively appealing and consistent with their understanding about the relative merits of prevention over treatment, and politicians, who like the ease with which cost effectiveness can be used as evidence that their funding decisions have the imprimatur of fiscal responsibility. As a result, cost-effectiveness estimates have been generated for a large number of discrete health interventions, particularly in healthcare systems in which the financing of services has been directly linked to proof of cost effectiveness, such as the United States and the United Kingdom. Increasingly, with the support of the international health community, cost-effectiveness criteria are informing decision making about support to health programs in developing countries [1, 2].

III. THE FAMILY OF ECONOMIC EVALUATION METHODS

Cost-effectiveness analysis often is confused with several other economic evaluation methods, and for that reason it is useful to define those as well.[1] Cost-minimization analysis, the simplest form of economic evaluation, compares the costs of two or more competing interventions; the cheapest one—regardless of differences in effectiveness—wins the competition for resources. This type of analysis is a sensible approach to allocating resources efficiently when the effectiveness of two interventions is identical, a rare circumstance.

Cost–benefit analysis (CBA) is an evaluation method in which the benefits of the health intervention are expressed in money terms, that is, a dollar value is placed on the life years gained, and thus a ratio of benefit to costs of less than 1 would imply that the intervention was not worth undertaking at all, whereas a benefit-to-cost ratio greater than 1 would suggest that the intervention is a good investment. Although CBA is a popular method for decisions about the advisability of allocating resources to investment projects, for example, building a hydroelectric dam, it is more problematic when evaluating investments in the health sector (or other social sectors). Placement of a dollar value on health benefits is an exercise fraught with conceptual and empirical difficulties; the sensitivity of the CBA result to assumptions about the value of life renders the method of limited practical value in the field of health policy.

Cost–utility analysis (CUA), which is often mistaken for cost-effectiveness analysis, attempts to incorporate the dimension of quality of life into the measurement of benefits. Benefits are measured as quality-adjusted life years, or QALYs, in which the gain in expected lifespan resulting from an intervention is weighted by the quality of that life, as assessed through some type of

---

[1]Many excellent texts are available for those interested in additional information about economic evaluation methods [3–7]. There are also a number of journal articles that compare and contrast the various methods of economic evaluation for vaccines and immunization programs [8–15].

systematic surveying of the affected (or general) population. Thus, an intervention that leads to a 10 year gain in life expectancy but implies considerable pain during those years might be estimated to have a lower QALY than an intervention that results in only an 8 year gain in years but with less pain during that period. Although there is considerable debate about the optimal ways to assess the subjective "quality" dimension, analysts generally agree that QALYs are closer to the fundamental concept of health benefits than are the standard physical measures used in cost-effectiveness analysis. Because many of the issues implied by CUA are similar to those with CEA and because CUA is increasingly popular, both are discussed in the following sections of this chapter.

IV.   IS IMMUNIZATION COST-EFFECTIVE?

Immunization with the core cluster of antigens (measles, polio, diphtheria, pertussis, tetanus, and BCG) is often considered to be the quintessential cost-effective intervention: the vaccines now are available at pennies per dose; the costs of delivering the services, particularly through existing health clinics, are relatively low; the diseases prevented would affect many children in the absence of an immunization program; and the health (and financial) consequences would be significant. Very few health interventions can claim as low a cost-to-effectiveness ratio as immunization, which is often estimated to cost about $15–25 per life year gained. This compares to hundreds or even thousands of dollars per life year gained for many common healthcare interventions. Thus, basic childhood immunization "wins" the cost-effectiveness contest against nearly all other possible uses of health resources. Even other preventive programs tend to be less cost effective because they are more costly to provide—often involving significant behavior change on the part of the patient population—and are less effective. In fact, unlike virtually all other health interventions, routine childhood immunization is so cheap and so effective that it is cost-saving to society.

However, the cost-effectiveness story for immunization is far from complete. A core question confronting policymakers and program managers is whether immunization programs should introduce one or more of the newer vaccines or vaccine combinations, which come at a high price relative to the traditional vaccines. In some cases, the newer vaccines prevent a set of diseases that are very costly to treat, but may affect only a narrow segment of the population (e.g., HepB). In other cases, the vaccines may prevent a prevalent disease, but one that may not always be high-cost in terms of lives or dollars (e.g., Hib). In addition, program managers and policymakers may have to consider whether the standard assumptions about cost effectiveness hold when the program tries to reach beyond the current population being covered.

Countless estimates have been made of the cost effectiveness of immunization; a small number of these are shown in Table 1. In general, these fall into three categories. First is estimates of the cost effectiveness of routine childhood immunization with the traditional package of antigens [16–19]. Many of these estimates have been generated not for the purposes of making specific resource allocation decisions but rather to demonstrate that childhood

TABLE 1 Selected Studies of Cost Effectiveness of Immunization

Vaccine	Country	Year and source	Vaccination strategy	C-E measure[a]
Measles	Developing countries with an average age of infection at 2 years	1999 [16]	Expanding routine infant immunization coverage from 50 to 80%	$2.53–5.06 per discounted life year gained (preliminary results)
Measles	Bangladesh	1999 [17]	Routine infant immunization	$288 per death prevented, $14.39 per life year gained
Measles	Developing countries	1993 [19]	Routine infant immunization	$2–15 per healthy life year gained
BCG, DTP, OPV, and measles	Guinea	1994 [18]	Routine infant immunization	$25 per life year saved
BCG, DTP, OPV, and measles	Low-income countries	1993 [19]	Routine infant immunization	$14–20 per healthy life year gained
BCG, DTP, OPV, and measles	Middle-income countries	1993 [19]	Routine infant immunization	$29–41 per healthy life year gained
BCG, DTP, OPV, HepB, yellow fever, vitamin A, and iodine supplements	Low-income countries	1990 [1]	Routine infant immunization	$12–17 per DALY gained
Hib	Developing countries with high mortality	1999 [16]	Routine infant immunization	$480 per discounted life year gained (preliminary results)
Hib	Africa	2000 [32]	Routine infant immunization	$21–22 per life year gained
Hib	Low-income Asia	1998 [31]	Routine infant immunization	$55 per life year gained
HepB	Bangladesh	1998 [28]	Routine infant immunization	$4809 per death prevented
HepB	Developing countries with high HBV prevalence	1999 [16]	Routine infant immunization	$219 per discounted life year gained (preliminary results)
HepB	Low-income countries with HBV prevalence less than 2%	2000 [32]	Routine infant immunization	$42–59 per life year gained
HepB	Low-income countries with HBV prevalence greater than 8%	2000 [32]	Routine infant immunization	$8–11 per life year gained

[a]Dollar amounts are given in U.S. dollars.

immunization is characterized by the unbeatable combination of low cost and high effectiveness. Toward that end, the literature includes a number of studies and policy papers emphasizing the highly cost-effective nature of immunization.

Second, estimates have been made of the relative cost effectiveness of different strategies to deliver routine immunization services, often comparing fixed-site services with mobile teams and/or mass campaigns. The outcome measure for these studies typically has been cost per fully immunized child, rather than a direct health benefit [20–27]. Occasionally, these studies have been used to influence the allocation of resources for specific immunization programs. More commonly, they have been used within the international health community to promote the idea of moving away from vertical programs and mass campaigns and toward the integration of immunization services into existing well-child health services in clinics.

Third, analysts have studied the cost effectiveness of new vaccines, particularly HepB and Hib, given specific epidemiological and demographic features of the population [28–32]. It is these studies that can claim to have directly influenced (rather than rationalized) decisions about how money is spent. In particular, the work on HepB has demonstrated that its introduction into routine infant immunization programs is likely to make sense in low- and middle-income countries only when the population-wide prevalence of the disease is relatively high.

Interestingly, work on the cost effectiveness of newer vaccines brings into sharp relief the subtleties of cost-effectiveness analysis and its interpretation. Although the range of estimates of the cost-effectiveness ratio for the traditional package of vaccines has been relatively narrow, the variation in estimates for the newer vaccines is strikingly wide: from $21 to at least $480 per life year saved for Hib and from $8 to at least $219 per life year saved for HepB. This variation demonstrates that the cost effectiveness of a health intervention is not an inherent and immutable property, like a chemical's specific gravity or a radioactive element's half-life. It is a characteristic that varies with the context—the structure and characteristics of the full health system, as well as the affected population—and the perspective underlying its measurement. This is easy to see in the following section, as we examine the elements of cost-effectiveness analysis.

V. WHAT DO CEA RESULTS REALLY MEAN AND FOR WHAT ARE THEY USEFUL?

To better understand the strengths and limitations of CEA (and CUA), it is worth decomposing both the numerator in a cost-effectiveness estimate, the cost part, and the denominator, the effectiveness part. As will be seen in the sections that follow, neither part is a straightforward measure that exists independent of the decision facing us or its context.

In estimating the cost effectiveness of a preventive health service such as immunization, as indicated earlier, the net cost refers to the total cost of the intervention *minus* the total cost of treating the illnesses that would occur in the absence of the intervention. If we put aside for the moment the issue of how easily costs can be measured, a basic question arises: whose "costs" do we care about? Is it the entity paying for the provision of the services, such as the national or local government or the insurer? Or is it society at large, including (but not exclusively) the individuals who must take an action to obtain the service and may incur both direct and indirect costs for transportation, lost work or school time, the effects of adverse events, and so forth? If we count only the costs to the agency providing the services, such as the health department, then it is very likely that the services that appear to be the least costly are those that require only marginal investments on top of existing activities, such as the integration of additional immunization services within a well-child program at a health clinic, rather than specialized activities, such as outreach programs in which "health visitors" travel to remote areas to offer in-home services to mothers and children. However, if the costs to all of those involved in service delivery and provision are included in the calculation of cost, the picture may change. It may be that obtaining services through a fixed site requires parents to pay for transportation, take time off from work or school, obtain childcare for other children in the household, and so forth, whereas the in-home services may imply none of these expenses. Thus, an intervention that is costly to the agency paying for services may be inexpensive to the consumer, and vice versa.

The net cost part of the cost-effectiveness analysis is complicated further by the need to subtract the costs *not incurred* because of the beneficial effects of the preventive intervention, for example, the costs of treatment for the measles cases that would have been incurred if those cases had not been prevented by routine childhood immunization. Treatment costs fluctuate significantly over time and across healthcare systems, implying large shifts in the cost-effectiveness calculations, even if the costs and effectiveness of the intervention itself do not change. Historically, in many countries the costs of treatment have escalated rapidly with the introduction of new technologies, increases in remuneration of health workers, and other factors. Thus, the inclusion of the averted costs—essential on a conceptual level—complicates the interpretation of cost-effectiveness estimates and the comparisons across countries. In addition, the averted costs are all future ones, so that they must be discounted to reflect the fact that a dollar available today is more valuable to society than a dollar available in the uncertain and distant future. Finally, just as for the costs of the intervention being evaluated, a decision must be made about whether the treatment costs should reflect the costs to the payer (government or insurer) or to society at large.

We now turn to the denominator in the cost-effectiveness (or cost-utility) estimate—the measure of health benefit (or quality-adjusted health benefit) that is associated with the intervention. In the measurement of the health benefits of immunization, we face a similar decision about what perspective to adopt—the individual's or the society's. This is reflected in the choice of a static or a dynamic model. A static estimate seeks to capture the health benefits among the population directly affected by the intervention during a fixed period, for example, the disability- or quality-adjusted life years gained by the children being vaccinated during a year. A dynamic estimate, on the other hand, attempts to take into account the fact that the health benefits from the prevention of a communicable disease are felt not just by the individuals receiving the preventive services but also by the individuals who are *not* exposed to diseases by those protected individuals and therefore who do not become ill themselves. Similarly, a truly ambitious analyst may attempt to capture the benefits of "herd immunity"—the direct immunological benefits obtained by unvaccinated individuals from being present in a community where vaccination coverage is high.

As will be immediately obvious, the multiplier effect of preventing a communicable disease can be immense and is, in fact, a large part of the justification for government spending on immunization services (the so-called "externalities" of the intervention). For each infection directly prevented, immunization can prevent a large number of secondary, tertiary, and other future infections. Estimation of the number of lives ultimately affected by the immunization of one child requires knowledge of disease prevalence, the degree of disease spread in populations with a given level of underlying susceptibility (due, for example, to nutritional status or other factors that affect immune response) and density, and the susceptibility, density, and size of the population (and possible changes in those factors over time). As with the estimates of future costs, the calculation of long-term and dynamic effects also requires the application of a discount factor: life years saved in the future are, in some sense, less "valuable" than life years lived today.

The decisions about what costs and benefits to include and exclude and how to measure them are largely determined by the resource allocation question at hand. As shown in Table 2, politicians may ask whether it is better for society to allocate resources toward preventive or curative health services. In that case, the analysis requires a broad brush, drawing in all social costs and a wide range of benefits. In contrast, a health service financier may have a narrower interest, wishing to know whether it makes economic sense to introduce HepB into the routine infant immunization program or to provide it only to high-risk populations. In that case, the costs to the households of obtaining the service are irrelevant to the decision (and may complicate interpretation of results); the focus is on the direct costs to the payer. With respect to benefits, the health service financier is mainly interested in the health benefits to the vaccinated population *unless*, for example, it is a social insurance entity that is liable for treatment costs across a wider population. Finally, both immunization program managers and economists may have an interest in understanding which immunization strategies make the best use of resources, but their perspectives are different: the program managers may wish to include only the direct costs to the program, whereas the economists may see the value of understanding the full set of social costs.

Although CEA provides useful information for priority setting, in several ways it is by no means a recipe for optimal resource allocation. First, the priority setting that results from a comparison of cost-effectiveness ratios may change when the budget constraint changes; an optimal mix under one budget may be different when the budget shrinks or grows [33]. Second, CEA assumes that health benefits to one group are equivalent in value to health benefits to another group—an assumption that may be inconsistent with broader public policy goals. In the allocation of resources for health, the government may wish to place a high priority on providing health benefits to vulnerable populations to achieve an equity objective. In general, cost-effectiveness analysis will not inform that choice. Third, politicians and their constituents may place a value on allocating resources in other ways that are not consistent with a mechanistic interpretation of cost-effectiveness results [34]. Being a pioneer in health care,

TABLE 2 Illustrative Resource Allocation Questions and Strategies for CEA

Illustrative question	Asked by	Perspective for CEA	CEA approach	Comments
Should we spend more on prevention?	Politician	Social	Include all social costs and dynamic model of health benefits.	Likely result is that preventive services are more cost-effective than curative services, particularly for those preventive interventions that imply few costs to the individual (e.g., do not require extensive behavior change) and/or for which treatment costs are high.
Should we include HepB in the routine infant immunization schedule or just provide it to high-risk populations?	Medical community–health agencies	Provider–payer	Include direct and indirect costs incurred by provider–payer. Use static (individual) model of health benefits unless payer covers community-wide curative health services; in that case, use dynamic model of health benefits.	Results will be determined primarily by prevalence and distribution of HepB, vaccination costs, and treatment costs.
Should services be delivered through fixed sites or outreach services?	Immunization program managers	Provider–payer	Include direct and indirect costs incurred by provider–payer. Use static (individual) model of health benefits; dynamic model will not add information useful to make the decision.	Results will be determined primarily by the direct costs of the different interventions.
	Economists	Society	Include direct and indirect costs incurred by provider–payer and households. Use static (individual) model of health benefits; dynamic model will not add information useful to make the decision.	If household costs are high, results may differ substantially from analysis done from the perspective of provider–payer alone.

spending "as much as it takes" to preserve an identifiable human life, and other objectives may be seen as having high value and may run in direct opposition to a pure cost-effectiveness perspective. Fourth, whereas the results of cost-effectiveness analyses may provide information about how funds can be invested well, they do not inform the question of what an insurer, health ministry, or society can *afford* to spend on health care or on any specific healthcare intervention. Finally, cost-effectiveness analysis does not generate information about who should pay for or provide a service—the government or the private sector.

VI. CURRENT QUESTIONS REQUIRING CEA

The immunization agenda at the beginning of the twenty-first century is full: traditional highly effective and very low cost vaccines do not yet reach a sizeable share of the world's children, and renewed international attention is being given to closing that gap. New vaccines and new formulations of old vaccines have entered the market, and efforts are underway to secure financing for their introduction into developing countries, which in the past have had to wait many years before new health technologies reached an affordable price. The eradication of smallpox, declared in 1980 and recognized as one of the great achievements of the modern era, has inspired the global health community to rally around the idea of eradicating poliomyelitis; in some quarters, there is talk of eradicating measles. Regrettably, threats of bioterrorism have led to the need to reconsider production and immunization against smallpox once again. CEA can provide part of the information required to make sound decisions about whether and how these efforts should proceed.

A.   How Can Cost-Effectiveness Analysis Be Used To Inform Decisions about Increasing Immunization Coverage?

By necessity, analyses of the cost effectiveness of immunization programs have been based on the costs of achieving current levels of coverage. Extension of coverage implies providing services to harder and harder to reach populations: those in remote areas with little access to health services and/or those with low (or even negative) demand[2] for services. There is little doubt that as coverage increases, the average costs will also increase. At the same time, it is entirely possible that the benefits will also be relatively greater, if the populations reached are more susceptible to diseases and/or are more costly to treat when ill because of the distance from health facilities or other factors. Thus, whether expanded immunization coverage is more or less cost-effective than current levels of immunization is impossible to predict without careful analytic study. The main question to address is as follows: Is there a point at which the cost effectiveness of coverage expansion is so low that immunization no longer is a "good buy" compared with competing uses of scarce healthcare resources?

[2]"Negative demand" pertains to populations who are skeptical about or hostile to the idea of vaccination and who will actively avoid it for themselves and their children.

The analytic results can be used—along with other information and opinions—to determine whether there is a point beyond which expansion should not be pushed.

B. What Does CEA Tell Us about Whether and How To Introduce New Vaccines, Such as Hib and HepB?

As countries (and donors) decide whether to fund the introduction of the relatively high-cost newer vaccines into national immunization programs, CEA can be a useful decision tool. In the case of Hib, the issue of how to measure the effectiveness and the treatment costs *averted* becomes central. Some of the illnesses that Hib prevents can be easily treated with low-cost antibiotics; thus, for settings in which modern health services are available, the cost effectiveness of Hib vaccination may be relatively low. At the same time, if health services are unavailable—or available only at a very high cost—then vaccination becomes an economically efficient strategy to reduce the burden of disease and the aggregate cost of its treatment.

In the case of HepB, another relatively costly new vaccine, the question to be addressed by CEA is slightly different, and the results are driven largely by the epidemiology and the high costs of treating the disease (including hepatocellular carcinoma). In some countries, the overall population prevalence of HepB is low (e.g., chronic carrier prevalence of less than 2%), but prevalence is very high among specific population groups, such as intravenous drug users, STD clinic patients, and others engaging in high-risk behaviors. In such settings, CEA generally yields the result that targeted programs are more economically efficient than routine infant immunization. In other countries, in which the endemic prevalence is 7–20%, the marginal costs of adding HepB to the infant immunization schedule may be more than offset by the society-wide gains in treatment costs averted. The main conclusion that can be drawn from the varied CEA on HepB is that the results are highly context-specific, and it is risky to generalize across settings.

C. Should Decisions about Funding Eradication Campaigns Be Made on the Basis of CEA?

As a report on the global impact of immunization programs elegantly stated, "The benefits of an eradication program are likely to be infinite, [so] the cost-effectiveness ratio for such programs will be near zero regardless of the cost in the numerator" [35]. Even with sensible adjustments of the discount rate to reflect an understanding that the long-term health benefits of the eradication program (e.g., deaths prevented) may be less valuable, on a unit basis, than health benefits that could be achieved in the nearer term with the same level of funding, it is likely that eradication will always "win" the cost-effectiveness competition when compared with routine immunization. A one-time investment that yields infinite returns, even over a very long time frame, is going to seem like a better deal than never-ceasing investments in a program that yields high but finite benefits. Thus, the decisions about whether to support the eradication of diseases such as polio and measles will have to be made on the basis of feasibility and affordability rather than on the theoretical grounds of long-term cost effectiveness. On the other hand, CEA can help managers decide among a variety of programmatic strategies by identifying the most efficient way to immunize a given population.

VII. WHAT REMAINS FOR COST-EFFECTIVENESS ANALYSIS?

Cost-effectiveness analysis is a well-regarded tool for economic analysis that has been widely used to inform or rationalize resource allocation decisions in the health sector. Immunization programs, with their highly effective, low-cost services, have been among the primary beneficiaries of increased attention to cost effectiveness: cost-effectiveness analyses have bolstered the case for greater spending on routine childhood immunization and have helped to steer immunization programs along a path of greater efficiency.

However, there is a vast and important agenda for improvement in and application of the methodologies. Some of the main elements of that agenda are highlighted next.

Systematically addressing the issue of uncertainty. By necessity, many of the most pressing questions for which cost-effectiveness analysis can be used require the estimation of variables that are unobservable and/or hard to measure, including future costs and health benefits. Increasingly, analysts are seeking ways to address the uncertainty in a systematic way. Among the newest efforts is the application of Monte Carlo and bootstrapping techniques to create a range of estimates of cost effectiveness based on an assumed distribution of the relevant variables [36–39]. The advantage of this approach is that it makes transparent the multiple assumptions that underlie the cost-effectiveness estimates and provides decision makers with a range of results based on known (or assumed) probabilities. Substantially more work is required to develop these methods, apply them to real-world questions, and effectively communicate analytic results to decision makers.

Identifying "moments" in the decision-making process when CEA is of the most benefit and introducing incentives for its use. CEA often is used to justify or reinforce a decision about resource allocation that has already been made. To better realize its potential for contributing to greater efficiencies within immunization programs, a more concerted effort is required to identify when those decisions are made—for example, when decisions are made about increasing coverage targets, introducing new vaccines, and/or changing the immunization program strategies—and how to apply the results of cost-effectiveness analysis. As part of those efforts, program planners may wish to consider both financial and nonmonetary incentives to encourage managers to undertake and use the results of cost-effectiveness analyses.

Standardization of methods for CEA of immunization programs. For Hib, HepB, and other vaccines for which the cost effectiveness in specific contexts is not yet known, new and better estimates of cost effectiveness are needed when decisions about their introduction are being made. To conduct those analyses, researchers will need standard methodologies that make clear the essential research questions, the analytic perspective, the ways of measuring costs and benefits, and the means of making uncertainty about both costs and benefits explicit in the development and presentation of cost-effectiveness results.

VIII. CONCLUSION

The cost effectiveness of traditional childhood vaccines is clear and undisputed, and the analyses that have demonstrated this feature have contributed to more general support for extending the benefits of national immunization programs. As new questions emerge regarding the correct approach to extend programs, in terms of both coverage and range of antigens, analyses of cost effectiveness have the potential to maximize the social benefits for every dollar invested. As shown in this part of Chapter 1, the appropriate use of cost-effectiveness analysis—or any other tool of economic evaluation—requires the analyst to understand the fundamental decisions at hand and the perspective that should be taken to examine them.

REFERENCES

1. World Bank (1993). "Investing in Health: World Development Report." Oxford, UK: Oxford University Press.
2. Murray, C. J. L., Evans, D. B., Acharya, A., and Baltussen, R. M. P. M. (2000). Development of WHO guidelines on generalized cost-effectiveness analysis. *Health Econ.* 9:235–251.
3. Jefferson, T. O., and Demicheli, V. (1996). "Elementary Economic Evaluation in Health Care." London: BMJ Books.
4. Drummond, M. F., O'Brien, B., Stoddart, G. L., and Torrance, G. W. (2001). "Methods for Economic Evaluation of Health Care Programmes," 2nd ed. New York: Oxford University Press.
5. Kobelt, G. (1996). "Health Economics: An Introduction to Economic Evaluation." London: Office of Health Economics.
6. Sloan, F., Ed. (1996). "Valuing Healthcare: Costs, Benefits and Effectiveness of Pharmaceutical and Other Medical Technologies." Cambridge, UK: Cambridge University Press.
7. Gold, M. R., Siegel, E. S., Russell, L. B., and Weinstein, M. C. (1996). "Cost-Effectiveness in Health and Medicine." New York: Oxford University Press.
8. Cutting, W. A. M. (1980). Cost–benefit evaluations of vaccination programmes. *Lancet* 20:634–635.
9. Demicheli, V., and Jefferson, T. (1996). Economic aspects of vaccination. *Vaccine* 14:941–943.
10. Van Damme, P., and Beutels, R. (1996). Economic evaluation of vaccination. *Pharmacoeconomics* 9(Suppl. 3):8–15.
11. Willems, J. S., and Sanders, C. R. (1981). Cost-effectiveness and cost–benefit analyses of vaccines. *J. Infect. Dis.* 144:486–493.
12. Creese, A. L., and Henderson, R. H. (1980). Cost–benefit analysis and immunization programmes in developing countries. *Bull. WHO* 58:491–497.
13. Creese, A. L., Sriyabbaya, N., Casabal, G., and Wiseso, G. (1982). Cost-effectiveness appraisal of immunization programmes. *Bull. WHO* 60:621–632.
14. Haaga, J. (1986). Cost-effectiveness and cost–benefit analyses of immunization programmes in developing countries. *In* "Advances in International Maternal and Child Health," D. B. Jelliffe and E. F. P. Jelliffe, Eds., Vol. VI. Oxford, UK: Clarendon Press.
15. Hinman, A. R. (1999). Economic aspects of vaccines and immunizations. *C.R. Acad. Sci. III* 322:989–994.

16. Edmunds, W. J., Gay, N. J., Gay, J. R., and Brisson M. (1999). "The Cost-Effectiveness of *Haemophilus influenzae* Type b (Hib), Hepatitis B Virus (HBV) and Measles Vaccination in Developing Countries." Preliminary report to WHO.

17. Walker, D., Khan, S., Akramuzzaman, S., Khan, M., Rushby, J. F., and Cutts, F. (1999). Cost-effectiveness analysis of measles control in Dhaka, Bangladesh. WHO (ref.no. HQ/98/4544190116328).

18. Jha, P., Bangoura, O., and Ranson, K. (1998). The cost-effectiveness of forty health interventions in Guinea. *Health Pol. Plan.* **13**:249–262.

19. Jamison, D. T., Mosley, W. H. O., Measham, A. R., and Bobadilla, J. L. (1993). "Disease Control Priorities in Developing Countries." Oxford, UK: Oxford University Press.

20. WHO (1980). Expanded programme on immunization: Economic Appraisal, Thailand. *Weekly Epidemiol. Record* **55**:289–296.

21. Phonboon, K., Shepard, D. S., Ramaboot, S., Kunasol, P., and Preuksaraj, S. (1989). The Thai expanded programme on immunization: Role of immunization sessions and their cost-effectiveness. *Bull. WHO* **67**:181–188.

22. Barnum, H. N., Tarantola, D., and Setaidy, I. F. (1980). Cost-effectiveness of an immunization programme in Indonesia. *Bull. WHO* **58**:499–503.

23. Robertson, R. L., and Qualls, N. (1985). Cost-effectiveness of immunization in the Gambia. *J. Tropical Med. Hygiene* **88**:343–351.

24. Shepard, D. S., Sanoh, L., and Coffi, E. (1986). Cost-effectiveness of the expanded programme on immunization in the Ivory Coast: A preliminary assessment. *Soc. Sci. Med.* **22**:369–377.

25. Creese, A. L., and Dominguez-Uga, M. A. (1987). Cost-effectiveness of immunization programs in Colombia. *Bull. PAHO* **21**:377–394.

26. Creese, A. L. (1984). Cost-effectiveness of alternative strategies for poliomyelitis immunization in Brazil. *Rev. Infect. Dis.* **6**:S404–S407.

27. Dominguez-Uga, M. A. (1988). Economic analysis of the vaccination strategies adopted in Brazil in 1982. *Bull. PAHO* **22**:250–268.

28. Disha, A., and Khan, M. M. (1998). Economic evaluation of hepatitis B vaccination in Bangladesh. Inter-Divisional Scientific Forum, ICDDR,B.

29. Hussey, G. D., Lasser, M. L., and Reekie, W. D. (1995). The costs and benefits of a vaccination programme for *Haemophilus influenzae* type B disease. *S. African Med. J.* **85**.

30. Beutels, P. (2001). Economic evaluations of hepatitis B immunization: A global review of recent studies (1994–2000). *Health Econ.* **10**:751–774.

31. Miller, M. (1998). An assessment of the value of *Haemophilus influenzae* type b conjugate vaccine in Asia. *Ped. Infect. Dis. J.* **17**:S152–S159.

32. Miller, M., and McCann, L. (2000). Policy analysis of the use of hepatitis B, *Haemophilus influenzae* type b, streptococcus pneumonia-conjugate and rotavirus vaccines in national immunization schedules. *Health Econ.* **9**:19–35.

33. Granata, A. V., and Hillman, A. L. (1998). Competing practice guidelines: Using cost-effectiveness analysis to make optimal decisions. *Ann. Intern. Med.* **128**:56–63.

34. Robinson, R. (1999). Limits to rationality: Economics, economists and priority setting. *Health Pol.* **49**:13–26.

35. England, S., Loevinsohn, B., Melgaard, B., Kou, U., and Jha, P. (2001). The evidence base for interventions to reduce mortality from vaccine-preventable diseases in low and middle-income countries. *CMH Working Paper Ser. WG5* **10**.

36. Fenwick, E., Claxton, K., and Sculpher, M. (2001). Representing uncertainty: The role of cost-effectiveness acceptability curves. *Health Econ.* **10**:779–787.

37. Baltussen, R. M. P. M., Hutubessy, R. C. W., Evans, D. B., and Murray, C. J. L. (2001). Uncertainty in cost-effectiveness analysis: Probabilistic uncertainty analysis and stochastic league tables. *WHO GPE Discuss. Paper Ser.* **34**.

38. Briggs, A. H. (1999). A Bayesian approach to stochastic cost-effectiveness analysis. *Health Econ.* **8**: 257–261.

39. Briggs, A. H., and Gray, A. M. (1999). Handling uncertainty in economic evaluation of healthcare interventions. *Br. Med. J.* **319**:635–638.

1 GLOBAL BURDEN OF DISEASE

PART C. Potential and Existing Impact of Vaccines on Disease Epidemiology

ELIZABETH MILLER

Immunization Division, Public Health Laboratory Service, Communicable Disease Surveillance Centre, London NW9 5EQ, United Kingdom

I. INTRODUCTION

Vaccines have the potential to produce major changes in disease epidemiology resulting both from direct protective effects on those vaccinated and from indirect herd immunity effects exerted on the unvaccinated. The latter may not necessarily be beneficial, with potential increases as well as decreases in disease incidence in the unvaccinated occurring depending on the biological properties of the organism, the epidemiology of the disease, the immunization strategy employed, the vaccine coverage achieved, and the mechanism of action of the vaccine. Only those vaccines that reduce disease transmission have the potential to generate herd immunity. The duration of vaccine-induced immunity and the extent to which it can be boosted by natural exposure are also major determinants of the impact of a vaccination program.

This part of Chapter 1 provides a conceptual framework that will help in understanding the epidemiological impact of existing vaccination programs and the factors that must be taken into account when designing a new program. The framework involves the use of mathematical models that can simulate the key stages of disease transmission in a population and can be used to predict the likely epidemiological impact of a new vaccine.

However, the starting point for those involved in evaluating an existing program or designing one for a new vaccine is to have a clear idea of the disease outcome that is being sought, as this has a major impact on choice of policy. The establishment of a surveillance system that can monitor progress toward achievement of the target outcome is also essential.

II. POTENTIAL EPIDEMIOLOGICAL OUTCOMES

Before vaccination policy can be determined, the aim in terms of disease control must be clearly defined. There are three main options: containment, elimination, and eradication [1].

A. Containment

This is the least ambitious of the options because it accepts that disease will remain endemic in the population and merely seeks to reduce morbidity to an "acceptably" low level. Inevitably, the concept of acceptable will change over time. For example when rubella vaccine was first introduced in the United Kingdom, a 90% reduction in the incidence of congenital rubella syndrome would probably have been considered acceptable. However, within two decades of its introduction, the birth of a single child with congenital rubella syndrome (CRS) was deemed unacceptable and a major policy revision was made in order to achieve the elimination of CRS in the United Kingdom [2].

B. Elimination

Disease elimination requires that there should be no endemic transmission of the infection in the target population, usually defined nationally or by a World Health Organization (WHO) region. This does not necessarily mean zero infections in that geographical locality because sporadic cases may still occur due to importations from outside. However, the spread from these exogenous cases should be sufficiently limited that endemic transmission is not reestablished [3]. This requires high immunity levels to be maintained in the population through continued vaccination.

Transition from disease containment to disease elimination may involve a stage at which the resources needed to prevent each additional case increase disproportionately. Extra resources are required both to support the cost of vaccine delivery, particularly for hard-to-reach groups or if the introduction of a second dose is necessary, and to establish the enhanced surveillance that is essential when the aim is elimination. Careful cost–benefit analyses need to be made, therefore, before elimination goals are set.

Measles elimination is currently the target of three WHO regions—the Americas, the eastern Mediterranean, and Europe—and a strategic framework within which elimination can be achieved has been defined [4, 5]. However, the wisdom of trying to achieve measles elimination in regions such as Africa, where there is difficulty in sustaining routine immunization programs, has been questioned [6]. Moreover, because of the highly infectious nature of measles, achieving and sustaining an elimination target elsewhere is proving difficult [7].

C. Eradication

Eradication is the most ambitious disease target because it requires the global destruction of the pathogen and allows for cessation of all control measures. As a result, there are huge cost savings for future generations. Eradication is only possible if there is no animal reservoir of the pathogen and if the vaccine is effective in preventing transmission. Smallpox is the only vaccine-preventable disease that has been eradicated so far, but global eradication of poliomyelitis is a target, although with some concerns about the ethics of pursuing this at the cost of other health interventions [8]. The prospects for measles eradication are being discussed [9].

Clearly, global eradication requires the pursuit of local elimination goals by all countries and regions, and it is this context in which measles elimination policies in poor countries with other calls on scarce healthcare resources could be justified [9].

III. SURVEILLANCE OF IMMUNIZATION PROGRAMS

A key requirement for achieving a successful immunization program is to have a surveillance strategy in place that can measure both the process (i.e., vaccine uptake) and the outcome (i.e., the epidemiological impact when compared with the prevaccination period) [10]. For the latter, information on the age-specific incidence of disease and, for those diseases for which there is a serological correlate of protection, on the age-specific distribution of immunity in the population before and after introduction of the immunization program is essential [11]. Surveillance of vaccine safety must also be established in order to assess the overall risk–benefit of the immunization program (see also Chapter 8).

Surveillance of the impact on disease incidence can involve a number of different data sources, including death notifications, clinical reporting schemes, and, where available, laboratory-based reporting [10]. Reliance on sentinel reporting derived from representative subsets of the population rather than nationally based reporting may be adequate when disease incidence is relatively high in the early stages of a new immunization program, but it is not sufficient if elimination policies are being pursued.

Laboratory confirmation of suspected cases becomes increasingly important as disease incidence declines. This is because the positive predictive value of a clinical case definition (i.e., the proportion of clinically suspected cases that are correctly diagnosed) will progressively decrease, becoming zero when the disease is eliminated. In the United Kingdom, for example, where measles transmission was interrupted in 1995, a noninvasive salivary immunoglobulin M (IgM) test is routinely offered for all clinically diagnosed cases of measles. Between July and September 2001, of the 338 measles cases notified in England and Wales on the basis of a clinical diagnosis and investigated by the salivary test, only 6 (1.7%) were confirmed as measles. Importations and limited spread from imported cases accounted for these few cases, rather than endemic transmission [12].

IV. IMMUNIZATION POLICY DETERMINATION

Although high-quality surveillance information is essential for achieving a successful immunization program, by itself it is not sufficient. The existence of a framework for making and implementing policy decisions in the light of surveillance information is also essential. The national vaccine advisory committees, such as the Advisory Committee on Immunization Practices (ACIP) in the United States, the National Advisory Committee on Immunisation (NACI) in Canada, and the Joint Committee on Vaccination and Immunisation (JCVI) in the United Kingdom, are examples of such policy-making committees. The key information inputs to the JCVI, mainly derived from the national Public Health Laboratory Service, and the mechanism for deciding policy are illustrated in Fig. 1.

Changes to an existing immunization program may be needed as more information emerges about the performance of the vaccine under conditions of routine use and as understanding grows about the epidemiological impact of the program. The use of mathematical models to assist in the interpretation of surveillance data for existing vaccination program, as well as exploring the potential impact of new programs, is an area of theoretical development that has seen increasing practical application [13].

V. BIOLOGICAL DETERMINANTS OF IMMUNIZATION PROGRAMS

The choice of an immunization strategy and its potential impact will be highly dependent on the biological properties of the vaccine, together with the epidemiological and biological characteristics of the natural infection. Some examples of the way in which these factors can interact are given in Table 1.

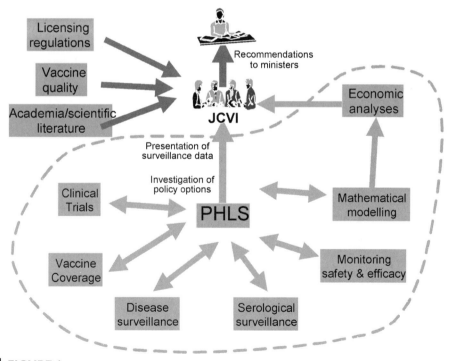

FIGURE 1   Role of JCVI in UK vaccine policy.

TABLE 1 Ways in Which the Biological Properties of the Disease and the Vaccine Impact on Immunization Strategy

Immunization program	Relevant biological characteristics of the disease	Relevant properties of the vaccine	Implications for vaccination program
Tetanus	Acquired from environment, no person-to-person transmission, maternal antibodies protect.	Effective in the presence of maternal antibodies, immunity wanes.	Eradication of pathogen not possible, disease elimination possible with universal coverage in infancy, regular boosters and maternal immunization if neonatal cases occur.
Rubella	Adverse consequences from infection in pregnancy, congenital infection can follow reinfection in pregnancy.	Highly effective against disease and transmission, less effective against reinfection, little evidence of waning immunity.	Elimination not possible with selective strategy as no herd effect and even with 100% coverage of childbearing population CRS cases due to reinfection can still occur.
Measles	Highly infectious, low average at infection.	Not effective in the presence of maternal antibodies, two doses needed to match immunity from disease, protects well against transmission.	Elimination–eradication possible but ideally needs vaccine that is effective in the presence of maternal antibody, but could still be achieved if high early coverage with a two-dose schedule can be sustained globally.
Varicella	Establishes latency and causes reactivation as shingles later, contact with varicella probably protects against shingles, risk of severe complications increases with age.	Vaccine establishes latency but risk of shingles reduced compared with natural infection, protects against transmission.	Potential to increase shingles in medium term by reducing booster effect from varicella exposure, potential for increasing cases of severe varicella by increasing average age of infection.
Pertussis	Immunity wanes, second attacks can occur.	Protects better against severe than mild disease, therefore difficult to stop transmission, vaccine-induced immunity wanes.	Elimination not possible with current vaccine, regular boosters probably required to maintain acceptable level of disease control.
Pneumococcal	More than 80 pathogenic serotypes, carriage more common than disease, serotype switching possible.	Highly effective against disease from vaccine serotypes, also reduces carriage of vaccine serotypes.	Containment is the only realistic goal but threatened by potential for replacement of vaccine with nonvaccine serotypes.

Although the implications of these biological and epidemiological characteristics can be understood intuitively, the ability to predict the likely outcome of a vaccination program on a more quantitative basis is desirable. This is essential if cost–benefit analyses are to be undertaken and if there are difficult choices about which strategy to adopt as, for example, with rubella vaccine (Table 2).

VI. DEVELOPMENT OF MATHEMATICAL MODELS

The ability to predict the likely population impact of a vaccination program and to select the most appropriate strategy to achieve the desired disease outcome has been considerably advanced with the development of sophisticated dynamic mathematical models that can simulate disease transmission in the population. The aim of building models is to simulate the key processes that underlie the interaction between the organism and the individual and that determine the behavior of the disease at the population level.

A. Model Structure

Such models typically have a compartmental structure representing the key sequential stages in the transmission of an infection. The simplest model involves progression from susceptibility (usually after loss of the passive immunity derived form maternal antibodies), to latency, when the individual has acquired the infection but is not yet infectious, to infectivity, and finally to recovery, when the individual is both immune and no longer capable of transmitting infection. The aim of vaccination is to move individuals from the susceptible to the immune state without entering the latent or infectious periods (Fig. 2a).

In reality, the biological processes determining these stages are usually more complex, particularly in relation to the duration and quality of natural

TABLE 2 Choice of Rubella Vaccination Strategy and Attendant Risks

Target population	Type of protection given to pregnant women	Risks
Universal vaccination of young children of both sexes	Indirect: Rubella-susceptible women protected from exposure	Continued occurrence of CRS cases and infections in pregnancy inevitable and may be considered unacceptable or that the program has been a failure.
Selective vaccination of girls and adult women	Direct: Women enter pregnancy already immune to rubella	Low coverage or waning immunity have the potential to increase the number of cases of CRS and infections in pregnancy.

or vaccine-induced immunity. An example of a more complex model that could be used to describe pertussis transmission is given in Fig. 2b. Provision exists not only for waning immunity with return to the fully susceptible state but also for boosting when partially immune such that return to the fully immune state can occur without becoming infectious.

Other refinements that may be made to improve correspondence with reality is to subdivide the population into groups, usually defined by age. This is particularly important if mixing within the population is not homogeneous but occurs preferentially within or between certain age groups. With measles, for example, one of the important determinants of contact is the opportunity for mixing that occurs in schools [14]. Measles models that allow for heterogeneous mixing between and within different age groups, as defined by school years, therefore have been developed [15–17].

B. Flows between Compartments

The flow of individuals from one compartment to the next can be represented by a set of differential equations that describe the rates of change of the population subgroups (susceptibles, infecteds, etc.) with respect to time. Given an initial set of conditions, such as the number of susceptible and infected individuals in the preimmunization era, the equations are used to derive the numbers of individuals in each state at any subsequent time interval. The

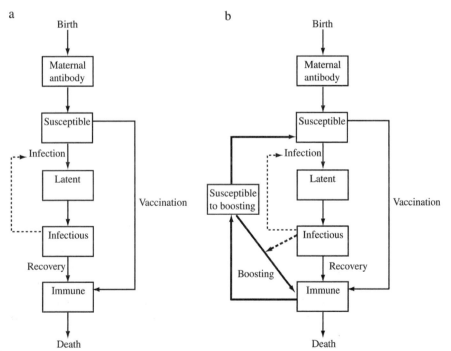

FIGURE 2 (a) Simple model that assumes that natural and vaccine-induced immunity is life-long. Solid lines indicate flows, dotted lines indicate influences. (b) Complex model that assumes that natural and vaccine-induced immunity wanes and is boosted by natural exposure. Solid lines indicate flows, dotted lines indicate influences.

interaction between susceptible and infectious individuals is the key equation and is derived from the "mass action" principle. In the simplest homogeneous mixing model, the rate at which new infections occur is equal to a constant (the transmission coefficient) multiplied by the product of the number of susceptible and infectious individuals in the population at any one time [18]. The transmission coefficient is disease- and population-specific because it depends on sociological and behavioral factors that influence the rate of contact between members of the population and on biological properties of the organism and host that influence the likelihood that a contact will result in transmission.

Other rates that must be defined are the rate of entry of new susceptibles into the population (a function of the birth rate and duration of maternal immunity) and the rate at which individuals exit through death (a function of the average life expectancy of individuals in the population and the case fatality rate of the disease). For simplicity, birth and death rates are usually assumed to be equal, a reasonable approximation for most developed countries. The rates at which individuals pass from the latent to the infectious compartment and from the infectious to the immune compartment will be directly determined by the biological properties of the organism, namely, the average duration of the latent and infectious periods.

C.   Parameter Estimation

Realistic estimates for the key model parameters are essential if valid models are to be built. Data that can be used to directly estimate parameters such as birth and death rates, the decay function for maternal antibody, and the average duration of the latent and infectious periods are usually available. However, this is not so for the transmission coefficient—the most difficult parameter to derive. Indirect methods for estimating this crucial parameter have been developed that utilize age-specific disease incidence data from the preimmunization era, the implicit assumption being that that mixing patterns do not alter appreciably over time. Suitable data can be obtained from prevaccination reports of incident cases as provided by disease surveillance systems, assuming that these are not biased with respect to completeness of reporting by age. Alternatively, and preferably, age-specific antibody prevalence data from the prevaccination period may be used [19, 20].

The parameter derived from such incidence data sets is called the "force of infection" and is the rate at which susceptible individuals acquire infection, from which the transmission coefficient can itself be derived. For the simple homogeneous mixing model, the force of infection is approximately equal to the reciprocal of the average age at infection. For age-stratified models, transmission coefficients for each age-group interaction are required. The so-called "who acquires infection from whom" (WAIFW) matrix specifies the transmission rates within and between each age group [15, 18]. For example, for a WAIFW matrix with five age groups, a total of 5^2 age-dependent transmission coefficients are required. Usually simplifications based on epidemiologically plausible assumptions can be made. For measles, for example, it is reasonable to assume that the contact rate between adults and 10- to 14-year-olds is unlikely to be substantially different from that between adults and 5- to 9-year-olds and that the matrix is symmetric [15].

D. Reproduction Number

A key summary statistic that is derived from other model parameters is the basic reproduction number of an infectious disease (R_0). This is defined as the average number of secondary cases that would be generated by a primary case in a totally susceptible population. It provides an overall measure of the potential for transmission of an infection within a population, and it is dependent not only on the transmission coefficient but also on the average duration of infectiousness. The higher the value of R_0, then the higher vaccine coverage must be to eliminate a disease [18].

In most real populations, not all individuals are susceptible to the disease. The (effective) reproduction number R takes into account the actual susceptibility of the population and is defined as the average number of secondary cases induced by a typical case. For diseases in endemic equilibrium, R will have an average value of 1. In reality R is constantly fluctuating. For example, the regular epidemic cycles seen with infections that confer permanent immunity and have no carrier state reflect natural cyclical changes in R as it oscillates around 1. There is a constant influx of new susceptibles into the population by births, and epidemics occur when this number reaches a critical threshold such that $R > 1$. The epidemic dies away when the number of susceptibles has been sufficiently reduced by infection to bring R below 1 again.

Elimination of an infectious disease by vaccination requires coverage rates that ensure R is maintained below 1. Interruption of transmission also requires that the average age at vaccination should be less than the average age at infection prior to the introduction of immunization; if disease elimination is being pursued, then vaccination must be applied at the youngest possible age. The effective reproduction number, like the average age at infection, can be estimated from serological data and has proven extremely useful in planning measles elimination policies [15].

E. Model Validation

Given the number of assumptions built into mathematical models and the inherent difficulty in estimating key parameters, model validation is essential. Computer simulations of a range of scenarios using different assumptions and parameter values are normally undertaken and the predictions compared with real epidemiological data. A simple comparison is whether the epidemic period of the disease (time between epidemics) and the age distribution of cases are accurately predicted by the model. For a simple disease transmission model, the epidemic period T can be derived from the average age at infection in the population A (obtained from serological or case notification data) and the serial interval between infections D (the sum of the latent and infectious periods) as follows:

$$T = 2(AD)^{1/2}$$

For the common vaccine-preventable infections such as measles, rubella, and pertussis, there is good correspondence between the observed and predicted epidemic period in most unvaccinated populations, suggesting that the simple model is a good approximation of reality [18].

For existing vaccination programs, model validation can utilize epidemiological data from the postvaccination period [21]. Introduction of a vaccine will tend to raise the average age at infection because it reduces the net rate of transmission in the population. The lengthening of the epidemic period and upward shift in the age distribution of cases predicted by the model using the vaccine coverage actually achieved can be compared with those observed in surveillance data.

VII.   EXAMPLES OF THE USE OF MATHEMATICAL MODELS TO GUIDE IMMUNIZATION POLICY

The use of mathematical models to guide immunization policy will be illustrated with three examples, two for existing vaccines (measles and rubella) and one with a relatively new vaccine (varicella).

A.   Measles

Measles presents a particular challenge to vaccinators because it is highly infectious with a very young average age at infection, and, as shown by experience in countries with single-dose programs, a two-dose schedule is essential if a disease elimination goal is being pursued (Table 1). There was much early optimism in the United States that a single-dose strategy with high coverage achieved through school entry laws could eliminate measles. Accordingly, the United States set a target date for this of October 1982, and an international conference was held in March 1982 to plan for global eradication [22]. Unfortunately this optimism was misplaced and sporadic measles outbreaks continued in the United States throughout the 1980s, culminating in the 1989–1990 resurgence that resulted in over 46,000 reported cases and 140 deaths. The focus of the resurgence was predominantly unvaccinated children under 5 years of age for whom immunization had been delayed until school entry and older, vaccinated school children who had received only a single dose [23].

The phenomenon of a prolonged period of low measles incidence—the so-called "honeymoon period"—followed by an explosive resurgence has now been observed in a number of countries with high coverage but only a single-dose program. During the honeymoon period when R is below 1, susceptible individuals accumulate in successive birth cohorts representing both those that fail to get vaccinated and those in whom vaccination fails. The latter proportion, though only around 5–10%, nevertheless reduces effective coverage substantially such that, with a 95% vaccination rate, the proportion susceptible in each birth cohort may still be as high as 10–15%. Inevitably over time the number of susceptible individuals reaches the critical threshold where R exceeds 1 and an epidemic ensues. The situation in the United States was compounded by the relatively late age at immunization of many children at the time of school entry.

It is now recognized that measles elimination requires high coverage at the earliest possible age and a two-dose program; delivery of the second dose may be through a routine program, as in the United States, Canada, and most European countries, or regular campaigns as in Latin America [24]. Targets for vaccine coverage, optimal ages at vaccination, and, for countries adopting the campaign approach, intervals between campaigns can now be set using predictions based on mathematical models. The preemptive use of a campaign to prevent an epidemic resurgence predicted by mathematical modeling has also been undertaken, first in the United Kingdom [15, 16] and subsequently elsewhere [25].

B. Rubella

When rubella vaccine first became available in the 1960s, there were two options for its use (Table 2). At the time, uncertainty about the duration of vaccine-induced immunity meant that the less risky option was to use it to render susceptible women of childbearing age immune rather than to seek to prevent the exposure of susceptible women during pregnancy by vaccinating young children. A second risk of a universal vaccination strategy, not appreciated at the time but subsequently revealed by mathematical modeling, was that at certain levels of vaccine uptake the rise in the average age at infection could result in an increase in rubella incidence in women of childbearing age compared with the prevaccination period [26]. Theoretical predictions were subsequently borne out by the experience in Greece, where two decades of moderate coverage with a universal infant program resulted in an overall increase in rubella susceptibility and incidence in women of childbearing age with an associated increase in cases of CRS [27]. Because the epidemiology of rubella differs between countries, even in the developed world, the coverage that is needed to avoid such perverse effects with a universal program will vary [28].

The advantage of a universal immunization program is that it is the only way to eliminate CRS, because even with 100% immunization of susceptible pregnant women, cases due to vaccine failure or reinfection in previously immune women would still occur. Models of rubella transmission have shown that maximum benefit with minimum risk can be achieved by a combined selective and universal strategy, and these were instrumental in the decision in the United Kingdom in 1988 to complement the existing strategy of vaccinating all school girls and susceptible adult women with universal infant immunization in the form of measles, mumps, rubella (MMR) vaccine [2, 29]. The epidemiological impact of the MMR program was immediate (Fig. 3), with a marked reduction in cases of CRS and abolition of the 3–4 yearly increases previously associated with the epidemic cycle of rubella.

The rubella policy now recommended by WHO emphasizes the importance of sustaining high coverage (>80%) if childhood vaccination is undertaken and supplementing this by vaccinating adult women to ensure high immunity levels in this critical group [30].

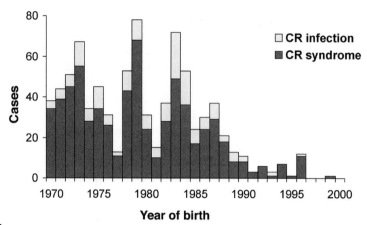

FIGURE 3 Cases of congenital rubella (CR) infection and syndrome, England and Wales, 1970–2000. From the National Congenital Rubella Surveillance Programme and the PHLS Communicable Disease Surveillance Centre.

C. Varicella

Varicella zoster virus causes chickenpox on primary exposure and can reactivate later in life to cause shingles. As primary infection is more serious in adults than in children, and as exposure to the virus might boost the immune response to both chickenpox and shingles, there are two main concerns regarding infant varicella vaccination: (1) that it could lead to an increase in adult disease as with the rubella example and/or (2) that it could lead to a temporary increase in the incidence of shingles. These potential adverse consequences have been investigated extensively prior to the introduction of the vaccine by using mathematical models [31–35].

The consensus view of these different modeling studies is that, although an increase in adult cases may occur, the overall varicella-associated burden is likely to decrease in the long term, regardless of the level of vaccine coverage. However, less attention has been paid to the potential for increasing zoster, but work in the United Kingdom suggests that it appears likely; the more effective vaccination is at preventing varicella, the larger the increase in zoster incidence [34]. Sensitive surveillance of both chickenpox and shingles is essential in countries that have implemented, or are about to implement, varicella vaccination.

VIII. CONCLUSION

The ability to understand and predict the potential epidemiological impact of a vaccination program has been greatly enhanced by the use of mathematical models of disease transmission. Of particular importance is the prediction of nonintuitive outcomes, because this can emphasize the need for enhanced surveillance in particular areas and ensure that safe vaccine coverage targets are set if a potentially risky option is undertaken.

The use of models to help understand the population biology of organisms such as the meningococci and pneumococci that have the potential to undergo serogroup–type switching and the relationship between carriage and disease-causing isolates is likely to be a major application in the future with the arrival of new polysaccharide conjugate vaccines. Similarly, for pathogens such as the herpes viruses that establish latency, models that can explore the outcome of altering by vaccination the complex relationship between primary infection and reactivation are likely to be of increasing utility for policymakers. However, underpinning all of the models must be sound baseline data on the epidemiology of the disease and its age-specific disease incidence and antibody prevalence profile in the population prior to embarking on any immunization program.

REFERENCES

1. Allwright, S. P. A. (1998). Elimination or reduction of diseases? *In* "Opportunities for Health Service Action in Europe," E. J. Silman and S. P. A. Allwright, Eds., pp. 5–7. Oxford, UK: Oxford Medical Publications.
2. Miller, E. (1986). Can congenital rubella be eliminated? *In* "Public Health Virology," P. Mortimer, Ed., pp. 36–49. Public Health Laboratory Service, London.
3. De Serres, G., Gay, N., and Farrington C. P. (2000). Epidemiology of transmissible diseases after elimination. *Am. J. Epidemiol.* **151**:1039–1048.
4. Ramsay, M. (1999). A strategic framework for the elimination of measles in the European region. Copenhagen: World Health Organization Regional Office for Europe.
5. Hersch, B. S., Tambini, G., Nogueira, A. C., Carrasco, P., and de Quadros, C. (2000). Review of regional measles surveillance data in the Americas, 1996–99. *Lancet* **355**:1943–1948.
6. Cutts, F. T., Henao-Restrepo, A. M., and Olive, J. M. (1999). Measles elimination: Progress and challenges. *Vaccine* S47–S52.
7. Gay, N. J. (2000). Eliminating measles—No quick fix. *Bull. WHO* **78**, 949.
8. Taylor, C., Cutts, F., and Taylor, M. (1997). Ethical dilemmas in current planning for polio eradication. *Am. J. Public Health* **87**:922–925.
9. Cutts, F. T., and Steinglass, R. (1998). Should measles be eradicated? *Br. Med. J.* **316**:765–767.
10. Begg, N., and Miller, E. (1990). Role of epidemiology in vaccine policy. *Vaccine* **8**: 180–189.
11. Osborne, K., Gay, N., Hesketh, L., Morgan-Capner, P., and Miller, E. (2000). Ten years of serological surveillance in England and Wales: Methods, results, implications and action. *Int. J. Epidemiol.* **29**:362–368.
12. http://www.phls.org.uk/publications/CDR%20Weekly/pages/immunisation.html#MMR27–39.
13. Nokes, D. J., and Anderson, R. M. (1988). The use of mathematical models in the epidemiological study of infectious diseases and the design of vaccination programmes. *Epidemiol. Infect.* **101**:1–20.
14. Fine, P. E. M., and Clarkson, J. A. (1982). Measles in England and Wales II. The impact of the measles vaccination programme on the distribution of immunity in the population. *Int. J. Epidemiol.* **II**:15–25.
15. Gay, N., Hesketh, L. M., Morgan-Capner, P., and Miller, E. (1995) Interpreting serological surveillance data for measles using mathematical models: Implications for vaccine strategy. *Epidemiol. Infect.* **115**:139–156.
16. Babad, H. R., Nokes, D. J., Gay, N. J., Miller, E., Morgan-Capner, P., and Anderson, R. M. (1995). Predicting the impact of measles vaccination in England and Wales: Model validation and analysis of policy options. *Epidemiol. Infect.* **114**:319–344.
17. Schenzle, D. (1998). An age-structured model of pre- and post-vaccination measles transmission. *IMA J. Math Appl. Med. Biol.* **1**:169–191.
18. Anderson, R. M., and May, R. M. (1991). "Infectious Diseases of Humans: Dynamics and Control," 2nd ed. Oxford, UK: Oxford University Press.

19. Grenfell, B. T., and Anderson, R. M. (1995). The estimation of age-related rates of infection from case notifications and serological data. *J. Hygiene (Cambridge)* **95**:419–436.
20. Farrington, C. P. (1990). Modelling forces of infection for measles, mumps and rubella. *Stat. Medicine* **9**:953–967.
21. Edmunds, W. J., van de Heijden, O. G., Eerola, M., and Gay, N. J. (2000). Modelling rubella in Europe. *Epidemiol. Infect.* **125**: 617–634.
22. Evans, A. S. (1985). The eradication of communicable diseases: Myth or reality? *Am. J. Epidemiol.* **122**:199–207.
23. Atkinson, W. L., and Orenstein, W. A. (1992). The resurgence of measles in the United States, 1989–90. *Annu. Rev. Med.* **43**:451–463.
24. de Quadros, C. A., Hersh, B. S., Nogueira, A. C., Carasco, P. A., and da Silveira, C. M. (1998). Measles eradication: Experience in the Americas. *Bull. WHO* **76**(Suppl 2):42–52.
25. Forrest, J. M., Burgess, M. A., Heath, T. C., and McIntyre, P. B. (1998). Measles control in Australia. *Communicable Dis. Intell.* **22**:33–36.
26. Knox, E. G. (1980). Strategy for rubella vaccination. *Int. J. Epidemiol.* **9**:13–23.
27. Panagiotopoulos, T., Antoniadou, I., and Valassi-Adam, E. (1999). Increase in congenital rubella occurrence after immunisation in Greece: Retrospective survey and systematic review. *Br. Med. J.* **319**:1462–1467.
28. Edmunds, W. J., Gay, N. J., Kretzschmar, M., Pebody, R. G., and Wachmann, H. (2000). The pre-vaccination epidemiology of measles, mumps and rubella in Europe: Implications for modelling. *Epidemiol. Infect.* **125**:635–650.
29. Anderson, R. M., and Grenfell, B. T. (1986). Quantitative investigations of different vaccination policies for the control of congenital rubella syndrome (CRS) in the United Kingdom. *J. Hygiene Cambridge* **96**:305–333.
30. World Health Organization (2000). Rubella vaccines: WHO position paper. *Weekly Epidemiol. Record* **75**:161–172.
31. Garnett, G. P., and Grenfell, B. T. (1992). The epidemiology of varicella–zoster virus infections: A mathematical model. *Epidemiol. Infect.* **108**:495–511.
32. Halloran, M. E., Cochi, S. L., Lieu, T. A., Wharton, M., and Fehrs, L. (1994). Theoretical epidemiologic and morbidity effects of routine varicella immunization of preschool children in the United States. *Am. J. Epidemiol.* **140**:81–104.
33. Garnett, G. P., and Ferguson, N. M. (1996). Predicting the effect of varicella vaccine on subsequent cases of zoster and varicella. *Rev. Med. Virol.* **6**:151–161.
34. Brisson, M., Edmunds, W. J., Gay, N. J., Law, B., and De Serres, G. (2000). Modelling the impact of immunisation on the epidemiology of varicella zoster virus. *Epidemiol. Infect.* **125**: 651–669.
35. Schuette, M. C., and Hethcote, H. W. (1999). Modeling the effects of varicella vaccination programs on the incidence of chickenpox and shingles. *Bull. Math Biol.* **61**:1031–1064.

2 IMMUNOLOGY

PART A. Basic Immunology of Vaccine Development

ROBERT A. SEDER AND JOHN R. MASCOLA

Vaccine Research Center, National Institute of Allergy and Infectious Diseases, National Institutes of Health, Bethesda, Maryland 20892

I. INTRODUCTION

The historical basis for the concept of immune memory following a primary exposure to an infection dates back to 430 BC when Thucydides described the plague of Athens: "It was with those who had recovered from the disease that the sick and the dying found most compassion. These knew what it was from experience, and had now no fear for themselves; for the same man was never attacked twice—never at least fatally." Approximately 2000 years later at end of the eighteenth century, in the seminal experiments by Jenner, it was conclusively shown that vaccination with cowpox virus could protect individuals following an infectious challenge with smallpox virus. This demonstrated that immunization could generate a protective immune response by mimicking specific aspects of an infectious agent. Some 80 years after Jenner's observations, Louis Pasteur developed the first vaccine in the laboratory. He produced a mixture of live and inactivated rabies virus as an effective rabies vaccine. Today, various vaccine formulations have been shown to be successful against a wide variety of viral and bacterial infections and have had a tremendous impact in reducing morbidity and mortality worldwide from infectious disease. Despite the enormous success of current vaccines, there are no uniformly effective vaccines for infections such as HIV, malaria, and *Mycobacterium tuberculosis*, which today account for a substantial proportion of the deaths worldwide from infection. Moreover, the substantial morbidity and mortality associated with Ebola virus infection and the potential threat of bioterrorism have made vaccine biology a major priority of scientific and medical investigation.

Although the goal of vaccination is to provide protection against disease caused by a microorganism, it is also important to develop a thorough understanding of the immune mechanism by which protection is achieved. This facilitates a more systematic approach to vaccine development that allows for the improvement of existing vaccines and the rational design of future vaccines. Thus, vaccine studies seek to define "immune correlates" of protection that can provide a useful guide in determining the type of immune response a vaccine should elicit. In this regard, immune responses can be divided into humoral (antibody) and cellular (T cell) immunity. The majority of licensed human vaccines work by inducing protective antibodies. Therefore, the immune correlates for most current vaccines are derived by measurement of the level of antibody elicited after immunization. It is also important to point out that, whereas antibodies are necessary and possibly sufficient for protection against certain infections, protective immunity against many infectious diseases involves a complex interplay of both the humoral and cellular immune responses. Indeed, a major impediment to the development of successful vaccines against HIV, *M. tuberculosis*, and malaria has been a lack of understanding of how to induce long-term protective cellular immune responses. This part of Chapter 2 will first review the basic immunology relevant to the goals of inducing and maintaining humoral and cellular immune responses. Later we will discuss how our current understanding of these immune mechanisms is being used to guide the design of vaccines capable of mediating protection against a variety of pathogens.

II. OVERVIEW OF THE IMMUNE RESPONSE

A. Innate and Adaptive Immunity

When a microbe enters the body, the immune system responds through a diverse set of mechanisms in an attempt to eliminate the infectious agent. These immune responses can be segregated into two major compartments designated the innate and the adaptive immune systems. The innate response relies on immediate recognition of antigenic structures common to many microorganisms by a selected set of immune cells with rapid effector function [1–4]. In contrast, the adaptive immune response is made up of B and T lymphocytes that have unique receptors specific to various microbial antigens. These antigen-specific receptors are encoded by genes generated during a complex process of gene rearrangement that occurs during the course of lymphocyte development. As each B and T lymphocyte contains a unique antigenic receptor, it allows for a large (10^{14} and 10^{18}, respectively) and diverse population of cells capable of recognizing a wide spectrum of pathogens. This is termed the lymphocyte repertoire. Although the size and diversity of the repertoire make it likely that there is an antigen-specific lymphocyte for any given pathogen, the frequency of these cells can be extremely low and normally will not be sufficient to protect the host against a primary infection. Following antigenic stimulation, there is activation and expansion of these antigen-specific cells. It is this clonal expansion, and ultimately the perpetuation of these antigen-specific cells (memory cells), that protects us against a secondary challenge. This initial or primary adaptive immune response to a foreign antigen takes 5–7 days to develop, during which time the infecting microbe can replicate in the body. Once primed, the secondary immune response is more rapid and robust. Thus, the goal of vaccination is to enhance the number of antigen-specific B and T cells against a given pathogen sufficient to provide protection following a secondary infectious challenge.

The other major arm of the immune system is the innate component. The innate immune response, which includes phagocytic cells, antimicrobial peptides, and complement, has been viewed to be primarily involved in the initial defense against infection; however, several recent discoveries have shown that the innate immune response has an additional important role in regulating the adaptive immune response. Although a thorough review of innate immunity is beyond the scope of the chapter, there are fundamental differences between the innate and adaptive immune responses that are important to review in order to understand how the innate response influences the adaptive response. Innate immune recognition of infectious pathogens is mediated by germline-encoded receptors that recognize a relatively limited number of highly conserved microbial structures termed pathogen-associated molecular patterns (PAMP) [1, 3]. These PAMPs interact with pattern recognition receptors, called toll-like receptors [5, 6], that are expressed on various cells of the innate immune response, including the major antigen-presenting cells (APCs) [macrophages, dendritic cells (DCs), and B cells] responsible for T-cell activation. It has been shown that pathogenic stimuli such as lipopolysaccharides [7], mycobacterial antigens [8], and specific sequences contained within bacterial DNA [9] can bind to toll-like receptors and stimulate the secretion of cytokines and chemokines from APCs. This innate recognition pathway leads to an increase in APC effector function. Stimulated APCs then play a key role in the initiation of the adaptive immune response. Thus, vaccines that can specifically activate the innate immune response could have a profound effect on the initial adaptive immune response generated by immunization. These observations on the interaction of innate and adaptive immunity will lead to a greater focus on how vaccines target APCs and stimulate the innate immune response to optimize the adaptive cellular immune response.

B. Adaptive Cellular Immunity

1. CD4⁺ T-Cell Responses

a. Antigen Presentation

The goal of immunization is to generate and sustain a population of antigen-specific immune cells that will mediate effector functions to prevent and/or control disease following exposure. The primary immune response is initiated by the presentation of antigen to naive CD4$^+$ T cells (Fig. 1). Each CD4$^+$ T cell expresses a unique receptor that recognizes antigen associated with major histocompatibility complex (MHC) class II molecules expressed on specific APCs. MHC class II expression has a relatively limited tissue distribution and is found primarily on DCs, macrophages, and B cells. Processing of

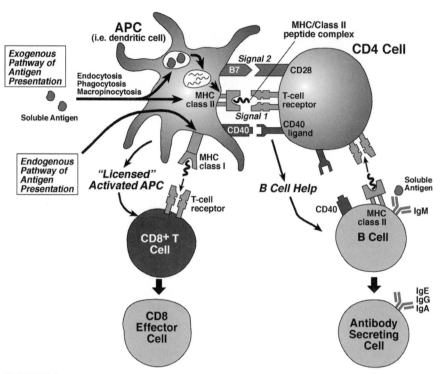

FIGURE 1 Initiation of primary humoral and cellular immune responses. Primary cellular immune responses are initiated when antigen is processed and presented by MHC class I and class II molecules to the antigen-specific T-cell receptor on CD8$^+$ and CD4$^+$ T cells, respectively (signal 1). In addition, engagement of CD28 on T cells with B7 molecules on APCs enhances the response (signal 2). Other costimulatory interactions between CD40L on T cells and CD40 on DCs facilitate activation of DCs, leading to activation of CD8$^+$ T cells. Similarly, interaction of CD40L on T cells with CD40 on B cells enhances B-cell activation and regulates isotype switching.

antigen by the MHC class II pathway occurs by the exogenous pathway, in which soluble antigens are taken up by APCs, enzymatically digested into peptides, and then associated with MHC class II molecules on the surface of the cell (Fig. 1). This MHC class II–peptide complex recognition of antigen by CD4$^+$ T cells is the first of two signals necessary to stimulate an effective CD4$^+$ T-cell response. APCs also express cell surface molecules such as B7-1 and B7-2, which interact with CD28 on T cells to provide further activation. This is referred to as the costimulatory pathway; it provides the second signal. For initiation of a functional primary T-cell response, both signal 1 and signal 2 are required, and the most potent and effective APC is the DC [10]. DCs are a complex group of myeloid- and lymphoid-derived leukocytes whose role is to initiate and modulate immune responses. DCs are found in lymphoid tissues and are also strategically located in potential pathogen entry sites, such as mucosal tissues and skin. Upon encountering antigen, immature DCs migrate to secondary lymphoid organs, where they initiate immune responses in both T cells and B cells. Of note, in addition to presenting antigen to T cells, DCs are capable of inducing a broad array of cytokines and chemokines with important roles in regulating the type of effector CD4$^+$ T cell generated.

Evidence suggests that there are specific subsets of DCs that can exert both stimulatory and inhibitory–regulatory effects on the immune response depending on the types of cytokines they produce. The type of APC, through its expression of costimulatory molecules and cytokine production, can markedly influence the type of immune response generated. Thus, in developing vaccines for infections requiring cellular and humoral immunity, substantial effort is aimed at targeting vaccine antigens to specific types of DCs that would lead to a potent immune response.

b. Effector Function of CD4⁺ T Cells

CD4⁺ T cells play a central role in several aspects of the immune response. The historical "helper" designation of CD4⁺ T cells referred to their ability to activate B cells and enhance the qualitative and quantitative aspects of antibody production (Fig. 1). The ability of CD4⁺ T cells to enhance B-cell activation and survival and to precipitate switching of the class of antibody secretion (described later) is mediated by the cognate interaction between CD40 ligand (CD40L) on T cells with CD40 on B cells [11]. Additionally, CD4⁺ T cells are important in sustaining or enhancing the effector function of CD8⁺ T cells [12–14]. This occurs through the production of cytokines such as interleukin-2 (IL-2) and/or through an indirect process of activating DCs. This latter function of DC activation by CD4⁺ T cells is mediated by CD40L on the CD4⁺ T cell interacting with CD40 on the DC and has been termed "licensing" of DC. Thus, upon stimulation by CD4⁺ helper cells, the licensed DC can directly stimulate CD8⁺ effector cells (Fig. 1). Together, these mechanisms highlight the critical role of CD4⁺ T cells in regulating B-cell and CD8⁺ T-cell responses. Thus, CD4⁺ T-cell activation is a critical component of successful vaccination.

In addition to the aforementioned functions, CD4⁺ T cells secrete a large number of cytokines that have both important immunoregulatory and direct biologic functions. Following activation, naive CD4⁺ T cells can differentiate into two relatively distinct subsets termed T helper 1 (Th1) and T helper 2 (Th2) type cells, on the basis of their characteristic production of specific cytokines [15, 16]. Th1 cells produce IL-2 and interferon-γ (IFN-γ), whereas Th2 cells produce IL-4, IL-5, IL-10, and IL-13. Th2 cells have been shown to be important in mediating the host response against extracellular parasites, such as worm infections. Th2-type responses are also strongly associated with mediating allergic and asthmatic disease. Because Th2 responses are often correlated with susceptibility to intracellular infections, immunization for such pathogens would not seek to elicit Th2 responses. On the other hand, Th1 cells are essential in mediating protection against a variety of intracellular infections, such as viral infections, and pathogens, such as *M. tuberculosis*, malaria, and *Leishmania major* (Fig. 2). Thus, as it relates to vaccine development, there is a strong interest in generating Th1 responses to protect against infections in which cellular immunity is a critical component of protection. The induction of Th1 responses for many infectious pathogens is also critically dependent on another cytokine, IL-12 [17]. IL-12 is produced by APC early in the course of an infection and acts on CD4⁺ T cells to promote a Th1 response. This again provides an example of the critical link between the innate and adaptive immune responses.

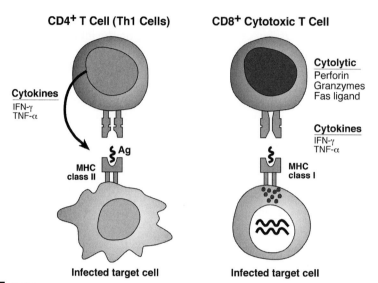

FIGURE 2   Effector function of Th1 and CD8$^+$ T cells. Th1 cells mediate much of their effector function through secretion of cytokines such as IFN-γ and TNF-α on infected cells expressing antigen in association with MHC class II molecules. CD8$^+$ T cells exert their effector function through cytokines and/or cytolytic mechanisms on infected target cells expressing antigen in association with MHC class I molecules.

2. CD8$^+$ T-Cell Responses

a. Antigen Presentation

CD8$^+$ T cells recognize antigens processed and presented by cells expressing MHC class I molecules (Fig. 1). With the exception of a small number of cells (red blood cells, spermatogonia), all cells in the body express MHC class I and are thus potentially able to present antigen to CD8$^+$ T cells and/or be targets for subsequent CD8$^+$ T-cell killing. In contrast to the way in which antigens are processed and presented in the MHC class II–CD4 pathway, MHC class I presentation to CD8$^+$ T cells is mostly limited to antigens processed through the endogenous pathway. In this pathway, antigen processing occurs within the cytosol (e.g. from viral infection of the cell), resulting in degradation of the antigen into small peptides that are loaded onto MHC class I molecules in the endoplasmic reticulum and then transported to the cell surface. It is important to note that this fundamental difference in antigen processing and presentation for CD4$^+$ and CD8$^+$ T cells has important ramifications in terms of how to design vaccines to elicit specific immune responses.

b. Effector Function

CD8$^+$ T cells mediate their effector function through the production of cytokines such as IFN-γ and tumor necrocis factor-α (TNF-α) and through a direct cytolytic (CTL) mechanism (Fig. 2). The mechanism of killing by the latter pathway occurs by the release of granule contents such as perforin and granzyme from activated CD8$^+$ T cells. Perforin is a complement-like protein that can produce physical disruption of a cell by the formation of pores in the lipid membrane. Granzymes are proteases that induce a process termed programmed cell death, or apoptosis. In addition, CD8$^+$ T cells can kill by a process of FAS-mediated lysis. Upon activation, CD8$^+$ T cells up-regulate the surface expression of FAS ligand (FAS-L), which binds to FAS on target cells and induces a cascade of events resulting in apoptosis of the target cell. CD8$^+$ T cells are thought to control viral as well as nonviral intracellular pathogens (fungi, protozoa, and certain bacteria) by lysis of infected target cells and/or by cytokine-induced inhibition of pathogen replication.

3. Cellular Memory

The ability to develop successful vaccines for diseases requiring cellular immunity (e.g., HIV, malaria, M. tuberculosis, L. major) has been hampered by a limited understanding of what is required to induce long-term cellular memory responses. Substantial advances in our understanding of cellular memory should aid the development of vaccines for a variety of diseases [18–20].

Induction of antigen-specific CD4$^+$ and CD8$^+$ T cells *in vivo* proceeds through three phases: expansion, death, and memory. During the initial expansion phase, there is a substantial increase in the frequency of antigen-specific cells following exposure to antigen [21, 22]. For CD8$^+$ T cells, the amount of antigen at the time of initial activation is critical for establishing the initial threshold of the response (burst size) that will ultimately determine the magnitude of the memory response [23, 24]. Moreover, it has been demonstrated that a brief (as little as 2 hr) encounter with a sufficient amount of antigen is enough to drive a naive CD8$^+$ T cell through multiple rounds of cell division to induce specific cytotoxic effector function [25–27]. Together, these observations highlight the importance of the amount of antigen at the initiation of the CD8$^+$ T-cell response to establish the optimal frequency of antigen-specific CD8$^+$ T cells. The expansion phase, which usually occurs over the first week, is followed by the death phase, in which >90% of the activated cells undergo activation-induced apoptosis. The memory phase of the CD8$^+$ T-cell response then ensues and can be stably maintained over a long period of time. It is notable that following a viral infection in mice, though the magnitude of the CD8 memory response remains stable for up to 3 years, there is a slow decline in the CD4$^+$ T-cell response [28]. These data are the first to suggest a demonstrative difference in the long-term stability of CD4$^+$ and CD8$^+$ memory cells; this may have important ramifications for vaccines requiring cellular immune responses.

a. Functional Differences between Naive and Memory T Cells

Memory CD4$^+$ and CD8$^+$ T cells differ from naive T cells in cell phenotype and biologic function. Memory cells can express different patterns of specific cell surface markers denoting prior activation. For example, mouse memory cells express surface levels of CD44 and low levels of CD62L (CD44^{hi} and CD62L^{lo}), whereas naive cells are CD44^{lo} and CD62L^{hi}. Expression of CD62L, however, can change over the course of time, making it a less specific marker for memory T cells in both mice and humans. In humans, there is also considerable overlap in phenotype between naive and memory T cells. In general, CD45RA, CD27^{hi}, and CD62L^{hi} are considered helpful in identifying naive T cells, whereas the markers CD45RO, CD11a, and CD95 are useful to define memory cells. In addition, the kinetics and magnitude of the memory immune response in terms of effector function (cytokine production and/or cytotoxic activity) are faster and of substantially higher magnitude following antigenic stimulation when compared with those of naive T cells [29–35].

b. Heterogeneity of the Memory Response: Interaction between Resting and Effector T Cells

In addition to being phenotypically and functionally different from naive T cells, it is important to note that memory T cells themselves are heterogeneous. Memory cells can be divided into "resting" and "effector" cells on the basis of a variety of phenotypic and functional properties. Resting memory cells reside in major lymphoid organs, are small in size, and require reactivation (e.g., exposure to antigen) to elicit effector functions. In contrast, effector memory cells reside in peripheral tissues, are usually larger in size, and can more rapidly produce effector cytokines and mediate direct cytolytic activity. Data show that resting and effector memory CD4+ and CD8+ T cells can be distinguished by their expression of surface markers known as lymph-node-homing receptors [36]. Resting memory cells in lymph nodes are CD62L^{hi} and CCR7+ and are also referred to as "central" memory cells [36]. By contrast, effector memory cells in peripheral tissues are CD62L^{lo} and CCR7−. Finally, there has been extensive investigation into the pathway of memory cell differentiation from naive T cells. A major question is whether resting memory T cells derive from effector cells ("linear" differentiation pathway) or arise independently from naive T cells. There is strong evidence from both *in vitro* and *in vivo* mouse studies that resting memory CD8+ T cells derive primarily from effector cells, strongly supporting the linear differentiation pathway for CD8+ T cells [37, 38]. It is not presently known whether the linear differentiation pathway exists for CD4+ effector T cells. In conclusion, the dynamics between resting and effector memory T cells and the factors controlling these cells are likely to be key determinants affecting whether immunization generates a memory immune response sufficient to provide protection.

c. The Requirement for Antigen in Maintaining Cellular Memory

Among human infectious diseases there are several examples of initial exposure to a microbial pathogen conferring lifelong immunity against reinfection. For certain infections, this immunity is maintained in the apparent absence of any possible reexposure to the primary infection. These observations raise two possibilities. The first is that long-term protective immunity is sustained in the absence of persistent infection and/or persistent antigen. Alternatively, long-term protective memory immunity could be sustained by the presence of small amounts of antigen from the primary infection and/or from cross-reactive self or environmental antigens. This question of whether antigen is required in sustaining long-term immunity (humoral and cellular) is critical for vaccine development. Although some initial data showed that antigen is required to maintain memory cells [39–41], more recent data in well-defined mouse models provide strong evidence that both memory CD4+ and CD8+ T cells can be maintained for long periods *in vivo* in an antigen-independent manner [35, 42–45]. Although antigen does not appear to be required to sustain memory CD4+ or CD8+ T cells per se, it is still possible that optimal immune protection against infection does require subsequent antigenic stimulation (i.e., vaccine booster immunization).

d. The Requirement for Cytokines in Maintaining Cellular Memory

The finding that antigen-specific CD8+ memory cells can be maintained at a stable number over a long period of time in the absence of antigen suggests that other factors, such as cytokines, might be involved in sustaining memory T cells. For CD8+ T cells, there is good evidence that IL-15, a potent T-cell growth factor, is important in homeostasis of CD8+ memory cells [18, 46–49]. A potent source of IL-15 is APCs such as DCs in response to a variety of stimuli, including type I interferons [50]. This again illustrates the importance of DCs for both inducing and sustaining cellular immune responses. In addition to IL-15, evidence in the mouse model suggests that, although the cytokine IL-7 is dispensable for antigen-induced expansion of CD8+ T cells, it is important for sustaining antigen-specific CD8+ T-cell memory response [51]. Finally, for CD4+ T cells, experimental mouse models of parasitic infection show that continuous IL-12 is required to mediate immune protection. It is not yet clear, however, whether IL-12 is essential to maintain the existence of memory Th1 cells or merely to sustain or enhance the production of IFN-γ from existing Th1 cells sufficient to control the infection [52, 53].

C. Humoral Immunity

1. Antibodies

Antibodies are a complex group of glycoproteins, termed immunoglobulins (Igs), that bind with high specificity to protein and polysaccharide antigens on microbial pathogens [54, 55]. Antibodies are present both in plasma and at mucosal surfaces and can mediate protective immunity by a variety of mechanisms. In fact, most current vaccines work by generating antibodies that prevent the invading organism from establishing infection in the host [56–58]. Binding of antibody to antigenic components of the invading pathogen can directly inactivate the microorganism or its toxin or can work in concert with complement proteins and cellular immune mechanisms to mediate protective immunity.

All Igs have the same basic structure consisting of four polypeptide chains held together by disulfide bonds (Fig. 3). Each antibody molecule contains two identical heavy chains (molecular weight of 55,000–75,0000) and two identical light chains (molecular weight of 25,000). The amino-terminal end of the Ig molecule contains two arms of highly variable protein sequences, termed variable domains, that recognize specific antigenic sequences. Each variable domain is made up amino acid sequences from one heavy and one light chain; the resulting amino acid diversity allows noncovalent binding to a vast number of conformational binding sites. Antibodies are termed "bivalent" because the two antigen-binding arms allow one antibody molecule to bind to two identical antigens. The more conserved carboxyl-terminal region of the antibody

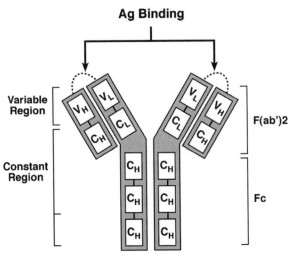

FIGURE 3   Ig structure. The basic structure of Ig molecules includes two identical disulfide-linked heavy chains that are each linked to an identical light chain. The antibody-binding region is formed by the combined sequence of the variable regions of the heavy (V_H) and light (V_L) chains. Together, the two variable domains create an antigen-specific binding site. Each of the two antibody-binding sites is referred to as a Fab region. Antibody class (isotype) and biologic activities such as complement fixation and binding to cell surface Ig receptors are determined by the constant (Fc) region of the Ig molecule.

includes the constant regions of the two heavy chains. It is these constant regions that bind to the Fc receptor of cells such as neutrophils and mononuclear phagocytes. Additionally, the constant regions define the five classes of antibody (IgG, IgA, IgM, IgD, and IgE) as well as the four subclasses of IgG and two subclasses of IgA [54]. As will be discussed later, Ig class and subclass are important determinants of the anatomic distribution and biologic function of antibodies.

2.   B Cells and Antibody Production

Antibodies are produced by B lymphocytes, or B cells, that express Ig of a single specificity on their cell surface. Thus, each B cell recognizes a unique antigenic structure and, upon encountering this antigen, is stimulated to secrete soluble antibody. It is estimated that the B lymphocytes are capable of producing up to 10^{14} different antibody variable regions, which are produced by an extraordinary process of splicing and recombining genes that encode the variable regions of Ig [59–61]. Although a detailed discussion of the mechanisms of gene rearrangements that occur in each B cell to produce specific antibodies is beyond the scope of this part of the chapter, it is important to understand the cellular events that lead to the high-level production of antibody and the generation of memory B cells.

The initial stimulation of a naive B cell occurs by binding of specific antigen to the surface-bound IgM receptor on the B cell. This antigen–antibody complex is internalized and processed by the B cell, and this initiates a process of B-cell proliferation and differentiation that results in the generation of either antibody-secreting cells (plasma cells) or memory B cells. These initial events account for the primary adaptive antibody response [56, 62]. Primary humoral immunity is composed mostly of IgM antibodies, which generally are not detectable until 5–10 days after antigen exposure [62, 63]; however, upon rechallenge with antigen, the presence of specific memory B cells and primed helper CD4+ T cells allows for a secondary antibody response that is more rapid and of greater magnitude than the primary response. Secondary antibody responses are characterized by antibodies that have higher affinity for the antigen and that are predominantly IgG, IgA, or IgE rather than IgM.

The cellular events responsible for initiation of the primary antibody responses involve interactions between DCs, CD4+ T cells, and B cells (Fig. 1). As noted previously, initial exposure to a foreign antigen results in antigen uptake and processing by DCs, which travel via the afferent lymphatics to the T-cell zone of draining lymph nodes where they stimulate CD4+ T cells. The presence of antigen-specific CD4+ T cells is a key component of an effective B-cell response [20, 56, 62–64]. Whereas surface-bound Ig on a naive B cell can recognize its cognate antigen, B cells require immunologic "help" from antigen-specific activated CD4+ T cells to most effectively differentiate into antigen-secreting plasma cells and memory B cells. For example, differentiation of B cells into antigen-secreting plasma cells is facilitated by the interaction of CD40 on B cells with CD40L on activated CD4+ T cells [11, 65]. Moreover, secretion of T-cell cytokines such as IL-2, IL-4, IL-5, IL-6, and IL-10 is important in regulating both qualitative (antibody classes) and quantitative aspects of antibody production [15, 16, 66]. Of note, it is the ability of B cells to also act as APCs that allows them to solicit T-cell help from antigen-specific T cells (Fig. 1). The B cell processes antigen and presents peptide–MHC class II complexes to antigen-specific CD4+ T cells that have been activated by a DC. The B cells then receive stimulation from CD4+ T cells by CD40L interacting with CD40 on B cells as well as through CD4+ T-cell cytokines. Thus, the CD4+ T cell and B cell can recognize different epitopes from the same antigen: the B cell recognizes native (unprocessed) antigen, whereas the T cell recognizes peptide antigens processed by the B cell. This interaction underlies the concept of conjugate vaccines.

3. T-Cell-Independent Antigens and Conjugate Vaccines

In contrast to the T-cell-dependent B-cell stimulation described previously, some antigens can directly stimulate B-cell proliferation and antibody secretion. This process of T-cell-independent B-cell activation takes place in response to highly polymerized antigens that can cause extensive cross-linking of B-cell receptors. An important example of this occurs with bacterial polysaccharides, which are repetitive carbohydrate determinants on the surface of some bacteria. The cellular events and outcomes associated with T-cell-independent B-cell activation are distinct from T-cell-dependent stimulation, and this has important implications for vaccine development. T-cell-independent B-cell stimulation produces a predominance of IgM and IgG_2 antibody-secreting cells, and the production of antigen-specific memory B cells is inefficient. Thus, even repeated immunization with a capsular polysaccharide produces low levels of antibodies and poor memory antibody responses upon challenge. This has been problematic for the development of vaccines to medically important bacteria such as *Haemophilus influenzae*, *Streptococcus pneumoniae*, and

Neisseria meningitidis [54, 67, 68]. To overcome these inherent limitations of polysaccharide antigens, new vaccine formulations have coupled bacterial polysaccharides to proteins that produce strong T-cell-dependent antibody responses. This allows a specific B cell to recognize the polysaccharide antigen and endocytose the coupled antigen complex. Some of the resultant helper peptides are loaded onto MHC class II on the surface of the B cell and provide for a cognate interaction between the antigen-specific CD4+ helper T cell and the B cell (Fig. 1). This approach has led to the development and licensure of highly effective vaccines against *H. influenzae* B (Hib) and *S. pneumoniae* that couple the bacterial capsular polysaccharide to known T helper antigens such as diphtheria or tetanus toxoid [67, 68].

4. Isotype Switching and Affinity Maturation

As mentioned, secondary exposure of memory B cells results in a rapid change in the class (also termed isotype) of antibody produced by antibody-secreting cells. This occurs by rearrangement of the genes encoding the constant regions of the Ig heavy chain, which confers the antibody isotype [59–61]. Gene rearrangement in antibody-secreting cells results in a switch from IgM antibodies to class IgG, IgA, or IgE antibodies. These antibodies retain the variable domains and, thus, the antigenic specificity of the original IgM antibodies. This process of isotype switching is regulated by the cognate interaction between CD40L on T cells with the CD40 receptor on B cells as well as by various T-cell cytokines. An important clinical example of the role of CD40L in isotype switching is the X-linked congenital genetic defect in the gene encoding CD40L. This defect leads to the hyper-IgM syndrome, which is characterized by the inability of B cells to switch from the IgM to the IgG antibody isotype [69]. Patients with this syndrome have elevated levels of IgM in their serum and almost no Ig of other isotypes. Cytokines produced by CD4+ T-cell cytokines also have a profound influence on the class of antibody produced. Thus, IL-4 made by Th2-type cells promotes the production of IgE, whereas TGF-β promotes the production of IgA. The fact that specific T-cell subsets (Th1 and Th2 cells), producing unique patterns of cytokine production, can alter the amount and the type of Ig produced may have relevance in terms of designing vaccines or using vaccine adjuvants that can elicit a particular type of T helper response that will lead to an appropriate antibody response [15, 16, 66].

In addition to isotype switching during early B-cell stimulation, the diversity of the B-cell repertoire is further increased by a process known as somatic hypermutation. In the spleen and secondary lymphoid tissues, unprocessed antigen can become trapped and retained on the surface of specialized cells, termed follicular DCs [70]. While in close contact with antigen-bearing follicular DCs in the germinal centers of lymphoid tissue, B cells experience extensive mutations of the variable regions of their Ig genes. By selection of mutant Ig with high affinity to cell-bound antigens, an efficient process of positive selection generates B cells that produce antibody with higher affinity for the antigen [71, 72]. Of note, follicular DCs are found in the germinal centers of lymph nodes and are distinct from the antigen-presenting DCs (Fig. 1) that initiate primary immune responses.

5. B-Cell Memory

Because antibody is a major component of protection against most microbial pathogens, it is likely that the continued presence of some level of antigen-specific antibody in the serum or at the mucosal surface is important to preserve long-term protection. Thus, an understanding of B-cell memory and how to generate and maintain long-term antibody responses is germane to the development of effective vaccines. The majority of serum antibody is produced by terminally differentiated antibody-secreting cells called plasma cells. These cells are distinct from memory B cells in that their main function is to continuously secrete large quantities of specific antibody [73, 74]. In contrast, memory B cells do not spontaneously secrete antibody but can proliferate and differentiate into antibody-secreting cells upon appropriate stimulation. Until relatively recently, it was believed that plasma cells had a relatively short lifespan (days to weeks) and that long-term antibody secretion required the continuous differentiation of memory cells into antibody-secreting cells; however, more recent data suggest that plasma cells may populate the spleen, bone marrow, and gut-associated lymphoid tissue and secrete antibody for years [73]. This would explain the lifelong presence of serum antibodies against some microbial antigens despite the lack of reexposure after initial infection. In addition, antibody to some well-known vaccine antigens such as diphtheria or tetanus toxoid appears to be maintained over many years without revaccination or reexposure. A possible alternative explanation for long-term antibody production is that initial antigen remains on follicular DCs and provides a constant source of antigenic stimulation that renews the population of antigen-secreting cells [24].

Whereas protective immunity is almost always due to a complex interplay of humoral and cellular immune factors, the amount of antibody required to confer protection against an invading pathogen is an important consideration in vaccine development [57, 58, 75, 76]. On one hand, antibody-mediated protection could result from preexisting serum or mucosal antibody present at a level sufficient to inactivate the inoculum of the invading pathogen. Alternatively, protection could result from a robust secondary immune response sufficient to clear the organism from circulation. Which of these mechanisms is most important in mediating protection in general is not known, and in some cases both may contribute [56]. It is therefore interesting to note that, for most licensed vaccines, successful immunization or maintenance of protective immunity is determined simply and accurately by measurement of the level of specific serum IgG [58]. For example, successful immunization against a respiratory tract virus such as influenza or an enteric virus such as poliovirus can be predicted by measuring the serum levels of neutralizing antibody present after vaccination. In these cases, it is possible that preexisting antibody completely prevents initial infection (sterilizing immunity), but it is more likely that the organism is eradicated after initial rounds of replication.

6. Mechanisms of Antibody-Mediated Protection

The fact that antibodies alone can protect against many infectious diseases has been clearly demonstrated in various animal models of infection [57, 77].

In addition, human studies have shown that specific antibodies can protect against disease caused by viruses, bacteria, and bacterial toxins. In the United States there are licensed antibody products that prevent infection by hepatitis A, hepatitis B, respiratory syncytial virus, varicella–zoster virus (VZV), rabies, cytomegalovirus (CMV), and vaccinia virus [78]. In addition, tetanus immune globulin is given to prevent tetanus that results from *Clostridium tetani* infection [54]. Prior to the development of effective vaccines against diseases such as polio and measles, passive administration of specific antibody was known to prevent infection with these viruses. In fact, until the licensure of the hepatitis A vaccine, it was common practice to treat travelers with immune serum globulin (IgG derived from healthy donors) to prevent hepatitis A infection.

To prevent infection, antibodies must act directly on cell-free bacteria or viruses to inactivate the infectious inoculum or to prevent early spread of infection. This is best illustrated by the action of IgG, IgA, or IgM antibodies that prevent cell-free viruses from infecting host cells. Thus, viral neutralization can be defined as the inactivation of viral infectivity due to reaction of the virus with specific antibody. Antibodies bound to viral determinants can interrupt viral infection of a cell by several mechanisms, including inhibition of cellular attachment, virus–cell fusion, and uncoating after viral entry [54, 77]. In addition to blocking cell infection, antibody coating of microbes can mediate phagocytosis of organisms when the antibody constant region is recognized by the Fc receptors of phagocytes such as neutrophils and macrophages. Specific antibody classes and subclasses have additional functions. For example, the Fc regions of IgM, IgG_1, and IgG_3 antibodies can bind and activate complement proteins that can facilitate phagocytosis or cause direct microbial death. IgA antibodies have been shown to block the intracellular transport of virus across mucosal epithelial cells and to cause extracellular viral aggregation that facilitates mucociliary clearance [79–81].

Once a virus or intracellular microbe has established infection within host cells, control of infection is generally mediated by cellular immune mechanisms; however, the humoral and cellular immune systems function together during a process known as antibody-dependent cellular cytotoxicity. Because viral infection of cells results in the surface expression of some viral antigens, infected cells can be recognized by specific antibody. The bound antibody then engages the Fc receptor of natural killer cells, which can lyse cells by the release of cytolytic cytokines and perforin molecules. A similar mechanism exists to protect against some parasitic infections: IgE-coated parasites are recognized and attacked by eosinophils that release molecules, such as major basic protein, that are toxic to the organism.

7. Immune Adjuvants

Vaccine development involves the identification of pathogen-specific antigens that can be administered in a formulation that will induce an immune response sufficient for protection. Whereas certain vaccines (e.g., recombinant protein or inactivated virus) have the ability to elicit the appropriate antigen-specific immune responses, the vaccines themselves are poorly immunogenic.

In such cases, immune adjuvants are used to alter the kinetics (induction and duration) as well as the qualitative and quantitative aspects of the immune response. Complete Freund's adjuvant (CFA), consisting of an emulsion of killed *Mycobacterium bovis* and oil, has been referred to as nature's adjuvant and remains a gold standard in rodent systems for the enhancement of immune responses. Unfortunately, due to its ability to cause inflammation and granulomatous reactions at the site of inoculation, it is precluded for use in humans. The efficacy of CFA is likely due to the potent immunostimulatory properties of the mycobacteria on innate immunity and antigen presentation, combined with the depot effect of the oil emulsification. At present, alum adjuvants consisting of aluminum salts are the only approved adjuvants for use in humans in the United States, although several newer adjuvants have been used in human clinical studies. Although alum can enhance the production of antibody, it has little effect on inducing Th1 or CD8+ T-cell responses. Moreover, there is evidence that alum can preferentially induce Th2 responses, which would not be preferred for a vaccination against diseases requiring cellular immune responses (e.g., *M. tuberculosis*, *L. major*, malaria, HIV). This has prompted the development of several newer classes of adjuvants that might be useful in eliciting strong cellular immune responses. These include bacterial adjuvants [monophosphoryl lipid A (MPL)], particulate adjuvants [immuno-stimulatory complexes (ISCOMS)], oil emulsion and emulsifier-based adjuvants (saponins), and synthetic adjuvants (synthetic lipid A, muramyl peptide analogs). Similar to CFA, it is likely that the mechanism of action for many of these adjuvant formulations is to cause an inflammatory reaction at the site of inoculation, which will serve to enhance innate immunity and APC function leading to activation of T cells.

As the cellular and molecular mechanisms of antigen presentation and T-cell activation have been more clearly elucidated, there has been substantial interest in using specific cytokines and/or costimulatory molecules as vaccine adjuvants. This approach could enhance or modulate immune responses by recruitment and activation of DCs as well as T cells. Thus, for vaccines in which Th1 responses are required (*M. tuberculosis* and *L. major*), the cytokine IL-12 can stimulate production of IFN-γ from CD4+ T cells following T-cell activation [52, 82]. In terms of enhancing APC function, costimulatory molecules such as CD40L or cytokines such as GM-CSF have been shown to be potent immune adjuvants in a variety of experimental mouse models [83]. In addition, a variety of pathogens are potent stimulators of APCs. One of the most actively studied and potentially useful immune adjuvants is specific nucleotide sequences of bacterial DNA. Compared with mammalian DNA, there are substantial differences in the number and methlylation pattern of specific nucleotide sequences (CpG motifs) within bacterial DNA [84]. These CpG motifs provide potent immune stimulation of APCs through binding to toll-like receptors. This results in the production of proinflammatory cytokines such as IL-12, IL-18, and TNF-α as well as increased expression of costimu-latory molecules on the surface of APCs [9]. As our understanding of protective immunity develops, immune modulators and vaccine adjuvants will be used in a more targeted manner to achieve the desired immune responses in a safe and tolerable manner.

III. DETERMINANTS OF VACCINE PROTECTION

A. Prevention of Infection versus Prevention of Disease

The primary goal of a vaccine is to protect the host against infectious complications of a pathogenic microorganism. This is accomplished by eliciting specific immune responses that will either inactivate the initial infection or eliminate the organism during its initial rounds of replication. Because vaccine efficacy is measured by prevention of disease manifestations, it is not always clear whether immunity is a result of the prevention of infection (i.e., complete inactivation of the infectious inoculum) or the prevention of disease due to rapid elimination of the organism after initial replication in the host. As an example, enteroviruses such as poliovirus infect through mucosal tissues of the gastrointestinal tract. Preexisting immunity, such as secretory IgA at the mucosal surface, could prevent initial infection. On the other hand, serum IgG antibodies likely neutralize and eliminate poliovirus before the virus has the opportunity to infect motor neurons (anterior horn cells) in the spinal cord. Although both mechanisms may play a role in protection, immunity against paralytic polio is best correlated with induction of serum antibodies that neutralize the specific serotype of the infecting poliovirus. As an example, immunization with poliovirus serotype 1 generates neutralizing antibodies that predominantly protect against type 1 poliovirus. Thus, to be broadly effective, current poliovirus vaccines include all three poliovirus serotypes.

Although neutralizing antibodies are the most common correlate of protection induced by current vaccines, there is growing appreciation for the role of the cellular immune response in mediating protection against a variety of infectious diseases. Indeed, for several viral diseases of humans, cellular immunity is known to be a critical component of protection. CD4 and CD8 T-cell responses are required to protect against clinical disease following initial exposure to many viral infections, such as measles, CMV, herpes simplex virus (HSV), VZV, and HIV-1 (HIV type 1). Furthermore, for diseases such as *M. tuberculosis* or *L. major*, there is little evidence that humoral immunity has any role in protecting against infection. Thus, a vaccine against these infections would need to induce a cellular immune response that would work by controlling the infection and preventing disease rather than preventing infection. This is supported by the observation that >90% of healthy individuals exposed to *M. tuberculosis* do not develop disease and are only at risk of developing clinical tuberculosis in the setting of immunosuppression. This central issue of preventing infection versus disease has relevance to the efforts to develop a vaccine against HIV-1. Although the goal of an HIV-1 vaccine is to completely prevent HIV infection through induction of neutralizing antibodies and a cellular immune response, it is possible that a vaccine will be successful in preventing or prolonging the development of disease by the induction of a predominantly cellular immune response.

B. Factors That Determine Vaccine Efficacy: Threshold, Time, and Place

Successful immunization requires the activation, replication, and differentiation of T and B lymphocytes leading to the generation of memory cells [56].

The majority of vaccines in current use require multiple immunizations to prime and maintain effective immunity. As previously noted, for a variety of bacterial and viral infections, there is a well-defined threshold for the amount of antibody required for protection. Furthermore, in animal models of diseases requiring cellular immune responses for protection, there also appears to be a certain threshold of effector CD8+ T cells [85]. The initial frequency and subsequent number of antigen-specific memory CD8+ T cells appear to be proportional to the initial antigenic load. In general, live infection induces a greater frequency of antigen-specific cells than immunizations with proteins, DNA vaccines, or recombinant nonreplicating viruses encoding specific antigen [23]. The magnitude of the memory phase is generally determined by the size of the initial clonal burst induced by immunization [21]. Moreover, the amount of antigen can also affect the qualitative aspect of CD8+ T-cell memory–effector responses [23]. Thus, for diseases that may require a high number of memory CD8+ T cells to mediate protection, it will be desirable to use a vaccine strategy that provides a sufficient antigenic load to maximize the initial burst size. Although it is clear that antigen is not required to sustain resting memory CD8+ T cells, it may be required to sustain optimal T-cell effector function. The central question that remains unresolved for vaccines against diseases requiring effector CD8+ T cells is whether persistent or periodic antigen (booster immunization) will be required to sustain a pool of circulating effector–memory CD8+ T cells that would be important in providing protection.

C. Immune Responses Elicited by Various Vaccine Formulations: Newer Approaches for Developing Vaccines for Diseases Requiring Cellular Immunity

Among currently licensed vaccines, only live attenuated vaccines allow for infection of host cells and efficient MHC class I presentation of antigen that can stimulate strong and persistent CD8+ T-cell responses (Table 1). Live attenuated vaccines best simulate natural immunity and generally produce the most persistent protection, making them the gold standard for diseases requiring both humoral and cellular immunity; however, from a practical and safety standpoint, attenuated whole organism vaccines may be precluded from widespread application for certain diseases (e.g., HIV-1 and malaria). Because the cellular immune response plays a major role in the control of HIV-1, *M. tuberculosis*, and malaria infection, there is an urgent need to develop vaccines for these diseases that optimize the induction and maintenance of cellular immune responses. The demonstration that DNA vaccines could induce broad cellular responses (CTL and Th1 responses) in small animals provided hope that they could become the future vaccine of choice for infections requiring cellular immunity in humans [83, 86]. Unfortunately, the induction of humoral and cellular immune responses following DNA vaccination in humans has been comparatively lower than in rodent models [87–89]. Nevertheless, the potential ability of DNA vaccines to induce both CD4 and CD8 responses in a safe and cost-effective manner has prompted investigators to attempt to optimize the delivery of DNA vaccination to enhance the immune responses in nonhuman primates and humans. These approaches include the use of needle-free injector devices to deliver the DNA, conjugation

TABLE 1   Comparative Immunogenicity of Various Types of Vaccines[a]

Vaccine	B cells/ antibody	CD4+T cells	CD8+T cells
Licensed vaccines			
Live attenuated[b] (measles, mumps, rubella; oral polio virus vaccine)	+++	+++	+++
Whole inactivated[c] (hepatitis A; influenza; inactivated polio virus vaccine; whole cell pertussis)	+++	++	+/−
Protein–subunit[c] (tetanus and diphtheria toxoids; hepatitis B)	+++	++	+/−
Polysaccharide (pneumococcal vaccine)	+	+/−	−
Polysaccharide–conjuate (*Haemophilus influenzae* type B conjugate vaccine)	+++	++	−
Novel vaccine formulations			
Plasmid DNA[d]	++	++	++
Live vectors[e]	++	++	++

[a]+ and − symbols indicate level of immunogenicity.

[b]Live attenuated vaccines may be precluded for use in immunocompromised patients and in patients with certain infections such as HIV.

[c]In some cases, whole inactivated and subunit antigens can access the endogenous MHC class I antigen processing pathway and elicit CD8+ T-cell responses.

[d]DNA immunization of rodents elicits robust antibody and CD4+ and CD8+ T-cell responses; however, DNA immunization has been less immunogenic in nonhuman primates and humans.

[e]As an example, pox viruses (e.g., vaccinia) engineered to express vaccine immunogens.

of DNA to polymers, and the use of various cytokines as adjuvants [90]. In addition, there have been several promising primate studies in which DNA vaccination is followed by a booster immunization with a recombinant viral vaccine. These studies demonstrate the striking enhancement of immune responses using DNA priming and boosting with a recombinant pox or adenoviral vector encoding the same foreign antigens [85, 91, 92]. This may be one of most promising vaccine approaches for several infectious diseases, including HIV–AIDS and malaria.

These newer vaccine approaches have benefited from an improved basic understanding of the immunology of antigen presentation and the factors important in sustaining memory immune responses. The continued improvement in our understanding of the nature of protective immunity against various infectious diseases should allow the rational design of vaccines and immunization strategies that generate protective immune responses against infectious diseases for which no vaccines yet exist.

REFERENCES

1. Hoffmann, J. A., *et al.* (1999). Phylogenetic perspectives in innate immunity. *Science* **284**:1313–1318.

2. Medzhitov, R., and Janeway, C. A., Jr. (1997). Innate immunity: The virtues of a nonclonal system of recognition. *Cell* **91**:295–298.

3. Medzhitov, R., and Janeway, C., Jr. (2000). Innate immunity. *New. England J. Medicine* **343**:338–344.

4. Medzhitov, R., and Janeway, C. A., Jr. (1997). Innate immunity: Impact on the adaptive immune response. *Curr. Opin. Immunol.* **9**:4–9.

5. Lemaitre, B., *et al.* (1996). The dorsoventral regulatory gene cassette spatzle/Toll/cactus controls the potent antifungal response in *Drosophila* adults. *Cell* **86**:973–983.

6. Medzhitov, R., Preston-Hurlburt, P., and Janeway, C. A., Jr. (1997). A human homologue of the *Drosophila* Toll protein signals activation of adaptive immunity. *Nature* **388**: 394–397.

7. Poltorak, A., *et al.* (1998). Defective LPS signaling in C3H/HeJ and C57BL/10ScCr mice: Mutations in *Tlr4* gene. *Science* **282**: 2085–2088.

8. Brightbill, H. D., *et al.* (1999). Host defense mechanisms triggered by microbial lipoproteins through toll-like receptors. *Science* **285**:732–736.

9. Hemmi, H., *et al.* (2000). A Toll-like receptor recognizes bacterial DNA. *Nature* **408**:740–745.

10. Banchereau, J., and Steinman, R. M. (1998). Dendritic cells and the control of immunity. *Nature* **392**:245–252.

11. Banchereau, J., *et al.* (1994). The CD40 antigen and its ligand. *Annu. Rev. Immunol.* **12**:881–922.

12. Ridge, J. P., Di Rosa, F., and Matzinger, P. (1998). A conditioned dendritic cell can be a temporal bridge between a CD4+ T-helper and a T-killer cell. *Nature* **393**:474–478.

13. Bennett, S. R., *et al.* (1998). Help for cytotoxic T-cell responses is mediated by CD40 signalling. *Nature* **393**:478–480.

14. Schoenberger, S. P., *et al.* (1998). T-cell help for cytotoxic T lymphocytes is mediated by CD40–CD40L interactions. *Nature* **393**:480–483.

15. O'Garra, A. (1998). Cytokines induce the development of functionally heterogeneous T helper cell subsets. *Immunity* **8**:275–283.

16. Seder, R. A., and Paul, W. E. (1994). Acquisition of lymphokine-producing phenotype by CD4+ T cells. *Annu. Rev. Immunol.* **12**:635–673.

17. Trinchieri, G. (1998). Immunobiology of interleukin-12. *Immunol. Res.* **17**:269–278.

18. Sprent, J., and Surh, C. D. (2001). Generation and maintenance of memory T cells. *Curr. Opin. Immunol.* **13**: 248–254.

19. Zinkernagel, R. M., *et al.* (1996). On immunological memory. *Annu. Rev. Immunol.* **14**:333–367.

20. Ahmed, R., and Gray, D. (1996). Immunological memory and protective immunity: Understanding their relation. *Science* **272**:54–60.

21. Hou, S., *et al.* (1994). Virus-specific CD8+ T-cell memory determined by clonal burst size. *Nature* **369**: 652–654.

22. McHeyzer-Williams, M. G., and Davis, M. M. (1995). Antigen-specific development of primary and memory T cells *in vivo*. *Science* **268**: 106–111.

23. Oehen, S., *et al.* (1992). Antivirally protective cytotoxic T cell memory to lymphocytic choriomeningitis virus is governed by persisting antigen. *J. Exp. Med.* **176**:1273–1281.

24. Ochsenbein, A. F., *et al.* (1999). A comparison of T cell memory against the same antigen induced by virus *versus* intracellular bacteria. *Proc. Natl. Acad. Sci. U.S.A.* **96**:9293–9298.

25. van Stipdonk, M. J., Lemmens, E. E., and Schoenberger, S. P. (2001). Naive CTL require a single brief period of antigenic stimulation for clonal expansion and differentiation. *Nat. Immunol.* **2**:423–429.

26. Wong, P., and Pamer, E. G. (2001). Cutting edge: Antigen-independent CD8 T cell proliferation. *J. Immunol.* **166**:5864–5868.

27. Kaech, S. M., and Ahmed, R. (2001) Memory CD8+ T cell differentiation: Initial antigen encounter triggers a developmental program in naive cells. *Nat. Immunol.* **2**:415–422.

28. Homann, D., Teyton, L., and Oldstone, M. B. (2001). Differential regulation of antiviral T-cell immunity results in stable CD8+ but declining CD4+ T-cell memory. *Nat. Med.* **7**:913–919.

29. Dutton, R. W., Bradley, L. M., and Swain, S. L. (1998). T cell memory. *Annu. Rev. Immunol.* **16**:201–223.

30. Veiga-Fernandes, H., *et al.* (2000). Response of naive and memory CD8⁺ T cells to antigen stimulation *in vivo*. *Nat. Immunol.* **1**:47–53.

31. Rogers, P. R., Dubey, C., and Swain, S. L. (2000). Qualitative changes accompany memory T cell generation: Faster, more effective responses at lower doses of antigen. *J. Immunol.* **164**:2338–2346.

32. London, C. A., Perez, V. L., and Abbas, A. K. (1999). Functional characteristics and survival requirements of memory CD4⁺ T lymphocytes *in vivo*. *J. Immunol.* **162**:766–773.

33. Pihlgren, M., *et al.* (1996). Resting memory CD8⁺ T cells are hyperreactive to antigenic challenge *in vitro*. *J. Exp. Med.* **184**:2141–2151.

34. Garcia, S., DiSanto, J., and Stockinger, B. (1999). Following the development of a CD4 T cell response *in vivo*: From activation to memory formation. *Immunity* **11**:163–171.

35. Tanchot, C., *et al.* (1997). Differential requirements for survival and proliferation of CD8 naive or memory T cells. *Science* **276**:2057–2062.

36. Sallusto, F., *et al.* (1999). Two subsets of memory T lymphocytes with distinct homing potentials and effector functions. *Nature* **401**:708–712.

37. Opferman, J. T., Ober, B. T., and Ashton-Rickardt, P. G. (1999). Linear differentiation of cytotoxic effectors into memory T lymphocytes. *Science* **283**:1745–1748.

38. Jacob, J., and Baltimore, D. (1999). Modelling T-cell memory by genetic marking of memory T cells *in vivo*. *Nature* **399**:593–597.

39. Gray, D., and Matzinger, P. (1991). T cell memory is short-lived in the absence of antigen. *J. Exp. Med.* **174**:969–974.

40. Kundig, T. M., *et al.* (1996). On the role of antigen in maintaining cytotoxic T-cell memory. *Proc. Natl. Acad. Sci. U.S.A.* **93**:9716–9723.

41. Bachmann, M. F., *et al.* (1997). Protection against immunopathological consequences of a viral infection by activated but not resting cytotoxic T cells: T cell memory without "memory T cells"? *Proc. Natl. Acad. Sci. U.S.A.* **94**:640–645.

42. Lau, L. L., *et al.* (1994). Cytotoxic T-cell memory without antigen. *Nature* **369**:648–652.

43. Murali-Krishna, K., *et al.* (1999). Persistence of memory CD8 T cells in MHC class I–deficient mice. *Science* **286**:1377–1381.

44. Swain, S. L. (1994). Generation and *in vivo* persistence of polarized Th1 and Th2 memory cells. *Immunity* **1**:543–552.

45. Swain, S. L., Hu, H., and Huston, G. (1999). Class II-independent generation of CD4 memory T cells from effectors. *Science* **286**:1381–1383.

46. Nishimura, H., *et al.* (2000). Differential roles of interleukin-15 mRNA isoforms generated by alternative splicing in immune responses *in vivo*. *J. Exp. Med.* **191**:157–170.

47. Kennedy, M. K., *et al.* (2000). Reversible defects in natural killer and memory CD8 T cell lineages in interleukin-15-deficient mice. *J. Exp. Med.* **191**:771–780.

48. Ku, C. C., *et al.* (2000). Control of homeostasis of CD8⁺ memory T cells by opposing cytokines. *Science* **288**:675–678.

49. Lodolce, J. P., *et al.* (1998). IL-15 receptor maintains lymphoid homeostasis by supporting lymphocyte homing and proliferation. *Immunity* **9**:669–676.

50. Waldmann, T. A., and Tagaya, Y. (1999). The multifaceted regulation of interleukin-15 expression and the role of this cytokine in NK cell differentiation and host response to intracellular pathogens. *Annu. Rev. Immunol.* **17**:19–49.

51. Schluns, K. S., *et al.* (2000). Interleukin-7 mediates the homeostasis of naive and memory CD8 T cells *in vivo*. *Nat. Immunol.* **1**:426–432.

52. Stobie, L., *et al.* (2000). The role of antigen and IL-12 in sustaining Th1 memory cells *in vivo*: IL-12 is required to maintain memory/effector Th1 cells sufficient to mediate protection to an infectious parasite challenge. *Proc. Natl. Acad. Sci. U.S.A.* **97**:8427–8432.

53. Gurunathan, S., *et al.* (1998). Vaccine requirements for sustained cellular immunity to an intracellular parasitic infection. *Nat. Med.* **4**:1409–1415.

54. Heinzel, F. P. (2000). Antibodies. *In* "Principles and Practice of Infectious Diseases," G. Mandell, J. Bennett, and R. Dolin, Eds., pp. 45–66. Philadelphia: Churchill Livingstone.

55. Frazer, J. K., and Capra, J. D. (1999). Immunoglobulins: Structure and function. *In* "Fundamental Immunology," W. E. Paul, Ed., pp. 37–74. Philadelphia: Lippincott-Raven.

56. Ada, G. (1999). The immunology of vaccination. *In* "Vaccines," S. A. Plotkin and W. A. Orenstein, Eds., pp. 28–39. Philadelphia: W. B. Saunders Company.

57. Krause, R. M., Dimmock, N. J., and Morens, D. M. (1997). Summary of antibody workshop: The role of humoral immunity in the treatment and prevention of emerging and extant infectious diseases. *J. Infect. Dis.* **176**(3):549–559.

58. Robbins, J. B., Schneerson, R., and Szu, S. C. (1995). Perspective: hypothesis: Serum IgG antibody is sufficient to confer protection against infectious diseases by inactivating the inoculum. *J. Infect. Dis.* **171**:1387–1398.

59. Tonegawa, S. (1983). Somatic generation of antibody diversity. *Nature* **302**:575–581.

60. Delves, P. J., and Roitt, I. M. (2000). The immune system. First of two parts. *New England J. Medicine* **343**:37–49.

61. Schwartz, R. S. (1995). Jumping genes and the immunoglobulin V gene system. *New. England J. Medicine* **333**:42–44.

62. Nossal, G. J. (1997). Host immunobiology and vaccine development. *Lancet* **350**:1316–1319.

63. Huston, D. P. (1997). The biology of the immune system. *J. Am. Med. Assoc.* **278**:1804–1814.

64. Garside, P., *et al.* (1998). Visualization of specific B and T lymphocyte interactions in the lymph node. *Science* **281**:96–99.

65. van Kooten, C., and Banchereau, J. (2000). CD40–CD40 ligand. *J. Leukoc. Biol.* **67**: 2–17.

66. Spellberg, B., and Edwards, J. E., Jr. (2001). Type 1/Type 2 immunity in infectious diseases. *Clin. Infect. Dis.* **32**:76–102.

67. Ward, J. I., and Zangwill, K. M. (1999). Haemophilus influenzae vaccines. *In* "Vaccines," S. A. Plotkin and W. A. Orenstein, Eds., pp. 183–221. Philadelphia: W. B. Saunders Company.

68. Fedson, D. S., Musher, D. M., and Eskola, J. (1999). Pneumococcal vaccine. *In* "Vaccines," S. A. Plotkin and W. A. Orenstein, Eds., pp. 553–607. Philadelphia: W. B. Saunders Company.

69. Geha, R. S., and Rosen, F. S. (1994). The genetic basis of immunoglobulin-class switching. *New England J. Medicine* **330**:1008–1009.

70. Liu, Y. J. (1997). Sites of B lymphocyte selection, activation, and tolerance in spleen. *J. Exp. Med.* **186**:625–629.

71. Foote, J., and Milstein, C. (1991). Kinetic maturation of an immune response. *Nature* **352**:530–532.

72. Jacob, J., *et al.* (1991). Intraclonal generation of antibody mutants in germinal centres. *Nature* **354**:389–392.

73. Slifka, M. K., *et al.* (1998). Humoral immunity due to long-lived plasma cells. *Immunity* **8**:363–372.

74. Manz, R. A., Thiel, A., and Radbruch, A. (1997). Lifetime of plasma cells in the bone marrow. *Nature* **388**:133–134.

75. Dittmer, U., Brooks, D. M., and Hasenkrug, K. J. (1999). Requirement for multiple lymphocyte subsets in protection by a live attenuated vaccine against retroviral infection. *Nat. Med.* **5**:189–193.

76. Zinkernagel, R. M. (1996). Immunology taught by viruses. *Science* **271**:173–178.

77. Parren, P. W., and Burton, D. R. (2001). The antiviral activity of antibodies *in vitro* and *in vivo*. *Adv. Immunol.* **77**:195–262.

78. Lutwick, L. I. (1996). Postexposure prophylaxis. *Infect. Dis. Clin. North Am.* **10**:899–915.

79. Mazanec, M. B., *et al.* (1992). Intracellular neutralization of virus by immunoglobulin A antibodies. *Proc. Natl. Acad. Sci. U.S.A.* **89**:6901–6905.

80. Bomsel, M., *et al.* (1998). Intracellular neutralization of HIV transcytosis across tight epithelial barriers by anti-HIV envelope protein dIgA or IgM. *Immunity* **9**:277–287.

81. Burns, J. W., *et al.* (1996). Protective effect of rotavirus VP6-specific IgA monoclonal antibodies that lack neutralizing activity. *Science* **272**:104–107.

82. Afonso, L. C., *et al.* (1994). The adjuvant effect of interleukin-12 in a vaccine against *Leishmania major*. *Science* **263**:235–237.

83. Gurunathan, S., Klinman, D. M., and Seder, R. A. (2000). DNA vaccines: Immunology, application, and optimization. *Annu. Rev. Immunol.* **18**:927–974.

84. Wagner, H. (2001). Toll meets bacterial CpG-DNA. *Immunity* **14**:499–502.

85. Schneider, J., *et al.* (1999). Induction of CD8[+] T cells using heterologous prime-boost immunisation strategies. *Immunol. Rev.* **170**:29–38.

86. Seder, R. A., and Gurunathan, S. (1999). DNA vaccines—Designer vaccines for the 21st century. *New England J. Medicine* **341**:277–278.

87. Wang, R., *et al.* (1998). Induction of antigen-specific cytotoxic T lymphocytes in humans by a malaria DNA vaccine. *Science* **282**:476–480.

88. Le, T. P., *et al.* (2000). Safety, tolerability and humoral immune responses after intramuscular administration of a malaria DNA vaccine to healthy adult volunteers. *Vaccine* **18**:1893–1901.

89. Calarota, S., *et al.* (1998). Cellular cytotoxic response induced by DNA vaccination in HIV-1-infected patients. *Lancet* **351**:1320–1325.

90. Barouch, D. H., *et al.* (2000). Control of viremia and prevention of clinical AIDS in rhesus monkeys by cytokine-augmented DNA vaccination. *Science* **290**:486–492.

91. Schneider, J., *et al.* (1998). Enhanced immunogenicity for CD8[+] T cell induction and complete protective efficacy of malaria DNA vaccination by boosting with modified vaccinia virus Ankara. *Nat. Med.* **4**:397–402.

92. Amara, R. R., *et al.* (2001). Control of a mucosal challenge and prevention of AIDS by a multiprotein DNA/MVA vaccine. *Science* **292**:69–74.

2 IMMUNOLOGY

PART B. Immunological Requirements for Vaccines To Be Used in Early Life

CLAIRE-ANNE SIEGRIST

WHO Collaborating Center for Neonatal Vaccinology, Departments of Pediatrics and Pathology, University of Geneva, 1211 Geneva 4, Switzerland

I. INTRODUCTION

Despite global infant immunization, 2.5 million infant deaths still result annually from acute respiratory and diarrheal infections [1] caused by a limited number of viral and bacterial pathogens for which vaccines are already available (pertussis, measles), could become available for global use [Haemophilus influenzae B (Hib), pneumococcal and meningococcal conjugates], or are currently being developed (rotavirus, respiratory syncytial virus). However, all of these vaccines, as well as novel vaccines against major killers such as malaria, tuberculosis, and HIV, have to meet a difficult challenge in order to be effective in early life: they must safely induce protective immune responses rapidly after birth, i.e., prior to pathogen exposure, despite the immaturity of the immune system and the inhibitory influence of antibodies of maternal origin.

II. EARLY LIFE VACCINE ANTIBODY RESPONSES

A. Antibody Responses to Current Infant Vaccines

The magnitude of immunoglobulin G (IgG) antibody responses that may be induced by a given vaccine is directly correlated to age at immunization. It has been previously observed that early rapid vaccination schedules (i.e., three vaccine doses given at 2, 3, and 4 months of age) result in lower antibody responses and a greater proportion of infants failing to respond to weaker antigens (such as diphtheria toxoid) than those schedules spanning a longer period (2, 4, and 6 months or 3, 5, and 12 months) [2]. This reflects not only immune immaturity but also a shorter time period available for affinity maturation of vaccine-induced B cells. In contrast, evaluation of antibody responses to a single vaccine dose allows precise evaluation of the immune maturation stage. As an example, a single dose of HIB conjugate vaccine elicits progressively higher serum anticapsular antibody concentrations when administered at 2–3 months, 4–6 months or 8–17 months [3]. Similarly, the rate of seroconversion and antibody concentrations steadily increase following measles [4–6] or mumps [7] immunization at 6, 9, 12, or 15 months of age. Analyses performed in infants without residual maternal antibodies at the time of immunization demonstrate that age at immunization is the main determinant of the magnitude of measles and mumps antibody responses <9 months of age. Infant antibody responses are not only of a weaker magnitude but also of a shorter duration than those elicited >12 months of age, requiring boosting in the second year of life for reactivation of memory B cells and maintenance of circulating antibodies. Accordingly, it has been suggested that early (<12 months) immunization could be associated with waning immunity against measles, which is not observed following immunization at an older age [8].

Whether early life immunization also results in antibodies of a different quality is an important question. Infant vaccines preferentially elicit IgG antibodies regardless of age at administration [9]. However, the infant IgM to IgG switch preferentially occurs toward IgG_1 and IgG_3 isotypes, and IgG_2 antibodies remain weak during the first 18 months of life [10] even for vaccines that induce preferential IgG_2 responses in adults. This limited switch to IgG_2 antibodies, which have a strong complement binding capacity, could contribute to infant susceptibility to infections with encapsulated bacteria (see later discussion). It is not yet clear whether early life immunization includes the risk of generating antibodies with lower avidity than those elicited later in life. In support of this hypothesis, age-related differences were demonstrated in the V region of antibody light chains following Hib conjugate vaccine administration [11], and limitations in antibody avidity were demonstrated years ago in mice following neonatal immunization with hapten conjugates. In contrast, the affinity maturation process was shown to be functional during the first year of life in infants [12], and neonatal murine immunization with clinically relevant infant vaccines resulted in an adult-like maturation process. Thus, the hypothesis that early life immunization is associated with suboptimal affinity maturation may not prove correct in infants and requires further studies.

In addition to limitations of antibody responses to protein antigens, infant and toddler responses to most bacterial capsular polysaccharides (PSs) are markedly limited, contributing to their high susceptibility to infections with encapsulated bacteria such as *H. influenzae b, Streptococcus pneumoniae*, and *Neisseria meningitidis*. The same limitations affect most polysaccharide vaccines, which remain poorly immunogenic before the age of 2 years and exhibit an age-dependent increase in vaccine efficacy between 2 and 10 years of age [13].

Thus, the first and major immunological requirement for vaccines to be effective earlier in life is to be able to induce stronger and faster antibody responses than those elicited by current infant vaccines.

B.　May Neonatal Immunization Induce Tolerance Rather Than Immunity?

Induction of tolerance rather than immunity by neonatal immunization is a frequent concern that has originated from many murine studies. However, in contrast to newborn mice and their profoundly immature immune system, human neonatal immunization induced either enhanced or conserved subsequent infant responses. Postnatal tolerance induction in humans has not yet been observed, despite a few reports in which neonatal immunization with whole-cell pertussis vaccines, a PRP-OMPc vaccine, or *N. meningitidis* group C polysaccharides (MenC PSs) resulted in impaired responses to subsequent vaccine doses. Lipopolysaccharide (LPS) containing whole-cell pertussis vaccine, OMPc, and MenC PSs all induce direct cross-linking of B-cell antigen receptors, which could lead to apoptosis of immature B cells in the absence of sufficient T-cell activation. Thus, neonatal tolerance is considered unlikely to occur with vaccine formulations inducing T-cell as well as B-cell responses in early life.

C.　Meeting the Challenge of Improving Early Life Antibody Responses

Factors that limit infant responses to bacterial PSs include (i) low complement activity, which limits the deposition of C3d on bacterial PS, (ii) weak expression of surface C3d receptors (CD21) on infant B lymphocytes, limiting synergy between B-cell receptor and complement receptor-mediated activation, and (iii) structural immaturity of the splenic marginal zone [14], to which C3d-bound PSs preferentially localize in adults. Altogether, these factors limit the capacity of marginal zone B cells and B1 cells to respond rapidly to particulate bacterial antigens [15]. In contrast, the glycoconjugate vaccines, which attach bacterial PS to a carrier protein and thus transform T-independent PSs into T-dependent antigens, are immunogenic and protective in young infants. However, despite their increased immunogenicity, glycoconjugate vaccines still require the administration of several doses, such that infections unfortunately may occur before completion of the vaccine series.

The mechanisms that limit early life antibody responses to protein antigens, whether present in protein, subunit, inactivated, or live attenuated vaccines, are not yet fully understood. Studies assessing the influence of age on vaccine antibody responses in neonatal, infant, and adult mice yielded observations very similar to those in human infants, provided that immunization was initiated ≥ 7 days of age (rather than in the immediate neonatal period) to compensate for the greater immaturity of newborn mice (reviewed in ref 16). Studies assessing the various stages of antigen-specific B-cell differentiation in early life have thus been initiated using these preclinical models of neonatal immunization. They have demonstrated that the slower and lower antibody responses to alum-adjuvanted protein vaccines reflect delayed and weaker induction of primary antibody-secreting cells (ASCs) compared to adults [17], which is associated with delayed germinal center (GC) induction. Studies are ongoing to identify the factors that limit early life GC and ASC responses. In addition to these limitations, there is limited establishment of the bone marrow pool of long-lived plasmocytes following neonatal murine immunization, which could explain the shorter kinetics of early life antibody responses [17].

In neonatal and infant mice, certain adjuvant formulations are able to significantly enhance early life vaccine responses, whereas others fail to do so despite their strong adjuvanticity in adult animals (reviewed in ref 16). However, none of the adjuvant formulations tested so far proved capable of correcting the delay in the induction of neonatal compared to adult antibody responses. Similarly, novel antigen delivery systems such as DNA vaccines did not enhance early life antibody responses compared to those elicited by conventional vaccines, despite the fact that neonatal DNA immunization did induce similar antibody responses in newborn and adult mice (reviewed in ref 18). Accordingly, DNA immunization of newborn–infant nonhuman primates against hepatitis B, HIV, or influenza also resulted in weak antibody responses [19], and sequential bleeding indicated a lack of antibody responses prior to 4 or 8 weeks of age after two or three vaccine doses [20]. Thus, currently available DNA vaccines unfortunately may not be the solution to limited antibody responses in early life (reviewed in ref 21).

In conclusion, vaccine formulations capable of rapidly inducing strong antibody responses in early life have not yet been identified, and additional studies are needed. In contrast, it was realized that induction of antigen-specific memory B cells can be achieved soon after birth, as illustrated in human infants by neonatal priming with oral polio and hepatitis B vaccines. Thus, certain vaccine antigens–delivery systems–adjuvants (but not others for as yet unknown reasons) are able to activate neonatal B cells and trigger their differentiation into memory B cells. This early priming is likely to prove essential to vaccine-mediated prevention in early life through early prime–later boost strategies.

III.   EARLY LIFE T-CELL RESPONSES

A.   Characteristics of Early Life T-Cell Responses

In contrast to the slow maturation of antibody responses, acquisition of antigen-specific T-cell responses is an early event. This was shown by *in utero* priming of fetal T-cell responses to allergens [22] and following neonatal bacille Calmette–Guérin (BCG) immunization with BCG [23]. The age-dependent maturation of T-cell vaccine responses is, however, as yet poorly characterized. T-cell proliferative responses following BCG were stronger when administration was delayed from birth until 2–6 months of age in some studies [24], whereas in contrast adult-like interferon-γ (IFN-γ) responses to neonatal BCG were reported in the Gambia [25]. Analyses of T-cell responses to measles and mumps immunization indicated similar antigen-specific T-cell proliferative and IFN-γ responses in infants immunized at 6, 9, or 12 months of age [7, 26]. However, proliferative and IFN-γ infant responses were significantly lower than those of adult controls [7, 26]. When compared with those of adults, infant T cells also showed a limited capacity to increase their IFN-γ release in response to IL-12 supplementation as compared to adults [26]. This could suggest that, although maturation is similar in 6-, 9-, and 12-month-old infants, it is not yet complete at 12 months of age despite the strong immune activation that follows the use of live replicating vaccines.

The understanding of early life CD8$^+$ cytotoxic responses (CTLs) is even more limited. Although infection-induced CTLs have been detected within the first weeks of life, CTL responses could be age-dependent [27]. CTLs were recovered in infants following influenza infection but not following immunization with a live influenza vaccine, suggesting that a certain immunogenicity threshold had not been reached in vaccinated infants [28]. Later in life, measles immunization at 9 months of age induced CTL responses similar to those found in children or adults with acute measles [29]. Thus, it seems likely that the maturation of CD8$^+$ cytotoxic responses will also prove age- and vaccine-dependent, although far more studies are required to clarify this important issue.

B.   Factors That Limit Early Life T-Cell Responses

Studies assessing the determinants of early life T-cell responses are limited to murine models of early life immunization. They indicate that antigen-specific T-cell responses may be readily elicited at an early stage, but that early immunization is associated with lower IFN-γ responses and higher IL-5, IL-4, and IL-13 responses to most conventional vaccines (reviewed in ref 16). Although T-cell maturation may be faster in lymph nodes than in the spleen [30], early life immunization results in limited induction of Th1-driven IgG$_{2a}$ antibodies and CTL responses [31, 32]. Altogether, these observations suggest a preferential differentiation of early life murine T-cell responses to viral–protein vaccines toward the Th2 pathway as a "default" developmental pathway (reviewed in ref 33).

This limited induction of early life Th1 responses is thought to reflect suboptimal APC–T-cell interactions rather than intrinsic T-cell defects. Evidence that neonatal APC function (assessed by IL-12 production) is indeed immature has been provided in mice. Similarly, human cord blood monocyte-derived cells (DCs) express limited amounts of IL-12 and IL-15, which was ascribed to a defect of IL-12 regulation in neonatal human APC [34]. Available data suggest that a limited IL-12–IFN-γ release capacity may persist during the first year of life [35]. Limited IL-12 production by APC was not corrected with the addition of CD40L trimers, supporting the existence of an intrinsic developmental immaturity of neonatal APC. Suboptimal APC–T-cell interaction could also result from deficient provision of T-cell help through limited CD154–CD40L expression. Transient differences in CD154 expression between neonatal and adult T cells have been reported, although more recent studies indicated that CD3-TcR-mediated activation of neonatal T cells did result in CD154 expression. Thus, the transiently limited expression of neonatal CD154–CD40L may not be a marker of intrinsically deficient early life T cells, but rather secondary to suboptimal T-cell activation. This would explain the successful induction of adult-like murine neonatal T-cell responses by adult dendritic cells [36] and vaccines or Th1 adjuvants capable of strong APC activation (reviewed in ref 16).

C.   Can Early Life Immunization Induce Adult-like CD4$^+$ and CD8$^+$ Vaccine Responses?

Studies in mice have clearly demonstrated that adult-like T-cell responses may be induced by novel delivery systems and/or adjuvants in the neonatal period. This has been achieved repeatedly by DNA immunization against a panel of vaccine antigens (reviewed in ref 18), and the induction of adult-like CD4+ and CD8+ early life T cells appears as a generic property of most DNA vaccines. This could result from prolonged antigenic exposure, allowing both prolonged immune stimulation and ongoing immune maturation to occur. However, induction of adult-like neonatal Th1 responses was also achieved by certain adjuvants. Among them, CpG motifs containing immunostimulating sequences enhanced early life INF-γ responses to adult levels, suppressed the neonatal burst of IL-5, and induced adult-like CTL responses [37, 38].

In conclusion, neonatal T cells may have greater requirements than adult T cells for costimulatory signals, such that the induction of Th1 and CTL neonatal responses essentially reflects the relative capacity of vaccines to activate neonatal APCs to thresholds sufficient for optimal T-cell activation to occur. Thus, mimicking (IL-12 supplementation) or triggering (CD40) optimal APC activation is sufficient to enhance IFN-γ and CTL neonatal responses. A first report assessing the capacity of a novel adjuvant, MF-59, to increase early life human T-cell responses demonstrated that MF-59 enhanced lymphoproliferative responses to recombinant HIV gp120 following immunization at birth, 2 weeks, 2 months, and 5 months [39]. Although cytokine responses could not be assessed, it seems reasonable to expect that certain adjuvant formulations or delivery systems will prove capable of enhancing early life Th1 responses, representing major progress in the control of early infections with intracellular pathogens.

IV. INFLUENCE OF MATERNAL ANTIBODIES ON INFANT VACCINE RESPONSES

It has long been recognized that residual maternal IgG antibodies (MatAbs) that are passively transferred during gestation may inhibit infant vaccine responses to measles [40] and oral poliomyelitis vaccine, and more recently it was noted that they may also affect responses to nonlive vaccines such as pertussis, tetanus and diphtheria toxoids, Hib conjugates, and hepatitis A. However, other clinical studies reported no influence of MatAb on responses to the same vaccines, requiring a better understanding of the main determinants of MatAb inhibition of infant responses.

A. Determinants of Maternal Antibody Inhibition of Infant Vaccine Responses

The main determinant of MatAb-mediated inhibition of antibody responses is the titer of MatAbs present at the time of immunization [41–43]. As passively transferred IgG MatAbs decrease with a half-life of a few weeks, a single month delay in vaccine administration is sufficient to reduce MatAb levels by 50%. Similarly, the use of higher doses of vaccine antigen increases the chance that some B-cell epitopes remain available for binding by infant B cells and may also provide the means to circumvent the inhibition. This was shown in pups of immune BALB/c mothers [42], for infant hepatitis A immunization, and indirectly following high-titer measles vaccine administration. Thus, strategies to circumvent MatAb inhibition of vaccine antibody responses have mainly included (i) delayed vaccine administration awaiting disappearance of MatAbs, (ii) use of higher vaccine doses, and (iii) development of mucosal vaccines, as concentrations of MatAbs reaching infant mucosae are significantly lower than those reaching serum.

However, it was observed that, although MatAbs may completely inhibit infant antibody responses if present at sufficient titers, they do leave CD4+ and CD8+ T-cell responses largely unaffected. This was first observed in mice under experimental conditions in which high titers of MatAbs were found to completely abrogate antibody responses but not to affect T-cell proliferative and cytokine responses [42, 43]. It is important to note that the presence of unaffected CD4+ responses was associated with a lack of inhibition of CTL responses by passively transferred antibodies [43–45]. In human infants, T-cell proliferative and INF-γ responses were also unaffected in 6- to 12-month-old infants immunized in the presence of MatAb concentrations that inhibited their antibody responses [5, 7, 26]. Accordingly, measles-specific T-cell responses were measured in 86.8% of 6-month-old infants immunized in the presence of MatAbs, whereas antibody responses were observed in only 36.7% [46].

B. Influence of Maternal Antibodies on Infant B- and T-Cell Responses

The inhibition of B-cell but not T-cell infant responses indicates the existence of distinct mechanisms, which can be hypothesized as follows (Fig. 1). Following the introduction of a vaccine antigen into a host with preexisting passive antibodies, MatAbs bind to specific B-cell vaccine epitopes. If the vaccine antigen/MatAb ratio is low, it prevents access of infant B cells to B-cell epitopes and therefore inhibits their differentiation into antibody-secreting cells. However, vaccine antigen–maternal antibody immune complexes are readily taken up by infant APCs, and processed, and their antigenic peptides are presented to CD4+ and CD8+ T cells. This allows CD4+–CD8+ priming to occur independently of the inhibition of B-cell responses (Fig. 1). Although a very high neutralizing MatAb/antigen ratio could completely neutralize a live attenuated viral vaccine load, thus preventing *in vivo* replication and reducing the effective immunizing dose to below that of the immunogenicity threshold required for induction of CD4+–CD8+ responses, this may occur only in a limited number of conditions because it was not observed following measles and mumps immunization. This early priming is likely to explain the reduced measles morbidity and mortality observed in vaccinated infants who failed to seroconvert due to the presence of maternal antibodies. It could also significantly facilitate prime–boost strategies in early life, as shown in mice [43].

C. Can Novel Vaccines Circumvent the Inhibition of Maternal Antibodies?

Novel vaccine formulations capable of prolonged immune stimulation could allow MatAb titers to decline and, thus, infant B cells to bind on recently synthesized–released vaccine antigens. Therefore, much attention was given to the immunogenicity of DNA vaccines in the presence of MatAbs. Depending on the experimental conditions used, DNA vaccines have been reported to either circumvent the inhibition of MatAbs or fail to do so (reviewed in ref 16). Compilation of available data strongly suggests that the same determinants of the influence of MatAb apply to both conventional and DNA vaccines. This

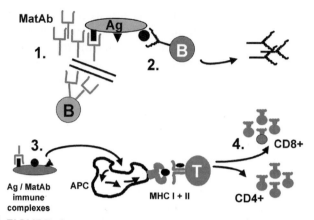

FIGURE 1 Influence of passively transferred antibodies on induction of vaccine responses. (1) Maternal antibodies (MatAbs) bind to their specific B-cell epitopes on the surface of vaccine antigens, thus preventing infant B cells from binding–activation. (2) Infant B cells that would be specific for B-cell epitopes against which MatAbs are low or absent (i.e., non-immunodominant maternal epitopes) may still bind vaccine antigens and differentiate into antibody-secreting cells. (3) Immune complexes formed by vaccine antigens–MatAbs are taken up by infant APCs and processed, and their peptides are presented on MHC class I–class II antigens at the cell surface. (4) This results in the induction of CD4+ and CD8+ responses, despite the inhibition of antibody responses.

includes MatAb titers at the time of immunization and the duration of antigen production and immune stimulation exceeding, or not, the period of MatAb persistence above inhibitory thresholds. Thus, it is expected that inhibition of B-cell responses to DNA vaccines will occur in the presence of high titers of MatAbs. Nevertheless, should DNA vaccines be able to induce sufficiently prolonged *in vivo* antigen delivery, allowing both ongoing immune maturation and the decline of MatAb concentrations, they would represent a most promising strategy for early life immunization in the presence of MatAbs. Alternatively, antigen presentation systems such as biodegradable polymer microspheres or ISCOMS could facilitate infant B-cell binding to vaccine antigens in spite of competing MatAbs. Finally, mucosally administered vaccines could avoid the influence of maternal antibodies, should the induction of vaccine responses occur at the mucosal level, avoiding contact with the higher levels of MatAb present in the spleen and lymph nodes.

V. CONCLUSION AND PERSPECTIVES

In order to be effective early in life, novel vaccines should be able to induce early and strong antibody responses. Available clinical and preclinical evidence suggests that this may not be easily achieved in the first weeks of life, even in the absence of maternal antibodies, until we gain a better understanding of the reasons for these limitations. In contrast, immune immaturity may not prevent the early induction of memory B cells. For protection against intracellular pathogens such as tuberculosis and malaria, certain infant vaccines will have to overcome a limited early life capacity for INF-γ and CTL responses. There is significant hope that this may be within reach, as these limitations apparently result from suboptimal APC–T-cell interactions rather than from intrinsic T-cell defects, i.e., they may be overcome by the use of specific adjuvants or delivery systems. The optimal immunogenicity–reactogenicity balance of these new vaccine formulations will have to be carefully defined in very young populations, calling for specific ethical and regulatory considerations.

In conclusion, strategies for early life immunization may, at present, optimally rely on an early priming of neonatal B cells, regardless of whether antibody responses develop, followed by boosting at a later age with a vaccine formulation enhancing antibody responses. As neonatally triggered Th2 responses are difficult to fully redirect toward Th1 responses, Th1-driving formulations should be used at the time of neonatal priming whenever the induction of Th1–CTL responses is required. Accordingly, a DNA priming–protein boosting strategy in infant mice was shown to result in strong Th1, CTL, and antibody responses against the F protein of respiratory syncytial virus [44]. Importantly, early primed T and B cells may both enhance subsequent responses and/or directly contribute to protection, even under conditions in which maternal immunity limits vaccine antibody responses.

ACKNOWLEDGMENTS

The author is most grateful to the support provided by the Fondation Mérieux and to all who participated in or supported the studies performed in the WHO Collaborating Center for Neonatal Vaccinology.

REFERENCES

1. World Health Organization (1996). Maternal Health and Safe Motherhood Programme. *MSM96.7*.
2. Booy, R., Aitken, S. J., Taylor, S., Tudor-Williams, G., Macfarlane, J. A., Moxon, E. R., Ashworth, L. A., Mayon-White, R. T., Griffiths, H., and Chapel, H. M. (1992). Immunogenicity of combined diphtheria, tetanus, and pertussis vaccine given at 2, 3, and 4 months versus 3, 5, and 9 months of age [see comments]. *Lancet* **339**:507–510.
3. Einhorn, M. S., Weinberg, G. A., Anderson, E. L., Granoff, P. D., and Granoff, D. M. (1986). Immunogenicity in infants of *Haemophilus influenzae* type B polysaccharide in a conjugate vaccine with *Neisseria meningitidis* outer-membrane protein. *Lancet* **2**:299–302.
4. Johnson, C. E., Nalin, D. R., Chui, L. W., Whitwell, J., Marusyk, R. G., and Kumar, M. L. (1994). Measles vaccine immunogenicity in 6- versus 15-month-old infants born to mothers in the measles vaccine era. *Pediatrics*, **93**:939–944.
5. Gans, H. A., Arvin, A. M., Galinus, J., Logan, L., DeHovitz, R., and Maldonado, Y. (1998). Deficiency of the humoral immune response to measles vaccine in infants immunized at age 6 months. *J. Am. Med. Assoc.* **280**:527–532.
6. Klinge, J., Lugauer, S., Korn, K., Heininger, U., and Stehr, K. (2000). Comparison of immunogenicity and reactogenicity of a measles, mumps and rubella (MMR) vaccine in German children vaccinated at 9–11, 12–14 or 15–17 months of age(*) [In Process Citation]. *Vaccine*, **18**: 3134–3140.
7. Gans, H., Yasukawa, L., Rinki, M., DeHovitz, R., Forghani, B., Beeler, J., Audet, S., Maldonado, Y., and Arvin, A. M. (2001). Immune responses to measles and mumps vaccination of infants at 6, 9, and 12 months. *J. Infect. Dis.* **184**:817–826.
8. Whittle, H. C., Aaby, P., Samb, B., Jensen, H., Bennett, J., and Simondon, F. (1999). Effect of subclinical infection on maintaining immunity against measles in vaccinated children in West Africa. *Lancet* **353**:98–102.

9. Ambrosino, D. M., Sood, S. K., Lee, M. C., Chen, D., Collard, H. R., Bolon, D. L., Johnson, C., and Daum, R. S. (1992). IgG1, IgG2 and IgM responses to two *Haemophilus influenzae* type b conjugate vaccines in young infants. *Pediatr. Infect. Dis. J.* **11**:855–859.

10. Plebani, A., Ugazio, A. G., Avanzini, M. A., Massimi, P., Zonta, L., Monafo, V., and Burgio, G. R. (1989). Serum IgG subclass concentrations in healthy subjects at different age: Age normal percentile charts. *Eur. J. Pediatr.* **149**:164–167.

11. Lucas, A. H., Azmi, F. H., Mink, C. M., and Granoff, D. M. (1993). Age-dependent V region expression in the human antibody response to the *Haemophilus influenzae* type b polysaccharide. *J. Immunol.* **150**:2056–2061.

12. Goldblatt, D., Vaz, A. R., and Miller, E. (1998). Antibody avidity as a surrogate marker of successful priming by *Haemophilus influenzae* type b conjugate vaccines following infant immunization. *J. Infect. Dis.* **177**:1112–1115.

13. De Wals, P., Dionne, M., Douville-Fradet, M., Boulianne, N., Drapeau, J., and De Serres, G. (1996). Impact of a mass immunization campaign against serogroup C meningococcus in the Province of Quebec, Canada. *Bull. WHO* **74**:407–411.

14. Timens, W., Boes, A., Rozeboom-Uiterwijk, T., and Poppema, S. (1989). Immaturity of the human splenic marginal zone in infancy. Possible contribution to the deficient infant immune response. *J. Immunol.* **143**:3200–3206.

15. Martin, F., Oliver, A. M., and Kearney, J. F. (2001). Marginal zone and B1 B cells unite in the early response against T-independent blood-borne particulate antigens. *Immunity* **14**:617–629.

16. Siegrist, C. (2001). Neonatal and early life vaccinology. *Vaccine* **19**:3331–3346.

17. Pihlgren, M., Schallert, N., Tougne, C., Bozzotti, P., Kovarik, J., Fulurija, A., Kosco-Vilbois, M., Lambert, P. H., and Siegrist, C. A. (2001). Delayed and deficient establishment of the long-term bone marrow plasma cell pool during early life. *Eur. J. Immunol.* **31**:939–946.

18. Bot, A. (2000). DNA vaccination and the immune responsiveness of neonates. *Int. Rev. Immunol.* **19**:221–245.

19. Prince, A. M., Whalen, R., and Brotman, B. (1997). Successful nucleic acid based immunization of newborn chimpanzees against hepatitis B virus. *Vaccine* **15**:916–919.

20. Bot, A., Shearer, M., Bot, S., Woods, C., Limmer, J., Kennedy, R., Casares, S., and Bona, C. (1999). Induction of antibody response by DNA immunization of newborn baboons against influenza virus. *Viral Immunol.* **12**:91–96.

21. Siegrist, C. A. (2000). Vaccination in the neonatal period and early infancy. *Int. Rev. Immunol.* **19**:195–219.

22. Prescott, S. L., Macaubas, C., Holt, B. J., Smallacombe, T. B., Loh, R., Sly, P. D., and Holt, P. G. (1998). Transplacental priming of the human immune system to environmental allergens:Universal skewing of initial T cell responses toward the Th2 cytokine profile. *J. Immunol.* **160**:4730–4737.

23. Marchant, A., Goetghebuer, T., Ota, M. O., Wolfe, I., Ceesay, S. J., De Groote, D., Corrah, T., Bennett, S., Wheeler, J., Huygen, K., Aaby, P., McAdam, K. P., and Newport, M. J. (1999). Newborns develop a Th1-type immune response to *Mycobacterium bovis* bacillus Calmette–Guerin vaccination. *J. Immunol.* **163**:2249–2255.

24. Pabst, H. F., Godel, J. C., Spady, D. W., McKechnie, J., and Grace, M. (1989). Prospective trial of timing of bacillus Calmette–Guerin vaccination in Canadian Cree infants. *Am. Rev. Respir. Dis.* **140**:1007–1011.

25. Vekemans, J., Amedei, A., Ota, M. O., D'Elios, M. M., Goetghebuer, T., Ismaili, J., Newport, M. J., Del Prete, G., Goldman, M., McAdam, K. P., and Marchant, A. (2001). Neonatal bacillus Calmette–Guerin vaccination induces adult-like IFN-γ production by CD4+ T lymphocytes. *Eur. J. Immunol.* **31**:1531–1535.

26. Gans, H. A., Maldonado, Y., Yasukawa, L. L., Beeler, J., Audet, S., Rinki, M. M., DeHovitz, R., and Arvin, A. M. (1999). IL-12, IFN-γ, and T cell proliferation to measles in immunized infants. *J. Immunol.* **162**:5569–5575.

27. Chiba, Y., Higashidate, Y., Suga, K., Honjo, K., Tsutsumi, H., and Ogra, P. L. (1989). Development of cell-mediated cytotoxic immunity to respiratory syncytial virus in human infants following naturally acquired infection. *J. Med. Virol.* **28**:133–139.

28. Mbawuike, I. N., Piedra, P. A., Cate, T. R., and Couch, R. B. (1996). Cytotoxic T lymphocyte responses of infants after natural infection or immunization with live cold-recombinant or inactivated influenza A virus vaccine. *J. Med. Virol.* **50**:105–111.

29. Jaye, A., Magnusen, A. F., Sadiq, A. D., Corrah, T., and Whittle, H. C. (1998). *Ex vivo* analysis of cytotoxic T lymphocytes to measles antigens during infection and after vaccination in Gambian children. *J. Clin. Invest.* **102**:1969–1977.

30. Adkins, B., Bu, Y., Cepero, E., and Perez, R. (2000). Exclusive Th2 primary effector function in spleens but mixed Th1/Th2 function in lymph nodes of murine neonates. *J. Immunol.* **164**:2347–2353.

31. Barrios, C., Brawand, P., Berney, M., Brandt, C., Lambert, P. H., and Siegrist, C. A. (1996). Neonatal and early life immune responses to various forms of vaccine antigens qualitatively differ from adult responses: Predominance of a Th2-biased pattern which persists after adult boosting. *Eur. J. Immunol.* **26**:1489–1496.

32. Barrios, C., Brandt, C., Berney, M., Lambert, P. H., and Siegrist, C. A. (1996). Partial correction of the TH2/TH1 imbalance in neonatal murine responses to vaccine antigens through selective adjuvant effects. *Eur. J. Immunol.* **26**:2666–2670.

33. Adkins, B. (2000). Development of neonatal Th1/Th2 function. *Int. Rev. Immunol.* **19**:157–171.

34. Goriely, S., Vincart, B., Stordeur, P., Vekemans, J., Willems, F., Goldman, M., and De Wit, D. (2001). Deficient IL-12(p35) gene expression by dendritic cells derived from neonatal monocytes. *J. Immunol.* **166**:2141–2146.

35. Chougnet, C., Kovacs, A., Baker, R., Mueller, B. U., Luban, N. L., Liewehr, D. J., Steinberg, S. M., Thomas, E. K., and Shearer, G. M. (2000). Influence of human immunodeficiency virus-infected maternal environment on development of infant interleukin-12 production. *J. Infect. Dis.* **181**:1590–1597.

36. Ridge, J. P., Fuchs, E. J., and Matzinger, P. (1996). Neonatal tolerance revisited: Turning on newborn T cells with dendritic cells [see comments]. *Science* **271**:1723–1726.

37. Kovarik, J., Bozzotti, P., Love-Homan, L., Pihlgren, M., Davis, H. L., Lambert, P. H., Krieg, A. M., and Siegrist, C. A. (1999). CpG oligodeoxynucleotides can circumvent the Th2 polarization of neonatal responses to vaccines but may fail to fully redirect Th2 responses established by neonatal priming. *J. Immunol.* **162**:1611–1617.

38. Brazolot Millan, C. L., Weeratna, R., Krieg, A. M., Siegrist, C. A., and Davis, H. L. (1998). CpG DNA can induce strong Th1 humoral and cell-mediated immune responses against hepatitis B surface antigen in young mice. *Proc. Natl. Acad. Sci. U.S.A.* **95**:15553–15558.

39. Borkowsky, W., Wara, D., Fenton, T., McNamara, J., Kang, M., Mofenson, L., McFarland, E., Cunningham, C., Duliege, A. M., Francis, D., Bryson, Y., Burchett, S., Spector, S. A., Frenkel, L. M., Starr, S., Van Dyke, R., and Jimenez, E. (2000). Lymphoproliferative responses to recombinant HIV-1 envelope antigens in neonates and infants receiving gp120 vaccines. AIDS Clinical Trial Group 230 Collaborators. *J. Infect. Dis.* **181**:890–896.

40. Albrecht, P., Ennis, F. A., Saltzman, E. J., and Krugman, S. (1977). Persistence of maternal antibody in infants beyond 12 months: Mechanism of measles vaccine failure. *J. Pediatr.* **91**:715–718.

41. Markowitz, L. E., Albrecht, P., Rhodes, P., Demonteverde, R., Swint, E., Maes, E. F., Powell, C., and Patriarca, P. A. (1996). Changing levels of measles antibody titers in women and children in the United States: Impact on response to vaccination. Kaiser Permanente Measles Vaccine Trial Team. *Pediatrics* **97**:53–58.

42. Siegrist, C. A., Cordova, M., Brandt, C., Barrios, C., Berney, M., Tougne, C., Kovarik, J., and Lambert, P. H. (1998). Determinants of infant responses to vaccines in presence of maternal antibodies. *Vaccine* **16**:1409–1414.

43. Siegrist, C. A., Barrios, C., Martinez, X., Brandt, C., Berney, M., Cordova, M., Kovarik, J., and Lambert, P. H. (1998). Influence of maternal antibodies on vaccine responses: Inhibition of antibody but not T cell responses allows successful early prime–boost strategies in mice. *Eur. J. Immunol.* **28**:4138–4148.

44. Martinez, X., Li, X., Kovarik, J., Klein, M., Lambert, P. H., and Siegrist, C. A. (1999). Combining DNA and protein vaccines for early life immunization against respiratory syncytial virus in mice. *Eur. J. Immunol.* **29**:3390–3400.

45. Bangham, C. R. (1986). Passively acquired antibodies to respiratory syncytial virus impair the secondary cytotoxic T-cell response in the neonatal mouse. *Immunology* **59**:37–41.

46. Pabst, H. F., Spady, D. W., Carson, M. M., Krezolek, M. P., Barreto, L., and Wittes, R. C. (1999). Cell-mediated and antibody immune responses to AIK-C and Connaught monovalent measles vaccine given to 6 month old infants. *Vaccine* **17**:1910–1918.

3 TRIAL DESIGN FOR VACCINES

PART A. Clinical Development of New Vaccines: Phase 1 and 2 Trials

W. RIPLEY BALLOU

Immunology and Infectious Diseases, Clinical Development, Medimmune, Inc., Gaithersburg, Maryland 20878

I. INTRODUCTION

The scientific basis for the development of new vaccines has accelerated greatly over the past 20 years. Major advances in the understanding of the pathophysiology of infectious diseases and a wealth of revolutionary technologies are expected to greatly enhance the feasibility of immunization against diseases for which vaccines do not currently exist [1]. Although relatively few experimental vaccines will make the transition from preclinical studies into trials involving human subjects, careful planning for clinical development activities can ensure that the most promising candidates are evaluated as efficiently as possible.

The clinical development of a new vaccine begins with a plan. The clinical development plan serves as a road map and time line and should lay out, in synopsis form, a series of clinical studies designed to provide the data needed for licensure. The plan should encompass phase 1 through phase 3 activities and identify issues that will need to be addressed in phase 4. The plan should begin with an explicit description of the proposed indications for which licensure will be pursued. The studies required to obtain the data supporting these indications should be described. Critical go–no-go milestones and alternative strategies should be identified. The majority of the clinical studies described in the development plan can be expected to occur in the phase 1 and phase 2 programs.

The phase 1 and phase 2 clinical programs should be expected to overlap with preclinical development. This will require an active dialogue between clinicians and research scientists. The preclinical phase provides the data that support the rationale for antigen selection, formulation, delivery route, and dose range to be employed in the clinic. Other preclinical objectives include the identification of an appropriate formulation capable of delivering the predicted dose range, an estimation of the potential need for booster doses, and an adequate preclinical safety database (usually including the site of injection toxicology studies that equal or exceed the doses and number of injections intended for human subjects). Additionally, well-characterized, high-throughput laboratory methods for assessing immunogenicity must be developed to support the clinical program.

Clear, succinct, and internally consistent protocols are essential for the success of a clinical development program. The clinical protocol is the definitive reference for study conduct [2]. Protocols should have a limited number of objectives (generally less than three), and these should be ranked as to their priority. The investigators must force themselves to stipulate the questions being asked, the data necessary to answer those questions, and methods that must be employed to collect those data. Integral to this process is the careful design of case report forms (CRFs), which will contain the entire set of data available for analysis at study completion. If the data are not in the CRF, they cannot be analyzed. Similarly, if the data are not going to be analyzed, they should not be collected in the CRF. Common mistakes among inexperienced investigators include poorly articulated objectives, too many objectives, and collection of data not required to achieve the objective. The tendency to collect data peripheral to the two or three most important objectives must be resisted. The practice of writing protocols collaboratively should be avoided as it tends to result in redundancies and inconsistencies that may lead to confusion among investigators.

Careful selection of investigators having previous experience with vaccine studies and access to suitable study populations and recruitment of a sufficient number of sites to insure rapid study enrollment are important keys to success. Investigator meetings are extremely valuable for insuring that each participating site is fully trained and understands the protocol-required procedures. The mechanics of data collection must be understood in great detail. Distribution and management of the investigational agent at the study sites require careful planning. Management of laboratory specimens for determination of vaccine safety and for measurement of immune responses must be coordinated, and systems must be established for reporting and analysis of study data.

The study investigators must be thoroughly familiar with the principles of good clinical practice (GCP) as described in the International Conference on Harmonization (ICH) Guidelines for Good Clinical Practice (Section E6) and Clinical Safety Data Management (Section E2A) [3]. These documents detail the responsibilities that the investigator, sponsor, and investigational review board assume when conducting studies involving human subjects, including the commitment of study investigators to conduct the study strictly according to the protocol. The ICH documents also provide essential information about what must be included in the clinical trial protocol, informed consent document, and investigator's brochure.

Vaccine development has a long tradition of investigator-driven research. Beginning with William Jenner and continuing until relatively recently, vaccinologists developed their own vaccines and tested them on themselves and their colleagues. Although current vaccine development still relies heavily on individuals who champion a vaccine strategy, there is now a far greater emphasis on understanding vaccine safety and therefore much greater regulatory oversight of all aspects of vaccine development. Investigators who wish to develop vaccines for the future must be committed to a strong multidisciplinary team approach that includes research and development, clinical development, investigational drug safety, regulatory affairs, biostatistics, and data management and that maintains a dialogue with governmental regulatory agencies. They must also insure that adequate organization and financial and personnel resources are available to drive the program through to completion.

II. PHASE 1: A PRIMARY FOCUS ON SAFETY

The phase 1 program is the period of clinical development during which a vaccine is introduced for the first time, usually into a study population consisting of healthy adult volunteers who do not have preexisting acquired immunity to the disease that the vaccine is designed to prevent. The choice of a naive population permits a detailed safety evaluation in the absence of preexisting immunity to the vaccine antigen and maximizes the interpretability of elicited immune responses. The evaluation of safety requires the establishment of a baseline prior to receiving the first dose of vaccine and the measurement of changes through serial clinical and laboratory evaluation after immunization.

The early clinical development of a new vaccine differs somewhat from that of small molecules and other biologicals where pharmacokinetic and pharmacodynamic studies are essential components of the development program. Because vaccine antigens are administered in discrete, widely spaced, and relatively small amounts, the distribution, accumulation, and elimination of the vaccine antigen are rarely described. In contrast, the local and general toxicity of the formulation and the immune responses generated by the vaccine are studied in great detail.

Phase 1 studies generally enroll fewer than 50 subjects and, thus, only permit a first-cut evaluation of vaccine safety using a systematic process of collecting adverse events (AEs). The number of volunteers exposed to the new vaccine will be expanded during the phase 2 program, in which several hundred volunteers can be enrolled per study. The safety data collection process is cumulative; summaries should be updated regularly and described in the investigator's brochure. ICH guideline E2A provides clear definitions for AEs and serious adverse events (SAEs). The emphasis should be on detecting important dose-limiting toxicities, recognizing that only the most frequent AEs and SAEs are likely to be detected in phase 1.

An important safety readout in phase 1 vaccine trials is reactogenicity, a measure of the local and general inflammatory responses to one or more components of the formulation. Although there is frequently no direct link between vaccine reactogenicity and immunogenicity, inflammatory processes at the cellular level clearly play a crucial role in the induction of immune responses. Reactogenicity may vary substantially among individuals who

develop equivalent immune responses to a vaccine. Reactogenicity is characterized by collecting solicited adverse events. Solicited AEs include local reactions
at the site of administration as well as selected general signs and symptoms.
They should be collected explicitly and systematically during the postvaccination observation period. Subjects should be monitored closely for 20–30 min
after vaccination for acute AEs (especially for allergic manifestations). For a
parenterally administered vaccine, reactions at the site of injection should be
quantitatively measured and graded for severity. For vaccines delivered by other
routes (oral, mucosal, transcutaneous), site-specific criteria may be required.
Typical periods for collecting solicited AEs are 4–7 days following each dose,
though these periods may be modified as additional reactogenicity data are generated for a given vaccine [4].

The occurrence of other AEs during a defined period after vaccination
should also be recorded. These should be collected in a nonleading (unso-
licited) fashion, for example, asking whether the general health status of the
volunteer has changed in any way since the previous visit. Where feasible, the
use of diary cards or other data collection tools can be helpful to obtain daily
information about AEs during the period after vaccination and the volunteer's
next follow-up visit. It is very important to train the volunteers in the standardized use of these tools, especially if quantitative data such as intensity of
AE and measurement of injection site reactions are requested. For each AE,
start and stop dates must be recorded, and the investigator must determine
whether a causal relationship exists between the administration of the vaccine
and the onset of the AE.

III. PRACTICAL CONSIDERATIONS FOR PHASE 1 STUDIES

The phase 1 program is characterized by one or more studies that have a
relatively small sample size and that ask a limited number of focused questions. The phase 1 program should be designed to select a vaccine dose that is
safe and sufficiently well-tolerated to justify exposure to larger study populations in phase 2. The study design should incorporate clear stopping rules for
individual subjects and for the study population as a whole that are based on
specific objective measurements and that permit either the principal investigator or the medical monitor to discontinue dosing pending review of the data
by an independent safety monitoring committee or regulatory authorities [5].

Phase 1 studies are principally designed to determine whether a vaccine has
an acceptable margin of safety and induces sufficiently robust and appropriate
responses to justify the considerable time and expense required to conduct
further clinical trials. It is, therefore, important to extract the most information from the study as efficiently as possible. Timely recruitment of studies is
an important part of this goal and is usually a challenge. Investigators routinely overestimate the availability of volunteers to participate in phase 1 and
phase 2 trials and, therefore, underestimate the time required to fully enroll
the studies. Enrollment rates can be enhanced by careful protocol design
that pays close attention to details of eligibility criteria, study complexity, and
time demands on study subjects. Sites may need to screen many volunteers

for each enrolled subject, and therefore advertisement or study announcement strategies designed to target a large potential audience must be developed. The use of multiple study sites is likely to reduce the enrollment period and can be cost-effective.

Dose selection for phase 1 must be determined on the basis of preclinical studies or experience with similar vaccines [6]. Vaccine dose may be limited by antigen availability, formulation requirements, or the sensitivity of assays used to characterize product stability. Although it is beyond the scope of this part of Chapter 3 to comment on every class of vaccines, certain generalizations can be made. The majority of licensed subunit vaccines contain 50 µg of protein antigen or less per adult dose, but optimal dose selection must be determined experimentally. Dose–response studies should include a threshold dose below which no or minimal responses are seen, a range over which increasing doses induce more consistent seroconversion or higher mean antibody titers, and a dose above which immune responses plateau. Vaccine antigens that are particulate in nature are generally more immunogenic than soluble proteins. Carbohydrate antigens may have unpredictable immunogenicity and often require conjugation to a protein for optimal immune responses. Adjuvants have been demonstrated to enhance the immune responses to virtually all classes of vaccines, but regulatory agencies are likely to insist that the requirement for their inclusion be proven in the clinic. In these studies, the vaccines formulated with and without adjuvant are tested side by side in a controlled, double-blind fashion. Studies of live-attenuated and DNA plasmid vaccines may require dose ranges extending over several orders of magnitude.

Phase 1 vaccine trials should routinely be conducted as blinded studies that include a control group as one of the study arms [7]. This design permits the most unbiased assessment of vaccine safety. Phase 1 studies frequently will evaluate doses extending over a considerable range, and it is desirable to demonstrate a dose response if possible. This typically requires the inclusion of a dose predicted to be less than optimally immunogenic, a dose representing the likely upper range considered practical, and one or more intermediate doses. Formulation of the vaccine antigen at a relatively high concentration permits dilution of the vaccine "at the bedside" or by the study pharmacist and allows a range of vaccine doses to be studied relatively simply. The initial choice of doses should be based upon preclinical immunogenicity studies, but it should recognize that doses administered to small experimental animals may not scale in a linear fashion and may be misleading when predicting response in humans. More importantly, immunogenicity in mice and rabbits can be notoriously poor as a predictor of immune response in humans, as they may greatly overestimate immunogenicity. Studies in nonhuman primates may be more informative, but the availability of suitable primates is often restricted [8].

Blood collection after vaccination is performed to assess vaccine safety and to measure immune responses. Blood collection for laboratory safety studies should occur just before and within 1 week after vaccination and should target a limited subset of laboratory assays selected for their clinical relevance. Serum or whole blood collection for analysis of immune responses should be collected at baseline, before booster doses are administered, and generally within 7–14 days following a dose. The amount of blood required for immunogenicity studies is usually underestimated, especially when assay development is still in progress. Because of their complexity, studies of cellular immune responses are frequently restricted to a subset of subjects or a limited number of sampling times, and explicit instructions and detailed technical training may be required for successful recovery and handling of viable peripheral blood cells. Sampling from other sites, such as collection of mucosal secretions, also requires the development of detailed study procedures. Great care must be taken to properly label, process, store, and ship laboratory samples.

The duration of follow-up for phase 1 vaccine studies should be at least 6 months after the final injection, but long-term follow-up to assess the duration of immune responses over several years is often valuable. Investigators should consider building in this option from the beginning to maximize volunteer participation.

IV. PROCEEDING TO PHASE 2

Once it is clear that the vaccine has acceptable reactogenicity and appears safe and immunogenic in healthy adults, it is appropriate to expand exposure to the vaccine to larger and more diverse populations in the phase 2 program. The evaluation of safety continues throughout the phase 2 program and employs methods that extend and refine the data collected in phase 1. Phase 2 generally includes additional dose–response studies, studies of different formulations, schedule optimization studies, and lot consistency studies. The introduction of a new vaccine into a pediatric population may require a parallel series of clinical trials, including age deescalation studies in which dose and schedules are optimized and studies in which the impact of concomitant immunization with existing pediatric vaccines is evaluated. Phase 2 also may include pilot efficacy trials that can be extremely valuable when planning for phase 3.

Schedule optimization involves establishing the optimal timing and number of doses required to induce protection. This process can be greatly facilitated if a correlate of protective immunity is known. Some live attenuated vaccines can induce fully protective responses after a single dose, but others in this class and most subunit vaccines require two or more doses to achieve optimal responses. Experience and experimental evidence suggest that booster injections may be more effective when delivered after immune responses have fallen from peak levels, but identification of the optimal timing for booster injections is often empiric [9].

Phase 2 trials typically enroll several hundred subjects per study, and they are commonly conducted at multiple centers. The scope of the phase 2 program will be driven by the complexity of the clinical questions to be resolved. If the vaccine target is a pediatric or otherwise vulnerable population, additional phase 1–2 safety and immunogenicity studies must be conducted in these groups. In the setting of childhood immunization, it is particularly important to assess the impact of the investigational vaccine on the immune responses to concomitant administration of licensed vaccines and vice versa [10].

One issue that arises during the design of dose escalation and formulation studies involving vaccines that include an adjuvant is whether to hold the amount of adjuvant constant and vary the antigen concentration or to hold the ratio of antigen and adjuvant constant and adjust the volume of vaccine administered. Both approaches have their merits and limitations, and the issue should be anticipated and explored during preclinical studies. Practical concerns such as blinding, minimum and maximum constraints on the amount of vaccine injected (generally 0.25–1.0 ml in adults), and the pluses and minuses of multiple vialings on the one hand and the potential for formulation "at the bedside" on the other should be considered well in advance. Because the emphasis on understanding the safety of the vaccine continues into phase 2, the inclusion of control groups is strongly recommended. The selection of a suitable control may be complex. Whenever a control group is used, the study should be designed as a randomized, double-blind trial.

Lot consistency studies are frequently included during the phase 2 program to insure that the safety and immunogenicity of the vaccine is equivalent from lot to lot prior to the start of costly pivotal phase 3 studies. As the demonstration of lot manufacturing and formulation consistency is required for licensure, it therefore may be prudent to conduct phase 2 studies with more than one lot of vaccine. Lot consistency studies may be designed as stand alone "shoot and bleed" safety and immunogenicity studies, or they may be bundled with dose-ranging or pilot efficacy studies. In either case, careful documentation of which lots are used in clinical studies is essential for proper analysis of lot consistency data.

Pilot efficacy studies should be strongly considered during phase 2 if the incidence of the disease is sufficiently high to obtain preliminary evidence of vaccine efficacy. In addition to reducing the risks associated with phase 3, they are invaluable for working out critical details such as refining end points and case definitions, improving study procedures and systems, evaluating study sites and grooming investigators, and permitting better sample size estimates for phase 3.

For certain infectious diseases, it may be possible to estimate the efficacy of a new vaccine by conducting experimental challenge studies [11]. In these studies, volunteers are recruited with the explicit understanding that an experimental challenge will be conducted once immunization has occurred. Such studies are extremely complex and pose extensive ethical and logistic concerns. They are limited to infectious diseases that are either fully treatable or self-limiting [for example, influenza, respiratory syncytial virus (RSV), rhinovirus, malaria, cholera, shigella, salmonella, gonorrhea, and possibly others]. Challenge doses must be well-characterized, reproducible, and highly standardized, and in certain settings they must be carried out under isolation or containment. Challenge studies must not include vulnerable populations such as children, prisoners, or anyone unable to provide fully informed consent [12]. Because of the greater risks inherent with challenge studies, great care must be taken to avoid inappropriate inducement to participate on the basis of volunteer payment.

Estimation of sample sizes for phase 2 pilot efficacy trials to be conducted in a field setting can be quite challenging. Often, investigators will be faced with less than complete data on disease incidence, which may have been collected in a nonintervention setting. Experienced researchers understand that the observed disease incidence during a study is almost always lower than expected. Decreased rates of disease may be due to factors such as the implementation of highly specific case definitions, the impact of improved case detection and treatment, and, of course, just bad luck. Although the statistician should always carry out formal sample size calculations, these should be considered estimates at best. A useful rule of thumb is to estimate the expected disease rates in the control arm using the best available data and then reduce this expected rate by up to 50%. Estimate the sample size based on the revised incidence rates and consider doubling the resultant sample size if resources permit. A phase 2 pilot efficacy study designed using this process is likely to be adequately powered and will permit reliable decisions to be made on the design and execution of pivotal phase 3 studies.

V. CONCLUSIONS

Phase 1 and phase 2 programs are designed to lay the foundation for proceeding to pivotal phase 3 trials. In the context of a vigorous laboratory research program, they provide for an understanding of the safety and immunogenicity of new vaccines, allow for the refinement of immunologic assays and the identification of clinical correlates of immunity, and provide an opportunity to establish proof of principle on a pilot basis. As such, they must be conducted rigorously and in the context of a dialogue with local and national regulatory agencies. The programs instruct the investigational team on the safety profile and immunogenicity of the vaccine and its optimal dose, formulation, route, and schedule. When feasible, properly executed pilot efficacy studies should help predict the likely outcome of large-scale pivotal trials of vaccine efficacy.

REFERENCES

1. Ellis, R. W. (2001). Technologies for the design, discovery, formulation and administration of vaccines. *Vaccine* **19**(17–19): 2681–2687. Review.
2. ICH Guidelines for Good Clinical Practice (E6), Section 4.5, Compliance with Protocol. Federal Register, May 9, 1997.
3. ICH Guidelines for Good Clinical Practice (E2A), Section 4.5, Clinical Safety Data Management: Definitions and Standards for Expedited Reporting. Federal Register, March 1, 1995.
4. Kester, K. E., McKinney, D. A., Tornieporth, N., Ockenhouse, C. F., Heppner, D. G., Hall, T., Krzych, U., Delchambre, M., Voss, G., Dowler, M. G., Palensky, J., Wittes, J. T., Cohen, J., Ballou, W. R. (2001). Efficacy of recombinant circumsporozoite protein vaccine regimens against experimental *Plasmodium falciparum* malaria. *J. Inf. Dis.* **183**:640–647.
5. Wittes, J. (1993). Behind closed doors: The data monitoring board in randomized clinical trials. *Stat. Med.* **12**(5–6):419–424.
6. Andre, F. E., and Foulkes, M. A. (1998). A phased approach to clinical testing: criteria for progressing from Phase I to Phase II to Phase III studies. *Dev. Biol. Stand.* **95**:57–60.
7. Schulz, K. F., and Grimes, D. A. (2002). Blinding in randomised trials: hiding who got what. *Lancet* **359**(9307):696–700.
8. Gordon, D. M., McGovern, T. W., Krzych, U., Cohen, J. C., Schneider, I., LaChance, R., Heppner, D. G., Hollingdale, M., Slaoui, M., Hauser, P., Voet, P., Sadoff, J. C., and Ballou, W. R. (1995). Safety, immunogenicity and efficacy of a recombinantly produced *Plasmodium falciparum* circumsporozoite protein / HBsAg subunit vaccine. *J. Inf. Diseases* **171**:1576–1585.

9. Middleman, A. B., Kozinetz, C. A., Robertson, L. M., DuRant, R. H., and Emans, S. J. (2001). The effect of late doses on the achievement of seroprotection and antibody titer levels with hepatitis b immunization among adolescents. *Pediatrics* **107**(5):1065–1069.

10. Schmitt, H. J., Zepp, F., Muschenborn, S., Sumenicht, G., Schuind, A., Beutel, K., Knuf, M., Bock, H. L., Bogaerts, H., and Clemens, R. (1998). Immunogenicity and reactogenicity of a Haemophilus influenzae type b tetanus conjugate vaccine when administered separately or mixed with concomitant diphtheria-tetanus-toxoid and acellular pertussis vaccine for primary and for booster immunizations. *Eur. J. Pediatr.* **157**(3):208–214.

11. Herrington, D. A., Clyde, D. F., Murphy, J. R., Baqar, S., Levine, M. M., do Rosario, V., and Hollingdale, M. R. (1988). A model for *Plasmodium falciparum* sporozoite challenge and very early therapy of parasitaemia for efficacy studies of sporozoite vaccines. *Trop. Geogr. Med.* **40**(2):124–127.

12. Hoffman, S. L. (1997). Experimental challenge of volunteers with malaria. *Ann. Intern. Med.* **127**(3):233–235.

3 TRIAL DESIGN FOR VACCINES

PART B. Phase 3 Studies of Vaccines

JOHN D. CLEMENS AND HYE-WON KOO
International Vaccine Institute, Seoul National University Campus, Seoul, Korea 151–742

I. INTRODUCTION

The past decade has witnessed a remarkable profusion of experimental vaccine candidates. New vaccine technologies and an improved understanding of protective immunity against infectious diseases, as well as novel methods of presenting antigens to elicit desired immune responses, have revolutionized our ability to prevent infections by vaccination. An indication of this phenomenal revolution in vaccine development is provided by the annual report on vaccine development for the year 2000 issued by the National Institute of Allergy and Infectious Diseases of the U.S. National Institutes of Health, which cited a pipeline of over 350 vaccine candidates now being developed against nearly 100 different infectious diseases [1].

This profusion of new candidate vaccines creates breathtaking opportunities for disease prevention. At the same time, it has created challenges for the proper evaluation of these agents. Rigorous and meticulous scientific evaluations are required before the introduction of a vaccine into public health practice. Phase 3 studies constitute the final evaluations of a new vaccine before its licensure. In this part of Chapter 3, we describe the goals and design of phase 3 studies, as well as several of the methodological challenges entailed in designing these studies.

II. PRELICENSURE PHASING OF VACCINE EVALUATIONS

The clinical evaluation of new vaccine is performed in a systematic and phased fashion. The phased approach is based on the sometimes unpredictable safety, immunogenicity, and protectivity of vaccine candidates. This unpredictability requires that initial evaluations be done on small numbers of subjects who are most likely to tolerate the side effects of a vaccine well. Thus, the first studies are done on small numbers of healthy adult volunteers, even if the vaccine is ultimately intended for use in another age group, such as infants. As confidence in the vaccine builds on the basis of these early studies, subsequent studies enroll progressively larger numbers, and move in steps to enroll the ultimate target group for the vaccine. Thus, for an infant vaccine, it is common for successive studies to move sequentially in successively lower age groups until the ultimate target group—infants—is enrolled. Enrollment of progressively larger numbers of subjects is also important in order to increase the statistical power of successive evaluations to detect infrequent vaccine side effects. For vaccines found to be suitably safe and immunogenic in phase 1–2 studies, phase 3 studies are conducted to provide rigorous evidence about vaccine protection against naturally occurring infections and to provide additional data on vaccine safety in larger numbers of persons. Phase 3 studies are constructed as experiments with clear hypotheses and are conducted in a population that normally experiences the disease against which the vaccine is targeted. Phase 3 studies thus constitute pivotal evaluations that provide the basis for decisions about whether to license a vaccine for use in public health practice.

III. PHASE 3 PARADIGM: THE RANDOMIZED, CONTROLLED CLINICAL TRIAL

A. Design

The randomized, controlled clinical trial (RCT) is now regarded as the gold-standard design for providing scientifically credible evidence about the clinical performance of vaccines [2, 3]. As a result, many national drug regulatory agencies require evidence of vaccine efficacy and safety from properly conducted RCTs before licensure of a new vaccine. In a two-group RCT of an experimental vaccine (Fig. 1), participants are recruited from a target population and enrolled for the study after acquisition of informed consent and ascertainment of eligibility. Prior to the study, a formal randomization scheme is devised to allocate subjects to the experimental vaccine group or to a comparison (control) group, and this scheme is applied to study participants who have been entered into the trial. Commonly, the comparison agent is selected such that investigators and participants are kept unaware of the identity of each dose administered in regard to whether the dose is the experimental vaccine or the comparison agent (double-blinding) [4]. This is done by choosing a comparison agent that, by virtue of having physical characteristics identical to those of the experimental vaccine and the same schedule and route of administration, can be given in a coded fashion that prevents both investigators and subjects from being able to guess the identity of the agent that is administered. Blinding can be accomplished with the use of either an inert placebo, such as normal saline, or an alternative active vaccine that has an appearance identical to that of the experimental vaccine under evaluation. In a trial of vaccine

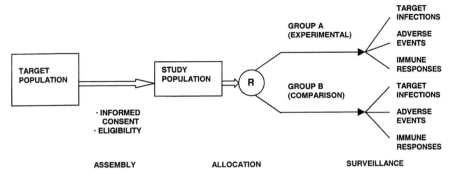

FIGURE 1   A simplified schematic of a phase 3 vaccine trial designed as a two-group, randomized, controlled trial. In such a trial, the study population is assembled from a target population and is then randomized to receive an experimental vaccine or a comparison agent. The experimental and comparison groups are followed longitudinally and concurrently to detect the comparative occurrence of target infections, adverse events, and immune responses in the two groups to assess vaccine protection, vaccine safety, and vaccine immunogenicity, respectively.

protection, this alternative vaccine must have no known ability to elicit immunological responses that protect against the target infection under study.

After randomization, study participants are followed longitudinally to detect target infections as well as adverse events that might be related to vaccination, including deaths, and the comparative occurrence of these events in the two groups provides the basis for assessing both vaccine protection and vaccine safety. The incidence of these events may be expressed as a *rate*, the number of detected events divided by the cumulative person–time at risk, or as a *risk*, the number of persons experiencing the event over some specified period of time divided by the number of persons at risk. Similarly, participants, or a subsample of participants, are typically assessed at baseline and at some point shortly after dosing to assess immune responses to the vaccine, and comparisons between the experimental vaccine group and the control group permit assessment of the proportion of apparent immune responders that can be attributed to receipt of the experimental vaccine.

Contrasts of the occurrence of events are typically expressed as rate ratios or risk ratios (RR), calculated as the risk or rate in the vaccine group divided by that for the comparison group. The conventional index of protection, *protective efficacy* (PE), is then calculated as $PE = (1 - RR) \times 100\%$[5], where RR is the risk or rate of the target infection in vaccinees divided by that for the comparison group. PE reflects the proportionate reduction of the incidence of the target infection in the experimental vaccine group relative to the comparison group. A value of 0% denotes no protection, that of 100% corresponds to complete protection, and negative values indicate a lower incidence in the comparison group than in the experimental vaccine group. Table 1 illustrates this type of calculation with the results for 1 year of follow-up of a phase 3 trial of two killed oral cholera vaccines in rural Bangladesh [6]. In this trial, one cholera vaccine under study contained cholera toxin B subunit together with killed whole cells, and the other contained only killed whole cells. The two vaccines were contrasted with a group that received an oral placebo consisting of killed *Escherichia coli* K12 strain whole cells.

TABLE 1   Field Trial of Killed Oral Cholera Vaccines in Matlab, Bangladesh: Analysis of Data for First Year of Surveillance[a]

| | Group[b] | | |
Developed cholera?	BS-WC	WC	K12
Yes	41	52	110
No	20,664	20,691	20,727
Total	20,705	20,743	20,837

[a]Protective efficacy (PE) for BS-WC: $[1-(41/20,705)/(110/20,837)] \times 100\% = 63\%$, $P<0.0001$; 95% confidence interval 46%, 74%. PE for WC: $[1-(52/20,743)/(110/20,837)] \times 100\% = 53\%$, $P<0.0001$; 95% confidence interval 33%, 66%. Data from ref 6.

[b]BS-WC, B subunit, killed whole-cell vaccine; WC, killed whole-cell vaccine; K12, *Escherichia coli* K12 strain placebo.

B. Strengths of the RCT Design in Phase 3 Trials of Vaccines

The RCT design of phase 3 trials of experimental vaccines is formulated with several objectives: to ensure ethical propriety, to protect against bias, to ensure statistically precise results, and to provide results that generalize suitably to the intended public health application of the vaccine.

Several specific features of RCTs serve these objectives. *Clear demarcation of the target population* for the trial defines a population to which the results of the trial ultimately can be extrapolated. *Eligibility criteria* for a trial are constructed not only to further define a group that is at high risk for the target infection and that can ultimately serve as a group to be vaccinated in public health practice but also to ensure that all participants in the trial have no absolute indications or contraindications to either the experimental vaccine or the comparison agent [7]. The latter feature is an ethical prerequisite for the use of randomization, a critical scientific feature of RCTs, because the use of randomization means that the assignment of a subject to one of the compared agents is totally out of the control of the investigator and is executed without reference to the needs or desires of the participants.

Acquisition of proper *informed consent* from potential subjects for their participation in RCTs is an ethical prerequisite. The particular requirements for obtaining informed consent may vary according to the age of the subject and the subject's legal capacity to give informed consent, as well as the particular *milieu* for the trial. In general, acquisition of informed consent indicates that subjects have agreed to participate after understanding the purposes and elements of the trial, as well as the possible benefits and risks of participation, and after being guaranteed that their decisions to participate and to continue participating are completely voluntary [8].

The RCT design also ensures that the enrollment of participants occurs without knowledge of what agent the participant will ultimately receive. This *independence of enrollment and allocation* is provided by suitable blinding of subjects and investigators to the identities of the compared agents, because such blinding ensures that a subject's choice to participate and an investigator's selection of a subject cannot be influenced by knowledge of what agent the subject is to receive. Without this feature, choices by subjects to participate and choices by investigators to enroll subjects could be affected by knowledge of the agent to be received, and these choices could distort the impartiality of randomized allocation [9, 10]. In lieu of blinding, independence of enrollment and allocation may be accomplished at least partially by allocating a subject to one of the compared agents only after the subject is fully enrolled in the trial. Of course, subjects could still withdraw prior to dosing in this situation, which would vitiate the protection against bias provided by randomization.

Randomized allocation of participants to the compared agents is the most powerful mechanism for balancing the compared groups in a trial with respect to features, such as age and socioeconomic status, that may be determinants of the risk of the target infection [11–13]. Use of randomization also provides a formal basis for using statistical tests to assess the role of chance in observed differences in the occurrence of outcomes in the compared groups [14].

Double-blinding in the administration of the compared agents protects against a bias that could occur if subjects decided to withdraw at the time of dosing on the basis of knowledge of what was to be received. It also protects against a bias that could occur if subjects used knowledge of what agent had been received to alter behaviors that might be related to the risk of the target infection. For example, if, in a cholera vaccine trial, subjects who received a cholera vaccine relaxed their attentiveness to personal hygiene whereas those receiving placebo did the reverse, the actual protective benefit of an effective vaccine could be partially or even completely nullified by this bias, which is sometimes referred to as *cointervention bias*.

Surveillance for target infections in a phase 3 trial is undertaken in a concurrent fashion for the compared groups to prevent secular biases in disease incidence that might occur if the calendar interval of follow-up were different for the groups. Surveillance is prospective with prearranged methods to ensure comprehensive detection of the infections of interest and with minimal losses to follow-up, so that the total impact of vaccination can be measured. If losses to follow-up occur through drop-outs from the study, out-migration, or deaths, these "competing" events are noted in the surveillance in order to evaluate the actual level of follow-up of the compared groups and to make statistical adjustments for any inequalities in follow-up that are observed [9, 10]. Surveillance is usually undertaken with prespecified methods for diagnostic evaluation of patients suspected of having the target infection and with clear clinical and microbiological definitions of the target infection to help ensure that all participants receive equivalent and appropriate diagnostic surveillance and to assist in explicitly defining the target illnesses prevented by the vaccine [4]. Deployment of quality control procedures to ensure that diagnoses are accurate and reproducible minimizes biases that might occur due to misclassification of clinical illnesses and helps guarantee that the true magnitude of a vaccine's preventive impact is measured. Formulation of these definitions prior to analysis of the trial data, rather than after the accumulated data have been inspected, helps to enhance the scientific credibility of the trial's findings. Moreover, double-blinded conduct of surveillance and suitably blinded procedures for making final diagnoses are important tools for ensuring that the trial's results are not distorted by detection bias [15]. Similar considerations apply to a trial's *surveillance for side effects and immune responses* to vaccination.

Finally, *sample size calculations* in planning phase 3 trials ensure that the trials have *adequate statistical power* to detect intergroup differences in outcome events of interest, including target infections, side effects, and immune responses.

IV. ADDITIONAL ISSUES FOR PHASE 3 TRIALS OF VACCINES

Although the basic paradigm of phase 3 RCTs of experimental vaccines is relatively straightforward, as outlined earlier, there are many complexities in designing, executing, and analyzing phase 3 vaccine trials. The following sections summarize several of these challenges.

A. Posing Research Questions for the Trial

As in any research project, posing the correct research question appropriately and in sufficiently specific terms is key to the successful outcome of a phase 3 vaccine trial, because the research questions posed guide the design, conduct, and analysis of the trial. In this regard, trials are typically designed to answer primary research questions, evidence bearing on which will be considered by regulatory agencies in deliberations about vaccine licensure, and secondary questions, which address additional important scientific issues of interest. In formulating the primary research questions, the following should be considered. The questions should clearly specify the target population for the study, the formulation and constituents of the vaccine, together with its mode and schedule of administration, the comparison agent to be used as a control agent, the target infections whose occurrence will serve as the basis for estimating vaccine protection, the adverse event outcomes whose occurrence will serve as the basis for evaluating vaccine safety, and the immune outcomes whose occurrence will serve as a basis for assessing vaccine immunogenicity. Failure to specify any of these features in a way that accurately reflects how a vaccine will be used in practice may impede the admissibility of a trial's results in regulatory decisions about licensure.

B. Determining the Type of Vaccine Protection To Be Measured

A key consideration in specifying the parameters for the primary study question is whether the trial is to measure direct or indirect vaccine protection. When a vaccine confers protection to an individual solely on the basis of vaccine-induced immunity in that individual, irrespective of whether other individuals in the same population have been vaccinated, it is called *direct protection* [16]. However, in practice vaccines are usually targeted to populations rather than individuals. For pathogens that are transmitted from person to person, targeting of a population rather than an individual for vaccination may have the added effect of reducing the intensity of transmission of the pathogen in the population, because the prevalence of immune individuals in the population, due to vaccination, reduces the probability that any individual will come into contact with the pathogen or increases the probability that, when an individual is exposed, the inoculum of exposure will be smaller. The protection resulting from this reduced intensity of transmission of the pathogen in a population is called *indirect protection* [16, 17].

A phase 3 trial of a vaccine can measure either direct protection per se or direct and indirect protection combined. To measure direct protection per se, allocation of subjects to vaccine and the comparison agents must not create high concentrations of vaccinees within the geographical units within which the pathogen is transmitted. This is usually accomplished by randomizing individuals rather than groups and by randomizing within blocks that balance the numbers of vaccinees and controls within epidemiologically relevant groups of individuals. To permit the estimation of direct and indirect protection combined, clusters of individuals rather than individuals per se must be the units of randomization [18]. One can visualize a hypothetical two-group trial in which clusters of individuals are randomized to an experimental vaccine or to an agent with no activity against the target pathogen. Contrast of the incidence of the target infection in vaccinees versus that in recipients of the comparison agent in such a trial provides a measure of the combination of direct protection and indirect protection of vaccinees, the latter due to any reduction of transmission within the vaccinated clusters. Comparison of the incidence in nonvaccinees within the vaccinated clusters versus that in nonrecipients of the comparison agent in the control clusters allows the estimation of indirect protection of nonvaccinated members of vaccinated groups ("herd immunity"). Finally, the overall comparison of the incidence of infection in groups allocated to the vaccine versus groups allocated to the comparison agent, regardless of whether subjects actually received the assigned agent, permits the estimation of the overall impact (direct plus indirect) of vaccination upon the entire group targeted for vaccination [19].

Most regulatory agencies require that phase 3 trials be individually randomized, presumably because of a preference for the demonstration of direct vaccine protection per se. However, a phase 3 trial of pneumococcal polysaccharide–protein conjugate vaccines given to Native American infants in the United States employed cluster randomization specifically because of a perceived need to demonstrate the indirect protective effects of this vaccine [20]. It is likely that we will see an increased use of cluster randomization in the future for vaccines from which strong indirect protective effects are anticipated.

C. Selecting an Appropriate Setting and Study Population

The success of a phase 3 trial is critically dependent on the choice of a setting and a study population. It is necessary to select a population in which the target infection occurs predictably and with a sufficient frequency to evaluate whether vaccination reduces the incidence of the target infection. Study populations may be defined on the basis of age, gender, and even behaviors. For example, an early trial of plasma-derived hepatitis B vaccine was conducted in homosexual men who resided in New York City [21], and ongoing trials of the gp120 vaccine against HIV in the United States and Thailand have enrolled homosexual men and intravenous drug users, respectively [22]. In some instances, the study population can be defined on the basis of documented recent exposure to the target pathogen, as in trials of newer generation vaccines against leprosy, which have focused upon family contacts of known patients with leprosy [23]. Regardless of the target population chosen for the trial, it is necessary to enumerate and characterize the population before the trial to enable later assessment of whether the final group for the study was representative of the intended target population.

Beyond these features, it is important that the trial have the informed and firm backing of local and, in some instances, national government authorities, as well as the backing of the local population from which the study population is to be recruited. Moreover, it is essential that the site have adequate research infrastructure, in terms of scientific and technical personnel who will work on the trial, as well as suitable laboratories for processing, storing, and testing specimens. Finally, even if a proposed study population has a high incidence of the target infection, a trial often will not be feasible if the population is not sufficiently stable, with minimal out-migrations, if the geography of the site is not propitious, allowing easy access to the study population, or if the site is politically unstable.

D.   Predicting the Incidence of the Target Expected in the Comparison Group

Basic to the planning of any RCT is suitable calculation of the size of the needed study population. One formula [24, 25] available for sample sizes required by two-group trials in which the groups are of equal sizes gives the size per group (N) as

$$N = \frac{(Z_a + Z_b)^2}{d^2}[P_1(1-P_1) + P_2(1-P_2)].$$

In this formula, P_1 is the expected incidence of the target infection in the comparison group, P_2 is the expected incidence in the group receiving the experimental vaccine, Z_a is the Z score taken as the threshold for declaring the $[P_1 - P_2]$ differences as "statistically significant" (e.g., 1.96 for a two-tailed P value of 0.05), and Z_b is the Z score for β error, the maximum tolerated probability of missing a significant difference when one really exists (e.g., 0.84 for a probability of 0.20). Because of the central importance of an adequate sample size for a trial, the variables in the formula provide an inventory of data that must be inspected and decisions that must be made prior to undertaking a phase 3 trial.

One of the most challenging aspects of planning a phase 3 trial is predicting P_1, the expected incidence of the target infection in the comparison group. The challenge of this prediction comes from several sources. First, because many infectious diseases display rather marked year-to-year fluctuations in incidence, it is relatively easy to be misled by inspecting one or even a few years of data on the incidence of the target infection in the population chosen for the study. Table 2, for example, shows the yearly incidence of cholera in Matlab, Bangladesh, the site of several trials of cholera vaccines. Over a 15-year period,

TABLE 2 Review of Cholera Cases and Hospitalization Rates Matlab, Bangladesh, 1966–1980[a]

Classical period			El Tor period		
Year	Cases	Annual rate ($\times 10^{-3}$)	Year	Cases	Annual rate ($\times 10^{-3}$)
1966	85	0.8	1973	63	0.2
1967	122	1.1	1974	1098	4.2
1968	344	1.3	1975	708	2.7
1969	376	1.4	1976	273	1.0
1970	465	1.8	1977	974	3.7
1971	387	1.5	1978	875	5.1
1972	355	1.3	1979	459	2.7
			1980	557	3.3

[a]Data from ref 26.

the annual incidence varied by over 25-fold [26]. This observation underscores the need to be conservative in predicting disease incidence.

A second factor requiring consideration is the effect of demographic variables, such as age, sex, and socioeconomic status, on the incidence of many infectious diseases. Substantial variations in incidence by these characteristics sometimes make average, population-wide figures for incidence relatively useless in predicting the incidence of a target infection for a trial, if the trial's study population is to be limited to certain subgroups within the overall population.

A third factor creating complexity results from the fact that investigators must usually rely on earlier incidence data derived from routine public health surveillance, rather than from the special, prospective surveillance system that is usually instituted for a trial. Routine surveillance often employs relatively loose clinical and microbiological criteria for defining infections, which tends to elevate the observed incidence. When a trial is begun, much stricter criteria typically are employed, which tend to depress the observed incidence. On the other hand, routine surveillance is usually based on routine diagnoses made in treatment settings, and many patients with the target infection may be missed either because they do not seek care or because they are not aggressively worked up with diagnostic procedures once they are seen. This feature would act to depress the incidence reported in routine surveillance. During a clinical trial, efforts usually are made to evaluate all patients who might have the target infection and to deploy diagnostic tests systematically on all such patients. This feature of surveillance during phase 3 trials would tend to elevate the observed incidence. Because of the unpredictability of the net balance between the factors that would inflate and factors that would diminish the incidence in a phase 3 trial, in relation to antecedent routine surveillance, investigators must be cautious in projecting antecedent incidence rates derived from routine surveillance to rates expected for a comparison group in a phase 3 trial.

A fourth consideration that must be taken into account in predicting the incidence of the target infection is that RCTs enroll volunteers from a target population, and volunteers generally are not representative of the population from which they are drawn. Generally, volunteers are more health-conscious and, as a result, have a lower than expected risk of infection. Table 3, for example, presents data on the occurrence of cholera and overall mortality in a field trial of killed oral cholera vaccines in rural Bangladesh. Strikingly, these data show a nearly 20% higher incidence of cholera in placebo recipients than in age- and gender-eligible persons who did not participate in the trial [27].

Yet another factor to be weighted is the anticipated duration of follow-up for the trial. A trial whose major purpose is to estimate 2-year vaccine efficacy in a population will have a much higher expected cumulative incidence of the target infection than a trial, in the same population, that aims at measuring 6-month efficacy. Of course, the duration of vaccine protection under study should not be altered arbitrarily merely to raise the expected incidence of disease; rather, the duration should be selected on the basis of what constitutes clinically meaningful vaccine protection.

Finally, anticipated subanalyses may affect the expected incidence. For example, many trials are designed to estimate vaccine protection not merely in the entire study population but in subgroups of the population. Table 4 shows, for example, meaningful differences in the protection conferred in Bangladesh by killed oral cholera vaccines in young children versus older persons. If such subgroups are of interest, the expected incidence of the target infection in each subgroup must be considered in designing a trial. Table 4 also illustrates another common objective in vaccine trials: measurement of vaccine protection against different phenotypic variants of the target pathogen. Table 4 shows markedly better protection by the killed oral cholera vaccine against *Vibrio cholerae* 01 of the classical biotype than of the El Tor biotype [28]. If phenotype-specific vaccine protection is to be measured, the incidence of each of the different phenotypes of interest must be considered separately in designing the trial and estimating sample size requirements.

TABLE 3 Occurrence of Cholera in Placebo Recipients and in Age- and Gender-Eligible Nonparticipants in the Field Trial of Killed, Oral Cholera Vaccines, Matlab, Bangladesh[a]

	Cholera		Relative risk[c]	
Group	Episodes	Incidence[b]	Crude	Adjusted
Placebo	249	3.54	1.00	1.00
Nonparticipant	453	4.21	1.19[d]	1.20[d]

[a]Data from ref 27.

[b]Per 10^4 person months of follow-up.

[c]Crude and adjusted (for confounding sociodemographic variables) relative risk of cholera in nonparticipants versus placebo recipients.

[d]$P<0.05$ (2-tailed) for cited elevation of relative risk.

TABLE 4 **Protective Efficacy of Killed Oral Vaccines against Cholera over 3 Years of Follow-Up in Bangladesh[a]**

Feature	Protective efficacy (%)[b]	
	BS-WC	**WC**
Age at vaccination (years)		
2–5	26[c]	22[c]
>5	63[e]	68[e]
Biotype of cholera		
Classical	58[e]	60[e]
El Tor	39[d]	40[d]

[a]Data from ref 28.
[b]BS-WC, B subunit, killed whole-cell vaccine; WC, killed whole-cell vaccine.
[c]$P<0.05$.
[d]$P<0.01$.
[e]$P<0.001$.

E. Ensuring That the Source Population Is Large Enough

The sample size formula provides a way to calculate the number of subjects finally analyzed for vaccine efficacy in a two-group RCT. Investigators sometimes fail to realize that the subjects at final analysis are drawn from a much larger population of subjects and sometimes represent a very small fraction of that population. Table 5 shows the process of progressive attrition from the original source population to the population recruited, dosed, and followed for 1 year in the Bangladesh trial of killed oral cholera vaccines [6]. In relation to the overall population of Matlab, Bangladesh, where the trial was conducted and where the entire population was assessed for eligibility and approached for informed consent, if deemed eligible, only about one-third of persons were fully dosed and followed for the first year of surveillance. If this latter group were to constitute the final group under analysis, the source

TABLE 5 **Assembly of Subjects in the Field Trial of Killed, Oral Cholera Vaccines in Matlab, Bangladesh[a]**

Group	Percent of total Matlab population (%)
Overall Matlab population	100
Randomized (after exclusion of ineligibles)	84
Participated (after exclusion of refusers and absentees)	48
Fully dosed (after exclusion of incomplete of incorrect doses)	33
Completely followed (after losses to follow-up in first year)	32

[a]Data from ref 6.

population for recruiting these participants therefore would have to be three times as large.

F.   Allocating and Administering the Compared Agents

Randomized allocation of subjects to the compared agents is a key safeguard against bias in RCTs of vaccines. However, even if randomized allocation is used, several issues still require attention. It must be decided whether randomized allocation is to create equal-sized or intentionally unequal-sized groups. Equal-sized groups are usually created in two-group trials to maximize the statistical efficiency of the design. When more than two groups are compared, unequal-sized groups may yield better statistical efficiency [29]. Moreover, on ethical grounds, there may be compelling reasons to minimize the number of participants who do not receive an active vaccine. If unequal sizes per group are desired, randomization procedures can easily be adapted to yield the desired allocation ratio.

Trials with small sample sizes may be susceptible to chance imbalances in important baseline characteristics in the compared groups. To prevent such imbalances, stratified allocation is sometimes used, wherein subjects are randomly allocated to different agents, usually in a balanced fashion ("blocking"), within subgroups defined by relevant risk factors for the target infection [30, 31]. Blocked allocation also helps to ensure that the compared groups will be of the desired relative size. For example, in a two-celled trial of vaccine versus placebo intended to compare groups of equal sizes, randomization might take place within blocks of every four consecutively assigned subjects to ensure that two subjects receive vaccine and two receive placebo [31].

Classically, in an RCT, randomization of a subject to one of the compared groups is done after the subject has been deemed eligible and has given appropriate informed consent. However, this prescribed sequence may not be readily accomplished in some settings, such as those in developing countries, in which communications may be inadequate for vaccination teams to be coordinated by a central randomization unit and in which it may not be optimal to entrust allocation procedures to the teams [15]. An alternative strategy in such circumstances is to prerandomize potential subjects before recruitment and ascertainment of eligibility and to safeguard against biased enrollment of subjects by double-blinding the trial. With this strategy, each member of an already characterized population can be randomly preassigned to one of the codes used to label the agents, or doses of agents can be ordered randomly and given consecutively to sequentially enrolled participants.

The unit of allocation must also be addressed in designing an RCT. It is statistically most efficient to randomize individuals rather than clusters of individuals [24]. However, randomization of individuals rather than groups may be difficult in trials of live vaccines that are transmitted from vaccinees to their contacts and that therefore might be transmitted from vaccinees to subjects in the comparison group. For example, in a trial of orally administered, live Ty21a typhoid vaccine conducted in Santiago, Chile, classrooms of students, rather than individual students, served as the units of allocation, partly to protect against transmission of the vaccine between vaccinees and controls [32].

Also, as noted earlier, cluster randomization may be required when the indirect effects of a vaccine are to be measured in a trial. When cluster randomization is used, sample size requirements for the trial are usually elevated, sometimes markedly, in comparison to those for an individually randomized trial, and special methods of statistical analysis are usually needed [33].

Several pragmatic issues require consideration in the packaging and labeling of agents in a vaccine trial. Although either single- or multiple-dose formulations can be tested in a field trial, the use of single-dose packaging minimizes wastage and safeguards against errors in measurement of the doses to be administered. Regardless of the choice of single- versus multiple-dose formulations, it is essential that the highest standards of sterility be maintained in the administration of injectable agents.

Several options are available for the coding of agents in a vaccine trial. Use of a unique code, such as a number, for each dose provides the greatest protection against the unblinding of a trial. However, the use of unique codes may create logistical difficulties in trials testing multidose regimens. Use of a limited number of codes is sometimes preferred; if a limited number is used, it is best to have several codes per treatment group. Moreover, because of the greater susceptibility of trials using a limited number of codes to inadvertent or intentional unblinding, it is useful to query subjects and investigators for their best guesses about the identities of the codes both during and after the trial. Moreover, if a limited number of codes is used, it is essential that the numbers of target outcomes accruing during a trial not be disaggregated by code because, for a highly protective vaccine, it may be readily apparent which codes correspond to the vaccine group and which correspond to the comparison group just by comparing the number of outcomes by code.

Regardless of the choice of a coding scheme, it is important to record what code was actually administered to each subject, because errors in administration are inevitable. For example, in the field trial of oral cholera vaccines in Bangladesh [34], 573 of 234,032 (0.2%) doses were not given as assigned, despite the use of a simple A–B–C coding system for the three compared agents. For agents packaged in single doses, it can be helpful to use removable, self-adhesive stickers giving the code letter on each vial, which can then be placed directly in the subject's dosing record. For agents that are not given by injection, it is also helpful to record the completeness of dosing. In the same Bangladesh field trial, for example, 3% of doses were judged to have been incompletely ingested due to vomiting, regurgitation, or refusal to ingest the entire volume of the agent.

Other aspects of the administration of the agents also require documentation during a trial. If the agents under investigation require a cold chain to preserve potency, it is essential to document storage temperatures throughout the itinerary of the agent from the producer to the ultimate recipient. A variety of thermal monitors are available for this purpose. Moreover, because of the need to be sure that vaccines administered in a field trial are fully potent at the time of administration, it is important to save a sample of doses that had been distributed for administration in the field for later potency testing and to assess immune responses to the compared agents in at least a sample of participants. If a vaccine fails to be as protective as expected in a trial, such information will be exceedingly helpful in investigating the reasons for the disappointing levels of protection.

G. Detecting and Defining Outcomes for Assessing Vaccine Protection

Investigators must choose an approach for detecting the target infections that occur in the study population. Two basic approaches are available. In *passive surveillance*, the investigator restricts surveillance to existing treatment settings and to patients who seek care in these settings. Diagnostic detection of the target infections then occurs as a part of the routine care offered to these patients. In *active surveillance*, the investigator visits or contacts the study population at regular intervals, irrespective of symptoms, to detect illnesses that are compatible with the target infections. Special arrangements are then required to conduct diagnostic tests on suspects to detect target infections. Of course, these two approaches represent extremes; the selected approach may represent a hybrid of the two extremes.

The choice between passive and active surveillance is often regarded as a simple trade-off. Passive surveillance, making use of existing treatment facilities and the patients who visit them, is logistically easier and less expensive, but will inevitably miss the portion of patients with target infections who do not seek care, yielding a lower detected incidence. This lower detected incidence, in turn, will translate into higher sample size requirements for the population enrolled into the trial. Active surveillance, which entails community-based visits to the entire study population, is logistically more complex and expensive. However, because active surveillance is geared to detect all, or nearly all, target infections occurring in the study population, the detected incidence is expected to be higher, which will lower sample size requirements.

What is less well-appreciated, however, is that the decision about the type of surveillance to be used will usually affect the clinical spectrum of detected target infections, with passive surveillance having a higher proportion of more severe illnesses among the target infections than active surveillance. If the vaccine under study has a greater impact on disease severity than on the occurrence of disease per se, the observed level of protection will differ using these two types of surveillance, and would be expected to be higher with use of passive surveillance. This phenomenon was nicely illustrated by the field trial of killed oral cholera vaccines in Bangladesh [35], where simultaneous active and passive surveillance conducted during the first year of the trial showed greater protective efficacy for both tested vaccines against cholera cases detected in passive surveillance (Table 6). This observation indicates that the choice of a mode of surveillance can influence the level of protection that is observed and that investigators must choose a type of surveillance that detects target infections of greatest clinical and public health relevance.

Regardless of whether passive or active surveillance, or a combination, is selected for a trial, a critical feature of any surveillance system is that it protects against detection bias, which can lead to erroneous estimates of vaccine protection due to unequal intensity of surveillance for target outcomes in the groups under comparison. This bias can occur in at least three ways. First, if there are unequal losses to follow-up of study participants in the compared

TABLE 6 Detection of Cholera with Passive versus Active Surveillance in the Field Trial of Killed, Oral Cholera Vaccines in Matlab, Bangladesh[a]

Method of detecting cholera	Group[b]		
	BS-WC (N = 1097)	WC (N = 1126)	K12 (N = 1062)
Passive surveillance	11 (57%)[c,d]	11 (58%)[d]	24
Active surveillance	28 (44%)[d]	31 (39%)[d]	48

[a]Data from ref 35.

[b]BS-WC, B subunit, killed whole-cell vaccine; WC, killed whole-cell vaccine; K12, *Escherichia coli* K12 strain placebo.

[c]Number of cases and (protective efficacy) for the cited group.

[d]$P < 0.05$. (2-tailed) for contrast between the cited vaccine group and the K12 placebo group.

groups, there may be an unequal opportunity to detect outcome events [15]. Standard statistical techniques, such as life-table analyses, can adjust estimates of protection for unequal periods of follow-up [9, 10]. Moreover, double-blinding helps ensure that losses will not occur on the basis of knowledge of the identity of the received agent. However, the most effective protection against this bias is to ensure a high level of follow-up, by selecting a study population that is geographically stable and that is compliant and by instituting vigilant efforts to maintain follow-up on as high a proportion of participants as possible. It is also essential that the losses that do occur be carefully documented, so that statistical adjustments can be made and so that the reasons for the losses can be tabulated and compared in the groups under study.

Second, even if very few subjects are lost to follow-up, a bias can still occur if subjects in one group are more or less likely to receive appropriate diagnostic tests for the target infection once they develop clinical indications diagnostic testing [15]. As noted earlier, employment of double-blinding is an effective safeguard against this type of bias. When double-blinding is not possible, alternative approaches to prevent and measure bias will be needed. For example, in trials of bacille Calmette–Guérin vaccine (BCG) against tuberculosis, double-blinding was impossible due to the scars left by BCG vaccination. To minimize unequal diagnostic testing of BCG vaccinees versus controls for pulmonary tuberculosis, one trial undertook periodic mass radiographic screening for pulmonary disease, irrespective of the presence of subjects' symptoms, and ensured blinded reading of these radiographs as a basis for initiating further diagnostic tests [36].

A third type of detection bias can occur if, in the process of making final diagnoses of the target infection in study subjects, diagnostic evidence is interpreted with knowledge of the agents received by these subjects. Again, double-blinding provides solid protection against this type of bias. In lieu of double-blinding, alternative strategies to prevent bias can be implemented. For example, to prevent biased diagnoses in trials of BCG vaccination against leprosy, investigators covered the injection sites of both vaccinees and nonvaccinees during physical examinations [37].

Sometimes it is useful to provide assurance that detection bias did not occur by conducting additional surveillance for an *indicator condition*. To serve as an indicator condition, the clinical features of the disease must be similar to those of the target infection and must be diagnosed using the same procedures, but the disease must not be preventable by the vaccine under evaluation. If the incidence of the indicator condition is found to be similar in the groups under comparison in a trial, this provides supportive evidence that detection bias was not operative. Paratyphoid A infections, for example, have served as useful indicator conditions for trials of certain vaccines against typhoid fever [38].

Finally, measures to protect against detection bias do not necessarily ensure that diagnoses of the target infection will be accurate; measures to prevent detection bias only ensure that the diagnostic process is equivalent in the compared groups. Diagnostic inaccuracies, even if they occur in a random fashion, may substantially distort estimates of vaccine protective efficacy [39]. It is, therefore, also essential that investigators arrange to collect diagnostic information in a systematic and accurate fashion and that they use appropriate and explicit diagnostic criteria for making final diagnoses.

H. Assessing Vaccine Safety

Detection of adverse events following dosing and comparison of rates of specific adverse events in the groups under study are essential components of a phase 3 trial. Indeed, phase 3 trials, because of their large size, offer a unique opportunity for evaluating vaccine safety before a vaccine is licensed. In years past, the approach taken by phase 3 trials was often to focus on documentation of frequent and expected adverse events occurring only during the first few days after dosing. The focus was primarily on documenting common side effects, generally of mild severity. Often these trials measured adverse events in only a small subsample of the trial population, as large sample sizes were not required to document frequent side effects, such as the occurrence of pain or erythema at an injection site. Some trials failed to conduct surveillance for adverse events altogether.

Several changes have occurred during the past several years in the approach to documenting vaccine safety in phase 3 trials. First, it is now well-recognized that not all adverse effects of vaccination are easily predicted. For example, intussusception following oral receipt of live rhesus rotavirus–reassortant vaccine, an event that is now well-documented, was not an expected side effect of this vaccine, at least at the time that phase 3 trials were undertaken [40]. It is, therefore, inappropriate in phase 3 trials to constrain the focus of surveillance to adverse events that can be readily predicted. Second, it is no longer acceptable to target surveillance only to adverse events that occur quite commonly, nor to place only a subsample of participants under surveillance for adverse events. Newer generation live oral rotavirus vaccines, for example, will be tested in phase 3 trials that, in aggregate, will be large enough to detect an attributable incidence of intussusception of approximately 1 case in 10,000 vaccine recipients. Finally, it is now appreciated that vaccine side effects can occur long after vaccine dosing. For example, quite unexpectedly, mortality rates during the third year of life were found to be elevated in female infants who received high-titered measles vaccines at 6 months of age [41]. As a result, national regulatory authorities are increasingly demanding that phase 3 trials be designed to capture adverse events of any grade of severity: those that are predictable as well as those that are not, those that occur at any point during the follow-up of subjects in the trial, not just the first several days after dosing, and those that occur infrequently. Expansion of the scope of surveillance in this fashion will have a major effect on increasing the size, cost, and complexity of future phase 3 trials.

I. Determining When to Terminate Surveillance

At the time that a phase 3 trial is designed, it is necessary to specify how long surveillance will be maintained. Of course, unforeseen dramatic events may force the termination of the trial earlier than anticipated. Among these are severe vaccine side effects that necessitate termination of enrollment, as well as an overwhelming benefit among vaccinees, such as a major reduction in mortality, that would require the termination of surveillance so that the beneficial vaccine can be offered to subjects who had been allocated to alternative agents.

In planning the duration of a trial, as well as in monitoring unforeseen adverse events and vaccine benefits, however, an investigator may be subjected to several conflicting pressures. First and foremost, there are the ethical mandates of protecting the human subjects enrolled in the trial and not depriving any of the subjects of an agent that has overwhelming benefit. Second, there is the intellectual desirability of continuing the trial until clinically and statistically meaningful results accrue. Third, from a public health perspective, there is a need for data on long-term vaccine efficacy, because the duration of vaccine protection is a major determinant of a vaccine's potential role as a public health tool. Fourth, because a large proportion of phase 3 trials are sponsored by commercial producers, there is the inevitable pressure to conclude a trial as soon as cogent benefit is demonstrable in order to accelerate licensure and marketing of the vaccine.

Because of the complexity of these considerations and pressures, an essential feature of modern phase 3 vaccine trials is that decisions about termination should be taken out of the hands of investigators and delegated to outside bodies lacking any conflicts of interest and having the expertise to make these decisions intelligently. Modern phase 3 trials are usually monitored by at least two bodies external to the investigative team: a Data and Safety Monitoring Board (DSMB) and an Institutional Review Board (IRB). The DSMB is typically composed of professionals with expertise in disciplines relevant to the trial, such as biostatistics, epidemiology, and clinical infectious diseases. These individuals cannot have any role as investigators in the trial. Typically, a DSMB has the authority to review and approve the final protocol for the trial. It also serves to monitor the trial periodically both by assessing the performance of the trial in meeting its process goals, such as subject enrollment, and by evaluating the adverse events and study end points that are observed among trial participants. In the last activity, they are responsible for recommending early termination of a trial to the sponsor of the trial in the event of unacceptable vaccine adverse reactions or overwhelming vaccine benefit. IRBs provide initial ethical clearance for a trial at all participating institutions and must be informed as well about significant adverse events occurring in the trial population. IRBs have the authority to rescind ethical clearance for a trial at any time during the conduct of the study.

J. Deciding on an Approach for Assessing Immune Responses to Vaccination

Immune responses to vaccination are measured in phase 3 vaccine trials with at least two distinct purposes. First, as noted earlier, it is essential that immune responses be measured to document that the vaccine given in the trial elicited the level of response expected on the basis of earlier studies. This is a prerequisite for any phase 3 trial because, if the trial yields disappointing results about vaccine efficacy, it is crucial to document whether the vaccine was as immunogenic as expected. Achievement of this goal is usually straightforward, because only a modest-sized sample of subjects needs to be evaluated immunologically to obtain statistically precise estimates of the magnitude of vaccine-induced responses. The ability to focus these evaluations on a small sample of participants is important to investigators, because immunological tests generally are performed on blood specimens and because bleeding of subjects often creates difficulties in recruiting participants for a vaccine trial.

A second potential use of measurements of immune responses to vaccination is the determination of an immunological correlate of vaccine protection. To define such a correlate, it is necessary to define a short-term immunological response that best discriminates between vaccinees who are protected versus vaccinees who the vaccine failed to protect. For example, analyses of serum anticapsular polysaccharide antibody responses to a capsular polysaccharide vaccine against *Haemophilus influenzae* type B (Hib) led to the conclusion that attainment of a serum titer of ≥ 1 µg/ml of anticapsular polysaccharide antibodies correlated with clinical protection against invasive Hib disease [42]. Definition of an immunological correlate of protection is a very important objective, because such a correlate permits future assessment of the protectivity of the tested vaccine and ones suitably similar to it, without a resort to full-scale efficacy trials. Indeed, phase 3 trials provide the only opportunity in the prelicensure sequence of studies for establishing such correlates.

Unfortunately, the establishment of an immunological correlate of vaccine protection can be very demanding, for both statistical and biological reasons. Statistically, in order to discern an immunological correlate of protection, it is necessary to contrast short-term immune responses to the vaccine in vaccinees who "succeeded" (e.g., those who did not develop the target infection) versus those who "failed" (e.g., those who did develop the target infection) during surveillance in the trial. Because very few vaccinees typically become infected, even in a large phase 3 trial, and because it is crucial to have a sufficient number of vaccine failures to define an immunological correlate of protection, it is desirable to obtain immunological measurements on as many vaccinees as possible. Unfortunately, at the time of vaccination, it is impossible to predict which subjects will become infected and, if the trial is double-blinded, it is also impossible to predict who received the vaccine of interest, as opposed to one of the other agents under evaluation. As a result, in order to define an immunological correlate of protection, it may be necessary to obtain suitable specimens from virtually every participant in the trial at the time of and shortly after dosing. As illustrated by the Bangladesh trial of killed oral cholera vaccines described earlier, phase 3 trials may enroll tens of thousands of participants, making the mere collection of adequate specimens a daunting task.

Even if an investigator is successful in obtaining the requisite specimens for immunological analysis, there is no guarantee that the immune parameters measured will correlate with clinical protection. In the late 1980s a field trial of two acellular pertussis vaccines was conducted in Sweden. One of the vaccines (JNIH-6) contained lymphocytosis promotion factor (LPF) and filamentous hemagglutinin, and the other (JNIH-7) contained LPF alone [43]. JNIH-6 conferred 69% protection against clinical pertussis, whereas the protective efficacy of JNIH-7 was 54%. Virtually all subjects in the trial were bled and assessed for short-term serum antibody responses to the defined antigens in the two vaccines. As shown in Table 7, however, geometric mean serum antibody titers to the vaccine antigens were virtually identical in vaccinees who developed pertussis versus those who did not, indicating that the measured antibodies were not correlates of protection.

K. Good Clinical Practice

Regulatory agencies have placed much emphasis on the concept of conducting vaccine trials leading to vaccine licensure with designs and procedures that conform to "good clinical practice." The essential elements of good clinical practice are outlined in Table 8. In essence, these elements are designed as basic criteria to ensure that trials are conducted in an ethically justifiable fashion, are scientifically sound, and are verifiable. The last feature refers to the need to create sufficient documentation during a trial that an independent auditor could verify that the findings of a trial accurately reflect the data actually collected.

TABLE 7 Antibody Levels in Sera Drawn 60–120 Days after Dosing in Vaccinated Infants with (Cases) or without (Noncases) Pertussis in a Swedish Trial of Acellular Pertussis Vaccines[a]

	JNIH-6 (LPF+FHA)		JNIH-7 (LPF)	
Antibody	Cases (N=17)	Noncases (N=122)	Cases (N=25)	Noncases (N=143)
Antitoxin[b]	118	81	211	164
IgG anti-LPF[b]	81	77	276	185
IgG anti-FHA[c]	20	25	2	2

[a]Data from ref 43. See text for explanation.

[b]Geometric mean reciprocal titer of antibodies to pertussis toxin or lymphocytosis promotion factor (LPF).

[c]Geometric mean ELISA units for antibody titers to filamentous hemagglutinin (FHA).

TABLE 8   Principles of Good Clinical Practice[a]

1. Clinical trials should be conducted in accordance with the ethical principles that have their origin in the Declaration of Helsinki and that are consistent with GCP and the applicable regulatory requirement(s).

2. Before a trial is initiated, foreseeable risks and inconveniences should be weighed against the anticipated benefit for the individual trial subject and society. A trial should be initiated and continued only if the anticipated benefits justify the risks.

3. The rights, safety, and well-being of the trial subjects are the most important considerations and should prevail over interests of science and society.

4. The available nonclinical and clinical information on an investigational product should be adequate to support the proposed clinical trial.

5. Clinical trials should be scientifically sound and described in a clear, detailed protocol.

6. A trial should be conducted in compliance with the protocol that has received prior Institutional Review Board (IRB)–Independent Ethics Committee (IEC) approval–favorable opinion.

7. The medical care given to and medical decisions made on behalf of subjects should always be the responsibility of a qualified physician or, when appropriate, a qualified dentist.

8. Each individual involved in conducting a trial should be qualified by education, training, and experience to perform his or her respective task(s).

9. Freely given informed consent should be obtained from every subject prior to clinical trial participation.

10. All clinical trial information should be recorded, handled, and stored in a way that allows its accurate reporting, interpretation, and verification.

11. The confidentiality of records that could identify subjects should be protected, respecting privacy and confidentiality rules in accordance with the applicable regulatory requirement(s).

12. Investigational products should be manufactured, handled, and stored in accordance with applicable good manufacturing practice (GMP). They should be used in accordance with the approved protocol.

13. Systems with procedures that assures the quality of every aspect of the trial should be implemented.

[a]From ref 50.

V. CONCLUDING REMARKS

The overall features of RCTs of vaccines are straightforward and, in principle, are identical to well-established guidelines for designing trials of clinical therapies [44]. Despite this apparent simplicity, however, the design and conduct of phase 3 RCTs of vaccines require an appreciation of many subtleties. The research questions posed by the trial must be clear, and the design of the trial must be crafted to ensure that the study population, compared agents, and outcomes address these questions. The design and implementation of a trial must provide safeguards against bias, while ensuring that vaccine protection, safety, and immunogenicity are measured with acceptable statistical precision and generalizability. Extensive background data are required to plan a trial with these features, and great care and considerable conservatism are needed in the interpretation of these data. In this part of Chapter 3, we have attempted to outline several of the complexities and challenges of evaluating vaccines in a fashion that provides cogent answers to clinical and public health questions and that also allows the generation of evidence sufficient for national regulatory bodies to grant licensure.

We have been intentionally selective in our choice of topics and have focused more on design issues than implementation issues, and we have barely touched on the analysis and reporting of these trials. A comprehensive treatise on phase 3 trials of vaccines would require in-depth discussions of ethical issues, regulatory and financial considerations, pragmatic issues in the implementation of field, data management, and laboratory activities of a trial, statistical issues in the monitoring and final analysis of trial results, and considerations related to the reporting of trials in the scientific literature. Moreover, additional challenges are raised when phase 3 trials are performed in a multicenter fashion. Other references provide useful discussions of many of these topics [31, 45].

There is a clear need to perform phase 3 trials of vaccines in less developed settings, as well as in industrialized countries. This derives not only from the occurrence of distinctive vaccine-preventable pathogens in less developed settings but also from the observation that vaccines are not always as protective when given to populations in developing countries [46]. Although the basic scientific prerequisites for phase 3 trials in industrialized countries are the same as those for trials completed in developing countries, trials in developing countries raise distinctive ethical challenges, which have been the source of several debates [47], and also may entail considerable complexities in project organization, regulation, and implementation [45].

Finally, experience underscores the need for vigilant and continued evaluation of the safety and protection conferred by vaccines even after they have been introduced into public health practice. Phase 3 trials, even of large size and impeccable quality, cannot provide complete assurance that vaccines will perform well when introduced into practice on a large scale and under routine conditions. Systems are in place to ensure timely and appropriate postlicensure evaluations in many industrialized countries [48]. Much work remains to be done in creating analogous systems in developing countries [49].

REFERENCES

1. National Institute of Allergy and Infectious Diseases (2000). Accelerated Development of Vaccines: The Jordan Report.
2. Byar, D. P., Simon, R. M., Friedewald, W. T., *et al.* (1976). Randomized clinical trials: Perspectives on some recent ideas. *New England J. Medicine* **295**:74–80.
3. Bailar, J. (1983). Introduction. *In* "Clinical Trials: Issues and Approaches," S. H. Shapiro and T. A. Louis, Eds. New York: Marcel Dekker.
4. Chalmers, T. C. (1983). The control of bias in clinical trials. *In* "Clinical Trials: Issues and Approaches," S. H. Shapiro and T. A. Louis, Eds., pp. 115. New York: Marcel Dekker.
5. Greenwood, M., and Yule, G. (1915). The statistics of anti-typhoid and anti-cholera inoculations. *Proc. R. Soc. Med.* **8**:113.
6. Clemens, J., Harris, J., Sack, D., *et al.* (1988). Field trial of oral cholera vaccines in Bangladesh: Results of one year of follow-up. *J. Infect. Dis.* **158**:60–69.
7. Clemens, J., and Horwitz, R. (1984). Longitudinal evaluation of therapy. *Biomedicine* **38**:440–443.
8. Levine, R. J., and Lebacqz, K. (1979). Ethical considerations in clinical trials. *Clin. Pharmacol. Ther.* **25**:728–741.

9. Peto, R., Pike, M. C., Armitage, P., *et al.* (1976). Design and analysis of randomized clinical trials requiring prolonged observation of each patient: I. Introduction and design. *Br. J. Cancer* **34**:585–612.

10. Peto, R., Pike, M. C., Armitage, P., *et al.* (1977). Design and analysis of randomized clinical trials requiring prolonged observation of each patient: II. Analysis and examples. *Br. J. Cancer* **35**:1–39.

11. Feinstein, A. R. (1973). Clinical biostatistics: XXII. The role of randomization in sampling, testing, allocation, and credulous idolatry (part 1). *Clin. Pharmacol. Ther.* **14**:601–615.

12. Feinstein, A. R. (1973). Clinical biostatistics: XXIII. The role of randomization in sampling, testing, allocation, and credulous idolatry (part 2). *Clin. Pharmacol. Ther.* **14**:898–915.

13. Feinstein, A. R. (1973). Clinical biostatistics: XXIV. The role of randomization in sampling, testing, allocation, and credulous idolatry (conclusion). *Clin. Pharmacol. Ther.* **14**:1035–1051.

14. Meier, P. (1983). Statistical analysis of clinical trials. *In* "Clinical Trials: Issues and Approaches," S. H. Shapiro and T. A. Louis Eds., pp. 115. New York: Marcel Dekker.

15. Clemens, J. D., Chuong, J. J., and Feinstein, A. R. (1983). The BCG controversy: A methodological and statistical reappraisal. *J. Am. Med. Assoc.* **249**:2362–2369.

16. Halloran, M., Haber, M., Longini, I., *et al.* (1991). Direct and indirect effects in vaccine efficacy and effectiveness. *Am. J. Epidemiol.* **33**:323–331.

17. Fine, P. (1993). Herd immunity: History, theory, practice. *Epidemiol. Rev.* **15**:265–302.

18. Clemens, J., Brenner, R., Rao, M., *et al.* (1996). Evaluating new vaccines for developing countries: Efficacy or effectiveness? *J. Am. Med. Assoc.* **275**:390–397.

19. Struchiner, C., Halloran, M., Robins J., and Spielman, A. (1990). The behaviour of common measure of association used to assess a vaccination programme under complex disease transmission patterns—A computer simulation study of malaria vaccines. *Int. J. Epidemiol.* **19**:187–196.

20. Moulton, L., O'Brien, K., Koeberger, R., *et al.* (2001). Design of a group-randomized *Streptococcus pneumoniae* vaccine trial. *Controlled Clin. Trials* **22f**:438–452.

21. Szmuness, W., Stevens, C., Harley, E., *et al.* (1980). Hepatitis B vaccine: Demonstration of efficacy in a controlled clinical trial in a high risk population in the United States. *New England J. Medicine* **303**:833–841.

22. Santiago, L. (1998). Large HIV vaccine trial begins. *GMHC Treat Issues* **12**:1–5.

23. Institute of Medicine. (1986). Diseases of importance in developing countries. *In* "New Vaccine Development. Establishing Priorities," Vol. II, pp. 241. Washington, DC: National Academy Press.

24. Donner, A. (1984). Approaches to sample size estimation in the design of clinical trials—A review. *Stat. Med.* **3**:199–214.

25. Armitage, P. (1971). "Statistical Methods in Medical Research." Oxford, UK: Blackwell.

26. Glass, R. I., Backer, S., Huq, M. I., *et al.* (1977). Endemic cholera in rural Bangladesh, 1966–80. *Am. J. Epidemiol.* **116**:959–970.

27. Clemens, J. D., Sack, D. A., van Loon, F. P. L., *et al.* (1992). Non-participation as a determinant of adverse health outcomes in a field trial of oral cholera vaccines. *Am. J. Epidemiol.* **135**:865–874.

28. Clemens, J., Sack, D., Harris, J., *et al.* (1990). Field trial of oral cholera vaccines in Bangladesh: Results from long-term follow-up. *Lancet* **335**:270–273.

29. Lachin, J. (1982). Statistical elements of the randomized clinical trial. *In* "The Randomized Clinical Trial and Therapeutic Decisions," N. Tygstrup, J. Lachin and E. Juhl, Eds., pp. 82. New York: Marcel Dekker.

30. Feinstein, A. R., and Landis, J. R. (1976). The role of prognostic stratification in preventing bias permitted by random allocation of treatment. *J. Chron. Dis.* **29**:277–284.

31. Pocock, S. (1983). "Clinical Trials: A Practical Approach." Chichester, UK: Wiley.

32. Levine, M. M., Ferreccio, C., Black, R.E., *et al.* (1987). Large-scale field trial of Ty21a live oral typhoid vaccine in enteric-coated capsule formulation. *Lancet* **1**:1049–1052.

33. Donner, A., and Donald, A. (1987) Analysis of data arising from a stratified design with the cluster as the unit of randomization. *Stat. Med.* **6**: 43–52.

34. Clemens, J. D., Sack, D. A., Harris, J. R., *et al.* (1986). Field trial of oral cholera vaccines in Bangladesh. *Lancet* **2**:124–127.

35. Clemens, J., Sack, D., Rao, M., *et al.* (1992). Evidence that inactivated oral cholera vaccines both prevents and mitigates *Vibrio cholerae* 01 infections in a cholera-endemic area. *J. Infect. Dis.* **166**:1029–1034.

36. Great Britain Medical Research Council. (1956). BCG and vole bacillus vaccines in the prevention of tuberculosis in adolescence and early life. *Br. Med. J.* **1**:413.

37. Brown, J. A., Stone, M. M., and Sutherland, I. (1966). BCG vaccination of children against leprosy: First results of a trial in Uganda. *Br. Med. J.* **1**:7–14.

38. Simanjuntak, C. H., Paleologo, F. P., Punjabi, N. H., *et al.*(1991). Oral immunisation against typhoid fever in Indonesia with Ty21a vaccine. *Lancet* **338**:1055–1059.

39. Fleiss, J. (1981). "Statistics for Rates and Proportions," 2nd ed., pp. 193. New York: Wiley.

40. Jacobsen, R., Adegbenro, A., Pankratz, V., and Poland, G. (2001). Adverse events and vaccination—The lack of power and predictability of infrequent events in a pre-licensure study. *Vaccine* **19**:2428–2433.

41. Expanded Programme on Immunization (1992). Safety of high titre measles vaccine. *Weekly Epidemiol. Record* **67**:357–362.

42. Peltola, H., Kayhty, H., Virtanen, M., and Makela, P. (1984). Prevention of *H. influenzae* type b bacteremic infection with a capsular polysaccharide vaccine. *New Engl. J. Med.* **310**:1566–1569.

43. Kallings, L., Olin, P., and Storsaeter, J. (1988). Protective effect of 2 acellular pertussis vaccines in a double-blind, placebo-controlled trial. *Lakartidningen* **85**:1994–1996.

44. Feinstein, A. R. (1985). "Clinical Epidemiology: The Architecture of Clinical Research." Philadelphia: Saunders.

45. Smith, P., and Morrow, R. (1996). Field Trials of Health Interventions in Developing Countries: A Toolbox. London: MacMillan.

46. Ward, J., Brenneman, G., Letson, G., *et al.* (1990). Limited efficacy of a *Haemophilus influenzae* type B conjugate vaccine in Alaskan Native infants. *New England J. Medicine* **323**:1393–1401.

47. Angell, M. (1997). The ethics of clinical research in the Third World. *New England J. Medicine* **337**: 847–849.

48. Chen, R., De Stefano, F., Davis, R., *et al.* (2000). The Vaccine Safety Datalink: Immunization research in health maintenance organizations in the USA. *Bull. WHO* **78**:186–194.

49. Clemens, J. (2000). Evaluating vaccine safety before and after licensure. *Bull.WHO* **78**: 218–219.

50. International Conference on Harmonisation (1996). E6: Guideline for Good Clinical Practice.

4 ETHICS AND VACCINES

RUTH MACKLIN* AND BRIAN GREENWOOD†

*Department of Epidemiology and Social Medicine, Albert Einstein College of Medicine, Bronx, New York 10461

†London School of Hygiene and Tropical Medicine, London WC1E 7HT, United Kingdom

I. ETHICAL PRINCIPLES

Ethical concerns related to preventive vaccines comprise three main categories: (1) design and conduct of vaccine research involving human beings; (2) guarantee of access to vaccines that research has demonstrated to be safe and effective; and (3) tension between the public health benefits of vaccines and respect for the autonomy of individuals who seek to refuse mandatory vaccination of themselves or their children. This chapter addresses these three categories and subsidiary concerns arising within them.

Several fundamental ethical principles governing research involving human subjects have gained virtually universal acceptance throughout the world. As stated in the Belmont Report, a document issued by the U.S. National Commission for the Protection of Human Subjects [1], three central principles are (1) respect for persons, (2) beneficence, and (3) distributive justice. Respect for persons requires that researchers obtain voluntary, informed consent from every prospective research subject or a legally authorized representative (for example, parents of minor children) and that the privacy and confidentiality of participants be adequately protected during and after the research process. Beneficence requires that researchers and ethics review committees seek to ensure that the anticipated benefits of a research project outweigh the risks to participants. Distributive justice requires that the benefits and burdens of research be distributed fairly among all groups, regardless of age, gender, social class, geographic location, race, or ethnicity.

In one form or another, these principles are embodied in numerous national and international regulations or guidelines. The most prominent international guidelines are those of the Declaration of Helsinki, first issued by the World Medical Association in 1964, with several relatively minor amendments since then and a more substantial revision in 2000 [2], and the International Ethical Guidelines for Biomedical Research Involving Human Subjects, issued by the Council for International Organizations of Medical Sciences (CIOMS) [3]. The CIOMS guidelines address issues specific to research conducted in developing countries.

II. VACCINE RESEARCH

Ethical aspects of research involving human subjects are well-known and widely acknowledged and will not be addressed in general terms here. Instead, the focus is on issues specific to research on preventive vaccines, with examples selected from different areas of vaccine research.

A. Risk–Benefit Considerations

As in all clinical trials, an ethical imperative exists to minimize risks to subjects. According to the predominant view, this concern is heightened in studies involving healthy human subjects. For example, the U.S. Food and Drug Administration (FDA) maintains that safety standards have to be set higher for vaccines than for other products because they are administered to healthy subjects. Moreover, the need to maximize safety may come at the cost of sacrificing the efficacy of an experimental vaccine. A prominent example exists in the development of preventive HIV–AIDS vaccines. Most scientists agree that a vaccine using live, attenuated virus is likely to be considerably more efficacious than vaccines using envelope antigens or recombinant virus vectors. However, the risks of using a live, attenuated virus in the preparation of HIV vaccines are, at least at present, thought to be too high to warrant development of this approach.

Arguably, an analogous situation exists in research in which there is anticipated therapeutic benefit to subjects. For example, highly toxic agents used in chemotherapy or high doses of radiation in cancer trials may have to be limited because of the probability that the experimental treatment will kill subjects before they die of cancer. However, that possibility is not ruled out in cancer trials. An examination of informed consent documents for studies involving patients with end-stage cancer or those whose disease has been refractory to other therapies reveals that death from the experimental drug is more than a remote possibility.

This points to another difference between standard clinical trials that may provide direct therapeutic benefit and vaccine research. In the latter, the benefit is not only uncertain but also remote in time. The very prospect of benefit to healthy subjects in vaccine trials may be nonexistent, as in cases in which the individuals are never actually exposed to the disease for which prevention is sought. In that situation, research subjects undergo the risks of participation in a trial but with no actual or potential benefit to themselves. Even if the research participants are eventually exposed to the disease, it may occur at a much later time than their participation in the trial, thereby making the benefits seem remote. These considerations indicate the need to maximize safety in vaccine trials, possibly to a greater extent than is called for by the usual prescription that research must minimize risks to human subjects and strive to maximize benefits.

Trials on HIV-preventive vaccines pose an ethical challenge that may be virtually unique in the research context. That is the ironic prospect that trial participants may be placed at a greater risk of acquiring HIV because they believe the vaccine will be successful in preventing infection, resulting in their tendency to engage in risky behavior or a failure to reduce risks they might otherwise seek to avoid. This prospect places a greater burden on counselors and the individuals who are responsible for obtaining the informed consent of volunteers in such trials.

B. Informed Consent of Research Participants

Problems and pitfalls in obtaining voluntary, informed consent from research subjects are well-known in all areas of research. These include complex and poorly designed consent documents, inadequate time and effort spent in the process of informing and ensuring that potential subjects truly understand what they are consenting to, and compromises in the voluntariness with which consent is granted when physicians enroll their patients in research and when subjects have an unwarranted expectation of therapeutic benefit. The first two problems are just as likely to exist in vaccine research as elsewhere, whereas the third issue is evidently not applicable. Specific concerns about informed consent have been noted in two different contexts. The first is that of research and development of vaccines for fertility regulation.

Since the late 1970s, the World Health Organization, in collaboration with other researchers, has sought to develop a fertility-regulating vaccine. However, in the late 1980s, a groundswell of opposition to antifertility vaccines arose among a coalition of feminists in different countries [4]. One (but only one) of the chief reasons for this strong opposition was the belief that antifertility vaccines hold out a serious prospect of abuse of women. This abuse stems from the likelihood that women's consent to accept an infertility vaccine, whether during research or after the research is concluded, is likely to be uninformed, nonvoluntary, or both. One argument in support of this view contended that women cannot understand a vaccination against pregnancy: "This is not an idea of women" [5]. Critics stressed the possibility of confusion between a disease-preventing vaccine and one designed to regulate fertility. One of the strongest critics of research and development of antifertility vaccines pointed to an information brochure designed by the Human Reproduction Program at WHO in conjunction with a Swedish research team preparing for a phase 2 trial of a candidate vaccine. The critic argued that the brochure had "serious shortcomings according to international standards of informed consent," noting that it did not stress sufficiently the novelty of immunocontraceptives, "which are compared to anti-disease vaccines which prevent 'illness...and even death'" [5, p. 107]. Against these criticisms, others have pointed out that problems regarding informed consent are no different, in principle, for antifertility vaccines than for any other type of contraceptive research [6].

A different set of concerns about informed consent has arisen in HIV-preventive vaccine research. A guidance document for HIV-preventive vaccine research issued by the Joint United Nations Programme on HIV–AIDS (UNAIDS) notes that "HIV preventive vaccine trials require informed consent at a number of stages" [7, p. 33]. These include eligibility screening prior to enrollment for those found to be eligible and thereafter throughout the trial. A purely informational matter is the need to make clear to research participants that they may test positive for HIV even if they are not infected. This is a concept whose understanding requires some sophistication and is of special concern with respect to HIV because of the risks of stigmatization and discrimination against individuals who test positive. Of greater concern is the need to emphasize in informed consent counseling and the consent document the need for continued risk reduction and the fact that the vaccine is not likely to be 100% effective in preventing HIV infection.

C. Justice in Research

An ethical principle that has been much neglected, especially when clinical trials of all sorts are conducted in developing countries, is that of distributive justice. This principle requires that the benefits and burdens of research must be distributed equitably. Equity requires that no one group—gender, racial, ethnic, geographic, or socioeconomic—receive disproportionate benefits or bear disproportionate burdens of the research.

An ethical requirement closely related to the principle of distributive justice is the view that research must be responsive to the health needs and priorities of the population in which it is conducted. In the 2000 revision of the Declaration of Helsinki, paragraph 19 states: "Medical research is only justified if there is a reasonable likelihood that the populations in which the research is carried out stand to benefit from the results of the research" [2].

The Declaration of Helsinki contains another provision that applies specifically to participants in research. Paragraph 30 states: "At the conclusion of the study, every patient entered into the study should be assured of access to the best proven prophylactic, diagnostic and therapeutic methods identified by the study." As applied to vaccine research, this provision requires that all participants who were in a control group and received a placebo or other substance not designed to prevent the disease under study should be given the successful vaccine product. Although this would satisfy the requirement of equity, it has a shortcoming from the point of view of scientific knowledge. There may be strong reasons to maintain an unvaccinated control group in order to determine for how long the protective effect of the vaccine lasts, as well as to detect the occurrence of rare side effects.

Another international document addresses the issue of justice in research. Guideline 8 of the 1993 CIOMS international ethical guidelines [3] states: "As a general rule, the sponsoring agency should ensure that, at the completion of successful testing, any product developed will be made reasonably available to inhabitants of the underdeveloped community in which the research was carried out."

The UNAIDS guidance document for preventive HIV vaccine research contains similar requirements, including an especially strong statement in guidance point 2: "Any HIV preventive vaccine demonstrated to be safe and effective...should be made available as soon as possible to all participants in the trials in which it was tested, as well as to other populations at high risk of HIV infection" [7]. These provisions in highly respected international guidelines are designed to prevent the exploitation of populations in developing countries by seeking to ensure that they enjoy the benefits, as well as experience the burdens, of serving as research participants. What these international guidelines fail to address, however, is the difficult question of whose responsibility it is to see that the products of successful research are made widely available in resource-poor countries. In such countries, even if a vaccine is cost-effective, the local Ministry of Health may not be able to afford to purchase and distribute the vaccine. Although the CIOMS guideline mentions the sponsoring agency, it does not specify that the sponsor has the sole responsibility for making the product available.

A convention has been established that vaccines developed for primary use in the developing world, such as a malaria vaccine, should be tested first in volunteers in the industrialized country of the sponsor before being administered to volunteers in the population for which the vaccine is intended. Although this practice was meant to limit the potential for exploitation in populations thought to be vulnerable, it has two ethical shortcomings. First, it appears to violate the principle that research should be responsive to the health needs of the population in which it is carried out. The need for a malaria vaccine in the north is limited to recreational travelers and workers who spend time in countries in the south. Second, it delays the time at which a vaccine will be available to those most in need, because later trials will have to be conducted in the developing countries in which the vaccine will be used. More recent thinking on this issue may call for a reversal of this convention in order to hasten the progress of vaccine development and distribution in developing countries.

III. PROVIDING ACCESS TO VACCINES PROVEN TO BE SAFE AND EFFECTIVE

Questions relating to justice and equity go beyond those addressed earlier, which pertain specifically to research. According to WHO, one in four children throughout the world did not receive routine immunization with the six basic vaccines against polio, diphtheria, whooping cough, tetanus, measles, and tuberculosis in 1998. The proportion of children immunized each year against these six diseases declined between 1990 and 1998 [8]. Moreover, the "vaccine gap" between children in industrialized countries and those in resource-poor countries has widened. Until a few years ago, children in Africa and in Europe received the same childhood vaccines. Today, children in industrialized countries typically receive eleven or twelve vaccines, whereas children in developing countries receive six or seven. This gap is likely to continue to widen due to the higher costs of newly developed vaccines.

Traditional vaccines have been relatively inexpensive. Products based on old technologies typically have cost only a few cents to manufacture. The total cost of vaccines in the past was around $1–2 in the United States. Newer vaccines, such as *Haemophilus influenzae* type B, meningococcal conjugate, and pneumococcal conjugate vaccines, are now being introduced into the childhood immunization programs of many industrialized countries at a cost of tens of dollars per vaccine. For example, the cost of a pneumococcal conjugate vaccine in the United States is around $50 per dose, and three or four doses are needed. Newer vaccines use techniques of biotechnology and, as a result, will be much more expensive to manufacture and, consequently, to purchase. Poor countries where these vaccines are needed most cannot afford them and will be unable to afford them in the future without assistance from wealthier nations or international organizations. If a gap already exists between vaccines that are widely available in industrialized countries and those to which people in resource-poor countries have access, the newer technologies provide a reason why the gap is likely to widen.

New developments in molecular biology, for example, delineation of the genomes of an increasing number of microorganisms, will result in the development of new vaccines in the coming years. If such vaccines are developed through the conventional pharmaceutical pathway, they are likely to be expensive, as research and development costs will be high, and private companies are committed to maximizing profits in order to maintain their obligations to stockholders. High costs for new vaccines will place them out of the reach of Ministries of Health in resource-poor countries and increase the "vaccine gap" between children in rich and poor countries. One approach to addressing this problem is to undertake the development of new vaccines through a public–private partnership in which some or all of the vaccine development costs are met by the public or philanthropic purse of wealthy countries, thus allowing vaccine prices to be lower than they would be if a biotechnology company were the sole sponsor. Nevertheless, despite this prospect, the outcome remains speculative. The partnership approach is being undertaken for the development of a group A meningococcal conjugate vaccine that will be used primarily in the Sahel countries of Africa, some of the poorest in the world. An alternative approach adopted by some of the major vaccine manufacturers is that of a tiered pricing system, which allows vaccines to be made available to poor countries at a lower cost than elsewhere.

A different ethical problem with the introduction of vaccines after research is concluded relates to the level of efficacy determined to be adequate. If a low-efficacy vaccine results from an early clinical trial of, say, a preventive HIV vaccine, there is likely to be great social pressure to introduce that vaccine rather than wait for one with higher efficacy. The main objection to the introduction of the low-efficacy vaccine is the problem already noted in the context of research: recipients of the vaccine may believe themselves to be adequately protected and, therefore, engage in risky behavior they otherwise would avoid. Economic disincentives may also exist, as the cost of manufacture and distribution could be unacceptable given the likelihood that more effective vaccines would render the less effective product useless in a short time.

Still another consequence of introducing a vaccine of low efficacy is that it will almost certainly affect the design of future vaccine trials. Once an effective vaccine is available outside a trial—even one of low efficacy—it will no longer be ethically acceptable to design future trials using placebos in a control arm. This could compromise the science of such future trials. Moreover, if the less effective preventive vaccine becomes widely available in industrialized countries but not in developing countries, the ethically problematic situation will arise in which researchers and sponsors seek to carry out placebo-controlled vaccine trials in the poor countries when that design would be prohibited in the industrialized countries.

Determination of the threshold of safety for a vaccine can also be problematic, especially because the risk–benefit balance may differ for industrialized and developing countries. An illustration of this problem is the discovery that a rotavirus vaccine, licensed after extensive trials in the United States for the prevention of childhood diarrhea, was associated with an unexpected incidence of intussusception that could be life-threatening. Following official recommendations, the vaccine was taken off the market in the United States. In industrialized countries such as the United States and those in western Europe, even severe childhood diarrhea can be successfully treated, so the risk of administering the vaccine could be judged to outweigh the risk to children of remaining unvaccinated. In developing countries, on the other hand, where diarrhea is responsible for one-fourth of all deaths in young children, the risk–benefit ratio is sharply different. As a public health measure, widespread administration of the rotavirus vaccine could save millions of lives [9].

Yet a quandary remains, when the situation is viewed from the perspective of the individual child. A healthy child in a developing country, who might never experience life-threatening diarrhea, could be exposed to a preventive vaccine that could cause a different life-threatening condition. Further complicating this situation is the dilemma confronting decision makers in developing countries. Once a vaccine with known adverse effects is withdrawn from the market in the United States, it could appear that leaders in developing countries are prepared to place their own populations at a risk found to be unacceptable in the industrialized world.

IV. PUBLIC GOOD VERSUS INDIVIDUAL AUTONOMY

An ethical issue pertaining to preventive vaccines that does not arise in the therapeutic arena is the tension between public health benefits and respect for individual autonomy. In medical practice, the ethical principle respect for persons is typically interpreted to allow individuals to refuse unwanted medical treatments, including life-saving treatments. Although this principle is not always honored and is rejected by certain religious faiths, recognition of individual autonomy supports the right of patients to refuse medical treatments and, in a more limited set of circumstances, the right of parents to refuse medical treatment for their children. In such cases, the health-related risks and benefits of administering or forgoing treatment pertain only to the individual patient. In the case of vaccines, however, the risks to individuals of receiving a vaccine in most cases would outweigh the risks to them of forgoing the vaccine. The benefits accrue to society in the form of public health protection.

Especially in a community in which everyone else has been vaccinated, the best option for an individual is to remain unvaccinated. For example, the risk of natural polio infection in many countries in which the transmission of polio has been eliminated is virtually nil, but vaccination with oral polio vaccine carries a risk of paralysis of about 1 in 1 million. The question, then, is whether individual parents should be granted the right to avoid placing their children at risk when the benefit is for the good of the community. A woman who founded an advocacy organization called the National Vaccine Information Center said, "It is immoral to write off an unknown number of children as expendable in the name of the greater good to justify public health policy" [10, p. 31].

In some places, childhood vaccination is not made compulsory by the state. Yet vaccination against various childhood diseases may be required for a child to enter school, which is the case in most localities in the United States. Although compulsory vaccination for adults has been less common (the exception being the military), some companies have insisted on vaccinating workers against influenza in order to cut down on absenteeism. Some workers respond by saying, "I'd rather get sick than be required to be vaccinated." The justification for the company's requirement in this situation is, however, different from that of protecting the health of the public. It is, rather, to protect the company's bottom line—a private rather than a public good—because workers can still become infected with and develop influenza outside the workplace. A public health rationale for requiring people to place themselves at risk from a vaccination is on more solid ethical ground than the economic interest of a company, as a strong presumption exists in free, democratic societies to respect the autonomy of individuals.

A special case of public good versus individual rights is the attempt to develop transmission-blocking malaria vaccines. If proven successful, such vaccines will not protect a vaccinated individual from illness but will stop an infected person from transmitting the infection to others. This could be viewed as a case of "pure altruism" on the part of individuals who choose to be vaccinated. Yet there may still be a benefit to individuals who have been vaccinated if a high proportion of the community is also vaccinated. In that case, overall transmission of malaria will decrease, and the vaccinated persons will be afforded some protection in the long term. Such vaccines will be effective only if whole communities, rather than individuals, agree to use them. Evaluation of such vaccines requires that the community be the unit of randomization, which raises ethical questions of how to obtain individual consent when the entire community must participate. If it is problematic to override individual autonomy by compulsory vaccination using proven products, it is virtually unprecedented to require individuals to receive an experimental vaccine in the absence of individual, voluntary, informed consent.

A conception of justice known as "compensatory justice" relates to what is owed to individuals who have been harmed. Most industrialized countries have sought a remedy for serious injury resulting from vaccination by providing compensation from public funds to the individual or family. In the United States, for example, the National Vaccine Injury Compensation Program is a "no-fault" compensation system that removes most of the liability burden from manufacturers for immunization-related injuries. In addition to the benefits the individual or family receives in the form of monetary compensation, programs such as this can have the desirable effect of helping to stabilize the supply and price of vaccines. Such programs do not exist in developing countries, adding to the disadvantages that already exist from undersupply and unaffordable cost of vaccines.

Public opposition to compulsory vaccination is not a new phenomenon. In 1904, Brazil experienced a public "revolution" against compulsory smallpox vaccination. However, even with all of the regulatory agencies for drugs and biologicals in place in industrialized countries, the number of advocacy organizations and spokespersons critical of vaccines appears to be on the rise. Despite the legitimate concern that the administration of vaccines can cause harm—sometimes serious—to individuals, the stunning achievements of the eradication of smallpox and elimination of polio in many parts of the world, including the WHO Pan American, European, and Western Pacific Regions, give testimony to the enormous public health benefits attained by preventive vaccines. Although it relies on a utilitarian calculation, the conclusion is unmistakable that the benefits of vaccines far outweigh the risks.

ACKNOWLEDGMENTS

R.M. thanks José Esparza, MD, Ph.D., for valuable assistance in the preparation of this chapter.

REFERENCES

1. National Commission for the Protection of Human Subjects of Biomedical and Behavioral Research (1979). "The Belmont Report: Ethical Principles and Guidelines for the Protection of Human Subjects of Research." Washington, DC.
2. World Medical Association (1964, 2000). Declaration of Helsinki, as amended by the WMA 52nd General Assembly, Edinburgh, Scotland, October 2000.
3. Council for International Organizations of Medical Sciences (CIOMS) (1993). "International Medical Guidelines for Biomedical Research Involving Human Subjects." Geneva, Switzerland: CIOMS.
4. Women's Global Network for Reproductive Rights (1993). "A Call for an Immediate Halt to Research on Antifertility Vaccines," unpublished manifesto.
5. Richter, J. (1996). "Vaccination against Pregnancy: Miracle or Menace?" London, New Jersey: Zed Books.
6. Round Table (1994). Vaccine contraceptives: Wisdom, optimism and combatting the potential for abuse." *Reprod. Health Matters* 4:107–114.
7. UNAIDS (2000). Ethical Considerations in HIV Preventive Vaccine Research. Geneva, Switzerland: UNAIDS.
8. WHO Vaccine Preventable Diseases Monitoring System (1999). Global Summary.
9. Melton, L. (2000). Lifesaving vaccine caught in an ethical minefield. *Lancet* 356:318.
10. Allen, A. (2001). "A Shot in the Dark," *New York Times Magazine*, May 6.

5 UNDERSTANDING MICROBIAL PATHOGENESIS AS A BASIS FOR VACCINE DESIGN

PART A. Bacteria

B. BRETT FINLAY
Departments of Microbiology and Immunology, Biochemistry, and Molecular Biology and Biotechnology Laboratory, University of British Columbia, Vancouver, British Columbia, Canada V6T 1Z3

I. INTRODUCTION

Bacterial pathogens utilize a wide variety of molecules (called virulence factors) to cause infections. Because these molecules are critical for disease, and because virulence factors are usually on the bacterial surface or secreted out of the pathogen, they make attractive targets for potential vaccine candidates. Capsules, toxins, and adhesins such as pili have been utilized in the past to develop successful vaccines. More recently, significant advances have been made in understanding the molecular basis of bacterial diseases, significantly increasing the number of potential targets. Much of this information has come by studying the host cell and bacterial pathogen in concert, focusing not just on the bacterial molecules but also on the host cell machinery that the virulence factor affects. This area of study has been termed cellular microbiology and has led to the realization that the interactions between many pathogens and their host cells are much more complex than previously thought. For example, in order for a toxin-mediated disease such as cholera to occur, the pathogen must also have factors that allow it to survive the low pH in the stomach, to penetrate the intestinal mucous, to adhere to the intestinal cell wall, and to persist within the gut while it secretes the toxin. We have also learned that the vibrio goes to great lengths to encode its adhesins and toxin in mobile genetic elements that can be moved between bacteria. Other pathogens that reside within host cells (intracellular pathogens), such as *Salmonella* and *Mycobacterium*, have developed a spectacular array of virulence factors that enable the organism to enter into host epithelial cells, penetrate through the epithelium, enter into phagocytic cells, and successfully survive within host cells that have evolved to normally kill bacteria. All of these virulence factors are potential targets for vaccines, and thus a solid understanding of the molecular basis of pathogenesis is a prerequisite for rational vaccine development.

In order for a virulence factor to be effective, it has to be produced in the right place at the right time. Often this is on the bacterial surface, secreted, or even delivered directly into the host cell. Much research has focused on studying the regulation of virulence factors, defining the conditions under which they are produced. As a general rule, virulence factors are usually tightly regulated and only expressed under conditions that mimic that particular stage of an infection. Because of this strict regulation, many virulence factors have gone undetected until recently. For example, molecules that allow a pathogen to enter into (invade) intestinal epithelial cells are usually expressed under conditions that mimic the gut. Virulence factors needed to mediate intracellular survival are again only expressed inside host cells. If one constructs strains that express these virulence factors constitutively, the strain is usually avirulent, emphasizing the necessity of virulence factor regulation. Thus, in order for a vaccine to be effective, one must know when and where the virulence factor is expressed and utilize this knowledge when developing vaccine strategies.

Significant research efforts have been focused on secretion systems, especially in gram-negative bacteria. The gram-negative cell wall is a complex structure, and movement of virulence factors across two membranes and a periplasm requires dedicated transport systems. Several types of secretion systems have been characterized (Table 1). Amazingly, at least two of these types of secretion systems (types III and IV) seem to be designed not only to transport virulence factors out of the bacterium but to inject them directly into the host cell. This has led to the uprecedented knowledge that several bacterial pathogens drive specific molecules into host cells, where they can interact directly with host cell signaling systems, the cytoskeleton, vesicular transport, and apoptotic pathways. However, this also implies that, in order to develop a vaccine against these molecules, a cell-mediated immunological strategy is required to access bacterial molecules found within the cytoplasm.

A comprehensive review of bacterial virulence factors and their mechanisms is beyond the scope of this part of Chapter 5, and readers are referred elsewhere for more complete reviews [1–3]. Instead, what is provided is an

TABLE 1 Gram-Negative Secretion Pathways

Type	Properties	Examples
I	Small number of components	*E. coli* hemolysin
II	General secretory pathway	*Pseudomonas* pili
III	Specialized virulence factor delivery system	*Salmonella* invasins, *Yersinia* antiphagocytosis
IV	Specialized system, related to DNA conjugation systems	Pertussis toxin, *Legionella* macrophage survival factors

overview of bacterial pathogenesis, focusing on the general types of virulence factors and how they work. Despite their large number, most virulence factors fall into one of a few categories when classified according to function. These include (a) toxins, which are secreted bacterial molecules that poison host cells by a variety of mechanisms following uptake by the host cell; (b) adhesins, which are bacterial surface structures that allow pathogens to adhere to host cells; (c) invasins, which are bacterial surface molecules that trigger host cells to enable bacterial uptake into them; (d) virulence factors, which mediate intracellular survival within host cells; (e) mechanisms to avoid phagocytosis, often encoded by surface structures such as capsules but including virulence factors injected directly into host cells to paralyze phagocytosis; and (f) mechanisms to alter host cell apoptosis (either trigger or inhibit it). Secretion systems and regulatory systems are also mentioned briefly, as these are essential for successful virulence factor production and useful for defining potential vaccine candidates and strategies.

II. TOXINS

Toxins are perhaps the best characterized virulence factors, because they are often the easiest to purify (concentrate the supernatant) and have distinct phenotypes. They have also been used extensively and very successfully in toxoid vaccines, as well as adjuvants. However, as virulence factors, they often have essential roles in altering and often killing host cells, usually by an enzymatic process [4, 5]. Exotoxins are usually proteins that are secreted by bacteria out into the supernatant. Endotoxin is bacterial lipopolysaccharide (LPS), which is principally carbohydrate, not protein, is nonenzymatic (higher doses needed), and is not normally secreted. Because LPS is a critical component of all gram-negative bacteria, generally it is not considered a bacterial virulence factor, although it stimulates toxic effects in the host by activating innate immune responses and cytokine production.

Exotoxins come in several types and specificities. They usually have two general functions, which can often be uncoupled: the ability to bind to a host cell receptor and enzymatic activity. For example, cholera, labile toxin (LT), and many others have five binding (B) subunits coupled to the active (A) enzymatic subunit. Several vaccines are made from the B subunits without the catalytic subunit, ensuring no toxin activity. The B subunits bind to host cell molecules such as carbohydrates (for example, cholera binds the GM1 ganglioside, a glycolipid on intestinal cell surfaces). Following binding, the A subunit then enters the cell, which is facilitated by the B subunits. Once inside the host cell, the A subunit modifies host cell molecules. For example, pertussis toxin modifies heterotrimeric G proteins and cholera toxin affects adenylate cyclase, whereas neurotoxins such as botulinum toxin cleave molecules essential for neurotransmitter release. The spectrum of host cell targets is extensive, including many molecules involved in cytoskeletal function, vesicular transport, protein synthesis, etc. We are only beginning to realize the vast diversity of toxin targets. We are also realizing that these toxins make excellent probes of normal cellular function [4].

Another class of toxins is capable of inserting directly into host cell membranes, thereby forming a pore and causing lysis of the host cell. For example, streptolysin O, *Staphylococcus* α-toxin, and *Escherichia coli* hemolysin all cause pore formation in host cells by assembling into oligomeric pores in the cell membrane. Because of the complexity of gram-negative cell envelopes (two membranes and a periplasm), most gram-negative toxins also have a dedicated secretion pathway to enable secretion. For example, hemolysin uses a type I secretion system, whereas pertussis toxin uses a type IV system.

III. ADHESINS

The ability to adhere to a particular host cell is usually central to a pathogen's ability to cause disease. Even pathogens that rely mainly upon a toxin-mediated event (such as cholera) require the organism to adhere to cell surfaces, thereby allowing bacterial growth and toxin production and secretion. Thus, blockage of adherence by vaccinating against adhesins makes a logical vaccine target. Veterinarian vaccines against *E. coli* fimbriae have been very effective against calf and pig scours.

Like toxins, there are many types of adhesins [6]. Because of the necessity to be on the bacterial surface to access host cells, many adhesins are at the tip of hairlike projections called pili or fimbriae. This presumably allows surface exposure, as well as flexibility upon initial contact, much like a harpoon on the end of a line. Some fimbriae, like those of uropathogenic E. coli (which causes urinary tract infections), have a dedicated tip adhesin. However, others such as *Pseudomonas aeruginosa* pili utilize the same protein as that that makes the pilus; however, because of pilin stacking, the adhesive domain is only exposed on the subunits that are at the tip, which are otherwise buried within the stalk. From a vaccine perspective, both the pilus tip and stalk are potential targets, as antibodies adherent to the tip would block adherence, whereas antibodies coating the stalk would enhance complement lysis and bacterial clearance. Like flagella and other surface organelles, adhesins have a complex base structure embedded in the outer membrane, plus several other chaperones and secretins that assist in pilus assembly.

In contrast to fimbriae, there are also many adhesins that are part of the bacterial surface, often embedded in the outer membrane. Collectively, these are called afimbrial adhesins. Some examples of these include pertussis filamentous hemagglutinin (FHA, binds several receptors), the opacity proteins (Opa) of *Neisseria*, which bind carcinoembryonic antigens (CEA), and several extracellular matrix binding proteins of gram-positive pathogens.

The receptors to which adhesins bind are varied but, not surprisingly, are usually cell surface molecules, including glycoproteins, glycolipids, and proteins. Generally, most adhesins target carbohydrates, which seems to give them much tissue and cell specificity if needed. *E. coli* type I pili bind mannose, which gives them a broad range of receptors given the high presence of mannose on cell surfaces. In contrast, uropathogenic *E. coli* binds globosides in a certain conformation only, which seems to specifically target the bladder epithelium.

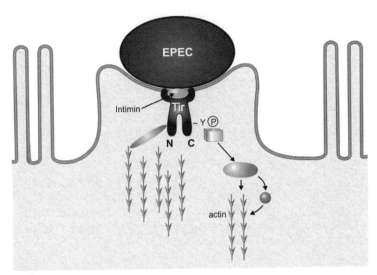

FIGURE 1 Enteropathogenic *E. coli* (EPEC) adherence to epithelial cells. EPEC uses a type III secretion system to inject Tir into host cell membranes, where it then serves as a receptor for intimin on the bacterial surface. Subsequently, actin cytoskeleton is concentrated beneath adherent organisms, raising the bacterium up onto a pedestal on the cell surface.

Most pathogens synthesize many adhesins whose expression is regulated differentially. For example, *Salmonella* has at least five adhesins that seem to contribute to adherence to the intestinal epithelium. Not surprisingly, mutation of a single adhesin does not seem to have any effect on virulence, yet mutation of several adhesins decreases virulence significantly. Because of this complexity and redundancy, definition of the contribution to virulence of particular adhesins has proven difficult. However, there is usually a correlation with the presence of particular adhesins among clinical isolates and disease severity.

A fascinating yet unexpected mechanism of adherence has been described for enteropathogenic *E. coli* (EPEC) and enterohemorrhagic *E. coli* (EHEC) [7], both significant causes of diarrhea, with EHEC also causing hemolytic uremic syndrome (HUS, commonly called hamburger disease). These pathogens utilize a type III secretion system to inject a bacterial molecule, Tir, directly into the host cell membrane (Fig. 1). Tir then functions as a receptor for an outer membrane molecule, intimin. Tir–intimin interactions bind these bacteria tightly to intestinal cells and also cause host cell cytoskeleton to accumulate in pedestals beneath adherent organisms, a process critical for disease. This is the first such example of a pathogen injecting its own receptor into host cells, and Tir and intimin make particularly appealing vaccine candidates, especially because one can vaccinate against both the adhesin and the receptor.

IV. INVASINS

Although most bacterial pathogens function as extracellular pathogens, several of the more serious pathogens actually reside inside host cells [8]. Some examples include *Salmonella, Mycobacterium, Listeria,* and *Chlamydia* species. Although perhaps the easiest way to enter a phagocytic cell is to be passively taken up by phagocytosis, most pathogens have derived specialized mechanisms for driving themselves into host cells, including nonphagocytic cells. This facilitates entry and penetration through intestinal epithelial cells and often places the microbe in a privileged and protected niche inside the host cell, thereby avoiding phagosome–lysosome fusion events.

There are two main mechanisms for pathogen-directed endocytosis (invasion) into host cells. The first involves a zipper-like uptake mechanism, whereby a bacterial surface molecule binds tightly to a receptor. The bacterial molecule is expressed over the entire bacterial surface, thereby allowing sequential binding of additional molecules to host receptors and "zippering" the pathogen into the host cell. This process requires the host cell actin cytoskeleton, although significant rearrangements of the host cell surface are not obvious. *Yersinia enterocolitica* and *Yersinia pseudotuberculosis* both encode an outer membrane protein called invasin, which binds to host cell surface β1-integrins (which normally bind matrix molecules such as fibronectin) very tightly, facilitating invasion. Expression of invasin in nonpathogenic *E. coli* enables these normally noninvasive organisms to enter nonphagocytic cells, such as epithelial cells. The gram-positive food-borne pathogen, *Listeria monocytogenes*, also has a family of invasins called internalins (Inl). InlA binds to E cadherins on the cell surface, mediating bacterial uptake via the cadherin–cytoskeletal linkage, whereas InlB binds to at least two host cell molecules to direct uptake.

The other mechanism invasive pathogens use to facilitate their uptake involves a significant activation of the actin cytoskeleton, causing significant membrane ruffling, macropinocytosis, and invasion. Pathogens that use such a mechanism include *Salmonella* and *Shigella* species (Fig. 2). These pathogens utilize a type III secretion system to inject bacterial effectors into host cells to alter host cell small GTPases such as rac, rho, and CDC42, whose activity controls actin polymerization levels. By promoting actin polymerization and interfacing directly with host cell signal transduction pathways, membrane ruffling occurs, which appears as a "splash" on the cell surface prior to invasion. Once these pathogens are inside the host cell, the cellular surface returns to normal, with the pathogen residing within a membrane-bound compartment inside the host cell.

V. INTRACELLULAR SURVIVAL

Once inside a host cell, invasive pathogens have derived several strategies to avoid the intracellular killing mechanisms that normally follow uptake. These include vacuole acidification and phagosome–lysosome fusion (which delivers toxic substances and enzymes to vacuoles containing ingested materials). However, this altered targeting within a host cell also contributes to the difficulty of developing vaccines against intracellular pathogens, because they often do not reside within the normal antigen presenting pathways used for cell-mediated immunity.

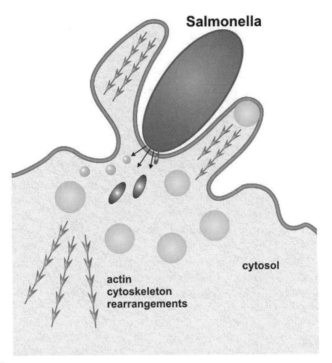

FIGURE 2 *Salmonella* invasion. *Salmonella* species use a type III secretion system to inject various effectors into the host cell, which interface with signaling and cytoskeleton function. This results in ruffles forming on the cell surface, which then engulf and internalize the bacteria. Once inside the cell, *Salmonella* species remain within a membrane-bound compartment, where they resist normal cellular killing mechanisms.

A small number of intracellular pathogens appear to thrive within phagolysosomes, apparently oblivious to lysosomal enzymes, low pH, etc. One such pathogen is *Coxiella burnetti*, the causative agent of Q-fever. How these pathogens overcome the hostile host intracellular defenses remains undefined. A similarly small number of pathogens lyse the vacuolar membrane once they are internalized, releasing them directly into the host cell cytosol where they flourish. *Shigella* species, *L. moncytogenes*, and *Rickettsia* have phospholipases that break down the vacuole membrane. Once free in the cytosol, additional bacterial proteins directly recruit the actin polymerization machinery, which then creates a comet trail of cytoskeleton behind the organisms, thereby propelling them through the cytosol and presumably directly into adjacent host cells, all without ever being exposed to the extracellular environment [1]. This is the primary reason that *L. moncytogenes* is used as a classic T-cell-mediated immune mechanism, as antibodies would never see the pathogen once it initially enters a host cell. From a vaccine perspective, cell-mediated immunity is necessary to control such pathogens, most likely by killing the host cell or by killing the pathogen within the host cell. It is surprising that only a few intracellular pathogens have chosen such a clever and effective intracellular survival mechanism.

Most intracellular pathogens choose to alter the targeting of the membrane-bound vacuole that surrounds the invading organism. Unfortunately, the mechanisms by which the pathogens achieve this are not well-characterized [8]. For example, *Salmonella* species utilize a type III secretion system to invade cells (see previous discussion) and then induce a second type III secretion system that delivers effectors into host cells to promote macrophage survival. Following invasion, these bacteria initially traffic along the normal cellular route, but then they diverge prior to lysosomal fusion. Some of the effectors include those that induce long filaments associated with the *Salmonella*-containing vacuole, whereas others cause actin to condense around the vacuole at later times, thereby affecting fusion with lysosomes. The *Mycobacterium* alters trafficking also, homing to early endosomes and preventing the vacuolar ATPase from fusing with the phagosome and thereby preventing vacuole acidification [9]. *Legionella pneumophila* and *Brucella abortus* use a similar strategy to avoid lysosomal fusion, but they employ a type IV secretion system for this effect. At present, the type IV effectors have not been identified, but this area and how pathogens affect intracelluar trafficking in general are under intensive investigation. As effector molecules are identified, they may make excellent vaccine candidates for a cell-mediated vaccine, although this remains to be tested.

VI. ANTIPHAGOCYTOSIS

An alternative strategy to avoid being killed by phagocytic cells is to prevent phagocytosis from occurring in the first place. Although only a few pathogens are currently recognized for this ability, their numbers have been increasing as additional pathogens are tested. The best studied antiphagocytic strategy is that utilized by the *Yersinia* species [10]. These pathogens also encode a type III secretion system, but this pathogen's injected effectors are designed to paralyze phagocyotosis. They include a potent tyrosine phosphatase, a serine–threonine kinase, and several actin poisons that collectively and effectively block phagocytosis. Enteropathogenic *E. coli* employs a similar strategy, utilizing its type III secretion system to inject as yet unidentified effectors that disrupt host cell phosphatidyl inositol-3 (PI-3) kinase, a signaling pathway essential for phagocytosis. Theoretically, if a vaccine could be designed to neutralize antiphagocytic effectors, pathogens such as these would then be susceptible to phagocytosis and subsequent intracellular killing. One precedent for this is the pneumococcal polysaccharide, which is slimy, making it difficult for phagocytic cells to ingest the organism. When antibodies, termed opsonins, are made to the capsule and bind to it, phagocytosis and intracellular killing rapidly ensue.

VII. APOPTOSIS

A theme that has emerged in bacterial pathogenesis is the ability to trigger apoptotic mechanisms that lead to programmed cell death of host cells [11]. This has a significant impact on the host immune system, as apoptotic death is designed not to incite inflammation. Although well-described for viral

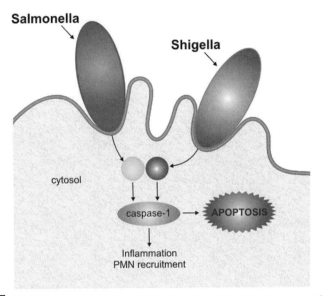

FIGURE 3 *Salmonella-* and *Shigella-*mediated induction of apoptosis. These pathogens inject factors into host cells via type III secretion systems that activate caspases, which then trigger cellular apoptosis and death. This "silent death" avoids the activation of inflammatory responses normally associated with cytolytic death, presumably providing the pathogen with an advantage.

pathogenesis, the ability to trigger or block apoptosis by bacterial pathogens is just being recognized. As is the case for most pathogenic mechanisms, triggering of apoptosis usually involves the delivery of effectors into host cells that interfere with normal signaling pathways. A good example of this are the *Salmonella* species (Fig. 3). It was observed that this pathogen triggers apoptosis in macrophages within the liver. It has also been found that there are at least three distinct mechanisms, two of which employ type III secretion systems, that *Salmonella enterica* serovar *typhimurium* employs to trigger apoptotic death in macrophages. In addition, it appears to have another mechanism that inhibits apoptotic death in epithelial cells by activating a prosurvival pathway. Whether these will represent potentially useful vaccine candidates remains to be determined.

VIII. CONCLUDING REMARKS

Microbial pathogenesis is a fascinating and complex phenomenon, with the pathogens utilizing a variety of virulence factors that all contribute to their pathogenic profile. However, blockage of any one of these key steps usually results in severe attenuation. Thus, vaccines that block a particular step should also block infection or tissue damage. Most vaccines have been developed against prominent bacterial surface structures, such as capsules, carbohydrates, toxins, and fimbriae. This is due to historical reasons, including their surface exposure and ease of purification. However, we now realize that many of the key virulence factors are tightly regulated and are expressed only at a particular tissue site at a particular stage of infection. It will be important to take this specific regulation into consideration when attempting to develop vaccines.

Finding a globally conserved virulence factor to vaccinate against is difficult. However, type III secretion systems play a central role in many gram-negative infections and are fairly conserved among the pathogens. In addition, they have components on their surface that would make logical vaccine targets. Type III effectors are quite pathogen-specific and usually are injected into host cells, both significant factors to consider when contemplating vaccine development. Similarly, development of vaccines against virulence factors that affect trafficking of intracellular pathogens will require clever vaccine strategies.

Within the past decade, significant strides have been made in our understanding of the molecular mechanisms of bacterial pathogenesis (type III systems were discovered less than a decade ago). In addition, we are now gaining significant insight into how virulence factors interface with host cellular machinery. The challenge ahead is to apply this sophisticated knowledge toward developing therapeutics and novel vaccines. Should we succeed, major inroads could indeed be made against many major infectious diseases.

REFERENCES

1. Finlay, B. B., and Cossart, P. (1997). Exploitation of mammalian host cell functions by bacterial pathogens. *Science* **276**:718–725.
2. Finlay, B. B., and Falkow, S. (1997). Common themes in microbial pathogenicity revisited. *Microbiol. Mol. Biol. Rev.* **61**:136–169.
3. Salyers, A. A., and Whitt, D. D. (1994). "Bacterial Pathogenesis: A Molecular Approach." Washington, DC: ASM Press.
4. Schiavo, G., and van Der Goot, F. G. (2001). The bacterial toxin toolkit. *Nat. Rev. Mol. Cell. Biol.* **2**:530–537.
5. Steele-Mortimer, O., Knodler, L. A., and Finlay, B. B. (2000). Poisons, ruffles and rockets: Bacterial pathogens and the host cell cytoskeleton. *Traffic* **1**:107–118.
6. Hultgren, S. J., Abraham, S., Caparon, M., Falk, P., St. Geme, J., and Normark, S. (1993). Pilus and nonpilus bacterial adhesins: Assembly and function in cell recognition. *Cell* **73**:887–901.
7. Kenny, B., DeVinney, R., Stein, M., Reinscheid, D. J., Frey, E. A., and Finlay, B. B. (1997). Enteropathogenic *E. coli* (EPEC) transfers its receptor for intimate adherence into mammalian cells. *Cell* **91**:511–520.
8. Meresse, S., Steele-Mortimer, O., Moreno, E., Desjardins, M., Finlay, B., and Gorvel, J. P. (1999). Controlling the maturation of pathogen-containing vacuoles: A matter of life and death. *Nat. Cell Biol.* **1**:E183–188.
9. Russell, D. G. (2001). *Mycobacterium tuberculosis*: Here today, and here tomorrow. *Nat. Rev. Mol. Cell. Biol.* **2**:569–586.
10. Ernst, J. D. (2000). Bacterial inhibition of phagocytosis. *Cell Microbiol.* **2**:379–386.
11. Weinrauch, Y., and Zychlinsky, A. (1999). The induction of apoptosis by bacterial pathogens. *Annu. Rev. Microbiol.* **53**:155–187.

5 UNDERSTANDING MICROBIAL PATHOGENESIS AS A BASIS FOR VACCINE DESIGN

PART B. Disease-Oriented Approach to the Discovery of Novel Vaccines

JEFFREY N. WEISER[*] AND ELAINE I. TUOMANEN[†]

[*]Departments of Microbiology and Pediatrics, University of Pennsylvania School of Medicine, Philadelphia, Pennsylvania 19104

[†]Department of Infectious Diseases and the Children's Infection Defense Center, St. Jude Children's Research Hospital, Memphis, Tennessee 38105

I. ADVANTAGES OF DISEASE-ORIENTED VACCINATION STRATEGIES

Existing vaccines confer protection from the effects of a single viral or bacterial agent. Decades of success in the development of new products have confirmed the promise of vaccination in the war on infectious disease. The number of immunizations and doses currently recommended for healthy children in particular, however, has strained the public's acceptance of additional vaccines and led to an effort to combine these products. Yet there are limitations to these efforts to simultaneously deliver multiple different vaccine components. Another approach to improve upon the current practice of "one vaccine for one pathogen" would be to develop a preventive strategy that is focused on a particular disease or organ system rather than a single agent. In many situations in which multiple pathogens can cause infection in one organ system, a single vaccine component would offer several clear advantages. In addition to ease of administration and fewer required doses, a disease-oriented vaccine could be made available more easily to at risk populations.

There are clear examples of sets of pathogens dominating the causes of infections of specific organ systems (Table 1). In some cases, this may reflect a common molecular strategy of infection and thereby indicate the existence of homologs with cross-protective potential for the agents causing pneumonia (see Section IV). This part of Chapter 5 will discuss examples of vaccine formulations in which one antigen could provide cross-protection against diverse etiologic agents. Apart from the clear-cut advantages of this vaccine design approach, there are limitations to its use. First, members of a set of bacteria causing an infection of organ A by a common molecular strategy may use very

TABLE 1   Predominant Causes of Bacterial Disease by Category (Organ System)

Disease	Bacteria
Sepsis (with meningitis)	*Streptococcus pneumoniae, Haemophilus influenzae* type B, *Neisseria meningitidis*
Sepsis (intensive care unit setting)	*Staphylococcus aureus, Enterococcus faecium, Enterococcus faecalis, Pseudomonas aeruginosa*
Sepsis (perinatal)	Group B Streptococcus, *Escherichia coli* K1
Pneumonia	*Streptococcus pneumoniae, Haemophilus influenzae, Mycoplasma pneumoniae*
Otitis media–sinusitis	*Streptococcus pneumoniae, Haemophilus influenzae, Moraxella catarrhalis*
Diarrhea	*Campylobacter jejuni, Yersinia enterocolitica, Vibrio cholera, Salmonellae, Shigella dysenterae*
Sexually transmitted disease	*Chlamydiae, Neisseria gonorrheae, Treponema pallidum*

diverse strategies to infect another organ B. For instance, the childhood pathogens *Streptococcus pneumoniae* and *Haemophilus influenzae* share mechanisms to infect the lungs, but differ significantly in how they infect the brain. Thus, protection against one disease may not offer protection against another disease caused by the same organism. Second, there are also multisystem infections of such great worldwide importance so as to justify pathogen-specific vaccines: tuberculosis, malaria, parasitic diseases, etc. Finally, there are infections that are caused predominantly by one bacterial species, arguing for a stand-alone vaccine. This is exemplified by urinary tract infection due to *Escherichia coli*, pharyngitis caused by *Streptococcus pyogenes*, and skin and bone infections due to *Staphylococcus aureus*.

An existing example of a disease-oriented approach is the effort to develop a multicomponent otitis media vaccine that would target several of the common etiologic agents causing that disease, including *Streptococcus pneumoniae* and *Haemophilus influenzae*, for universal administration to all children in the first months of life. These are also among the most common causes of serious bacterial infection of the lower respiratory tract in adults, raising the possibility that a similar product could be used to extend the spectrum of the current 23-valent pneumococcal vaccine as an updated preventive strategy against pneumonia. A single pneumonia vaccine could, for instance, be targeted to persons with known cardiopulmonary disease. Moreover, acute respiratory infection with bacterial pneumonia as the most serious manifestation remains among the leading causes of death worldwide according to the World Health Organization. Likewise, if were vaccines available to prevent common public health problems such as sepsis, diarrhea, or sexually transmitted disease, they could be targeted to the specific populations at greatest risk. This strategy would probably be more acceptable than individual products that each prevent one of many causes of these entities.

II. IDENTIFICATION OF COMMON TARGETS OF AGENTS CAUSING SIMILAR DISEASES

The challenge in any disease-oriented approach to vaccine discovery is the varied nature of the microbes capable of causing a similar spectrum of disease. Of course, a vaccine for a particular disease could be constructed of components from the multiple pathogens to be included. If, on the other hand, there were common protective antigens among the organisms causing related disease, it would greatly simplify development as well as administration. In most cases, however, the pathogens to be targeted are highly diverse. A bacterial pneumonia vaccine, for instance, optimally would need to provide coverage against at least a gram-positive species (*S. pneumoniae*), a gram-negative species (*H. influenzae*), and a mollicute (*Mycoplasma pneumoniae*) [2]. Unfortunately, the bacterial cell surface is generally considered to consist of structures that differ in composition from species to species, with few known examples of common antigens that have been revealed through the process of biochemical analysis. An example of an antigen common to two otherwise diverse species, both etiologic agents of meningitis, is the capsular polysaccharide of *Escherichia coli* K1 and *Neisseria meningitidis* type b [23]. These polysaccharides are composed of immunologically indistinguishable polysialic acid, which is poorly immunogenic, and a "self-antigen." The latter characteristic raises the hypothetical concern that immunization could lead to autoimmunity. Other examples are the conserved oligosaccharide structures, such as the digalactoside, galα1–4gal, on the lipopolysaccharide (LPS) of both *Haemophilus* and *Neisseria*. This structure, however, is also an example of molecular mimicry of the host and, therefore, not a promising vaccine candidate [21].

Two novel approaches to be discussed in this part of Chapter 5 offer improved prospects for the identification of common disease-specific vaccine targets. First, the availability of a large number of whole microbial genome sequences offers the potential to reveal previously unknown homologies among surface antigens. A methodology to identify these elements will be outlined in the next section. The second approach to be addressed later takes advantage of our increasing understanding of the molecular mechanisms of pathogenesis to find critical pathways for host–microbe interactions used by different organisms causing a similar spectrum of disease.

III. GENOMICS APPROACH TO IDENTIFICATION OF ANTIGENS CONSERVED AMONG ORGANISMS CAUSING SIMILAR DISEASES

Rather than depending on the traditional approach of biochemical characterization of microbial epitopes, whole genomes may now be efficiently compared to identify common features and to explore the hypothesis that there will be similar features among pathogens causing similar patterns of infection. An example of this approach is the whole genome comparative algorithm developed in this laboratory by Fueyo and Crabtree [9]. Initially, genes or amino acid sequences in two or more genomes are compared, and their relative homology is ranked according to the BLAST and PSI-BLAST algorithms. In a second

step, genes or amino acid sequences present in other genomes are then sub-
tracted from the data set. This allows for the identification of genes or amino
acid sequences common to organisms with a similar feature, such as the ability to occupy a particular host environment, that is not found among microbes
with other lifestyles. This algorithm has been applied to bacteria that cause
upper and lower respiratory tract infections (Table 2). In this example, the
publicly available whole genomes of pathogens that are otherwise dissimilar
but live in the same host environment and cause similar patterns of disease
were compared. Two respiratory tract pathogens found in nature only in
the human nasopharynx, *H. influenzae* and *S. pneumoniae*, were compared
by searching all *H. influenzae* open reading frames (ORFs) for homologs in
S. pneumoniae. Many of the homologous genes include "housekeeping genes,"
which are found widely among both respiratory and nonrespiratory tract bacterial species. The data set obtained was then made more specific by subtract-
ing homologous genes–sequences found in two representative species that are
not found in the human respiratory tract, the gram-positive *Bacillus subtilis*
and the gram-negative *E. coli*. A BLAST cutoff score of $P < 1.0 \times e^{-20}$ was used
as the criterion for homology in the data shown in Table 2, which shows genes
(based on the *H. influenzae* ORF number) found in two respiratory tract
pathogens, *H. influenzae* and *S. pneumoniae*, without homologs in either
B. subtilis or *E. coli*. The genes identified by this process could be important
for survival in this particular environment. In addition, this data set reveals
potential antigens based on a similar sequence that are common to distantly
related pathogens such as *S. pneumoniae* and *H. influenzae*. A similar process
can be applied to other databases to find common features among organisms
causing other diseases. When the 12 genes identified in the example shown in
Table 2 (with a BLAST cutoff score of $P < 1.0 \times e^{-20}$) were then compared to
the entire set of publicly available microbial genomes, using PSI-BLAST, the
majority, including HI0095, HI00131, HI659, HI660, HI1453, HI1537,
HI1538, and HI1540, had homologs only in other organisms found predominantly or exclusively in the respiratory tract, such as *Actinobacillus actinomycetemcomitans*, *Streptococcus pyogenes*, *Mycobacterium tuberculosis*, and
Porphyromonas gingivalis. This observation demonstrates the specificity of
the approach in finding sequences conserved among a group of organisms with
a common target organ. These ORFs could encode candidate antigens for disease-oriented vaccine discovery directed at respiratory tract infection.

IV. CANDIDATE ANTIGEN CONSERVED AMONG BACTERIAL PATHOGENS OF THE RESPIRATORY TRACT

Most of the genes identified by the preceding algorithm are of unknown function. Four of the genes (HI1536–1540), however, comprise an operon-like
locus in *H. influenzae* that is required for the uptake of environmental choline,
its phosphorylation, and its incorporation onto the surface of the LPS oligosaccharide in the form of choline phosphate or phosphorylcholine (ChoP) [25].
A similar set of genes in *S. pneumoniae* appears to be responsible for the
substitution of ChoP on its lipo- and cell-wall-associated teichoic acid [26].

TABLE 2 Genes Conserved among Respiratory Tract Pathogens Identified by Whole Genome Analysis[a]

HI ORF	S. pneumoniae	E. coli	B. subtilis	Species with homolog[b]	P-score	Type[c]	Function[d]
0095	$1.6 \times e^{-72}$	0.00019	$6.8 \times e^{-10}$	*Actinobacillus actinomycetemcomitans*	e^{-128}	R	Unknown
				Mycobacterium tuberculosis	$4.0 \times e^{-24}$	R	
0131	$6.1 \times e^{-29}$	$3.5 \times e^{-7}$	0.50	*Actinobacillus pleuropneumoniae*	$3.6 \times e^{-191}$	R	Iron binding (p)
220.2	$3.1 \times e^{-72}$	0.49	0.90	*Actinobacillus actinomycetemcomitans*	$4.0 \times e^{-85}$	R	Unknown
				Streptococcus pyogenes	$5.0 \times e^{-69}$	R	
				Acetobacter aceti	$8.7 \times e^{-48}$	NR	
				Vibrio cholerae	$2.0 \times e^{-28}$	NR	
659/660[e]	$3.6 \times e^{-38}$	0.05	0.10	None			Unknown
1038	$3.6 \times e^{-62}$	0.30	0.0081	*Porphyromonas gingivalis*	$8.0 \times e^{-55}$	R	Unknown
				Pyrococcus horikoshii	$5.0 \times e^{-39}$	NR	
				Methanococcus jannaschii	$2.0 \times e^{-33}$	NR	
1244	$1.4 \times e^{-25}$	0.65	0.13	*Clostridium acetobutylicum*	$9.0 \times e^{-20}$	NR	Unknown
1453	$7.2 \times e^{-31}$	$4.6 \times e^{-7}$	$1.4 \times e^{-10}$	*Neisseria meningitidis*	$6.4 \times e^{-20}$	R	Thioredoxin(p)
1537	$1.5 \times e^{-34}$	0.20	$7.3 \times e^{-3}$	None			Choline kinase(p)
1538	$4.7 \times e^{-31}$	$3.7 \times e^{-5}$	$4.9 \times e^{-10}$	None			Choline permease (p)
1539	$1.1 \times e^{-40}$	$4.9 \times e^{-4}$	$1.3 \times e^{-4}$	*Treponema pallidum*			Choline nucleoside transferase(p)
1540	$5.6 \times e^{-35}$	0.87	0.14	None		NR	NDP-choline transferase(p)

[a]BLAST and PSI-BLAST comparisons to *Haemophilus influenzae* open reading frames (HI ORFs) based on the genomes listed here were performed using the genomic sequences available from www.pseudomonas.com, NCBI's PSI-BLAST server at http://www.ncbi.nlm.nih.gov/blast/psiblast.cgi, and Kyoto University's BLAST server at http://www.kyoto/kegg/ad.jp. Annotated Complete Genomes Searched: *Aeropyrum pernix, Aquifex aeolicus, Archaeoglobus fulgidus, Bacillus subtilis, Borrelia burgdorferi, Campylobacter jejuni, Chlamydia pneumoniae CWL029, Chlamydia pneumoniae AR39, Chlamydia pneumoniae J138, Chlamydia muridarum, Chlamydia trachomatis D/UW-3/CX, Deinococcus radiodurans R1, Escherichia coli, Haemophilus influenzae, Helicobacter pylori 26695, Helicobacter pylori J99, Methanobacterium thermoautotrophicum, Methanococcus jannaschii, Mycobacterium tuberculosis, Mycoplasma genitalium, Mycoplasma pneumoniae, Neisseria meningitidis MC58, Neisseria meningitidis Z2491, Pyrococcus abyssi, Pyrococcus horikoshii, Rickettsia prowazekii, Synechocystis PCC6803, Thermotoga maritima, Treponemm pallidum, Ureaplasma urealyticum, Xylella fastidiosa.* Completed Genomes Searched (annotation in progress): *Bordetella pertussis, Clostridium acetobutylicum, Clostridium tetani, Corynebacterium glutamicum, Escherichia coli O157:H7, Halobacterium sp. NRC1, Lactococcus lactis, Mycobacterium tuberculosis, Neisseria gonorrhoeae, Pasteurella multocida, Pseudomonas aeruginosa PA01, Pyrobaculum aerophilum, Pyrococcus furiosus, Rhodobacter capsulatus, Sulfolobus tokodaii, Streptococcus pyogenes, Vibrio cholerae.* Incomplete Genomes Searched: *Actinobacillus actinomycetemcomitans, Caulobacter crescentus, Candida albicans, Chlamydia trachomatis MOPN, Chlorobium tepidum, Enterococcus faecalis, Mycobacterium bovis, Neisseria meningitidis serogroup A, Pseudomonas putida, Porphyromonas gingivalis, Salmonella typhimurium, Shewanella putrefaciens, Staphylococcus aureus COL, Staphylococcus aureus NCTC 8325, Streptococcus mutans, Streptococcus pneumoniae, Thiobacillus ferrooxidans, Trypanosoma brucei rhodensiense, Vibrio cholerae serotype 01, Yersinia pestis CO-92 biovar.*

[b]Homology based on $P < 1.0 \times e^{-20}$.

[c]Type refers to whether this species is associated primarily with the respiratory tract (R) or other sites (NR).

[d](p) refers to putative function.

[e]HI659 and HI660 are a single ORF in *S. pneumoniae.*

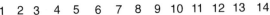

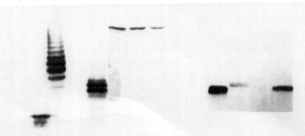

FIGURE 1   Composite Western blot summarizing reactivity of six species of mucosal pathogens with MAb TEPC-15 recognizing phosphorylcholine (ChoP). Lane 1, nontypeable *H. influenzae* strain H233 with ChoP − LPS; lane 2, nontypeable *H. influenzae* strain H233 with ChoP + LPS; lane 3, *S. pneumoniae* strain R6 lipoteichoic acid; lane 4, *A. actinomycetemcomitans* LPS of strain 4; lane 5, *A. actinomycetem-comitans* LPS of strain Jp-2; lanes 6–10, *P. aeruginosa* grown at 26.5, 30.0, 33.5, 37.0, 39.5°C, respectively; lane 11, purified *N. meningitidis* pilin; lanes 12–14, piliated *N. gonorrhoeae* strain MS11 variants with and without the ChoP epitope. Not shown are the commensal *Neisseria* and *H. somnus*, which display the phosphorylcholine epitope on their LPS, and the polar lipid with ChoP from *Mycoplasma* species, which does not transfer in Western blot analysis.

Choline, a highly unusual structural feature in prokaryotes, therefore, is on the cell surface of two major pathogens residing in the human respiratory tract, suggesting that it may act to promote survival in this niche. Further evidence to support this hypothesis comes from screening other species for the expression of ChoP (see Fig. 1). Bacterial ChoP is recognized by the natural IgA monoclonal antibody with the T15 idiotype, TEPC-15, and a number of other species have been shown to have structures that are recognized by this antibody [10, 12]. These include pili of *Neisseria meningitidis*, a temperature-regulated cell envelope protein in *Pseudomonas aeruginosa*, the LPS of *A. actinomycetemcomitans* (the etiologic agent of juvenile periodontitis), the LPS of commensal *Neisseria* species, the LPS of the bovine pulmonary pathogen, *Haemophilus somnus*, and a polar membrane lipid in various *Mycoplasma* species, including *M. pneumoniae* [6, 8, 15, 17, 22]. This list includes the most common etiologic agents of two major diseases: the bacterial causes of community-acquired pneumonia (*S. pneumoniae*, *H. influenzae*, and *M. pneumoniae*) and the encapsulated pathogens that are responsible for the overwhelming majority of cases of bacterial meningitis (*S. pneumoniae*, *H. influenzae*, and *N. meningitidis*). Moreover, a representative survey of common bacterial pathogens failed to identify the ChoP epitope on organisms that are not, at least in part, associated with the respiratory tract.

What then is the role of ChoP in allowing bacteria to survive in this particular host environment? Many of the species that display the cell-surface ChoP epitope undergo on–off switching or phase variation in its expression [11, 22, 25]. This suggests that there may be advantages and at other times disadvantages to the display of this unusual cell-surface prokaryotic structure. Bacterial ChoP, for instance, is recognized by an innate immune mechanism involving C-reactive protein that, in the presence of complement, is bactericidal to *H. influenzae* [19, 24]. The molecular mechanism of phase variation of choline incorporation in *H. influenzae* involves slip-stranded mispairing of tandem DNA repeats in a putative choline kinase (LicA) [24]. By direct determination of the number of repeats in specimens isolated from the human respiratory tract, it was possible to demonstrate that >90% of bacteria resident in this site express ChoP. ChoP, therefore, appears to be particularly important to the organism while resident in the natural host on the mucosal surface, a necessary initial step in the pathogenesis of respiratory tract infection. This may be because ChoP acts to promote adherence to host cells. For *S. pneumoniae*, *H. influenzae*, and *A. actinomycetemcomitans*, it has been shown that ChoP mimics platelet activating factor (PAF, a host molecule that also contains phosphorycholine), allowing the bacteria to both adhere to and gain entry into host epithelial cells by binding to its receptor (rPAF) [7, 15, 18]. Because the ability to adhere to and possibly invade host cells is a key step in colonization and persistence on the mucosal surface, this is an example of a common pathway in pathogenesis utilized by different organisms causing a similar spectrum of disease. Furthermore, the conservation of expression of ChoP among the most successful pathogens that reside primarily in the respiratory tract points to its importance to these species. ChoP is also known to be immunogenic and, at least in some cases, the target of protective antibody. Together these observations suggest that ChoP could function as an effective single antigen for the prevention of diseases such as otitis media, sinusitis, and pneumonia, if innate defenses targeting this structure were augmented by acquired immunity.

This approach is less likely to be as effective in preventing meningitis, even though the spectrum of pathogens is similar, because ChoP is down-regulated on pathogens circulating in the bloodstream and does not appear to be as dominant a mechanism for crossing the blood–brain barrier as it is for crossing respiratory epithelia. This example points out two lessons: (1) an understanding of bacterial pathogenesis can reveal vaccine antigens assembled by multiple (and even diverse) genes rather than one gene–one gene product–one vaccine antigen; and (2) innate immunity may point to surface structures of pathogens that share shape and function, but not necessarily sequence. Libraries of shapes arising from efforts in structural biology may reveal important new common epitopes for organ-based vaccines.

What then are the prospects for inducing protective immunity against ChoP? Despite the fact that choline phosphate is also found on host membrane lipids as a component of phosphatidylcholine, ChoP is antigenic, and a naturally acquired antibody recognizing bacterial ChoP is not uncommon [1, 5, 14, 16]. Antibody of the T15 idiotype is protective when given passively against systemic pneumococcal infection in a murine model, and mice carrying the transgene with the T15 idiotype are more resistant to infection by *S. pneumoniae* [3, 4, 13]. Following intranasal immunization of ChoP linked to a protein carrier, mice developed antibody to ChoP and were protected from fatal pneumococcal infection [20]. It remains unknown whether the antibody produced by humans in response to ChoP is functionally equivalent to the T15 idiotype antibodies generated in mice. Antibody recognizing ChoP may be detected in human serum, but these studies thus far have failed to establish whether they are protective or whether there is a relationship between antibody levels and incidence of pneumococcal infection [14]. It is, therefore, still unclear at this point whether induction of an immune response to ChoP in humans would confer protection against *S. pneumoniae* or other respiratory tract pathogens expressing this structure in a disease-oriented approach to vaccine development.

REFERENCES

1. Andersson, I., Rosen, V., Hakansson, A., Aniansson, G., Hansson, C., Andersson, B., Nylen, O., Sabharwal, H., and Svanborg, C. (1996). Antibodies to pneumococcal polysaccharides in human milk: Lack of relationship to colonization and acute otitis media. *Ped. Infect. Dis. J.* 15:498–507.
2. Bartlett, J., and Mundy, L. (1995). Community-acquired pneumonia. *New England J. Medicine* 333:1618–1624.
3. Briles, D. E., Forman, C., Hudak, S., and Claflin, J. L. (1982). Anti-phosphorylcholine antibodies of the T15 idiotype are optimally protective against *Streptococcus pneumoniae*. *J. Exp. Med.* 156:1177–1185.
4. Briles, D. E., Forman, C., Hudak, S., and Claflin, J. L. (1984). The effects of idiotype on the ability of IgG1 anti-phosphorylcholine antibodies to protect mice from fatal infection with *Streptococcus pneumoniae*. *Eur. J. Immunol.* 14:1027–1030.
5. Briles, D. E., Scott, G., Gray, B., Crain, M. J., Blaese, M., Nahm, M., Scott, V., and Haber, P. (1987). Naturally occurring antibodies to phosphocholine as a potential index of antibody responsiveness to polysaccharides. *J. Infect. Dis.* 155:1307–1314.
6. Cox, A., Howard, M., Brisson, J., van der Zwan, M., Thibault, P., Perry, M., and Inzana, T. (1998). Structural analysis of the phase-variable lipooligosaccharide from *Haemophilus somnus* strain 738. *Eur. J. Biochem.* 253:507–516.
7. Cundell, D. R., Gerard, N. P., Gerard, C., Idanpaan-Heikkila, I., and Tuomanen, E. I. (1995). *Streptococcus pneumoniae* anchor to activated human cells by the receptor for platelet-activating factor. *Nature* 377:435–438.
8. Deutsch, J., Salman, M., and Rottem, S. (1995). An unusual polar lipid from the cell membrane of *Mycoplasma fermentans*. *Eur. J. Biochem.* 227:897–902.
9. Fueyo, J., Crabtree, J., and Weiser, J. (2000). A genomic approach to identify bacterial genes required for survival in specific host environments. International Conference on Microbial and Model Genomes, Paris, France.
10. Gillespie, S. H., Ainscough, S., Dickens, A., and Lewin, J. (1996). Phosphorylcholine-containing antigens in bacteria from the mouth and respiratory tract. *J. Med. Microbiol.* 44:35–40.
11. Kim, J., and Weiser, J. (1998). Association of intrastrain phase variation in quantity of capsular polysaccharide and teichoic acid with the virulence of *Streptococcus pneumoniae*. *J. Infect. Dis.* 177:368–377.
12. Kolberg, J., Holby, E. A., and Jantzen, E. (1997). Detection of the phosphorylcholine epitope in streptococci, *Haemophilus*, and pathogenic *Neisseriae* by immunoblotting. *Microb. Pathogen.* 22:321–329.
13. Lim, P. L., Choy, W. F., Chan, S. T. H., Leung, D. T. M., and Ng, S. S. M. (1994). Transgene-encoded antiphosphorylcholine (T15+) antibodies protect CBA/N (xid) mice against infection with *Streptococcus pneumoniae* but not *Trichinella spiralis*. *Infect. Immun.* 62:1658–1661.
14. Nordenstam, G., Andersson, B., Briles, D., Brooks, J. W. J., Oden, A., Svanborg, A., and Eden, C. S. (1990). High anti-phosphorylcholine antibody levels and mortality associated with pneumonia. *Scand. J. Infect. Dis.* 22:187–195.
15. Schenkein, H., Barbour, S., Berry, C., Kipps, B., and Tew, J. (2000). Invasion of human vascular endothelial cells by *Actinobacillus actinomycetemcomitans* via the receptor for platelet-activating factor. *Infect. Immun.* 68:5416–5419.
16. Scott, M. G., Briles, D. E., Shackelford, P. G., Smith, D. S., and Nahm, M. H. (1987). Human antibodies to phosphocholine. IgG anti-PC antibodies express restricted numbers of V and C regions. *J. Immunol.* 138:3325–3331.
17. Serino, L., and Virji, M. (2000). Phosphorylcholine decoration of lipopolysaccharide differentiates commensal *Neisseriae* from pathogenic strains: Identification of *licA*-type genes in commensal *Neisseriae*. *Mol. Microbiol.* 35:1550–1559.

18. Swords, W., Buscher, B., Ver Steeg Ii, K., Preston, A., Nichols, W., Weiser, J., Gibson, B., and Apicella, M. (2000). Non-typeable *Haemophilus influenzae* adhere to and invade human bronchial epithelial cells via an interaction of lipooligosaccharide with the PAF receptor. *Mol. Microbiol.* **37**:13–27.

19. Szalai, A. J., Briles, D. E., and Volanakis, J. E. (1995). Human C-reactive protein is protective against fatal *Streptococcus pneumoniae* infection in transgenic mice. *J. Immunol.* **155**:2557–2563.

20. Trolle, S., Chachaty, E., Kassis-Chikhani, N., Wang, C., Fattal, E., Couvreur, P., Diamond, B., Alonso, J., and Andremont, A. (2000). Intranasal immunization with protein-linked phosphorylcholine protects mice against a lethal intranasal challenge with *Streptococcus pneumoniae*. *Vaccine* **118**:2991–2998.

21. Virji, M., Weiser, J., Lindberg, A., and Moxon, E. (1990). Antigenic similarities in lipopolysaccharides of *Haemophilus* and *Neisseria* and expression of a digalactoside structure also present on human cells. *Microb. Pathogen.* **9**:441–450.

22. Weiser, J. N., Goldberg, J. B., Pan, N., Wilson, L., and Virji, M. (1998). The phosphorylcholine epitope undergoes phase variation on a 43 kD protein in *Pseudomonas aeruginosa* and on pili of pathogenic *Neisseria*. *Infect. Immun.* **66**:4263–4267.

23. Weiser, J. N., and Gotschlich, E. C. (1991). Outer membrane protein A (OmpA) contributes to serum resistance and pathogenicity of *Escherichia coli* K-1. *Infect. Immun.* **59**:2252–2258.

24. Weiser, J. N., Pan, N., McGowan, K. L., Musher, D., Martin, A., and Richards, J. C. (1998). Phosphorylcholine on the lipopolysaccharide of *Haemophilus influenzae* contributes to persistence in the respiratory tract and sensitivity to serum killing mediated by C-reactive protein. *J. Exp. Med.* **187**:631–640.

25. Weiser, J. N., Shchepetov, M., and Chong, S. T. H. (1997). Decoration of lipopolysaccharide with phosphorylcholine: A phase-variable characteristic of *Haemophilus influenzae*. *Infect. Immun.* **65**:943–950.

26. Zhang, J.-R., Idanpaan-Heikkila, I., Fischer, W., and Tuomanen, E. (1999). Pneumococcal *licD2* gene is involved in phosphorylcholine metabolism. *Mol. Microbiol.* **31**:1477–1488.

5 UNDERSTANDING MICROBIAL PATHOGENESIS AS A BASIS FOR VACCINE DESIGN

PART C. Immunological Memory and Vaccines against Acute Cytopathic and Noncytopathic Infections

ROLF M. ZINKERNAGEL
Institute of Experimental Immunology, University Hospital, 8091 Zurich, Switzerland

I. INTRODUCTION

Resistance to infections is based primarily on nonspecific mechanisms, such as interferons, defensins, natural immunity, including complement, natural antibodies, and natural killer cells, activated phagocytes, and many additional mechanisms [21]. These nonspecific resistance mechanisms are responsible for the major part [45] (>95%) of host defense. For example, the absence of interferon receptors increases the susceptibility of mice or humans to viral infections by several orders of magnitude. Specific immunity is phylogenetically a rather new, additional fine tuner of resistance, emerging as a result of co-evolution between hosts and infectious agents.

The two arms of the immune system, humoral and cellular, fulfill the following major functions. The immunoglobulin (Ig) receptor of B cells and secreted antibodies directly recognizes complex folded proteins or carbohydrates. Protective antibodies inactivate and block the action of infectious agents or toxins by covering them and/or by facilitating their phagocytosis. IgM and IgG protect against antigens in blood and the lymphatic system, IgA protects on mucosal membranes [28, 41], and IgE triggers mast cells and basophils in skin and mucosae. In contrast to B cells and antibodies, T cells recognize small peptides presented on the cell surface by major histocompatibility (MHC) antigens [64]. Cytotoxic CD8+ T cells are specific for fragments of proteins synthesized by the cell itself and are presented by MHC class I (HLA-A, -B, -C) antigens; this pathway includes not only self-peptides but also viral, intracellular bacteria or tumor antigens. Phagocytosed antigens are processed in phagolysosomes and are presented by MHC class II antigens

(HLA-D, -P, -Q). Dendritic cells, which either are infected themselves or are able to take up infectious, foreign, or decaying self-antigen, transport antigens to organized lymphatic tissues. They are therefore of key importance in inducing T-cell responses. Although antibodies act directly where they are released or transported, T cells actively emigrate into peripheral solid tissues. T cells can act via direct contact or by specific release of immune mediators, such as interferon or tumor necrosis factors, or they can act nonspecifically via recruitment and activation of macrophages. Cytopathic viruses or bacteria that cause an acute disease or a lethal infection are, in general, controlled most efficiently by soluble diffusable factors, including T-cell-dependent cytokines [such as interferon-γ (IFNγ) and tumor necrosis factor (TNF)] and by specific neutralizing antibodies. Noncytopathic intracellular organisms usually cause no direct cell or tissue damage and, therefore, no direct disease, even though they tend to persist. In this case, immune control is mediated by perforin-dependent, cytotoxic, and cytokine-releasing T cells, which cause inflammation and tissue damage [15, 29]. Because the immune system cannot distinguish a priori between cytopathic and noncytopathic infections, it cannot really "foresee" its beneficial and detrimental effects on the host; it merely responds to antigen. Therefore, protection by immunity represents an equilibrum between optimal resistance against the various cytopathic infections and avoidance of excessive immunologically mediated tissue damage. Clinical examples of unbalanced immunity against poorly or noncytopathic infections causing disease by immunopathology are tuberculoid leprosy, fulminant aggressive hepatitis B virus (HBV), hepatitis C virus (HCV), or HIV infections leading to AIDS.

The following general rules define T- and B-cell responses and their dependence upon antigen structure, antigen localization, dose, and how long antigen is available [65, 67]. (1) Conventional immune responses of T and B cells can only be induced in organized secondary lymphoid organs, such as lymph nodes, Peyer's patches, and the spleen. (2) T cells react against cell-associated antigens that are localized in secondary lymphoid organs in sufficient amounts and for a period of at least 2–5 days. Accordingly, antigens that always stay outside of secondary lymphoid organs, including antigens that reach secondary lymphoid organs below the essential time and quantity thresholds, are ignored by the immune system. Self-antigens, which are localized strictly extralymphatically, are immunologically ignored. Because the thresholds are high with respect to both the dose and the time during which the antigen must be available for the induction of an immune response, ignorance or indifference to self-antigens is an important mechanism for avoiding autoimmunity. At the other end of the spectrum, antigens that are always in primary or secondary lymphoid organs—such as serum proteins—induce and delete all potentially reactive T cells. This process is called negative selection [44]. (3) B cells react against highly repetitive, rigidly ordered antigenic determinants with short-lived IgM responses independently of T help, particularly if combined with a polyclonal activator. These antigens are called T-independent type I. Other multimeric, but nonrigid antigens, including those on cell surfaces, will also induce B-cell IgM responses [50] if presented together with indirect (or unlinked) T help (T-independent type II antigens). B cells usually react against monomeric or oligomeric antigens only if structurally linked specific T help is provided. It is important to note that all B-cell responses become dependent upon linked T help if antigen doses are limited [50]. Also, the switch to long-lived IgG and the maintenance of IgG responses are usually dependent upon linked carrier-specific T help.

As stated, the highly repetitive paracrystalline identical determinants on most infectious agents distinguish them from the usually mono-oligomeric self-antigens accessible to B cells [3, 34]. This repetitive structure, combined with the relatively large size and the local concentration of the pathogen, induces T-independent B-cell responses efficiently. Induction is enhanced by polyclonal B-cell activators such as lipopolysaccharides or by maximal cross-linking of immunoglobulin receptors on B cells. During the early phase of infection, IgM B cells expand rapidly and produce early neutralizing IgM antibodies. The expanded specific B-cell pools are efficient targets for T helper cells, which have been induced by phagocytosed and processed antigen by day 4–6 after infection [4, 26]. Many B cells therefore are readily switched to IgG-producing plasma cells. This switch is very important because IgM is short-lived (<2 day half-life) and, due to its size, cannot diffuse readily into solid tissues; in particular, only IgG may reach the central nervous system and, importantly, the fetus.

The tight T-cell control of IgG responses means that, in general, B cells have not to be efficiently deleted by self-antigen. Although they may not react against monomeric antigens, nevertheless they are potentially self-reactive. However, such auto-reactive B cells are not readily induced to produce IgM or even switch to IgG responses because highly repetitive ordered self-antigens normally do not exist in the lymphatic system or in blood [5]. Even if they were triggered to produce some auto-reactive IgM transiently, this would be of little consequence because of the short half-life of IgM. The switch to IgG requires linked functional T help [63], but self-specific T help is generally deleted by negative selection and, therefore, is not available for long-lasting IgG auto-antibody responses. Thus, antigen structure and lack of T help are key guarantors for B-cell unresponsiveness against self-antigens, such as thyroglobulins or insulin. In this context, it is important to remember that sugar antigens are poor immunogens because they fail to induce necessary helper T cells, particularly in infants. The difficulty to develop adequate vaccines against carbohydrates has been overcome (e.g., against *Haemophilus influenzae* [52]) by chemically coupling such bacterial carbohydrate antigens to proteins that are capable of inducing appropriately linked T help.

Specificity is a key issue in any discussion about immune protection, immunological memory, and vaccines. Immune specificity has evolved in response to pressure from infectious agents. It can be defined operationally as the capacity of B or T cells to discriminate between individual pathogens. This capacity is dependent on the kinetics of the host–parasite interaction and the initial dose of antigen. The specificity of immunity is measured most directly by protection or cross-protection *in vivo*, i.e., protection by immunity against poliovirus strain I is absent against a subsequent infection with poliovirus II (Table 1) [47] or similarly by influenza virus HA1 against HA2. Because both cytotoxic and helper T cells against serotype-defined virus groups are shared between the various serotypes, the obvious lack of cross-protection between

TABLE 1 Protection against Serotype-Defined Infections: Essential Role of Neutralizing Antibodies But Not of T Cells

Virus	Serotypes	Specificity of CD4⁺T and CD8⁺T	Neutralizing Antibodies
Influenza	Various	Largely shared	Highly specific
Polio	I, II, III	Largely shared (nonessential for protection)	Highly specific (essential for protection)

Experimental evidence		Response	
Immunization	Challenge		
Serotype A Influenza virus or rhabdo virus	Serotype B	Secondary anti-A and secondary anti-B	Primary anti-B

serotypes (e.g., poliovirus I, II, III) in human populations indicates that only preexistent neutralizing antibodies and not primed helper or cytotoxic T cells are responsible for protection (Table 1) [26, 36, 57].

Infectious agents that exhibit various serotypes are often highly cytopathic and cause acute diseases. Innumerable specificities of antibodies are usually induced by virus infection, but only neutralizing antibodies are protective [3, 19]; other antibodies, particularly those against internal viral antigens, are virtually irrelevant for protection. Those infections that tend to persist, including many viruses [15, 40, 61] such as HBV, HCV, and HIV, facultative intracellular bacteria such as mycobacteria [8, 9, 33, 37], and other intracellular parasites, are usually controlled initially by T cells.

II. THE ROLES OF EARLY EVENTS AND PARASITE VARIABILITY IN HOST–PARASITE RELATIONSHIPS

Most cytopathic agents leave little leeway between death and survival. In contrast, a wide range of relationships is possible between low or noncytopathic infections and the host. Many parameters, such as interferon production and susceptibility of the infection to interferons, influence this equilibrium. In addition, mutation rates of the pathogens play a major role. Variability of antigens permits infectious agents to escape T cells or neutralizing antibodies, in the individual as well as at the population level. The variability of the parasite genome plays a major role in these selections. Selection of mutations is driven by epidemiological forces, either in the whole population, such as antigenic drift and shift in acute influenza viruses, or within an individual host for persistent viruses, such as HBV and possibly HIV or LCMV in mice. Therefore, RNA viruses that have an error-prone genome are particularly variable and can escape host immune responses efficiently. The games between absolute numbers of infectious agents, mutation frequencies, and efficiency of defense influence the subtle overall balance that results in either early efficient control [e.g., asymptomatic tuberculosis (TB), short-term HBV infection, asymptomatic long-term survivors of HIV infection] or chronic immunopathological disease that may evolve over a long time (e.g., organ-specific TB, AIDS, liver cirrhosis, perhaps diabetes type 2, some forms of arteriosclerosis and arthritis). Thus, early events set the stage and the equilibrum of host–parasite relationships that will determine the eventual outcome of slowly developing, immunopathologically mediated disease. Although this has become more obvious with a better understanding of HIV infections, other infections exhibit similarly vast spectra of differing diseases; this includes TB, leprosy, HBV, HCV (LCMV in mice), probably infections with coxsackie- or measles viruses [including subacute sclerosing panencephalitis (SSPE)], and most classical parasite infections.

III. HOW TO MEASURE IMMUNITY

Which measurements of immune responses, including T-cell and neutralizing antibody responses, correlate best with protective immunity or immunological memory to predict the efficiency of vaccines? Neutralizing antibody responses are readily measured *in vitro* by a virus plaque-reduction assay or by neutralization of the activity of bacterial toxins. Neutralizing antibodies, usually of the IgG type, possess an overall avidity of around $\geq 10^9 \ M^{-1}$ and need to be available in serum at concentrations of around $10^{-8} \ M$ ($\geq 1 \ \mu g/ml$) to be protective in tightly controlled murine model infections [4, 56]. Nonneutralizing antibodies are quite useful parameters to monitor the infection history of a patient, and neutralizing titers often correlate with ELISA titers, but not always (e.g., against non- or poorly cytopathic infections, such as HBV, HCV, HIV, LCMV, or TB).

The protective capacity of cytotoxic T cells correlates reasonably well with direct measurable lytic activity of lymphocytes in a 4- to 5-hr *in vitro* assay tested against infected target cells or target cells pulsed with T-cell peptides at concentrations of around 10^{-9}–$10^{-10} \ M$. A peptide concentration of $10^{-6} \ M$ yields results that tend to overestimate activities and relative precursor numbers of T cells lysing targets or T cells restimulated *in vitro* with high concentrations of peptides on antigen-presenting cells to reveal intracellular interleukin staining. These assays indicate priming reliably and measure the number of cells responding in buffered saline *in vitro*, but they yield only indirect correlates of their activation state and protective immunological activity *in vivo* (Tables 2 and 3).

T-cell-mediated immunity can be monitored by the injection of antigen intracutaneously in some infectious diseases. This skin test measures delayed type hypersensitivity (DTH) mediated by T cells. This DTH reaction empirically reveals an immune status only for TB, leprosy, and perhaps sarcoidosis. Importantly, DTH reactions against most viruses cannot assess the immune status [35], because on the one hand the DTH antigen is not antigenic or is degraded. On the other hand, an alternative explanation, documented by experiments in mice, indicates that the preactivated T cells are often absent that are needed for a DTH to develop within 48 hr. Thus, readily inducible DTH probably signals the persistence of antigen linked to an active infectious process (Tables 2–4).

TABLE 2   T-Cell Memory and Persistence of Infection: Infection Immunity

	Localization–spread of infection–vaccine	Persistence	Presence of activated T cells	
			<3 years	>70 years
BCG	Subcutaneous lesion regional lymph node	1–3 years	+	–
TB	Granuloma in lung + lymph node	Life-long	+	±
Leprosy	Inapparent tuberculoid lepromatous	Life-long	+ ++	± ++

IV. IMMUNOLOGICAL MEMORY

Immunological memory is defined as follows. A host, having survived an infection or primed with an antigen, will react more rapidly and with higher titers of antibodies or T cells to a second infection or antigen exposure. It is often explained as a special quality of individual T or B cells: they have acquired special "memory" characteristics when compared *in vitro* to naive cells or

TABLE 3   Rules of Memory T- and B-Cell Responses

	B Cells		T Cells	
	Resting	Activated (antibody-producing) plasmacytes	Resting (with effector function)	Activated
Location	Blood Lymph node Spleen	Germinal center Bone marrow Lymph node Spleen	Blood Lymph node Spleen	Blood Lymph node Spleen Migrate to periphery
Status–function	Increased frequency? Altered cellular parameters?	Antibody production	Increased frequency? Altered cellular parameters?	Emigration into solid organs
Protection	Slow, not much different from nonimmune	Immediate and efficient	Slow, not much different from nonimmune	Immediate and efficient
Antigen dependence	No	Yes	No	Yes
Interpretation	Theoretical memory is antigen-independent	Low-dose antigen-driven response for protective memory	Theoretical memory is antigen-independent	Low-dose antigen-driven response for protective memory

TABLE 4 Adoptive Transfer of Primed T and B Cells without Antigen Cannot Protect Recipients against Disease

Adoptively transferred	Inert antigen added during transfer	Neutralizing antibody titer on day 3 after transfer	Protection against disease by cytopathic virus
Unprimed T + B	−	<1/40	−
	+	<1/100	−
Primed T + B	−	<1/40	−
	+	<1/100	−
Serum from primed donors	−	>1/1000	++

		Increased CTLp on day 0	Protection by CD8+ T cells against immunopathological consequences of infection
Primed CTLs	−	10–30×	−
	+	30–100×	++

activated effector lymphocytes (Tables 2–4). This memory status correlates with increased precursor frequencies and enables the system to respond quickly and efficiently to a second exposure. The nature of the "memory status" correlates with the acquisition of numerous surface molecules on lymphocytes, but overall memory is still poorly understood [1, 24, 39, 53, 67]. An alternative possibility is that immunity depends on a low-level antigen-driven immune response keeping T cells activated and maintaining protective antibody titers; therefore, this would mean that protection by immunological memory disappears when antigen disappears. Obviously these two views—inherent special quality versus antigen-driven response—differ fundamentally. It would be very important to understand how protective memory functions over time against infectious agents or tumors so that we can improve vaccines.

A. Why and What Kind of Immunological Memory?

Children that have been infected with measles, pox, polio, or numerous other viruses are subsequently resistant to the same infection [22, 57] (Tables 3–5). Many years of immunological research have been spent on immunological memory [6, 24, 47, 67]. The period of life before and after birth may, from an evolutionary point of view, be the key to understanding immunity [65, 67]. During this period of physiological immuno-incompetence, adoptively transferrable maternal immunological memory is essential for the survival of the offspring and of the species. We may speculate that a naive adult host who does not survive a first infection will not need immunological memory, whereas a host having survived a first infection is fit to survive another infection

TABLE 5   Vaccines and Immunity

Available–successful		Not available–
Efficient	**Not completely satisfactory**	**not successful**
Smallpox	Measles	TB
Poliovirus I–III	Mumps	Leprosy
Rubella	RSV	Parasitic diseases
Tetanus		HCV
Diphtheria		HIV
Pertussis		Solid peripheral antigenic tumors
Protective means: neutralizing antibodies	Neutralizing antibodies plus T cell	Activated T cells (plus neutralizing antibodies in some cases)

with the same agent later. Other, but less directly life-limiting, benefits of functional immunological memory include improved fitness and herd immunity (see later discussion) [43, 47].

B.   Immunological Memory by Neutralizing Antibodies

How can protection be provided to an immuno-incompetent host during this particularly critical period? Passively acquired antibody-mediated protection is absolutely required pre- and postnatally. Therefore, transferable immunity is probably an essential precondition in vertebrates (fishes, birds, and mammals) for maturation of the immune system.

Co-evolution of infectious agents and MHC polymorphism has prevented easy selection of highly cytopathic mutants capable of evading MHC-restricted T-cell recognition. On the other hand, MHC polymorphism has endangered immunological maternal–fetal relationships during ontogeny. The danger of graft versus host or host versus graft reactions between mother and offspring is reduced by the lack of MHC antigen expression in the placental contact areas, by general immunosuppression of the mother, and by virtually complete immunodeficiency of the offspring until birth [10, 11, 68, 69]. Protective antibodies in the serum of the mother are passively transmissible, soluble forms of immunological experience. They protect the offspring for as long as it needs to develop its own T-cell competence and generate its own T-helper-cell-dependent protective and long-lived neutralizing IgG antibody responses. Here, we argue that the preexistence or co-evolution of transmissible antibodies has offered a basis to develop MHC polymorphism and MHC-restricted, T-cell-mediated immunity. This would signify that the development of cytopathic agents that could not be controlled efficiently by adoptively transferred antibodies during this critical period of immuno-incompetence would not have been possible because such infections would have endangered the survival of the species. Adoptively transferable immunological experience by antibodies from the mother to the immuno-incompetent offspring can be seen in agammaglobulinemic patients or mice and newborn calves. Infants incapable of generating their own immunoglobulins will be protected by maternal antibodies for the first 3–9 months after birth [23]. Serum IgG antibody is transferred via placenta (but not also via milk as in mice) to the serum in humans. Importantly, human milk antibodies are active within the gut and influence the gut flora, at least before weaning. Interestingly, in contrast to humans and mice, calves are born without serum immunoglobulins (this is why we use fetal calf serum for lab work), because maternal immunoglobulin cannot be transported through the completely double-layered placenta. During the first 18 hr after birth, they take up colostral maternal immunoglobulins via the gut [10]. Gut epithelia transport immunoglobulins to the blood only during this time period. If calves are not given colostral milk early enough, they remain without maternal protective antibodies and die of various infections during the next few weeks. Their own still immature immune system cannot act efficiently enough to mount protective immune responses.

The next important question is how antibody levels in human serum and milk are kept high enough to provide protection for the offspring [64, 65, 68, 69]. One way would be to induce antibody levels that cover all relevant infectious diseases during the 270 days of a human pregnancy or the 20 days of a mouse pregnancy. Such infections would, however, threaten the survival of the embryo and of the newborn. In fact, cytopathic infections during pregnancy must be avoided as they often cause abortion or developmental abnormalities [54]. Although infection during pregnancy is rare, high levels of immunity in most members of a species (i.e., herd immunity) usually prevent this problem. Overall, herd immunity depends on the infectious agent (acute or persistent), level of immunity (neutralizing antibody titers or/and activated T cells), population density and migration, and animal reservoirs. Thus, immunological memory at the individual level depends on and is in equilibrium with herd immunity at the population level.

All life-threatening acute infections must be survived by mothers before puberty, and classical childhood diseases represent the co-evolutionarily balanced infectious disease experience before procreation commences. Thus, immunological memory represents accumulated immunological experience and protection before pregnancy, including herd immunity [43, 47, 68, 69]. In contrast to Hollywood films, one can only die once from infections in real life, and that—at least in developing countries—is mostly through infections during the early period after birth; in western, vaccinated populations, it is mostly in old age when immune surveillance wanes. Both males and females have a comparable immune system, but immunological memory transmissible from mother to offspring is also regulated by important hormonal influences; they improve overall antibody responsiveness in females compared to males. The evolutionary cost is, as a consequence, that the ratio of auto-antibody-dependent autoimmune disease is 5 times more frequent in females than in males [68, 69].

V. IMMUNITY BY T CELLS

Immunological experience transferred via antibodies from mother to offspring is crucial for survival. What about T-cell memory? Neutralizing antibody responses against related but serotypically distinct viruses are limited, not by primed T helper cells or CTLs but by the precursor frequency of the specific B cells (Tables 2 and 3). Memory T cells cannot be transmitted from mother to offspring because of mutual immunological rejection. Therefore, why should long-term, cell-mediated memory be needed? Two aspects must be discussed here: (1) the role of specific T-cell-mediated protective immunity and (2) the important role of immunity that depends on ongoing low-level infections, which includes the so-called specific infection immunity and less specific concomitant immunity [33, 38, 67]. Immunological memory cannot be sustained by IgM antibody because of its very short half-life of only 1–2 days. In addition, because of the lack of receptors and its large molecular size, IgM cannot be transmitted to offspring via placenta or milk. The switch from IgM to IgG requires primed T helper cells, and IgG has a half-life of about 3 weeks. Additionally, IgG is more diffusible and transportable via various Fc receptors, and this includes transport to the offspring. As stated earlier, adoptively transferable maternal antibody memory is key to species survival, and therefore it is not surprising that all vaccines that provide efficient protection working today do so via neutralizing antibodies (Table 5).

VI. IS PROTECTIVE MEMORY ANTIGEN-DEPENDENT OR -INDEPENDENT?

B cells cannot mature to become antibody-producing plasma cells in the absence of antigen [25, 51]. To receive signals from specific T helper cells, B cells process antigen bound to surface Ig receptors in order to present the relevant peptides on MHC class II on their surface. This process is necessary for B cells' maturation to plasma cells, but it is not sufficient to prime naive T cells. T helper cells are only induced efficiently in secondary lymphatic organs by antigen-presenting cells (APCs), including dendritic cells (DCs) offering helper peptides via MHC class II. After priming, increased precursor frequencies of specific T and B cells are readily demonstrated in humans or mice [1, 58, 67], but primed T and B cells without specific antigen are *not* protective by themselves, as shown in adoptive transfer experiments [50, 59] (Table 3). Protection requires preexistent high neutralizing antibody titers that are antigen-dependent by B cells that mature to plasma cells. Some experiments have suggested that perhaps plasma cells may have a very long half-life, up to 150–300 days [42, 58]. However, this evidence unfortunately is flawed because antibody responses against nonprotective antigens composed of multiple undefined determinants have been evaluated instead of neutralizing antibody responses, which are specific for the tip of viral glycoproteins. The results show that increased antibody titers depend upon antigen-driven plasma cell maturation and responses. In fact, protective antibody titers usually decrease over time (e.g., against diptheria, tetanus toxins, or measles vaccines [27, 49]). Although ELISA binding measures binding qualities of $\leq 10^7$ M^{-1}, protection against reinfection and *in vitro* neutralization require affinities of $\geq 10^8$ M^{-1}. All of these observations strongly suggest that the maintenance of protective neutralizing or opsonizing antibody titers is antigen-dependent (Tables 3 and 4).

How does cell-mediated immune protection and protective T-cell-mediated memory work, how is it maintained, and what is its role? Many experiments in mice have demonstrated that adoptively transferred CD8[+] T cells protect against acute infections with noncytopathic viruses or tumors. Under special conditions, such protection experiments have also been successfully performed with cytopathic viruses, such as with influenza virus in mice [20]. However, as stated earlier, it is a fact that neither humans nor mice are efficiently protected against distinct serotypes of viruses, despite primed memory CD8[+] and CD4[+] T-cell specificities. This strongly indicates that such T-cell responses cannot efficiently protect across distinct serotypes. Nevertheless, if T cells are acutely activated, they can exhibit a protective phenotype during the period of activation, but usually this period only lasts for about 3 weeks [2, 26]. It is therefore necessary to emphasize that primed memory T cells are only reactivated to become protective effector T cells by antigen in lymphatic organs. Therefore, any experimental protocol that brings great amounts of antigen into spleen and lymph nodes (such as by infection of mice iv or ip with 10^6 PFU of nonlytic virus; a highly cytopathic virus would kill the host) reactivates many effector T cells within 8–16 hr. This reactivated T-cell response is then rapidly protective. In contrast, the same virus infecting the same primed host initially with a physiologically low dose strictly extralymphatically (e.g., skin, mucosa, or olfactory nerve) will require either pre-existing primed T cells to quickly eliminate newly infected peripheral cells or antibodies to prevent mucosal infection or systemic spread (Table 4). Strictly peripheral extralymphatic reinfections are the physiological route of a natural challenge infection and reveal the relatively slow activation kinetics of primed T cells more adequately. Experience with many infectious diseases, including tuberculosis, leprosy, and perhaps a few seronegative HIV–AIDS-resistant patients, demonstrates that T-cell memory provides efficient protection against "reinfection" from within and without. As experienced, this protection is clearly antigen-driven and is relevant for the individual host, but only indirectly important for the offspring via herd immunity. Infections controlled crucially by T cells are largely non- or low-cytopathic or variably so (e.g., herpes viruses) [15, 31, 66], are often slow in kinetics, and have a tendency to persist. These infections will not kill the host rapidly, but rather tend to establish a balanced state of infection–immunity as defined by Mackaness. This term describes the coexistence of low numbers of infectious agents usually in granulomas together with an active immune antibody and T-cell response (for salmonellosis, TB, or leprosy [38]). Similar conditions are offered by low-level infections of peripheral, nonlymphatic cells or organs, e.g., neurons for herpes virus, kidney cells or lung epithelial cells for cytomegalovirus (CMV), and perhaps β-islet cells or heart myocytes for coxsackievirus [14, 30]. The few infectious foci are well-checked by an active, ongoing immune response in the host. This response against TB or leprosy is seemingly exclusively T-cell-mediated, but against salmonella, coxsackie-, or herpes viruses, arena viruses, and probably rubella, it also includes neutralizing antibodies (reviewed in refs 18, 31, 46).

Many non- or poorly cytopathic agents, such as HBV, HCV, or HIV, are transmitted before (transplacentally, such as with LCMV in mice) or mainly during birth via maternal blood. Alternatively, they are transmitted orally early in life via peripheral mucosal or epithelial infections (herpes, CMV). Because the offspring are immuno-incompetent and because maternal immune defenses against these agents obviously have failed (or are well balanced for herpes viruses), viruses are best transmitted during this period of immuno-incompetence without endangering survival of the offspring and, therefore, also not of the host species. Some of the persistent noncytopathic infections may eventually cause serious disease late in the host, such as primary liver cell carcinomas 40 years after HBV infections or perhaps some chronic autoimmune diseases or chronic immunopathologies (e.g., AIDS) by others. These consequences of chronic persistent infections usually show up much later than necessary for the species to procreate and survive [68, 69]. Although this may not yet be true for HIV_1, probably because mutual adaptation with humans is still insufficient, it may already apply to SIV in certain monkey species and HIV-2 in humans.

VII. HOW IS PROTECTIVE MEMORY MAINTAINED BY INFECTIOUS AGENTS OR VACCINES?

As pointed out, sufficiently high neutralizing protective antibody titers, primarily in mothers but for herd immunity also in males, are essential for survival of the offspring and of the species. Maintenance of high neutralizing antibody titers may be achieved via the following: (1) reexposure to the antigen from external sources, a route typically used by poliovirus (Tables 1, 3, 4, and 5) [62]. Usually, spread of the Sabin vaccine strains within households, schools, or via public swimming pools keeps immunity boosted. (2) Reexposure from antigen sources within the host. This mechanism is essential for understanding immunity against TB, leprosy, HBV, HCV, HIV, many parasites, and also against measles virus. Measles virus persists in the host not in a replication-competent form, but as a crippled virus apparently often missing a functional matrix protein [7]. From this point of view, SSPE represents an extreme form of persistence of measles virus in central nervous tissue. In fact, PCR testing using one single probe revealed a positive signal in 25% of autopsies of patients >60 years of age [32] that had been exposed to wild-type measles virus during childhood. Similarly, HBV virus [55] in men or LCMV virus in mice [16] persists at very low levels and boosts the immune responses of T and B cells repeatedly [13, 17, 48, 60]. (3) Antibody–antigen complexes on follicular dendritic cells are maintained for long periods of time and boost both antigen-specific B cells directly as well as T helper cells indirectly. Because cross-priming and cross-processing [12] of inert antigens can only exceptionally access the MHC class I pathway even in dendritic cells, these antigen depots in general are neither capable of maintaining activated CD8 T cells nor, as a beneficial consequence, are they reduced or eliminated by CTLs. In the absence of antigen boosts, antibody responses will eventually dwindle.

VIII. CONCLUSION

T- and B-cell responses are largely regulated by structure, dose, localization, and duration of the availability of antigen. Without antigen reaching secondary lymphoid organs, there is no immune response and also no long-lasting immunity. Thus, immunological memory provides immunity (i.e., survival advantage) via low-level antigen-driven immune responses. Preexistent high titers of neutralizing antibodies represent immunological experience and transmissible protection to initially immuno-incompetent offspring. Activated T cells protect the host together with antibodies against T-cell-mediated immunopathology potentially triggered by persisting poorly or noncytopathic infections. Because antigen drives high antibody titers and activation of T cells, any long-term protecting vaccine must conform to these rules.

REFERENCES

1. Ahmed, R., and Gray, D. (1996). Immunological memory and protective immunity: Understanding their relation. *Science* **272**:54–60.
2. Albert, M. R., Ostheimer, K. G., and Breman, J. G. (2001). The last smallpox epidemic in Boston and the vaccination controversy, 1901–1903. *New England J. Medicine* **344**:375–379.
3. Bachmann, M. F., Kalinke, U., Althage, A., Freer, G., Burkhart, C., Roost, H., Aguet, M., Hengartner, H., and Zinkernagel, R. M. (1997). The role of antibody concentration and avidity in antiviral protection. *Science* **276**:2024–2027.
4. Bachmann, M. F., Kündig, T. M., Hengartner, H., and Zinkernagel, R. M. (1997). Protection against immunopathological consequences of a viral infection by activated but not resting cytotoxic T cells: T cell memory without "memory T cells?" *Proc. Natl. Acad. Sci. U.S.A.* **94**:640–645.
5. Bachmann, M. F. and Zinkernagel, R. M. (1997). Neutralizing antiviral B cell responses. *Annu. Rev. Immunol.* **15**:235–270.
6. Beverley, P. C. (1990). Human T cell memory. *Curr. Top. Microbiol. Immunol.* **159**:111–122.
7. Billeter, M. A., Cattaneo, R., Spielhofer, P., Kaelin, K., Huber, M., Schmid, A., Baczko, K., and M.ter, V. (1994). Generation and properties of measles virus mutations typically associated with subacute sclerosing panencephalitis. *Ann. N.Y. Acad. Sci.* **724**:367–377.
8. Blanden, R. V., Mackaness, G. B., and Collins, F. M. (1966). Mechanisms of acquired resistance in mouse typhoid. *J. Exp. Med.* **124**:585–600.
9. Bloom, B. and Ahmed, R. (1998). Immunity to infection. *Curr. Opin. Immunol.* **10**:419–421.
10. Brambell, R. W. R. (1970). "The Transmission of Immunity from Mother to Young." Amsterdam: North Holland Publishing Corp.
11. Brent, L. (1996) "A History of Transplantation Immunology." New York: Academic Press.
12. Carbone, F. R., Kurts, C., Bennett, S. R., Miller, J. F., and Heath, W. R. (1998). Cross-presentation: A general mechanism for CTL immunity and tolerance. *Immunol. Today* **19**:368–373.
13. Carman, W. F., Zanetti, A. R., Karayiannis, P., Waters, J., Manzillo, G., Tanzi, E., Zuckerman, A. J., and Thomas, H. C. (1990). Vaccine-induced escape mutant of hepatitis B virus. *Lancet* **336**:325–329.
14. Cherry, J. D. (1986). "Coxsackie Viruses B," pp. 1322–1369. Philadelphia: W.B. Saunders.
15. Chisari, F. V. and Ferrari, C. (1995). Hepatitis B virus immunopathogenesis. *Annu. Rev. Immunol.* **13**:29–60.
16. Ciurea, A., Klenerman, P., Hunziker, L., Horvath, E., Odermatt, B., Ochsenbein, A. F., Hengartner, H., and Zinkernagel, R. M. (1999). Persistence of lymphocytic choriomeningitis virus at very low levels in immune mice. *Proc. Natl. Acad. Sci. U.S.A.* **96**:11964–11969.
17. Ciurea, A., Klenerman, P., Hunziker, L., Horvath, E., Senn, B. M., Ochsenbein, A. F., Hengartner, H., and Zinkernagel, R. M. (2000). Viral persistence in vivo through selection of neutralizing antibody-escape variants. *Proc. Natl. Acad. Sci. U.S.A* **97**:2749–2754.
18. Collins, F. M., Mackaness, G. B., and Blanden, R. V. (1966). Infection–immunity in experimental salmonellosis. *J. Exp. Med.* **124**:601–619.
19. Dimmock, N. J. (1993). Neutralization of animal viruses. *Curr. Top. Microbiol. Immunol.* **183**:1–149.

20. Doherty, P. C., Topham, D. J., Tripp, R. A., Cardin, R. D., Brooks, J. W., and Stevenson, P. G. (1997). Effector CD4+ and CD8+ T-cell mechanisms in the control of respiratory virus infections. *Immunol. Rev.* **159**:105–117.

21. Fearon, D. T. (2000). Innate immunity—beginning to fulfill its promise? *Nat. Immunol.* **1**:102–103.

22. Fenner, F. (1949). Mousepox (infectious ectromelia of mice): A review. *J. Immunol.* **63**:341–373.

23. Good, R. A., and Zak, S. J. (1956). Disturbances in gamma globulin synthesis as experiments of nature. *Pediatrics* **18**:109–149.

24. Gray, D. (1993). Immunological memory. *Annu. Rev. Immunol.* **11**:49–77.

25. Gray, D., and Skarvall, H. (1988). B cell memory is short-lived in the absence of antigen. *Nature* **336**:70–73.

26. Gupta, S. C., Hengartner, H., and Zinkernagel, R. M. (1986). Primary antibody responses to a well-defined and unique hapten are not enhanced by preimmunization with carrier: Analysis in a viral model. *Proc. Natl. Acad. Sci. U.S.A.* **83**:2604–2608.

27. Guris, D., McCready, J., Watson, J. C., Atkinson, W. L., Heath, J. L., Bellini, W. J., and Polloi, A. (1996). Measles vaccine effectiveness and duration of vaccine-induced immunity in the absence of boosting from exposure to measles virus. *Pediatr. Infect. Dis. J.* **15**:1082–1086.

28. Holmgren, J., Brantzaeg, P., Capron, A., Francotte, M., Kilian, M., Kraehenbuhl, J. P., Lehner, T., and Seljelid, R. (1996). European Commission COST/STD Initiative. Report of the expert panel VI. Concerted efforts in the field of mucosal immunology. *Vaccine* **14**:644–664.

29. Kagi, D., Ledermann, B., Bürki, K., Zinkernagel, R. M., and Hengartner, H. (1996). Molecular mechanisms of lymphocyte-mediated cytotoxicity and their role in immunological protection and pathogenesis in vivo. *Annu. Rev. Immunol.* **14**:207–232.

30. Kaplan, M. H., Klein, S. W., McPhee, J., and Harper, R. G. (1983). Group B coxsackievirus infections in infants younger than three months of age: A serious childhood illness. *Rev. Infect. Dis.* **5**:1019–1032.

31. Kapoor, A. K., Nash, A. A., and Wildy, P. (1982). Pathogenesis of herpes simplex virus in B cell-suppressed mice: The relative roles of cell-mediated and humoral immunity. *J. Gen. Virol.* **61**(Pt. l):127–131.

32. Katayama, Y., Hotta, H., Nishimura, A., Tatsuno, Y., and Homma, M. (1995). Detection of measles virus nucleoprotein mRNA in autopsied brain tissues. *J. Gen. Virol.* **76**:3201–3204.

33. Kaufmann, S. H. E. (1993). Immunity to intracellular bacteria. *Annu. Rev. Immunol.* **11**:129–163.

34. Kundig, T. M., Bachmann, M. F., DiPaolo, C., Simard, J. J., Battegay, M., Lother, H., Gessner, A., Kuhlcke, K., Ohashi, P. S., and Hengartner, H. (1995). Fibroblasts as efficient antigen-presenting cells in lymphoid organs. *Science* **268**:1343–1347.

35. Kündig, T. M., Althage, A., Hengartner, H., and Zinkernagel, R. M. (1992). Skin test to assess virus-specific cytotoxic T-cell activity. *Proc. Natl. Acad. Sci. U.S.A.* **89**:7757–7761.

36. Liang, S., Mozdzanowska, K., Palladino, G., and Gerhard, W. (1994). Heterosubtypic immunity to influenza type A virus in mice. Effector mechanisms and their longevity. *J. Immunol.* **152**:1653–1661.

37. Mackaness, G. B. (1964). The immunological basis of aquired cellular resistance. *J. Exp. Med.* **120**:105–120.

38. Mackaness, G. B. (1969). The influence of immunologically committed lymphoid cells on macrophage activity in vivo. *J. Exp. Med.* **129**:973–992.

39. Mackay, C. R. (1993). Immunological memory. *Adv. Immunol.* **53**:217–265.

40. Mackay, I. R. (1983). Immunological aspects of chronic active hepatitis. *Hepatology* **3**:724–728.

41. Macpherson, A. J., Gatto, D., Sainsbury, E., Harriman, G. R., Hengartner, H., and Zinkernagel, R. M. (2000). A primitive T cell-independent mechanism of intestinal mucosal IgA responses to commensal bacteria [In Process Citation]. *Science* **288**:2222–2226.

42. Manz, R. A., Lohning, M., Cassese, G., Thiel, A., and Radbruch A. (1998). Survival of long-lived plasma cells is independent of antigen. *Int. Immunol.* **10**:1703–1711.

43. Mims, C. A. (1987). "Pathogenesis of Infectious Disease." London: Academic Press.

44. Moskophidis, D., Laine, E., and Zinkernagel, R. M. (1993). Peripheral clonal deletion of antiviral memory CD8+ T cells. *Eur. J. Immunol.* **23**:3306–3311.

45. Müller, U., Steinhoff, U., Reis, L. F., Hemmi, S., Pavlovic, J., Zinkernagel, R. M., and Aguet, M. (1994). Functional role of type I and type II interferons in antiviral defense. *Science* **264**:1918–1921.

46. Narayan, K. M. and Moffat, M. A. (1992). Measles, mumps, rubella antibody surveillance: Pilot study in Grampian, Scotland. *Health Bull.(Edinburgh)* **50**:47–53.

47. Nathanson, N. (1990). Epidemiology. *In* "Virology," B. N. Fields and D. M. Knipe, Eds., pp. 267–291. New York: Raven Press.

48. Nossal, G. J. V., Austin, C. M., and Ada, G. L. (1965). Antigens in immunity: VII. Analysis of immunological memory. *Immunology* **9**:333.

49. Ochsenbein, A. F., Karrer, U., Klenerman, P., Althage, A., Ciurea, A., Shen, H., Miller, J. F., Whitton, J. L., Hengartner, H., and Zinkernagel, R. M. (1999). A comparison of T cell memory against the same antigen induced by virus versus intracellular bacteria. *Proc. Natl. Acad. Sci. U.S.A.* **96**:9293–9298.

50. Ochsenbein, A. F., Pinschewer, D. D., Odermatt, B., Ciurea, A., Hengartner, H., and Zinkernagel, R. M. (2000). Correlation of T cell independence of antibody responses with antigen dose reaching secondary lymphoid organs: Implications for splenectomized patients and vaccine design. *J. Immunol.* **164**:6296–6302.

51. Ochsenbein, A. F., Pinschewer, D. D., Sierro, S., Horvath, E., Hengartner, H., and Zinkernagel, R. M. (2000). Protective long-term antibody memory by antigen-driven and T help-dependent differentiation of long-lived memory B cells to short-lived plasma cells independent of secondary lymphoid organs. *Proc. Natl. Acad. Sci. U.S.A* **97**:13263–13268.

52. Peltola, H. (2000). Worldwide *Haemophilus influenzae* type b disease at the beginning of the 21st century: Global analysis of the disease burden 25 years after the use of the polysaccharide vaccine and a decade after the advent of conjugates. *Clin. Microbiol. Rev.* **13**:302–317.

53. Rajewski, K. (1989). Evolutionary and somatic immunological memory. *In* "Progress in Immunology VII," F. Melchers, *et al.*, Eds., pp. 397–403. Berlin: Springer-Verlag.

54. Rawls, W. E. (1974). Viral persistence in congenital rubella. *Prog. Med. Virol.* **18**:273–288.

55. Rehermann, B., Ferrari, C., Pasquinelli, C., and Chisari, F. V. (1996). The hepatitis B virus persists for decades after patients' recovery from acute viral hepatitis despite active maintenance of a cytotoxic T-lymphocyte response. *Nat. Med.* **2**:1–6.

56. Roost, H. -P., Bachmann, M. F., Haag, A., Kalinke, U., Pliska, V., Hengartner, H., and Zinkernagel, R. M. (1995). Early high-affinity neutralizing anti-viral IgG responses without further overall improvements of affinity [see comments]. *Proc. Natl. Acad. Sci. U.S.A.* **92**:1257–1261.

57. Sabin, A. B. (1985). Oral poliovirus vaccine: History of its development and use and current challenge to eliminate poliomyelitis from the world. *J. Infect. Dis.* **151**:420–436.

58. Slifka, M. K. and Ahmed, R. (1998). Long-lived plasma cells: A mechanism for maintaining persistent antibody production [In Process Citation]. *Curr. Opin. Immunol.* **10**:252–258.

59. Steinhoff, U., Müller, U., Schertler, A., Hengartner, H., Aguet, M., and Zinkernagel, R. M. (1995). Antiviral protection by vesicular stomatitis virus-specific antibodies in alpha/beta interferon receptor-deficient mice. *J. Virol.* **69**:2153–2158.

60. Tew, J. G., Kosco, M. H., Burton, G. F., and Szakal, A. K. (1990). Follicular dendritic cells as accessory cells. *Immunol. Rev.* **117**:185–211.

61. Walker, B. D., Chakrabarti, S., Moss, B., Paradis, T. J., Flynn, T., Durno, A. G., Blumberg, R. S., Kaplan, J. C., Hirsch, M. S., and Schooley, R. T. (1987). HIV-specific cytotoxic T lymphocytes in seropositive individuals. *Nature* **328**(6128):345–348.

62. Watanabe, M. (2001). Polio outbreak threatens eradication program. *Nat. Med.* **7**:135.

63. Weigle, W. O. (1973). Immunological unresponsiveness. *Adv. Immunol.* **16**:61–122.

64. Zinkernagel, R. M. (1996). Immunology taught by viruses. *Science* **271**:173–178.

65. Zinkernagel, R. M. (2000). On immunological memory. *Phil. Trans. R. Soc. London* **355**:369–371.

66. Zinkernagel, R. M. (2000) What is missing in immunology to understand immunity? *Nature Immunol.* **1**(3):181–185.

67. Zinkernagel, R. M., Bachmann, M. F., Kündig, T. M., Oehen, S., Pircher, H. P., and Hengartner, H. (1996). On immunological memory. *Annu. Rev. Immunol.* **14**:333–367.

68. Zinkernagel, R. M. (2002). On natural and man-made vaccines. *Annu. Rev. Immunol.* In press.

69. Zinkernagel, R. M. (2001). Maternal antibodies, childhood infections and autoimmune diseases. *N. Engl. J. Med.* **345**:1331–1335.

5 UNDERSTANDING MICROBIAL PATHOGENESIS AS A BASIS FOR VACCINE DESIGN

PART D. Parasitic Diseases, with an Emphasis on Experimental Cutaneous Leishmaniasis

PASCAL LAUNOIS,* HEIKE VOIGT,* ALAIN GUMY,* ABRAHAM ASEFFA,*
FABIENNE TACCHINI-COTTIER,* MARTIN RÖCKEN,† AND JACQUES A.
LOUIS*

*World Health Organization Immunology Research and Training Center, Institute of
Biochemistry, University of Lausanne, 1066 Epalinges, Switzerland

†Department of Dermatology and Allergy, Ludwig-Maximilians University, 80337 Munich,
Germany

I. INTRODUCTION

Parasites have to survive in their vertebrate host during a sufficiently prolonged period of time to achieve their life cycle through successful transmission via insect vectors. In their vertebrate hosts, parasites are often confronted
by vigorous effector immune responses that they have to subvert somehow to
be able to outlast and be successfully transmitted.

The immune response comprises several components in terms of effector
cells, antibodies, and signaling molecules such as cytokines. It is now well-recognized that not all components of the immune response triggered during the
process of parasitism have antiparasite effector functions. In this context, one
strategy devised by *Leishmania* has been to utilize some components of the host
immune response to its own benefit. Therefore, an effective anti-leishmania
vaccine should only induce an effector mechanism(s) ultimately leading to
destruction of the parasites and abstain from triggering a component(s) of the
immune response favoring parasite survival. Parasites have also devised a variety of ways by which to escape the effector components of the immune
response that they have elicited in their mammalian hosts. Therefore, the balance between the components of the immune response with effector functions
and those irrelevant for protection on the one hand and the effector immune
response and the escape processes induced by parasites on the other largely
determines the outcome of infection. Thus, elucidation of the mechanisms
involved in the progression of parasitic diseases and definition of the relevant
immune effector mechanisms are equally important for the identification of
potential targets for beneficial interventions. Indeed, the specific triggering of

potent effector responses without activating the immune responses' components leading to pathology together with perverting the escape mechanisms evolved by parasites are the aims of vaccination.

The importance of cytokines, as antiparasite effector molecules and as mediators of pathology, has been reinforced by the description of two functionally different CD4[+] T-cell populations that can be distinguished on the basis of the pattern of cytokines they produce. In both mice and humans, T helper 1 (Th1) cells produced interleukin-2 (IL-2) and interferon-γ (IFN-γ), whereas Th2 cells produced IL-4, IL-5, and IL-13 [1, 2]. Th1 cells are responsible for cell-mediated immune reactions, whereas Th2 cells are involved in humoral immunity [3]. Resistance and susceptibility to infection with several pathogens have been analyzed with respect to the Th1–Th2 paradigm. Resistance to intracellular pathogens such as *Leishmania major* is associated with the development of a polarized Th1 response [4, 5]. Although resistance to extracellular parasite is now known to correlate with the generation of Th2 cells, the respective roles of polarized Th1 or Th2 effector responses in protective immunity against *Schistosoma*, an extracellular pathogen, are not yet fully understood [6]. Furthermore, expression of immunity in some parasitic diseases, such as malaria, cannot simply be analyzed in view of the development of polarized Th1 or Th2 responses because infection with this parasite is a multifocal process [7].

In this part of Chapter 5, we discuss, with the prospect of vaccine development, some components of the immune response triggered during infection with three parasites (with particular emphasis on the murine model of infection with *L. major*) and some aspects of the mechanisms devised by parasites to escape the host's effector immune response.

II. SOME ASPECTS OF IMMUNITY AGAINST MALARIA

Malaria is due to infection with *Plasmodium* species and remains a major infectious disease in tropical and subtropical regions, inducing a high burden of morbidity and mortality. In their mammalian hosts, malaria parasites ultimately develop in the erythrocytes, where membrane insertion of parasite-derived proteins takes place. Infected erythrocytes thus are one of the key targets of immune effector responses triggered during infection with *Plasmodium* [8]. However, a primary survival mechanism of the malaria parasite lies in its capacity to change the composition of its antigens expressed on the surface of infected erythrocytes [9].

The pathology associated with *Plasmodium falciparum* infection is due in part to the adherence of infected red blood cells to endothelial cells, inducing dysfunction of many organs including the brain (7). Analysis of the mechanisms underlying adherence of infected erythrocytes to blood vessels has shown that the membrane-inserted parasite protein PfEMP1 (*P. falciparum* erythrocyte membrane protein 1) is one of the important factors [9, 10]. More than 10 host receptor molecules for the binding of erythrocytes infected with *P. falciparum* have been identified, and each parasite clone has the ability to bind to only a subset of them [9]. The fact that malaria parasites bind to diverse receptors might be a strategy used by the parasites to survive in different human hosts. Indeed, because spleen is able to eliminate mature erythrocytes, it was postulated that adhesion of infected red blood cells to endothelium protects parasites from splenic clearance, maintaining them in the vasculature [11]. Potential anti-adherence vaccines should then be designed to reverse these interactions between the parasite and the host *in vivo*. In this context, a pregnancy malaria vaccine might be rapidly developed because, in the placenta, infected erythrocyte parasites bind to a specific receptor, chondroitin sulfate A (CSA) [12]. Indeed, anti-adhesion antibodies against CSA-binding parasites correlate with human resistance to placental malaria [13].

Large families of variant antigens (e.g., the *var* and *rif* families, Stevor and Clag) give the parasite the ability to escape immune responses, thus ensuring their transmission. PfEMP1 has been implicated not only in adhesion of infected erythrocytes to endothelial cells but also in antigenic variation [9]. Therefore, an effective anti-malaria vaccine targeting infected erythrocytes theoretically should elicit antibodies that are reactive with most variant antigens expressed at the surface of infected erythrocytes and thus, most isolates of *P. falciparum*. In this context, the use of T-cell epitopes conserved among most variant antigens might prime the host for the induction of effective variant-specific antibody responses following infection [10].

Although inflammatory cytokines are important defense mechanisms against several microorganisms including *Plasmodium*, overproduction of cytokines such as tumor necrosis factor-α (TNF-α) and IFN-γ can have severe pathological consequences. During malaria, the contribution of TNF-α and IFN-γ, produced in large amounts, to the severity of disease has been documented [14]. As TNF-α and IFN-γ have been shown to up-regulate the expression of some receptors on endothelial cells, these cytokines might play a role in the adhesion of infected erythrocytes to endothelial cells, resulting in the sequestration of infected erythrocytes that has been documented during severe malaria. Thus, identification of the antigens of *Plasmodium* that trigger an overproduction of these inflammatory cytokines should be instrumental for the design of measures preventing the development of severe disease.

III. SOME ASPECTS OF IMMUNITY TO *SCHISTOSOMA MANSONI*

Mechanisms of anti-*Schistosoma* immunity have been analyzed extensively within the framework of polarized Th1 and Th2 responses in various animal models and humans. Strikingly, depending upon the model under study, distinct Th cell subsets were found to be essential for the manifestations of protective immunity. The majority of results obtained in human and rat studies have clearly established the crucial role of Th2 responses in protective immunity to schistosomes [6]. Indeed, in humans, acquisition of immunity to schistosome correlates, in several studies, with the production of parasite-specific and nonspecific immunoglobolin E (IgE) [6]. In both humans and rats, it has been shown that the killing of *Schistosoma* larvae is mediated by an antibody-dependent, cell-mediated cytotoxicity (ADCC) mechanism, which involves IgE, eosinophils, and platelets [15–17]. Given the IL-4 dependence of IgE production on the one hand and the role of IL-5 in the production of eosinophils on the other, it appears logical that the generation of Th2 responses should be the aim of vaccination against *Schistosoma*. In sharp contrast, several studies in mice have shown that Th1 responses are required for the induction of protective immunity to *Schistosoma* (18). By studying the mechanism of resistance to infection following immunization with irradiated cercariae, it was clearly shown that the induction of CD4+ T-cell response with a Th1 functional phenotype was essential [19]. The lack of expression of functional receptors for IgE on mouse eosinophils has provided a rational basis for the absence of IgE-dependent effector mechanisms in mice [20].

Confronted with these opposing results, choice of the nature of the polarized Th response to be triggered by efficient immunoprophylactic measures is rather difficult. However, findings have revealed that, after two immunizing doses of irradiated cercariae, mice develop mixed Th1–Th2 responses that contribute to protection against infectious challenge [21]. Furthermore, results obtained in double-deficient mice for cytokines (either IL-4–IL-10 or IL-10–IL-12), which mounted polarized Th1 and Th2 responses, respectively, and were equally protected by vaccination, suggest that, although distinct antiparasite effector mechanisms operate in these mice, each response is able to induce protective immunity [22]. Thus, as proposed, protective immunization against schistosomes might best aim at inducing both Th1 and Th2 responses [22]. Such a task might be difficult to achieve given the mutually cross-regulatory activities of Th1 and Th2 responses. In this context, more information on the various parameters that influence polarization of Th cells *in vivo* is needed, underlying the practical importance of basic studies on this issue.

IV. IMMUNITY TO *LEISHMANIA*

Distinct features of the spectrum of clinical manifestations observed in humans infected with various *Leishmania* species have been successfully reproduced in inbred mice of different genetic backgrounds following infection with a single strain of *Leishmania*, i.e., *Leishmania major*. This murine model of infection has thus been used extensively in attempts to correlate component(s) of the immune response associated with either healing of lesions or severe pathological manifestations. Admittedly focusing on results from our laboratory, we will summarize the results pertaining to the dissection of (1) the immune effector mechanisms restricting the growth of leishmania, (2) the immune mechanisms leading to progressive disease, and (3) the mechanisms mediating the induction of these disparate responses with opposite effects on disease progression.

A. Murine Model of Infection with *L. major*

After infection with *L. major*, mice from the majority of inbred strains (i.e., C57BL/6) develop lesions that resolve spontaneously, leaving the animals immune to reinfection. These mice are referred to as genetically resistant to *L. major*. In contrast, mice from a few inbred strains (i.e., BALB), referred to as genetically susceptible, develop progressive lesions that do not heal and they do not become immune to reinfection. This genetically determined resistance and susceptibility to *L. major* in mice has been demonstrated to result from the development of CD4+ T-cell responses with distinct functional characteristics. The first evidence for the development, *in vivo*, of polarized CD4+ Th1 and Th2 responses has been obtained precisely in this murine model of infection. Thus, genetically determined resistance and susceptibility to infection with *L. major* were clearly related to the development of polarized Th1 and Th2 responses, respectively [4, 5]. The dominant role of IFN-γ produced by CD4+ Th1 cells in the control of infection with *L. major* has been well-documented [23, 24]. In the mouse, IFN-γ renders macrophages, the host cells for *Leishmania*, microbicidal through the synthesis of the inducible nitric oxide synthase (iNOS), leading to the production of toxic reactive nitrogen radicals [25]. In addition, apoptotic death of macrophages induced by CD4+ Th1 cells through the Fas–Fas L pathway also appears to be involved in resistance, but to a lesser degree than IFN-γ [26, 27]. Susceptibility to *L. major* is also governed by Th2-derived cytokines with macrophage-deactivating properties. IL-4, IL-10, and IL-13 have been shown to interfere with the induction of iNOS that is triggered by IFN-γ [3, 28, 29].

B. Mechanisms underlying Genetically Determined Differences in Th Cell Differentiation following Infection with *L. major*

Differentiated Th1 and Th2 cells arise from a common CD4+ T-cell precursor [30]. Several parameters have been reported to be able to influence the pathway of maturation of CD4+ T-cell precursors [3]. Among the several stimuli influencing the development of distinct CD4+ Th effector cells, cytokines themselves critically regulate this process [31]. By using CD4+ T cells transgenic for a unique T cell receptor (TCR) α–β receptor, it was demonstrated that IL-12 and IL-4 are important factors for Th1 and Th2 cell maturation, respectively [32–38].

1. Th1 Cell Development in Resistant Mice following Infection with *L. major*

The crucial role of IL-12 in Th1 cell maturation in resistant mice following infection with *L. major* has been demonstrated using either anti-IL-12 neutralizing antibodies or mice with disruption of the IL-12 gene [39, 40]. Furthermore, treatment of BALB/c mice with exogenous IL-12 redirected Th1 cell maturation and resistance to *L. major* in these otherwise highly susceptible mice [41, 42]. Although the direct role of IFN-γ in Th1 cell development *in vivo* is still a matter of debate [23, 24], IL-12, besides its direct effect on activated CD4+ T cells, could promote Th1 cell development by enhancing IFN-γ production via a variety of cell types, including natural killer (NK) cells and CD4+ T cells [42, 43]. Other cytokines, such as IL-1α or IL-18, have also been described as cofactors in Th1 cell development [31]. In this context, IL-18 was reported to participate in resistance to *L. major* through its ability to increase IFN-γ production by differentiated Th1 cells rather than through an effect on Th1 differentiation [44, 45].

2. Th2 Cell Development in Susceptible Mice following Infection with *L. major*

The dominant role of IL-4 in directing Th2 cell effector cell differentiation in BALB mice following infection with *L. major* is supported by many experimental results. More than 10 years ago, the ability of neutralizing anti-IL-4 antibodies in redirecting protective Th1 cell development in BALB/c mice was established [46].

a. IL-4 Is Required for Th2 Cell Development in BALB/c Mice Infected with L. major

We have demonstrated a burst of IL-4 mRNA expression in the draining lymph node of BALB/c mice within 16 hr after infection with *L. major* [47]. Importantly, this IL-4 production occurred during the period in which neutralizing IL-4 antibodies were capable of redirecting protective Th1 development in BALB/c mice [46, 48]. After this initial IL-4 mRNA burst, IL-4 mRNA expression returned to baseline values before the occurrence of a second and permanent wave of IL-4 transcripts, which reflects the establishment of a Th2 response. The cognate recognition of a single epitope of the *Leishmania* homolog of mammalian RACK1, designated LACK [49], was demonstrated to drive this early IL-4 response by a restricted population of MHC class II restricted CD4+ T cells that express the Vβ4–Vα8 TCR chains [50].

Superantigen(s) reacts with all T cells expressing a particular Vβ TCR chain [51]. By taking advantage of the fact that, after initial local stimulation of the T cells bearing the corresponding Vβ TCR chain, superantigen leads to their systemic deletion, we have been able to construct BALB/c mice lacking this particular subpopulation of Vβ4–Vα8 CD4+ T cells. The 3′ long terminal repeats of the mouse mammary tumor viruses (MMTV, Swiss IBM Moro [SIM]) encode a superantigen that ultimately leads to the systemic deletion of CD4+ T cells expressing the Vβ4 TCR chain [52]. The importance of Vβ4–Vα8 CD4+ T cells for subsequent Th2 cell development following infection with *L. major* was studied in BALB/c mice rendered deficient in Vβ4 cells as a result of neonatal exposure to MMTV (SIM). Vβ4 CD4+ T-cell-deficient BALB/c mice did not exhibit early IL-4 mRNA expression after inoculation with *L. major*, and Th2 cell differentiation was abrogated in these mice [50]. Strikingly, the presence of these cells was required for the establishment of progressive disease [50]. Similar to Vβ4-deficient mice, BALB/c mice rendered tolerant to LACK as a result of the transgenic expression of this molecule under MHC class II promoters in the thymus were resistant to *L. major* and developed a Th1 response [53].

The requirement of LACK-reactive Vβ4–Vα8 CD4+ for Th2 cell development and susceptibility to *L. major* in BALB/c mice stems from their capacity to produce IL-4 rapidly in response to infection. Indeed, administration of recombinant IL-4 during the first 64 hr after infection to resistant Vβ4-deficient BALB/c mice redirected Th2 cell development and susceptibility to *L. major* [54]. Furthermore, the induction of a specific unresponsive state in LACK-reactive Vβ4–Vα8 CD4+ T cells following treatment of BALB/c mice with altered LACK proteins that differ by a single amino acid from the natural I-A^d-restricted epitope antagonized early IL-4 response to the wild-type

LACK epitope, inhibited Th2 cell development, redirected Th1 cell maturation, and resulted in long-term protection [55]. The results imply that the role of these Vβ4–Vα8 CD4+ T cells is limited to provide the IL-4 necessary for Th2 maturation and that they are not essential at the effector phase of the Th2 response.

The rapidity of IL-4 production by these Vβ4–Vα8 CD4+ T cells following infection with *L. major* would suggest, at first glance, that these cells were activated before exposure to LACK. In this vein, it has been suggested that, as a result of prior stimulation with cross-reactive epitopes, LACK-reactive memory cells could be the source of this early IL-4 response in BALB/c mice [56]. Because activated T cells have been shown to become irreversibly committed to the expression of a particular cytokine pattern [57], our data revealing the functional plasticity of these cells in terms of cytokine production could suggest that they are not differentiated memory cells [58]. Only the possibility to visualize LACK-reactive Vβ4–Vα8 CD4+ T cells with appropriate molecular probes will permit us to know whether these cells express activation markers prior to exposure to *L. major*.

Early IL-2 responses following infection of BALB/c mice with *L. major* were also observed in our laboratory, and kinetic analysis of this response revealed that it precedes the early peak of IL-4 transcription. Strikingly, the early IL-4 response by LACK-reactive Vβ4–Vα8 CD4+ T cells was found to be IL-2-dependent. The cellular origin of this early IL-2 response necessary for the rapid expression of IL-4 transcripts is currently being determined.

Analysis of the accessory cell requirement for the expression of IL-4 transcripts in Vβ4–Vα8 CD4+ T cells following infection has further and unexpectedly demonstrated that B cells are essential. A B-cell-dependent enhancement of IL-4 production by T cells has also been described in other systems [59] and B-cell-deficient mice have been shown to mount enhanced Th1 responses [60]. Experiments aimed at determining whether the early IL-4 response by Vβ4–Vα8 CD4+ T cells requires cognate recognition of their specific epitope on B cells are in progress. The hypothesis that simultaneous recognition by T cells of their specific epitope on B and professional antigen presenting cells (APC) (i.e., dendritic cells) could signal B cells to produce cytokine(s) able to interfere with the Th1-inducing signals from DCs in the close vicinity is attractive.

b. IL-12 Unresponsiveness and Th2 Cell Development following Infection with L. major

The IL-12 receptor (IL-12R) comprises two components, the IL-12Rβ1 and IL-12Rβ2 subunits, which are expressed only on activated T cells [61]. Analysis of the kinetics of IL-12Rβ1 and IL-12Rβ2 mRNA expression in CD4+ T cells from BALB/c and C57BL/6 mice revealed that, whereas both chains of the IL-12 receptor were expressed 24 hr after infection, the IL-12Rβ2 chain was no longer detectable 48 hr after infection only in BALB/c mice [62]. The absence of the IL-12Rβ2 chain at this time was correlated to a state of unresponsiveness to IL-12 in CD4+ T cells, in terms of IFN-γ production [48]. Strikingly, CD4+ T cells from BALB/c mice treated with neutralizing anti-IL-4 antibodies at the onset of infection maintained IL-12Rβ2 chain expression and responsiveness to IL-12 [62]. Further results showed that this IL-4 induced down-regulation of IL-12Rβ2 chain expression could be rescued by exogenous IFN-γ [62]. Together these results confirm *in vivo* data obtained *in vitro*, which had shown the requirement of the IL-12Rβ2 chain for IL-12 signaling and the capacity of IL-4 to down regulate IL-12Rβ2 chain expression [63, 64]. By losing the expression of the IL-12Rβ2 chain, CD4+ T cells from BALB/c mice, induced to differentiate toward the Th2 pathway by IL-4, become refractory to the Th1-differentiating signals of IL-12, stabilizing their Th2 commitment. The ability of IFN-γ to interfere with the IL-4-induced down-regulation of the IL-12Rβ2 subunit could account for the necessity of IFN-γ in the IL-12-induced generation of Th1 cell maturation.

It has been shown that BALB/c mice with a transgenic expression of the IL-12Rβ2 gene remained susceptible to *L. major* [65]. Furthermore, transgenic expression of the IL-12Rβ2 gene on Th2 cells differentiating *in vitro* did not result in decreased IL-4 production, despite restoration of IL-12-mediated STAT4 signaling [65, 66]. Thus, because IL-18 receptors are only expressed on Th1 cells [31], it is possible that IL-18 signaling is also required for interference with Th2 cell maturation.

c. Affinity of the TCR and Th2 Cell Development following Infection with L. major

Results from elegant experiments using mice transgenic for the β–chain of the LACK-specific TCR and multivalent I-A^d-dominant LACK peptide–MHC class II molecules to visualize reactive cells have revealed that T cells, although expanding similarly in susceptible and resistant mice following infection with *L. major*, express low-affinity TCR only in BALB/c mice [67]. Other results also suggest that the pathway of Th cell maturation is influenced by the affinity of the specific TCR [68, 69]. In the leishmania system, however, it is not known whether the selection of low-affinity T cells in BALB/c mice stems from a T-cell intrinsic mechanism or is unrelated to the T-cell compartment. Indeed, following infection with *L. major*, the load of parasite antigens reaching the draining lymph nodes has been shown to be significantly higher in BALB/c mice [70]. It would be of interest to contrast the relative influence of the affinity of TCR in Th differentiation with the pressure of other stimuli. In this context, interventions redirecting Th1 cell maturation in BALB/c mice were not followed by expansion of high-affinity T cells [71].

d. IL-4 Can Induce either Th1 or Th2 Cell Development Depending on Its Cellular Target

Data suggest that, paradoxical to its well-known effect on Th2 cell development, IL-4 may favor Th1 cell differentiation. In this context, it has been demonstrated that IL-4 induces IL-12 production by DC *in vitro* [72, 73]. Data from our groups have revealed that IL-4 is capable of inducing the maturation of Th1 cells and establishing resistance to *L. major* in susceptible BALB/c mice [74]. These effects require that the availability of IL-4 be restricted to the initial period of DC activation that precedes T-cell priming. Extension of IL-4 availability to the period of T-cell activation promotes Th2 development. This paradoxical effect of IL-4 on Th1 cell differentiation results from IL-4-induced IL-12 production by DC [74]. These results highlight the dual and opposite roles that IL-4 can exert on Th cell maturation and demonstrate that these opposing effects depend on the nature of the cells targeted for IL-4 signaling.

C. Why Are C57BL/6 Mice Unable to Mount Early IL-4 Responses following Infection with *L. major*?

In contrast to susceptible BALB/c mice, C57BL/6 and other resistant mice do not mount an early IL-4 response following infection with *L. major* or injection of LACK [47]. Therefore, it has been hypothesized that the LACK-specific Vβ4–Vα8 CD4+ T cells represent a unique subpopulation in BALB/c mice that produces great amounts of IL-4. Alternatively, a higher frequency of these cells in BALB/c mice could account for the ability of the initial IL-4 response to LACK to exceed the threshold required for Th2 development [50]. Both of these hypotheses are unlikely because of our findings demonstrating that neutralization of either IL-12 or IFN-γ in C57BL/6 mice at the initiation of infection allows the expression of this rapid IL-4 response to *L. major*–LACK. Strikingly, this early IL-4 response in C57BL/6 mice also occurs in CD4+ T cells that express the Vβ4–Vα8 TCR chains. In addition, similar frequencies of LACK-specific CD4+ T cells were observed in the lymph nodes of susceptible and resistant mice transgenic for the β-chain of the LACK-specific TCR [67]. Together with previous data showing that treatment of BALB/c mice with recombinant IL-12 suppresses the early IL-4 response to *L. major* [47], these results indicate that an impairment in the mechanism(s) that down-regulates the early IL-4 response by Vβ4–Vα8 T cells might underlie the susceptibility of BALB/c mice. Of interest, evidence was obtained showing that the increased susceptibility resulting from treatment of resistant mice with anti-IL-12 resulted from the IL-4 produced during the early stage of infection.

V. CONCLUDING REMARKS

Containment of several pathogenic microorganisms by their host depends upon the differentiation of naive CD4+ T cells into effector cells secreting cytokines decisive for protection. Over the past years, our understanding of immunity to parasitic infections has been significantly enhanced by the discovery of distinct CD4+ cell populations. In particular, the murine model of infection with *L. major* not only has permitted us to validate *in vivo* the existence of distinct CD4+ T-cell subpopulations but has also demonstrated their crucial role on the outcome of infectious diseases. This model is now revealing itself as a powerful tool to understand the cellular and molecular mechanisms operating in the selective development of peripheral CD4+ T cells *in vivo*. A thorough definition of these mechanisms is a prerequisite for the development of novel interventional strategies for the prevention and treatment of serious infectious diseases.

ACKNOWLEDGMENTS

The experiments from our groups reported in this part of Chapter 5 were supported by grants from the Swiss National Science Foundation, Sandoz Research Foundation, Deutsche Forschungsgemeinschaft (RO 764/8-1, SFB 217, and SFB 456), and the French Ministry of Research.

REFERENCES

1. Mossmann, T. R., and Coffman, R. L. (1989). Th1 and Th2 cells: Different patterns of lymphokine secretion lead to different functional properties. *Annu. Rev. Immunol.* 7:145–173.
2. Zurawski, G., and de Vries, J. E. (1994). Interleukin 13, an interleukin 4-like cytokine that acts on monocytes and B cells, but not on T cells. *Immunol. Today* 15:19–26.
3. Abbas, A. K., Murphy, K. M., and Sher, A. (1996). Functional diversity of helper T lymphocytes. *Nature* 383:787–793.
4. Lockskey, R. M., Heinzel, F. P., Sadick, M. D., Holaday, B. J., and Gardner, K. D. (1987). Murine cutaneous leishmaniasis: Susceptibility correlates with different expansion of helper T cell subset. *Ann. Inst. Pasteur/Immunol.* 138:744–749.
5. Reiner, S. L., and Locksley, R. M. (1995). The regulation of immunity to *Leishmania major*. *Ann. Rev. Immunol.* 13:151–177.
6. Capron, A., Dombrowicz, D., and Capron, M. (1999). Regulation of the immune response in experimental and human schistosomiasis: The limits of an attractive paradigm. *Microbes Infect.* 1:485–490.
7. Chen, Q., Schlichterle, M., and Walgreen, M. (2000). Molecular aspects of severe malaria. *Clin. Microbiol. Rev.* 13:439–450.
8. Bull, P. C., Lowe, B. S., Kortok, M., Molyneux, C. S., Newbold, C. I., and Marsh, K. (1998). Parasite antigens on the infected red cell surface are targets for naturally acquired immunity to malaria. *Nature Med.* 4:358–360.
9. Graig, A., and Scherf, A. (2001). Molecules on the surface of the *Plasmodium falciparum* infected erythrocyte and their role in malaria pathogenesis and immune evasion. *Mol. Biochem. Parasitol.* 115:129–143.
10. Duffy, P. E., Graig, A. G., and Baruch, D. I. (2001). Variant proteins on the surface of malaria-infected erythrocytes developing vaccines. *Trends Parasitol.* 17:354–356.
11. Hommel, M., David, P. H., and Oligino, L. D. (1983). Surface alterations of erythrocytes in *Plasmodium falciparum* malaria. Antigenic variation, antigenic diversity and the role of the spleen. *J. Exp. Med.* 157:1137–1148.
12. Fried, M., and Duffy, P. E. (1996). Adherence of *Plasmodium falciparum* to chondroitine sulfate A in the human placenta. *Science.* 109:330–342.
13. Fried, M., Nosten, F., Brockman, A., Brabin, B. J., and Duffy, P. E. (1998). Maternal antibodies block malaria. *Nature* 395:851–852.
14. Miller, H. L., Good, M. F. and Milon, G. (1994). Malaria pathogenesis. *Science* 264:1878–1883.
15a. Capron, M., and Capron, A. (1994). Immunoglobulin E and effector cells in schistosomiasis. *Science* 264:1876–1877.
15b. Butterwoth, A. E., Sturrok, R. F., Houba, V., Mahmoud, A. A., Sher, A., and Rees, P. H. (1975). Eosinophils as mediators of antibody dependent damage to schistosomula. *Nature* 256:727–729.
16. Gounni, A. S., Lamkhioued, B., Ochiai, K., Tanaka, Y., Delaporte, E., Capron, A., Kinet, J. P., and Capron, M. (1994). High affinity IgE receptor on eosinophils is involved in defence against parasites. *Nature* 367:183–186.
17. Capron, M., Bazin, H., Joseph, M., and Capron, A. (1981) Evidence for IgE-dependent cytotoxicity by rat eosinophils. *J. Immunol.* 126:1764–1768.
18. Pearce, E. J., and Sher, A. (1991). Functional dichotomy in the CD4+ T cell response to *Schistosoma mansoni*. *Exp. Parasitol.* 73:110–116.

19. Wynn, T. A. (1999). Immune deviation as a strategy for schistosomiasis vaccines designed to prevent infection and egg-induced immunopathology. *Microbes Infect.* 1:525–534.
20. de Andres, B., Rakasz, E., Hagen, M., McCormik, M. L., Mueller, A. L., Elliot, D., Metwali, A., Sandor, M., Britigan, B. E., Werinstock, J. V., and Lynch, R. G. (1997). Lack of Fc-epsilon receptors on murine eosinophils: Implications for the functional significance of elevated IgE and eosinophils in parasitic infections. *Blood* 89:3826–3836.
21. Caulada-Benedetti, Z., Al-Zamed, F., Sher, A., and James, S. (1991). Comparison of Th1- and Th2-associated immune reactivities stimulated by single versus multiple vaccination of mice with irradiated *Schistosoma mansoni* cercariae. *J. Immunol.* 146:4489–4494.
22. Wynn, T. A., and Hoffmann, K. F. (2000). Defining a Schistosomiasis vaccination strategy— Is it really Th1 versus Th2? *Parasitol. today* 16:497–501.
23. Wang, Z. E., Reiner, S. L., Zheng, S., Dalton, D. K., and Locksley, R. M. (1994). CD4$^+$ effector cells default to the Th2 pathway in interferon γ-deficient mice infected with *Leishmania major*. *J. Exp. Med.* 179:1367–1371.
24. Swihart, K., Fruth, U., Messmer, N., Hug, K., Behin, R., Huang, S., Del Giudice, G., Aguet, M., and Louis, J. A. (1995). Mice from a genetically resistant background lacking the interferon γ receptor are susceptible to infection with *Leishmania major* but mount a polarized T helper cell 1-type CD4$^+$ T cell response. *J. Exp. Med.* 181:961–971.
25. Assreuy, J., Cunha, F. Q., Epperlein, M., Noronha-Dutra, A., O'Donnell, C. A., Liew, F. Y., and Moncada, S. (1994). Production of nitric oxide and superoxide by activated macrophages and killing of *Leishmania major*. *Eur. J Immunol.* 24:672–676.
26. Conceiçao-Silva, F., Hahne, M., Schröter, M., Louis, J., and Tschopp, J. (1998). The resolution of lesions induced by *Leishmania major* in mice requires a functional Fas (APO-1, CD95) pathway of cytotoxicyty. *Eur. J. Immunol.* 28:237–245.
27. Huang, F. P., Xu, D., Esfandiari, E. O., Sands, W., Wei, X. Q., and Liew, F. Y. (1998). Mice defective in Fas are highly susceptible to *Leishmania major* infection despite elevated IL-12 synthesis, strong Th1 responses and enhanced nitric oxide production. *J. Immunol.* 160:4143–4147.
28. Paludan, S. R., Lovmand, J., Ellermann-Eriksen, S., and Mogensen, S. C (1997). Effect of IL-4 and IL-13 on IFN-γ induced production of nitric oxide in mouse macrophages infected with herpes simplex virus type 2. *FEBS Lett.* 414:61–64.
29. Ohmory, Y., and Hamilton, T. A. (1997). IL-4-induced STAT6 suppresses IFN-γ-stimulated STAT1-dependent transcription in mouse macrophages. *J. Immunol.* 159:5474–5482.
30. Röcken, M., Saurat, J. H., and Hauser, C. (1992). A common precursor for CD4$^+$ T cells producing IL-2 and IL-4. *J. Immunol.* 148:1031–1036.
31. O'Garra, A. (1998). Cytokines induce the development of functionally heterogeneous T helper cell subsets. *Immunity* 8:275–283.
32. Hsieh, C. S., Macatonia, S. E., Tripp, C. S., Wolf, S. F., O'Garra, A., and Murphy, K. M. (1993). Development of TH1 CD4$^+$ T cells through IL-12 produced by *Listeria*-induced macrophages. *Science* 260:547–549.
33. Macatonia, S. E., Hosken, N. A., Litton, M., Vieira, P., Hsiesh, C. S., Culpepper, J., Wysocka, M., Trinchieri, G., Murphy, K. M., and O'Garra, A. (1995). Dendritic cells produce IL-12 and direct the development of Th1 cells from naive CD4$^+$ T cells. *J. Immunol.* 15:5071–5079.
34. Seder, R. A., Gazzinelli, R., Sher, A., and Paul, W. E. (1993). Interleukin 12 acts directly on CD4$^+$ T cells to enhance priming for interferon-γ production and diminishes interleukin 4 inhibition of such priming. *Proc. Natl. Acad. Sci. USA* 90:10188–10192.
35. Le Gros, G., Ben-Sasson, S. Z., Seder, R., Finkelman, F. D., and Paul, W. E. (1990). Generation of interleukin 4 (IL-4) producing cells *in vivo* and *in vitro*: IL-2 and IL-4 are required for *in vitro* generation of IL-4-producing cells. *J. Exp. Med.* 172:921–929.
36. Seder, R. A., Paul, W. E., Davis, M. M., and Fazekas de St Groth, B. (1992). The presence of interleukin 4 during *in vitro* priming determines the lymphokine-producing potential of CD4$^+$ T cells from T cell receptor transgenic mice. *J. Exp. Med.* 176:1091–1098.
37. Swain, S. L., Weinberg, A. D., English, M., and Huston, G. (1990). IL-4 directs the development of Th2-like helper effectors. *J. Immunol.* 145:3796–3806.
38. Seder, R. A., and Paul, W. E. (1994). Acquisition of lymphokine-producing phenotype by CD4$^+$ T cells. *Annu. Rev. Immunol.* 12:653–673.
39. Heinzel, F. P., Rerko, R. M., Ahmed, F., and Pearlman, E. (1995). Endogenous IL-12 is required for control of Th2 cytokine responses capable of exacerbating Leishmaniasis in normally resistant mice. *J. Immunol.* 155:730–739.

40. Mattner, F., Magram, J., Ferrante, J., Launois, P., Di Padova, K., Behin, R., Gately, M. K., Louis, J. A., and Alber, G. (1996). Genetically resistant mice lacking interleukin-12 are susceptible to infection with *Leishmania major* and mount a polarized Th2 cell response. *Eur. J. Immunol.* **26**:1553–1559.

41. Heinzel, F. P., Schoenhaut, D. S., Rerko, R. M., Rosser, L. E., and Gately, M. K. (1993). Recombinant interleukin 12 cures mice infected with *Leishmania major*. *J. Exp. Med.* **177**:1505–1509.

42. Scharton-Kersten, T., Afonso, L. C. C., Wysocka, M., Trinchieri, G., and Scott, P. (1995). IL-12 is required for natural killer cell activation and subsequent T helper 1 cell development in experimental leishmaniasis. *J. Immunol.* **154**:5320–5330.

43. Wakil, A. E., Wang, Z. E., Ryan, J. C., Fowell, D. J., and Locksley, R. M. (1998). Interferon γ derived from CD4+ T cells is sufficient to mediate T helper cell type 1 development. *J. Exp. Med.* **188**:1651–1656.

44. Ohkusu, K., Yoshimoto, T., Takeda, K., Ogura, T., Kashiwamura, S., Iwakura, Y., Akira, S., Okamura, H., and Nakanishi, K. (2000). Potentiality of interleukin-18 as a useful reagent for treatment and prevention of *Leishmania major* infection. *Infect. Immun.* **68**:2449–2456.

45. Monteforte, G. M., Takeda, K., Rodriguerz-Sosa, M., Akira, S., David, J. R., and Satoskar, A. R. (2000). Genetically resistant mice lacking IL-18 gene develop Th1 response and control cutaneous *Leishmania major* infection. *J. Immunol.* **164**:5890–5893.

46. Sadick, M. D., Heinzel, F. P., Holaday, B. J., Pu, R. T., Dawkins, R. S., and Locksley, R. M. (1990). Cure of murine leishmaniasis with anti-interleukin 4 monoclonal antibody. Evidence for a T-cell-dependent, interferon-γ-independent mechanism. *J. Exp. Med.* **171**:115–127.

47. Launois, P., Ohteki, T., Swihart, K., MacDonald, H. R., and Louis, J. A. (1995). In susceptible mice, *Leishmania major* induce very rapid interleukin-4 production by CD4+ T cells which are NK1.1⁻. *Eur. J. Immunol.* **25**:3298–3307.

48. Launois, P., Swihart, K., Milon, G., and Louis, J. A. (1997). Early production of IL-4 in susceptible mice infected with *Leishmania major* rapidly induces IL-12 unresponsiveness. *J. Immunol.* **158**:3317–3324.

49. Mougneau, E., Altore, F., Wakil, A. E., Zheng, S., Coppela, T., Wang, Z. E., Waldmann, R., and Locksley, R. M. (1995). Expression cloning of a protective *Leishmania* antigen. *Science* **268**:563–566.

50. Launois, P., Maillard, I., Pingel, S., Swihart, K., Xenarios, I., Acha-Orbea, H., Diggelmann, H., Locksley, R. M., Mac Donald, H. R., and Louis, J. A. (1997). IL-4 rapidly produces by Vβ4 Vα8 CD4+ T cells in BALB/c mice infected with *Leishmania major* instructs Th2 cell development and susceptibility to infection. *Immunity* **6**:541–549.

51. White, J., Herman, A., Pullen, A. M., Kubo, R., Kappler, J. W., and Marrack, P. (1989). The Vβ-specific superantigen staphylococcal enterotoxin B: Stimulation of mature T cells and clonal deletion in neonatal mice. *Cell* **56**:27–35.

52. Maillard, I., Emy, K., Acha-Orbea, H., and Diggelmann, H. (1996). A Vβ4-specific superantigen encoded by a new exogenous mouse mammary tumor virus. *Eur. J. Immunol* **2**:1000–1006.

53. Julia, V., Rassoulzadegan, M., and Glaichenhaus, N. (1996). Resistance to *Leishmania major* induced by tolerance to a single antigen. *Science* **274**:421–423.

54. Himmelrich, H., Launois, P., Maillard, I., Biedermann, T., Tacchini-Cottier, F., Locksley, R. M., Röcken, M., and Louis, J. A. (2000). In BALB/c mice, IL-4 production during the initial phase of infection with *Leishmania major* is necessary and sufficient to instruct Th2 cell development resulting in progressive disease. *J. Immunol.* **164**: 4819–4825.

55. Pingel, S., Launois, P., Fowell, D. J., Turck, C. W., Southwood, S., Sette, A., Glaichenhaus, N., Louis, J. A., and Locksley, R. M. (1999). Altered ligands reveal limited plasticity in the T cell response to a pathogenic epitope. *J. Exp. Med.* **189**:1111–1120.

56. Julia, V., McSorley, S. S., Malherbe, L., Breittmayer, J. P., Girard-Pipau, F., Beck, A., and Glaichenhaus, N. (2000). Priming by microbial antigens from the intestinal flora determines the ability of CD4+ T cells to rapidly secrete IL-4 in BALB/c mice infected with *Leishmania major*. *J. Immunol.* **165**:5637–5645.

57. Bird, J. J., Brown, D. R., Mullen, A. C., Moskowitz, N. H., Mahowad, M. A., Sider, J. D., Gajewski, T. F., Wang, C. R., and Reiner, S. L. (1998). Helper T cell differentiation is controlled by cell cycle. *Immunity* **9**:229–237.

58. Maillard, I., Launois, P., Himmelrich, H., Acha-Orbea, H., Diggelmann, H., Locksley, R. M., and Louis, J. A. (2001). Functional plasticity of the LACK-reactive Vβ4–Vα8 CD4+ T cells normally producing the early IL-4 instructing Th2 cell development and susceptibility to *Leishmania major* in BALB/c mice. *Eur. J. Immunol.* **31**:1288–1296.

59. Skok, J., Poudrier, J., and Gray, D. (1999). Dendritic cell-derived IL-12 promotes B cell induction of Th2 differentiation: A feedback regulation of Th1 development. *J. Immunol.* **163**:4284–4291.

60. Moulin, V., Andris, F., Thielemans, K., Maliszewski, C., Urbain, J., and Moser, M. (2000). B lymphocytes regulate dendritic cell (DC) function *in vivo*: Increased interleukin-12 production by DCs from B cell-deficient mice results in T helper cell type 1 deviation. *J. Exp. Med.* **192**:475–482.

61. Presky, D. H., Yang, H., Minetti, L. J., Chu, A. O., Nabavi, N., Wu, C. Y., Gately, M., and Gubler, U. (1996). A functional interleukin 12 receptor complex is composed of two β-type cytokine receptor subunits. *Proc. Natl. Acad. Sci. USA* **93**:14002–14007.

62. Himmelrich, H., Parra-Lopez, C., Tacchini-Cottier, F., Louis, J. A., and Launois, P. (1998). The IL-4 rapidly produced in BALB/c mice after infection with *Leishmania major* down-regulates the IL-12 receptor β2 chain expression on CD4+ T cells resulting in a state of unresponsiveness to IL-12. *J. Immunol.* **161**:656–661.

63. Szabo, S. J., Jacobson, N. G., Dighe, S., Gubler, U., and Murphy, K. M. (1995). Developmental commitment to the Th2 lineage by extinction of IL-12 signaling. *Immunity* **2**:665–675.

64. Szabo, S. J., Dighe, A. S., Gubler, U., and Murphy, K. M. (1997). Regulation of the interleukin (IL)-12R β2 subunit expression in developing T helper (Th1) and Th2 cells. *J. Exp. Med.* **185**:817–824.

65. Nishikomori, R., Gurunathan, S., Nishikomori, K., and Strober, W. (2001). BALB/c mice bearing a transgenic IL-12 receptor β2 gene exhibit a nonhealing phenotype to *Leishmania major* infection despite intact IL-12 signaling. *J. Immunol.* **166**:6776–6783.

66. Heath, V. L., Showe, L., Crain, C., Barrat, F. J., Trinchieri, G., and O'Garra, A. (2000). Ectopic expression of the IL-12 receptor β2 in developed and committed Th2 cells does not affect the production of IL-4 or induced production of IFN-γ. *J. Immunol.* **164**:2861–2865.

67. Malherbe, L., Filippi, C., Julia, V., Foucras, G., Moro, M., Appel, H., Wucherpfennig, K., Guéry, J. C., and Glaichenhaus, N. (2000). Selective activation and expansion of high affinity CD4+ T cells in resistance mice upon infection with *Leishmania major*. *Immunity* **13**:771–782.

68. Pfeiffer, C., Stein, J., Southwood, S., Ketelaar, H., Sette, A., and Bottomly, K. (1995). Altered peptide ligands can control CD4 T lymphocyte differentiation *in vivo*. *J. Exp. Med.* **181**:1569–1574.

69. Constant, S., Pfeiffer, C., Woodard, A., Pasqualini, T., and Bottomly, K. (1995). Extent of T cell receptor ligation can determine the functional differentiation of naive CD4+ T cells. *J. Exp. Med.* **182**:1591–1596.

70. Laskay, T., Diefenbach, A., Rollinghoff, M., and Solbach, W. (1995). Early parasite containment is decisive for resistance to *Leishmania major* infection. *Eur. J. Immunol.* **25**:2220–2227.

71. Malherbe, L. (2001). Visualisation des cellules T CD4+ anti-parasites au cours de l'infection de souches de souris syngéniques par le parasite intra cellulaire *Leishmania major*. Ph.D. thesis, University of Nice-Sophia Antipolis.

72. Kalinski, P., Smits, H. H., Schuitmaker, J. H., Vieira, P. L., van Eijk, M., de Jong, E. C., Wierrenga, E. A., and Kapsenberg, M. L. (2000). IL-4 is a mediator of IL-12p70 induction by human Th2 cells: Reversal of polarized Th2 phenotype by dendritic cells. *J. Immunol.* **165**:1877–1881.

73. Hochrein, H., O'Keefe, M., Luft, T., Vandenabeele, S., Grumont, R. J., Maraskovsky, E., and Shortman, K. (2001). Interleukin (IL) 4 is a major regulatory cytokine governing bioactive IL-12 production by mouse and human dendritic cells. *J. Exp. Med.* **192**:823–833.

74. Biedermann, T., Zimmermann, S., Himmelrich, H., Gumy, A., Egeter, O0., Sakrauski, A. K., Seegmüller, I., Voigt, H., Launois, P., Levine, A. D., Wagner, H., Heeg, K., Louis, J. A., and Röcken, M. (2001). Interleukin-4 instructs Th1 responses and resistance to *Leishmania major* in susceptible BALB/c mice. *Nature Immunol.* **2**:1054–1060.

6 DISEASE STATES AND VACCINES: SELECTED CASES

PART A. Introduction

STANLEY A. PLOTKIN
Aventis Pasteur, Doylestown, Pennsylvania 18901

An objective observer looking back at the 200-year history of vaccines or looking forward to the licensure of many experimental vaccines developed using modern technology would have good reasons for viewing vaccination as a success story. Yet for all of the successful conquest of epidemic diseases, and despite the bloom of a thousand flowers in the laboratory, there is a sense of malaise in the world of vaccines. As discussed here, this malaise derives from technical and social issues that impede the fulfillment of the goal of vaccine development—to protect populations against endemic and epidemic disease.

I. ISSUES IN PEDIATRIC VACCINATION

A. Pertussis

The keystone of infant vaccination is a combined vaccine against diphtheria, tetanus, and pertussis. Because of troubling local and systemic reactions in children, developed countries have mostly replaced whole cell pertussis with acellular pertussis vaccines based on purified protective components of the bacterium. This change has been accompanied by a satisfying drop in complaints, but comparative data suggest that many of the acellular vaccines are less efficacious than well-prepared whole cell vaccines[1]. Thus, there is a dilemma for developing countries: whether to remain with the less reactogenic, more efficacious whole cell vaccines or to switch to the safer, but possibly less efficacious acellular vaccines. For the moment, the cost of the latter has rendered the choice academic, but eventually the issue will become acute, particularly in light of the fact that some whole cell vaccines made by national institutes may not be of good quality.

B. Polio

Polio is supposed to be eradicated from the globe in the next few years, although stubborn foci still exist in the Indian subcontinent and Africa [2]. The best way to handle the last phase of eradication is in dispute because the oral attenuated vaccine is capable of causing sporadic and epidemic polio due to reverse mutations. Without attempting a discussion of this contentious issue, one possible way of handling the end game would be to switch to inactivated vaccine when there is no longer any known circulation of virulent virus, in order to provide safety against occult wild or reverted strains. This strategy would be based on the use of IPV (inactivated polio vaccine) in combination vaccines (see Section I.E).

C. Hib

Protein-conjugated *Haemophilus influenzae* B (Hib) polysaccharide has been superlatively successful in eliminating Hib syndromes, including meningitis, epiglottitis, and pneumonia, but also has reduced pharyngeal carriage of the organism. Hib conjugate has also been included in combination vaccines.

D. Hepatitis B

The need to protect infants from acquiring infection from their mothers is universally acknowledged. Many countries have adopted a policy of infant vaccination, although a few argue that the incidence in their populations is too low to justify general vaccination. Be that as it may, the success of hepatitis B vaccine in preventing infection and its later neoplastic consequences in the liver [3] has induced its inclusion in combination vaccines.

E. Combinations

Combination vaccines have been developed containing six valences: diphtheria, tetanus, acellular pertussis, IPV, Hib, and hepatitis B. This did not prove to be easy, both for pharmacological and for immunological reasons. Now there are additional antigens that one might wish to add to a combination for use in infants, such as pneumococcal, meningococcal, respiratory syncytial virus, parainfluenza viruses, and hepatitis A. However, because of difficulties, the future of combination vaccines is probably one of fission into syndrome-related vaccines, rather than the construction of 7-, 8-, or 10-valent products.

F. Pneumococcal

Invasive and pneumonic infections due to pneumococci are important throughout the world, although mortality in developed countries is low due to medical intervention. A protein-conjugated vaccine containing seven specific polysaccharide serotypes has been shown to be highly efficacious [4]. However, the number of serotypes will have to be augmented to immunize children in developing countries, where other serotypes are also common.

G. Meningococcal

The United Kingdom has had great success with a new meningococcal group C conjugate vaccine, and vaccines against three other serogroups are in the offing. Unfortunately, for technical reasons a group B vaccine based on capsular polysaccharide is not available, but there is hope for an outer membrane protein-based vaccine.

H. Carrier Proteins

The outstanding successes of bacterial polysaccharides conjugated with carrier proteins have encouraged their combination or association in infant vaccination. However, most licensed vaccines thus far have used tetanus or diphtheria proteins, and the potential results of injecting large amounts are increased reactions and decreased immunogenicity. Thus, it is urgent to find new carrier proteins, particularly as there are nosocomial pathogens for which this type of vaccine might be employed (see later discussion).

I. Rotavirus

The withdrawal of rotavirus vaccine from routine use in the United States was a significant event both for the vaccine and for vaccine safety in general. Although the association between administration of the vaccine strain and intussusception [5] rendered the vaccine unusable in developed countries, the high mortality rate due to rotavirus infection in the tropics argues for further study of new orally administered, attenuated strains. The comforting point is that the efficacy of repeated infection of the intestine with attenuated strains against wild viruses is beyond doubt, and one can be optimistic about the eventual availability of a rotavirus vaccine.

J. RSV (Respiratory Syncytial Virus) and PIV (Parainfluenza Viruses)

The great unsolved problem of pediatric vaccination is how to prevent respiratory infections in young infants, the most significant of which are due to RSV and parainfluenza type 3. In both cases, development of attenuated strains is far advanced, but other avenues are being actively explored. Aside from safety issues, the major difficulty is inducing protection at a very early age.

K. MMR

The three valences contained in this vaccine—measles, mumps, and rubella—have been highly successful in reducing or eliminating disease. Scandinavia [6] and North America have largely eliminated the three diseases, and Latin America has launched a campaign aimed at similar success. Aside from the usual difficulty of raising immunization coverage to a high level in developing countries, the paradoxical situation is that in Europe immunization continues to lag because of unrealistic safety concerns. In addition, the inclusion of rubella and mumps in measles-containing vaccines for developing countries has been slowed by cost of the combined vaccine.

II. ISSUES IN ADULT VACCINATION

A. Influenza

With the exception of AIDS, influenza is the last pandemic disease still unconquered, although the widespread use of inactivated vaccine has done much to moderate its lethal effects. However, the efficacy of the inactivated vaccine is only moderate in high-risk individuals and is specific to the strains in the vaccine. Moreover, the reservoir of infection in children has not heretofore been touched.

Fortunately, a live attenuated vaccine given by aerosol may soon be licensed. This vaccine is highly efficacious in children [7] and may also be useful in adults to augment immunity induced by the inactivated vaccine. Moreover, the live vaccine appears to give a broader immune response than the inactivated vaccine, which should reduce the year-to-year variation in efficacy of protection.

B. Pneumococcal

A multivalent polysaccharide vaccine against pneumonia has long been available, but its efficacy remains controversial. This situation could improve if protein-conjugated vaccines are made for adults or if one of the pneumococcal membrane proteins fulfills the promise of protection crossing serotypes.

C. Zoster

Recrudescent varicella, also known as zoster, may be prevented or ameliorated in severity by vaccination of adults with the current live varicella vaccine, in order to boost the cellular immunity against the virus, which wanes with age [8]. The results of an efficacy trial will be available soon.

D. Boosters

A large area of uncertainty in current vaccinology is how to interpret data concerning immunologic memory and how often booster vaccination is necessary. Constant maintenance of antibody titers appears to be necessary for toxin-mediated diseases, whereas memory appears to be sufficient to protect against a long-incubation-period disease such as hepatitis B. A great many infections and vaccines lie in between on this continuum, but the location is unclear for many of them. Only the accumulation of additional long-term data will allow rational decisions to be made about the need for and timing of boosters.

III. ISSUES CONCERNING NEWER TECHNIQUES FOR VACCINE DEVELOPMENT [20]

A. Cellular Immunity

Any analysis of the vaccines currently in use discloses that, with the exception of bacille Calmette–Guérin (BCG), all protect completely or predominantly by the induction of antibodies [9]. In most cases, the antibodies act at the bacteremic or viremic stages of the infection.

Now we are confronted by pathogens for which the protective mechanisms are wholly or predominantly those mediated by lymphocytes and macrophages. Examples include the herpes viruses, papillomaviruses, HIV, tuberculosis, and probably malaria as well. Live vaccines have traditionally been the best means of inducing cellular responses, but they may not be appropriate for many of these infections. Thus, new techniques have been sought to induce cellular responses. The three major approaches to this goal have been live vectors, nucleic acid vaccines, and lipopeptides.

B. Vectors

A vector is, by definition, a living organism into which the genes of a pathogen have been introduced and that has been attenuated naturally or in the laboratory. When introduced into the host, the vector expresses the foreign genes and induces a protective immune response against the pathogen. The main vectors in experimental use at the moment are poxviruses [9A], alphaviruses [10], adenoviruses, BCG, and salmonella (administered orally).

The most advanced development of vectors has been in the field of AIDS, where poxviruses and adenoviruses have been shown to induce good CTL (cytotoxic T lymphocyte) responses against HIV proteins such as *gag*. Another example is the induction of CTL against cytomegalovirus (CMV) in human volunteers by a canarypox containing the gene for the major matrix protein of CMV [11].

Plants have captured the imagination as another possible vector. The foreign genes are inserted either into the plant genome or into the genome of plant viruses, which then replicate in the plant before the plant material is administered as oral vaccines [12].

C. Nucleic Acids

The discovery that inoculation of foreign DNA inserted into plasmids could immunize animals has revolutionized antigen discovery, but so far it has not yielded a practical vaccine [13]. Part of the problem lies in the poorer response of humans than of mice, part in the need for an adjuvant to reduce the dose of DNA required, and part in the fears evoked by the injection of genetic material.

However, progress is being made in this area through the use of "prime–boost" protocols [14], in which the DNA primes the immune system for later recall using a more conventional approach, and through the development of adjuvants that potentiate the DNA.

The future of DNA immunization at the moment is entwined with AIDS vaccines: several clinical trials are now testing whether this approach will work.

D. Adjuvants

Although current techniques can produce a protein of high purity, they cannot guarantee immune responses to that protein. Depending on the native immunogenicity of the particular protein, an adjuvant may be needed to obtain a protective level of antibodies. Until recently, aluminum salts were the only acceptable adjuvants, but newer substances are under consideration to drive the immune system toward either antibodies (Th2) or cellular immunity (Th1) [15].

E. Genomics

The genetic sequences for various microbial genomes are pouring out of molecular biology laboratories. The genomes are then scanned for open reading frames [ORFs] that translate into protective proteins. Organisms such as *Chlamydiae pneumoniae* and group B meningococcus have yielded up numerous putatively protective candidate genes. However, no vaccine developed by genomics has yet made its way to the clinic or proven its superiority over previously known candidate antigens.

F. Route of Administration

Although new combination vaccines will reduce the number of injections needed, it is evident that we are reaching the practical limit for parenteral immunization, especially in pediatrics. Therefore, administration by other routes becomes a priority. The oral route is already in use for polio, rotavirus, cholera, and typhoid vaccines. The intranasal influenza vaccine was mentioned earlier. Intranasal vaccination is also attractive for sexually transmitted disease (STD) vaccines because of its ability to induce responses in the genital as well as nasal mucosa.

The idea of transcutaneous immunization is also exciting. Early studies suggest that antigen absorption through the skin is possible, especially using weak electrical currents, but the efficacy of this route remains to be demonstrated [16].

G. New Targets

It is easy to name the three big targets for this century: vaccines against HIV, malaria, and tuberculosis. Prior efforts to develop vaccines against these diseases have failed in part because of inadequate funding, a situation that has been remedied in the case of HIV, but the technical issues should not be underestimated. In each of the three cases the principal problem has been uncertainty about the immunological correlate of protection. If that were known, it would probably be possible to find a way of inducing the correct response.

In addition to the traditional targets of epidemic and endemic infectious diseases, new targets are being found in the diseases of advanced societies: nosocomial infections and chronic diseases.

Antibiotic resistance of resident bacteria has created an acute problem in hospitals, as infections due to staphylococci, enterococci, and pseudomonas become difficult or impossible to treat. Therefore, vaccine development is in full swing against those pathogens [17]. Although not, strictly speaking, a nosocomial phenomenon, resistant pneumococci are examples of pathogens that now may be controlled by immunization, at least in children.

The recognition that certain chronic diseases, such as diabetes mellitus, atherosclerosis, and multiple sclerosis, may be the result of chronic infections or of autoimmune responses to infection operating in the background of genetic disposition has led to the development of experimental vaccines to prevent or modify those abnormal responses [18].

In addition, chronic infections such as HIV are now the subject of attempts to develop therapeutic vaccines that will suppress the pathogens by eliciting stronger cellular responses.

Table 1 lists currently available vaccines and those likely to be licensed soon.

TABLE 1 **Vaccines Currently Available and Those in Advanced Clinical Trials**

Attenuated organisms grown *in vivo*	Killed Whole Organisms
Smallpox[a]	Typhoid
Yellow fever	Cholera
Rabies (neural tissue)	Plague
Attenuated organisms grown *in vitro* or *in ovo*	Pertussis, whole cell
	Polio (injected)
BCG	Japanese encephalitis
Polio (oral)	Hepatitis A
Measles	Tick-borne encephalitis
Mumps	
Rubella	Purified proteins
Adenovirus[a]	Diphtheria
Typhoid (oral)	Tetanus
Varicella	Hepatitis B[c]
Rotavirus[a,b]	Pertussis, acellular
Influenza (nasal)[a,b]	Lyme disease[c]
RSV[a]	HIV[a,c]
Dengue[a]	Meningococcus, group B[a]
Subunits	Vectored antigens
Influenza (injected)	Papilloma virus[a]
Rabies (cell culture)	HIV[a,c]
Anthrax	
Bacterial polysaccharides	Protein-conjugated polysaccharides
Pneumococcus	*Haemophilus influenzae* type B
Meningococcus (A, C, W, Y)	Pneumococcus
Typhoid (Vi)	Meningococcus group C
	Streptococcus group B[a]

[a]Unlicensed or not in use.
[b]Reassortants.
[c]Produced by recombinant technology.

IV. CHALLENGES FACING USE OF VACCINES

A. Safety

The twentieth century saw the gradual control of many infectious diseases through clean water supplies, antibiotics, and vaccination. However, the disappearance of infectious diseases has had the paradoxical effect of weakening enthusiasm for vaccination among people in developed countries. Fear of vaccines to some extent has replaced fear of infectious diseases, as reactions to vaccines, real and imaginary, have become more numerous than cases of the disease being prevented [19].

It is easy to show that in the absence of vaccination the situation would be worse, but from a purely selfish point of view the best strategy is not to be vaccinated oneself but to depend on one's neighbors being vaccinated. Thus, increasing numbers of parents prefer "natural" disease to vaccination. Although they often change their minds once their child suffers a vaccine-preventable infection, this group of people is a threat to disease control. In fact, the situation is worse in Europe than in the United States. Whereas philosophical objections to vaccination impact on less than 1% of American children, parental refusal has significantly reduced vaccine coverage in the United Kingdom and France. One ironic result is that importations of measles into the United States are coming more often from prosperous Europe than from poor countries.

Nevertheless, one must recognize that vaccine safety will be a predominant issue in the future. The withdrawal of rotavirus vaccine due to its association with intussusception is a concrete example of a safety debacle. Although other posited associations, such as multiple sclerosis with hepatitis B vaccine and autism with measles vaccines, appear to be without factual basis, there are good reasons to study such phenomena as allergy and stress in relation to vaccination. Vaccine safety must be a paramount consideration.

B. Vaccine Cost

Cost of vaccine is problematic in both industrialized and poor countries. In the former, the realization that vaccine manufacture is no longer a cheap process, but that on the other hand it is in the interest of the state to promote vaccination as a form of health prevention, requires the appropriation of major budgets for vaccine purchase. Although there is nothing inherently wrong with spending money on vaccines rather than on armaments, economics are beginning to impinge on decisions to recommend vaccination. The licensed pneumococcal conjugate vaccine is an example. Four recommended doses exceed $200 in cost when purchased on the private retail market. Although the vaccine is highly efficacious, reluctance to use it arose because of the price.

For poor countries the problem is worse. With minuscule health budgets and poor infrastructure, many countries are unwilling or unable to buy modern vaccines. This problem is now receiving attention by many prominent groups, and the idea is gradually growing that rich countries and donors, such as the Gates Foundation and the World Bank, must support vaccination in poor countries, for both humane and selfish reasons.

C. Vaccine Production

The revolution of rising expectations and the economic advances in Asia have generated a demand for vaccines that is unprecedented. However, at the moment there are only four major vaccine manufacturers capable of adherence to modern standards, and their total capacity is inadequate for future needs. Indeed, in some particular situations, such as influenza vaccine for 2000–2001, their capacity was shown to be inadequate for present needs.

Where will new capacity for vaccine production come from? There would appear to be only a limited number of possibilities: increased construction of facilities by the four majors; growth of biotechnology companies into major vaccine manufacturers; growth of regional small manufacturers in countries such as Brazil, Cuba, India, Korea, and Japan; partnerships between regional and major manufacturers; and development of new institutions to make vaccines. Which of these solutions will be adopted remains to be seen, but it is certain that vaccine production must be expanded to fill the growing need.

REFERENCES

1. Plotkin, S. A., and Cadoz, M. (1997). The acellular pertussis vaccine trials: An interpretation. *Ped. Infect. Dis. J.* **16**:508–517.
2. Dowdle, W., and Birmingham, M. (1997). The biologic principles of poliovirus eradication. *J. Infect. Dis.* **175**(Suppl. 1):S286–S292.
3. Chang, M. H., *et al.* (1997). Universal hepatitis B vaccination in Taiwan and the incidence of hepatocellular carcinoma in children. *New England J. Medicine* **336**(26):1855–1859.
4. Black, S., Shinefield, H., Fireman, B., *et al.* (2000). Efficacy, safety and immunogenicity of heptavalent conjugate pneumococcal vaccine in children. *Ped. Infect. Dis. J.* **19**:187–195.
5. Murphy, T. V., Gargiullo, P. M., Massoudi, M. S., *et al.* (2001). Intussusception among infants given an oral rotavirus vaccine. *New England J. Medicine* **344**:564–572.
6. Peltola, H., Heinonen, O. P., Valle, M., *et al.* (1994). Elimination of indigenous measles, mumps and rubella from Finland by a 12-year, two-dose vaccination program. *New England J. Medicine* **331**:1397–1402.
7. Belshe, R. B., Mendelman, P. M., Treanor, J., *et al.* (1998). The efficacy of live attenuated, cold-adapted, trivalent, intranasal influenzavirus vaccine in children. *New England J. Medicine* **338**:1405–1412.
8. Hayward, A. R., Buda, K., Jones, M., *et al.* (1996). Varicella zoster virus-specific cytotoxicity following secondary immunization with live or killer vaccine. *Viral Immunol.* **9**:241–245.
9. Plotkin, S. A. (1999). Vaccination against the major infectious diseases. *C.R. Acad. Sci., Paris* **322**:943–951.
9a. Tartaglia, J. (1998). Recombinant poxvirus vaccine candidates: Update and perspectives. *Res. Imm.* **149**:79–82.
10. Caley, I. J., Betts, M. R., Davis, N. L., *et al.* (1999). Venezuelan equine encephalitis virus vectors expressing HIV-1 proteins: Vector design strategies for improved vaccine efficacy. *Vaccine* **17**:3124–3135.
11. Berencsi, K., Gyulai, Z., Gonczol, E., *et al.* (2001). A canarypox vector expressing cytomegalovirus phosphoprotein 65 induces long-lasting cytotoxic T cell responses in HCMV seronegative volunteers. *J. Infect. Dis.* **183**:1171–1179.
12. Thanavala, Y., Yang, Y.-F., Lyons, P., *et al.* (1995). Immunogenicity of transgenic plant-derived hepatitis B surface antigen. *Proc. Natl. Acad. Sci. U.S.A.* **92**:3358–3361.
13. Cichutek, K. (2000). DNA vaccines: Development, standardization and regulation. *Intervirology* **43**:331–338.
14. Excler, J. L., and Plotkin, S. (1997). The prime–boost concept applied to HIV preventive vaccines. *AIDS* **11**(A):S127–S137.
15. Aguado, T., Engers, H., Pang, T., and Pink, R. (1999). Novel adjuvants in clinical testing, November 2–4, 1998, Fondation Mérieux, Annecy, France: A meeting sponsored by the World Health Organization. *Vaccine* **17**:2321–2328.

16. Glenn, G. M., Scharton-Kersten, T., and Arving, C. R. (1999). Advances in vaccine delivery: Transcutaneous immunization. *Exp. Opin. Invest. Drugs* 8:797–805.

17. Naso, R., and Fattom, A. (1996). Polysaccharide conjugate vaccines for the prevention of gram-positive bacterial infections. *In* "Novel Strategies in Design and Production of Vaccines," J. Cohen and A. Shafferman, Eds. New York: Plenum Press.

18. Flanders, G., Graves, P., and Rewers, M. (1999). Prevention of type 1 diabetes from laboratory to public health. *Autoimmunity* 29:235–246.

19. Gellin, B. G., and Schaffner, W. (2001). The risk of vaccination—The importance of "negative" studies. *New England J. Medicine* 344:372–373.

20. Plotkin, S. A. (2001). Vaccines in the 21st century. *Infect. Dis. Cl. N.A.* 15(1):307–327.

6 DISEASE STATES AND VACCINES: SELECTED CASES

PART B. Polio

CIRO A. DE QUADROS
Pan American Health Organization, Washington, DC 20037

I. INTRODUCTION

Polio was depicted in stiles in ancient Egypt during the eighteenth dynasty (1580–1350 BC). It is also believed that epidemics of "clubfoot" described by Hippocrates, the father of medicine, and Galen, the Roman physician, may have been due to poliomyelitis, as they referred to both congenital clubfoot as well as clubfoot that occurs in early infancy.

The first modern clinical description of the disease was made by the Italian physician Giovanni Battista Monteggia in 1813. Subsequently, the German physician Jacob von Heine published a most detailed description of the disease in 1840. Karls Oskar Medin, a Swedish pediatrician, described several characteristics of the disease during a major epidemic in Scandinavia, including the milder, atypical form without paralytic illness, which is nonetheless important in the chain of transmission of the disease.

The search for the causative agent of the disease spanned over 50 years and was completed in 1908 when Karl Landsteiner, an Austrian immunologist, and his assistant, E. Popper, reproduced the disease in monkeys (*Cynocephalus hamadrias* and *Macaca mullata*), showing lesions in their spinal cord exactly like the ones observed in human poliomyelitis. By 1909, the entire scientific community was certain of the viral etiology of the disease, and by 1912 it was generally agreed that poliomyelitis was an infectious disease with epidemic characteristics. The work of Landsteiner gave origin to the idea that a vaccine could prevent the disease, and he was awarded the Nobel prize for his work.

In the United States, initial efforts to fight polio dealt mainly with the rehabilitation of its crippling aspects after Franklin D. Roosevelt was paralyzed by polio in the early 1920s. The search for a vaccine was intensified when the National Foundation for Infantile Paralysis (NFIP) was formed in 1938 led by Basil O'Connor, a former law partner of President Roosevelt. The NFIP was an offspring of the President's Birthday Ball Commission (PBBC), and its main focus was to raise funds to fight poliomyelitis, with over 10% of funds being targeted at research. In the late 1940s, the newly created World Health Organization established an Expert Committee on Poliomyelitis, which further outlined critical areas for research [1].

A major breakthrough came when John F. Enders, Thomas H. Weller, and Frederick C. Robbins cultivated strains of the three serotypes of poliovirus in a variety of extraneural human tissues. Their work earned them the Nobel prize and opened the way for the development of vaccines against poliomyelitis.

II.   THE DISEASE AND ITS EPIDEMIOLOGY

The causative agent of poliomyelitis is an enterovirus of the family *Picornaviridae*. The genome is a single linear molecule of single-stranded RNA, appearing smooth and round in outline by electron microscopy. There are three antigenic types of polioviruses called types 1, 2, and 3. All three can cause paralysis, although type 1 causes paralysis most often, type 3 less frequently, and type 2 rarely does so. Most epidemics are due to type 1. Humans are the only known reservoir [2].

Infection is spread from person to person. The mouth is the usual site of entry, and it first multiplies in the lymph nodes in the pharynx and gastrointestinal tract. Once it has entered the body, it invades local lymphoid tissue, enters the bloodstream, and then may invade certain types of nerve cells. In the process of intracellular multiplication, the virus damages or completely destroys these nerve cells. Fecal–oral transmission is most common in developing countries where sanitation is poor, whereas oral–pharyngeal transmission is likely to be more common in industrialized countries and during outbreaks.

Most persons infected with the wild poliovirus exhibit symptoms that are associated with minor illnesses, such as mild fever, muscle pains, headache, nausea, vomiting, stiffness of the neck and back, and, less frequently, signs of aseptic (nonbacterial) meningitis. Unapparent (subclinical) infections are common, ranging between 100:1 and 1000:1, depending upon the strain of the poliovirus. The incubation period, from exposure to the virus to the onset of first symptoms, averages 7–10 days and ranges from 4 to 30 days.

The paralysis is typically flaccid, more commonly affecting the legs than the arms, and asymmetric, developing rapidly after onset, usually within 4 days. Sequelae are generally present 60 days after the onset of paralysis. Fever is usually present before the onset of paralysis. The case–fatality rate varies between 2 and 20% among persons who develop the paralytic form of the disease and up to 40% if there is bulbar or respiratory involvement [3].

Laboratory confirmation is critical to the final diagnosis, as many other clinical conditions may cause similar symptoms. Stool specimens collected for virus culture from cases of acute flaccid paralysis (AFP) and their contacts are the most sensitive and effective way to rule out wild poliovirus transmission. Cerebrospinal fluid (CSF), throat specimens, or blood are not likely to yield virus, and therefore collection of these materials is not routinely recommended. In the event of death, a definite diagnosis of polio can be made or rejected by examining the spinal cord.

Among the many conditions that may be confused with polio are Guillain–Barré syndrome (GBS) and transverse myelitis. Other conditions that may be clinically similar to paralytic poliomyelitis include traumatic neuritis, tumors, meningitis–encephalitis (less frequent), and illnesses produced by a variety of toxins. The most prominent difference between poliomyelitis and other causes of AFP is that for polio the paralytic sequelae are generally severe and permanent, whereas for many other causes of AFP paralysis tends to resolve or improve within 60 days of onset [4].

All nonimmunized persons are susceptible to polio. Epidemiological evidence shows that infants born to mothers with antibodies are protected naturally against paralytic disease for a few weeks. Immunity is obtained from infection with the wild virus and from immunization. Immunity following natural infection (including unapparent and mild infections) results in both humoral and local intestinal cellular responses. Such immunity is thought to be lifelong and can serve as a block to infection with subsequent wild viruses, therefore helping to break chains of transmission. There is thought to be little, if any, cross-immunity between poliovirus types.

III. POLIO VACCINES

Currently two effective polio vaccines are available. The first is inactivated polio vaccine (IPV), which was developed by Jonas Salk and tested successfully in a field trial conducted by Thomas Francis in 1954 in which over 200,000 children (the "polio pioneers") received the vaccine without any serious reactions or accidents, becoming available in 1955. Since then it has been used successfully in several industrialized countries [5].

Subsequently, Albert Sabin developed the live, attenuated oral poliomyelitis vaccine (OPV), which was first used in mass campaigns in 1959 [6]. The OPV is one of the most heat-sensitive vaccines in common use and should be kept frozen or at a temperature below 8°C at all times. A number of key issues favor the choice of OPV for use in developing countries as well as for an eradication program. These include the development of intestinal immunity and ability to reduce intestinal spread of wild virus, duration of immunity, ease of administration in both routine and mass campaigns, and cost. Probably the most critical issue relates to the effect of the vaccine on wild poliovirus transmission. It has been well-documented that the use of OPV can successfully interrupt wild poliovirus transmission in both developed and developing countries. Vaccination with the inactivated virus vaccine (IPV) protects against clinical disease and suppresses pharyngeal excretion of the virus, conferring good humoral immunity, but it has less effect on intestinal excretion. Thus, vaccination with IPV, though avoiding the possibility of a reduced number of vaccine-associated paralytic cases (VAPP) due to the use of OPV, does not provide resistance to the carriage and spread of wild virus in the community and would have comparatively little effect on the transmission of the wild poliovirus in developing countries, where transmission is primarily by the fecal–oral route.

The overall risk in the United States for paralytic polio associated with OPV in vaccine recipients is 1 case per 5.2 million doses distributed. The risk of VAPP cases in vaccine recipients for the first dose is 1 case per 1.3 million doses. The experience in Latin America has been similar: the overall risk in vaccine recipients in Latin America is 1 case per 4.2 million doses administered, whereas the risk for recipients of first doses is 1 per 1.5 million doses distributed. Vaccine-associated cases are usually due to types 3 and 2.

Under ideal conditions in temperate countries, a primary series of three doses of OPV produces seroconversion to all three virus types in over 95% of vaccine recipients and is thought to have a clinical efficacy of nearly 100%. Three properly spaced doses of OPV should confer lifelong immunity. In developing, tropical countries, the serologic response to OPV may be only 85%. This may be due to breaks in the cold chain, interference with the vaccine's ability to produce intestinal infection because of the presence of other enteroviruses, presence of diarrhea that causes excretion of the virus before it can attach to the mucosal cell, and other factors.

Although the schedule may vary in some countries, for routine services it is recommended that three doses of trivalent OPV be applied at 4- to 8-week intervals, beginning at 6 weeks of age. It should be administered orally, that is, directly into the mouth. Each single dose consists of two or three drops (approximately 0.1 ml) of live oral poliovirus vaccine or the dosage recommended by the manufacturer. Polio vaccine may be given simultaneously with any other childhood immunization.

IV. POLIO AND ITS GLOBAL ERADICATION

A. History and Strategy

In 1985, the Pan American Health Organization (PAHO) launched an initiative to eradicate poliomyelitis from the western hemisphere by the year 1990. In August 1991, the last case of poliomyelitis in the western hemisphere was detected in Peru, and in September 1994, an international certification commission declared that indigenous transmission of wild poliovirus was interrupted in this region. The experiences in the Americas, led by PAHO, originated the initiative for the global eradication of poliomyelitis, launched by the World Health Assembly in May 1988. The strategies used in the polio eradication program are based on the epidemiology of the disease, the available vaccines, and their effective use.

The major components of the strategy used to interrupt transmission of wild poliovirus in the Americas were the achievement and maintenance of high levels of vaccination coverage in children less than 5 years of age, development of a surveillance system for the timely detection and investigation of persons with AFP ("probable" poliomyelitis cases), and creation of a surveillance system for detection of wild poliovirus supported by a regional network of diagnostic virology laboratories.

To achieve and maintain high levels of vaccination coverage in the population at risk, routine vaccination services have been supplemented by annual national immunization days (NIDs) in all countries in the region in which wild poliovirus was circulating in 1985. Mass vaccination campaigns have been conducted twice a year, with an interval of 4–6 weeks between campaigns, over a short period of time (usually 1 or 2 days) in which one dose of OPV is administered to all children in the target age group (generally children 0–4 years of age), regardless of prior vaccination history.

Following completion of the NIDs, vaccine coverage and surveillance data identified areas that either had low coverage and/or wild poliovirus transmission persisted. These areas were targeted for "mop-up" activities in the form of "house-to-house" vaccination of all children under 5 years of age, regardless of previous vaccination status. The same strategy has been used for the global eradication of the disease (see Table 1) [7].

B. Progress to Date

On October 2001, over 10 years had elapsed since a 2-year-old boy living in Junin, Peru, suffered from poliomyelitis; this was the last case of poliomyelitis associated with wild poliovirus isolation (type 1) in the Americas. On September 29, 1994, the International Commission for the Certification of Poliomyelitis Eradication (ICCPE), after thoroughly reviewing regional polio vaccination coverage, AFP surveillance, and laboratory data, concluded that wild poliovirus transmission had been interrupted in the Americas. The certification of the interruption of wild poliovirus transmission in the Americas is a major milestone in the global effort to eradicate poliovirus [8].

In 1988, the World Health Assembly resolved to eradicate polio globally by 2000. Although substantial progress toward this objective has been reported from all 6 WHO regions, 20 countries were considered to still have endemic transmission of wild poliovirus by the end of 2000. Additional efforts are needed to complete the eradication task. WHO has prepared a global plan of action that anticipates certification of polio eradication in 2005 (3 years after poliovirus circulation has been interrupted globally).

The reported incidence of poliomyelitis globally in 2000 was 2,880 confirmed cases reported to date (provisional), compared with 35,000 cases in 1988 (see Fig. 1). For the first 10 months of the year 2001, only 412 cases have been reported to WHO [9].

It is now 4 years since the last polio case was identified in the western Pacific region in Cambodia in 1997, and the region was certified as polio-free in the year 2000. The last case in the European region was in 1998 in southeastern Turkey. Poliovirus is now largely confined to a few countries in the eastern Mediterranean, southeast Asia, and African regions [10].

In the eastern Mediterranean region there were 261 wild-virus-confirmed cases in 2000 (175 from Pakistan). Wild poliovirus transmission remains widespread in Afghanistan and Pakistan. Following an outbreak during 1999, Iraq has not detected wild poliovirus since January 2000, despite increasingly

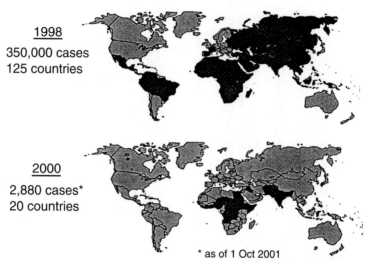

1998
350,000 cases
125 countries

2000
2,880 cases*
20 countries

* as of 1 Oct 2001

FIGURE 1   Polio-endemic countries, by number of poliovirus isolates from AFP cases, 2000. Source: WHO.

sensitive surveillance. Wild poliovirus is still endemic in Sudan where 4 virus-confirmed cases were found, but surveillance there is not yet sufficient to accurately assess the intensity of virus transmission. Improved surveillance in Somalia identified an outbreak of type 1 poliovirus (41 cases) in the Mogadishu area. Wild virus transmission continues in upper Egypt, with 3 virus-confirmed cases reported in the first half of 2000 and early 2001.

Only 272 virus-confirmed cases were reported from the southeast Asia region in 2000, compared to 1,161 cases in 1999. This rapid decrease was mainly attributable to a significant reduction in the number of virus-confirmed cases reported from India, the largest country in the region (264 virus-confirmed cases in 2000, compared to 1,126 in 1999). Transmission in Bangladesh was greatly reduced, with only 1 wild-poliovirus-associated case reported near Dhaka City in September 2000.

In the African continent, the quality of AFP surveillance has improved in 2000, with a regional nonpolio AFP rate of 1.5 per 100,000 compared to 0.8 per 100,000 in 1999. The proportion of AFP cases from which adequate stool samples were collected increased from 31% in 1999 to 52% in 2000. Wild poliovirus transmission is largely confined to Nigeria, central Africa, and the Horn of Africa. No virologically confirmed polio cases are being reported from north Africa and the countries of southern Africa. However, high-level surveillance has demonstrated the continued circulation of wild poliovirus in Egypt. At least 2 cases have been confirmed in 2001. Although poliovirus is not being found in east African countries, surveillance is not yet developed sufficiently to be confident that poliovirus has been eradicated there. Virological surveillance demonstrates major separate reservoirs of poliovirus in west Africa, Nigeria, and the Democratic Republic of Congo. Surveillance is not sufficiently developed in many countries to provide a clear picture of the circulation of poliovirus on the continent [11].

■ **TABLE 1 Global Polio Eradication Highlights**

1908: Karl Landsteiner and Edwin Popper discover the causative agent of polio in Vienna.

1938: March of Dimes launched by President Roosevelt.

1949: Enders, Weller, and Robbins cultivate the poliovirus in tissue culture.

1954: Killed inactived poliomyelitis vaccine developed by Jonas Salk.

1958: Attenuated oral poliomyelitis vaccine developed by Albert Sabin.

1985: The Pan American Health Organization (PAHO) launches an initiative to eradicate polio from the Americas by 1990.

1988: The World Health Assembly approves a resolution calling for the global eradication of polio by the year 2000, adopting the strategies developed by PAHO.

1991: Last case of poliomyelitis in the Americas is detected in a 3-year-old boy in Peru.

1994: Americas are certified as polio-free.

1997: Last case of polio reported in the western Pacific region, in Cambodia.

1998: Last case of polio reported in the European region, in Turkey.

1999: WHO Director General intensifies campaign by declaring "war on polio."

2000: Western Pacific region certified as polio-free.

2000: Vaccine-derived polio outbreak documented in Hispaniola, Americas.

Once the transmission of wild poliovirus in human populations around the world has been stopped, laboratories worldwide will represent the only remaining source of wild poliovirus. These laboratories may be storing specimens from known cases of poliomyelitis or other materials unknowingly infected with wild poliovirus (i.e., stool samples collected at a time and place of wild poliovirus circulation and stored under conditions known to preserve polioviruses). Global certification of polio eradication will require documentation that laboratory containment of wild poliovirus has been implemented in each country worldwide. During 2000, countries of the three polio-free WHO regions and selected polio-endemic countries began to create national inventories of laboratories containing wild poliovirus.

The ultimate goal of an eradication program is the discontinuation of the control measures, in this case, the cessation of immunization. In the region of the Americas, beginning July 12, 2000, and continuing into 2001, an outbreak caused by vaccine-derived poliovirus occurred on the island of Hispaniola in Haiti (7 laboratory-confirmed cases) and in the Dominican Republic (14 laboratory-confirmed cases). All cases were either unvaccinated or inadequately vaccinated. Genetic sequencing of the virus showed that the outbreak was caused by a vaccine-derived type 1 Sabin poliovirus that appeared to have acquired neurovirulence and the transmission characteristics of wild poliovirus. In response, the Dominican Republic and Haiti implemented nationwide OPV immunization campaigns, and transmission appears to have been interrupted, with the last confirmed cases reported in the Dominican Republic in January 2001 and in Haiti in July 2001. Occurrence of this phenomenon was observed once before, in Egypt between 1983 and 1993, and after Hispaniola in the Philippines in 2001 [12].

This phenomenon poses a major challenge for the cessation of immunization and requires urgency for the implementation of a research agenda to address the gaps in scientific knowledge regarding this issue. Therefore, the strategy for the cessation of immunization against polio can only be decided once there is a clear indication of how populations will be protected. This includes data on the possibility of circulation of vaccine-derived polioviruses and how the risk of potential circulation of a reintroduced wild virus, either by accident, from a laboratory, or from a chronic shedder, can be minimized.

Regardless of these important questions, the eradication of wild poliovirus, together with the eradication of smallpox, will be written in history as one of the greatest accomplishments of medicine, demonstrating the tremendous impact that vaccines can have on improving the lives and the well-being of humans.

REFERENCES

1. Paul, J. R. (1971). "A History of Poliomyelitis." Yale University Press, New Haven and London.
2. White, D. O., and Fenner, F. J. (1994). "Medical Virology," 14th ed. Academic Press.
3. Long, E. R. (1928). "A History of Pathology." Baltimore: Williams & Wilkins.
4. Pan American Health Organization (1988). "Polio Eradication Field Guide." Technical Paper No. 6.
5. Pan American Sanitary Bureau (1959). "Live Poliovirus Vaccines." Scientific Publication No. 44.
6. Pan American Health Organization (1971). "International Conference on the Application of Vaccines Against Viral, Rickettsial, and Bacterial Diseases of Man." Scientific Publication No. 226.
7. de Quadros, C. A. (1997). Global eradication of poliomyelitis. *Int. J. Infect. Dis.* 1(3):127–129.
8. Daniel, T. M., and Robbins, F. C. (1997). "Polio." Rochester, NY: University of Rochester Press.
9. World Health Organization (1997). Polio: The beginning of the end. WHO/EPI/GEN/97.03.
10. World Health Organization (1998). Resolution WHA41.28, WHA41/1988/REC/1, p. 72.
11. World Health Organization (2000). Weekly Epidemiological Record (WER), No. 49, 397–398.
12. World Health Organization (2001). Weekly Epidemiological Record (WER), No. 17, 126–131.

6 DISEASE STATES AND VACCINES: SELECTED CASES

PART C. Rubella

JENNIFER M. BEST* AND SIOBHAN O'SHEA†

*Department of Infection, Guy's, King's, and St. Thomas' School of Medicine, Kings College London, St. Thomas' Campus, London SE1 7EH, United Kingdom

†Department of Infection, Guy's and St. Thomas' Hospital Trust, London SE1 7EH, United Kingdom

I. DESCRIPTION OF THE DISEASE

A. Postnatally Acquired Rubella

Rubella is usually a mild disease when acquired in childhood. It is spread via the respiratory route, and, although efficient in crowded situations, it is not as infectious as measles. The incubation period is 13–20 days, following which a macular–papular rash may appear. The rash appears first on the face and spreads to the trunk and limbs; it seldom lasts for more than 3 days. Adults may experience a prodromal phase with malaise and fever and more severe symptoms. Lymphadenopathy may be present for 1 week before and up to 14 days after onset of the rash. The suboccipital, postauricular, and cervical lymph nodes are most frequently affected. Joint pains (arthralgia–arthritis) usually start as the rash disappears and last for 3–4 days or occasionally for up to a month. They occur in up to 70% of postpubertal females, but are less frequent in prepubertal females and males. Although it has been suggested that rubella may be associated with chronic joint disease, there is no convincing evidence to support this [1]. Other rare complications are hemorrhagic conditions, including thrombocytopenia (1 per 3000 cases) and encephalitis (1 per 5000 cases).

Up to 30% of cases of rubella are subclinical. It may also present atypically with minimal lymphadenopathy and a rash that fades quickly. Rubella cannot be diagnosed clinically as other virus infections (e.g., enteroviruses and measles) may induce a typical rubelliform rash, whereas parvovirus B19, Dengue, Chikungunya, and Ross River viruses may induce a rubelliform rash with arthralgia. Thus, it is important that laboratory techniques be used if a definitive diagnosis is required. This is particularly important when a pregnant woman develops a rubelliform rash or is in contact with a rubella-like illness. A more detailed description of clinical rubella and laboratory diagnosis is given elsewhere [2–4].

B.   Risk of Rubella in Pregnancy

If acquired in the first 12 weeks of pregnancy, rubella virus invariably crosses the placenta to infect the fetus and will cause congenital malformations in more than 80% of cases. Figure 1 shows the relationship between the clinical manifestations of congenital rubella and gestational age at maternal infection. Spontaneous abortion may occur in up to 20% of cases when rubella occurs during the first 8 weeks of pregnancy. The incidence of congenital malformations declines rapidly after 12 weeks of gestation, with an approximately 17% risk between 13 and 16 weeks of gestation and 6% between 17 and 20 weeks [3]. Current evidence suggests that there is no risk of congenital rubella infection if maternal rubella occurs within the 11 days after the last menstrual period and before conception [6]. In most countries, termination of pregnancy is considered when rubella infection is confirmed during the first trimester.

Rubella reinfection may occasionally occur in populations in which rubella continues to circulate. It is usually asymptomatic, but experimental studies have suggested that reinfection may occasionally be accompanied by a viremia. Prospective studies on women who have had laboratory-confirmed rubella reinfection in the first trimester of pregnancy suggest that the risk of congenital abnormalities is no more than 5%, significantly less than that of following primary infection [3].

C.   Congenital Rubella

Rubella virus interferes with organogenesis, and if infection occurs in early pregnancy it may be associated with a number of abnormalities (Table 1), the

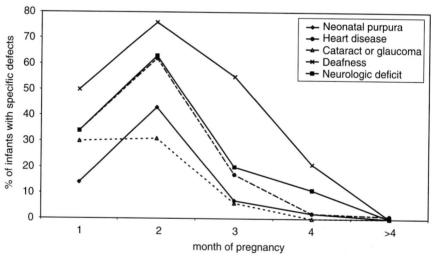

FIGURE 1   Relationship between the gestational age at time of maternal infection and clinical manifestations of congenital rubella, extrapolated with permission from Cooper et al. [5].

TABLE 1   Common Congenital Manifestations of Congenital Rubella

Permanent	Developmental	Transient
Sensorineural deafness	Sensorineural deafness	Low birth weight
Cataract	Peripheral pulmonary stenosis	Hepatosplenomegaly
Retinopathy	Mental retardation	Meningoencephalitis
Microphthalmia	Central language defects	Thrombocytopenic purpura
Patent ductus arteriosus	Diabetes mellitus	Bone lesions
Pulmonary valcular stenosis		
Ventricular septal defect		
Microcephaly		

most common being cataracts, sensorineural deafness, and heart defects [7]. Infected infants may excrete rubella virus for at least a year, and virus may persist in such sites as the brain and lens for even longer. Delayed manifestations of congenital rubella may become apparent in adolescence [3].

II. EPIDEMIOLOGY

A. Prevaccine

Before the introduction of rubella vaccination programs, rubella had a worldwide distribution. There is no animal reservoir. In temperate climates it was usual for outbreaks to occur in the spring, with an increased incidence every 3–4 years. The average age at infection was 9 years, and rubella was uncommon in preschool children. Approximately 80% of women of childbearing age were immune to rubella. Susceptible women were at risk of acquiring infection from their own children and from occupational exposure.

In some tropical countries, rubella is acquired in early childhood, and more than 90% of women of childbearing age are immune. However, there may be considerable differences in seroprevalence between rural and crowded city populations (Fig. 2). In some large countries, such as India, there may also be significant geographical differences in rubella susceptibility as a result of localized epidemics [8, 9]. Women in island communities in which rubella is not endemic are at particular risk from importation of the virus. Seroprevalence among such women was shown to be as low as 32% in some Caribbean islands before vaccination was introduced [9]. Until recently, rubella vaccine had not been extensively used in developing countries.

B. Postvaccine

The main purpose of rubella vaccination is to prevent women from acquiring rubella during pregnancy in order to protect the unborn child. As a consequence of successful vaccination programs, which are described in Section IV, rubella is now a rare disease in the United States, Finland, and Sweden.

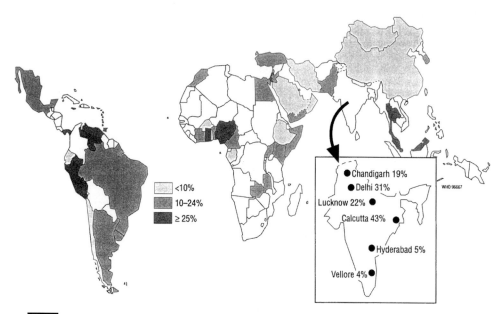

FIGURE 2   Rubella susceptibility in women of childbearing age: selected studies in developing countries prior to the use of vaccine [9] (reproduced with permission from the World Health Organization. The designations employed and the presentation of material on this map do not imply the expression of any opinion whatsoever on the part of the World Health Organization concerning the legal status of any country, territory, city or area or of its authorities, or concerning the delimitation of its frontiers or boundaries. Dotted lines represent approximate border lines for which there may not yet be full agreement.).

However, outbreaks have occurred in unvaccinated population groups (e.g., Hispanics and Amish) in the United States [10]. The United Kingdom saw a substantial reduction in the number of cases during the early 1990s, but outbreaks occurred among unvaccinated boys and young men in 1993 and 1996 (Fig. 3). Some outbreaks have occurred as a result of the importation of virus from countries that do not have effective vaccination programs. Thirty-one women acquired rubella during pregnancy between 1991 and 1996. Nineteen of the women had not received rubella vaccine as they were not born in the United Kingdom, while eight acquired rubella outside the United Kingdom [11]. Although most European countries have adopted universal childhood vaccination with measles–mumps–rubella (MMR) [12, 13], seroepidemiological studies show that more than 5% of 2- to 9-year-old children remain susceptible to rubella in most countries [14].

III.   RUBELLA VACCINES

A.   Development of Rubella Vaccines

The first rubella vaccines were developed in the 1960s and licensed in 1969–1972 [15, 16]. Ten vaccine strains have been licensed (Table 2). The RA27/3 vaccine

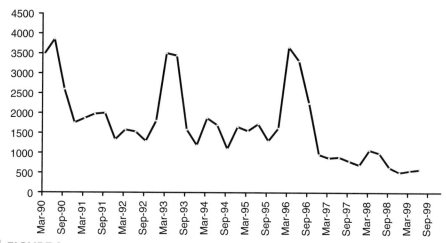

FIGURE 3 Incidence of rubella in England and Wales, 1990–1999 (courtesy of the Communicable Disease Surveillance Centre, London).

strain is now used worldwide with the exception of Japan, where locally produced vaccines are used [18, 19]. As the attenuated vaccines are safe and induce durable immune responses in the majority of vaccinees, recombinant or subunit vaccines have not been developed.

The RA27/3 vaccine is free of animal serum, but it contains 0.4% human albumin and 25–50 µg/ml neomycin. It is lyophilized and it should be stored at 2–8°C and protected from light before reconstitution with sterile distilled water provided by the manufacturer [18]. At room temperature there is a significant loss of potency after 3 months, and this is accelerated at higher temperatures. Potency is maintained long term when lyophilized vaccine is stored frozen. A cold chain is therefore required when conducting immunization programs in tropical areas. The RA27/3 vaccine dose should be at least 1000 plaque-forming units (PFU) of virus delivered subcutaneously. This strain may also be administered intranasally, when 95% seroconversion rates have been obtained. Intranasal vaccination has not been adopted for routine use, however, due to the higher dose required (10,000 PFU) and the difficulty of delivering the vaccine successfully by this route to small children.

B. Vaccine Efficacy and Immune Responses

All rubella vaccines have been shown to induce an immune response in approximately 95% of susceptible vaccinees. Neutralizing and hemagglutination–inhibition (HAI) antibodies are associated with protection. Antibodies induced by vaccination are usually at a lower concentration than those induced by naturally acquired infection. The RA27/3 strain is now used by most manufacturers as it induces an immune response most comparable to that following natural infection [19]. Reinfection, as demonstrated by a significant rise in rubella immunoglobulin G (IgG) and a transient rubella IgM response, occurs more commonly in vaccinees than in those with naturally acquired immunity.

TABLE 2   Attenuated Rubella Vaccines

Vaccine	Strain derivation	Passage history (no. of passes)
Vaccine Strains in Current Use[a]		
RA27/3	Kidney of rubella-infected fetus	Human embryonic kidney (4) WI-38 fibroblasts (17-25)
BRD-2	Child with postnatal rubella in China, 1980	Human diploid cells (30)
TO-336	Pharyngeal secretion from a case of postnatal rubella, Toyama, 1967	Vervet monkey kidney (7) Primary guinea pig kidney (20) Primary rabbit kidney (3)
Takahashi	Throat washing from a case of postnatal rubella, Matsue, 1968	Vervet monkey kidney (4) Rabbit testicle cells (36) Rabbit kidney cells (1)
Matsuba	Throat washings from a patient with rubella in Kumamoto, 1969	Vervet monkey kidney (1) Swine kidney cells (60) Rabbit kidney (6)
TCRB19	Throat swab from patient with rubella in Tokyo, 1967	Vervet monkey kidney (1) Bovine kidney (20) Primary rabbit kidney (3)
Other Vaccine Strains		
Cendehill	Urine from a case of postnatal rubella, 1963	Vervet monkey kidney (3) Primary rabbit kidney (51)
HPV77	From an army recruit with rubella, U.S.A., 1961	Vervet monkey kidney (77)
HPV77.DE5	As above	Vervet monkey kidney (77) Duck embryo fibroblasts (5)

[a]Information on the Japanese vaccine strains from Shishido and Ohtawara [17].

Antibodies usually develop 10–28 days after vaccination, but testing should be delayed until 8 weeks if seroconversion is to be confirmed, as the response may sometimes be delayed. Individuals who have failed to seroconvert should be retested 3 months after vaccination, and anyone still susceptible should be revaccinated. The major classes of rubella-specific antibodies can be detected, including serum and nasopharyngeal IgA. Specific IgM can be detected in most vaccinees 3–8 weeks after immunization and may occasionally persist at low levels for a year or more. Antibodies are long-lasting in the majority of vaccinees, but in approximately 10%, concentrations may decline to low levels (<15 IU/ml) within 5–8 years [19]. Lymphoproliferative responses are generally low following immunization, which may contribute to why reinfection is more common in vaccinees than in those with naturally acquired immunity.

C. Adverse Reactions

Rubella vaccines are generally well-tolerated. Lymphadenopathy, rash, and joint symptoms (arthralgia–arthritis) may occur 10 days to 4 weeks after vaccination.

Joint symptoms may occur in up to 50% of adult women but in less than 10% of children, usually lasting for 3–4 days. Persistent joint symptoms are rare [3, 18] and there is no convincing evidence that rubella vaccine is associated with chronic joint disease [1].

D. Contraindications

As rubella is a live attenuated vaccine, it should not be given to immunocompromised patients, with the exception of HIV-positive individuals, or to women who are pregnant. Vaccine strains cross the placenta and infect the fetus, although they do not appear to cause fetal damage (see later discussion). Pregnancy should be avoided for 1 month after vaccination. Other contraindications are listed in Table 3 and covered in more detail elsewhere [3, 18]. Susceptible women who have received anti-D immunoglobulin may be vaccinated postpartum, but they should be tested for seroconversion 2–3 months later.

E. Vaccination in Pregnancy

Although rubella vaccination is contraindicated in pregnancy, it is inevitable that some pregnant women are inadvertently vaccinated. In such cases, a therapeutic abortion is not indicated, as current evidence indicates that vaccine strains will not damage the fetus. This evidence has been obtained from prospective studies in the United States, United Kingdom, Germany, and Sweden (Table 4). Although 13 out of 417 (3.1%) infants had laboratory evidence of congenital infection, only one had an abnormality compatible with congenital rubella. However, only 20% (135 of 661) of the women were known to have been vaccinated during the high-risk period, between 1 week before and 6 weeks after conception.

F. Measles, Mumps, and Rubella (MMR) Vaccine

Monovalent rubella vaccine is offered to susceptible adult women, but in many countries it is usually given to children as a combined measles, mumps, and rubella (MMR) vaccine. Antibody responses to the rubella component are similar to those obtained with rubella vaccine alone.

TABLE 3   Contraindications to Rubella Vaccine

Immunosuppression

Congenital immune deficiency

Pregnancy

Other live vaccine within previous 3 weeks

Febrile illness

Thrombocytopenia

Hypersensitivity to antibiotics contained in vaccine preparation (e.g., neomycin)

Passively acquired antibodies (human immunoglobulin, blood transfusion, maternal antibodies in children <12 months of age)

TABLE 4 Combined Data for Risk of CRS in Infants Born to Susceptible Women Whose Pregnancies Were Complicated by Rubella Immunization[a]

| Country | Live births to women receiving rubella immunization | | No. with evidence of infection/no. tested (%) | No. with abnormalities compatible with CRS |
	Within 3 months of conception or during pregnancy	In the high-risk period[b]		
U.S.A.	321	113	6/222 (2.7)	0/321
Germany[c]	260	NK[d]	3/144 (2.1)	0/260
Sweden	5	NK	0/5	0/5
UK[e]	75	22	4/46 (8.7)	1[g]/75
Totals	661	135 (20%)	13/417 (3.12)	1[g]/661

[a]Adapted from Best and Banatvala [3] with permission from John Wiley & Sons Ltd.
[b]U.S.A. used between 1 week before and 4 weeks after conception. UK used 1–6 weeks after LMP.
[c]G. Enders, personal communication, 1998.
[d]NK, not known.
[e]P. Tookey, personal communication, 2000.
[f]Three of the rubella IgM-positive infants were born to mothers who were inadvertently immunized within 6 weeks after LMP.
[g]One rubella IgM-positive infant had a heart murmur which had resolved by 2 months of age.

G. Safety of MMR Vaccine

MMR vaccines have been licensed since 1971, are used in 93 countries, and have been given to >500 million children worldwide. However, in 1998 Wakefield and colleagues [20] suggested that MMR immunization was associated with inflammatory bowel disease and autism. This study described 12 children whose parents or general practitioner reported that symptoms had appeared soon after MMR immunization. However, the article showed only a temporal association without evidence of causation, as discussed in the accompanying editorial [21]. Unfortunately, this study attracted considerable media attention and generated significant adverse publicity for the vaccine. However, several longitudinal studies have shown no increased incidence of inflammatory bowel disease or autism among MMR vaccinees [22–24]. The safety of licensed MMR vaccines has also been confirmed by many agencies [25–29]. The UK Department of Health [25] has advised that children should not be given monovalent measles, mumps, and rubella vaccines, instead of MMR, as they will remain at risk of acquiring infection during the intervals between each vaccine. There is also the risk that unlicensed monovalent vaccines might be used.

IV. RUBELLA VACCINATION PROGRAMS AND THEIR IMPACT

A. Vaccination Programs

In 1969, the United States adopted a universal rubella vaccination program, with vaccine offered to preschool children of both sexes. The aim was to interrupt the transmission of rubella and reduce the risk of pregnant women being exposed to the virus. The United Kingdom and other European countries adopted a selective vaccination program, which aimed to protect women of childbearing age. Vaccine was offered to 10- to 14-year-old girls and susceptible adult women who had been identified by rubella antibody screening. However, neither program was entirely successful. In the United States, vaccination of preschool children resulted in an increase in the average age at which infection was acquired. As a consequence, rubella outbreaks occurred among susceptible adolescents and young adults in the late 1970s. A vigorous vaccination campaign was therefore directed at adolescents and susceptible adult women who had missed vaccination in childhood, resulting in a further decline in the incidence of rubella and congenital rubella (Fig. 4). Although outbreaks have continued to occur in unvaccinated groups, only 20 cases of rubella were reported in the United States during the first 10 months of 2001 [30; S. Reef, personal communication, 2001].

In the United Kingdom, 90% uptake of rubella vaccine was achieved among prepubertal girls, and only 2–3% of women of childbearing age were susceptible to rubella by 1986–1988. However, as susceptible pregnant women continued to acquire rubella, primarily from their own children, this program was augmented in 1988 by offering MMR to all children at 12–15 months of age [31]. This was followed in 1994 by mass vaccination of all 5- to 16-year-old children with measles and rubella (MR) vaccine, in order to

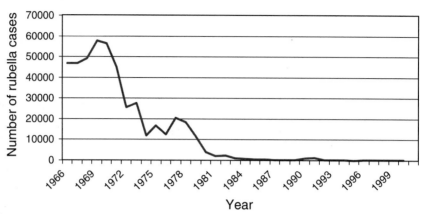

FIGURE 4   Reported rubella in the United States, 1966–2000 (courtesy of the Centers for Disease Control, Atlanta, Georgia).

prevent a predicted measles epidemic. As a consequence, the schoolgirl vaccination program was discontinued in 1996, but susceptible adult women, as
identified by antibody screening, continue to be vaccinated. It is particularly
important that women born in countries without rubella vaccination programs are targeted.

Children in the United Kingdom now receive a second dose of MMR
before they enter school at age 5. Two doses of MMR are also given to children in the United States and many other European countries [13, 31]. A study
has suggested that few countries in Europe have sufficient coverage with
MMR or rubella vaccine to prevent further outbreaks of rubella [14]; the two-
dose strategy should help to reduce the number of susceptible children.

B.   Current Initiatives

Valuable lessons have been learned from the experiences in the United States
and Europe. It is now recognized that vaccine uptake among children must be
maintained above 80% in order to prevent an alteration in rubella transmission dynamics, which may lead to increasing susceptibility among women of
childbearing age and an increase in the incidence of congenital rubella. Ideally,
susceptible adult women should be identified and vaccinated in parallel with
childhood vaccination. Alternatively, mass vaccination of adults and/or children may be employed.

In 1995–1996 the WHO Global Program on Vaccines and Immunization
conducted a worldwide survey on rubella, congenital rubella, and use of
rubella vaccination. This resulted in two reviews published in 1997, which
demonstrated that congenital rubella was a problem that had been largely
overlooked in many developing countries, in which it is a significant cause of
deafness, blindness, and mental retardation [9, 13]. These data prompted
WHO to organize a meeting in 2000 to review the global status of congenital
rubella and its prevention. This meeting recommended that all countries
should assess their rubella situation, make plans to introduce rubella vaccine,

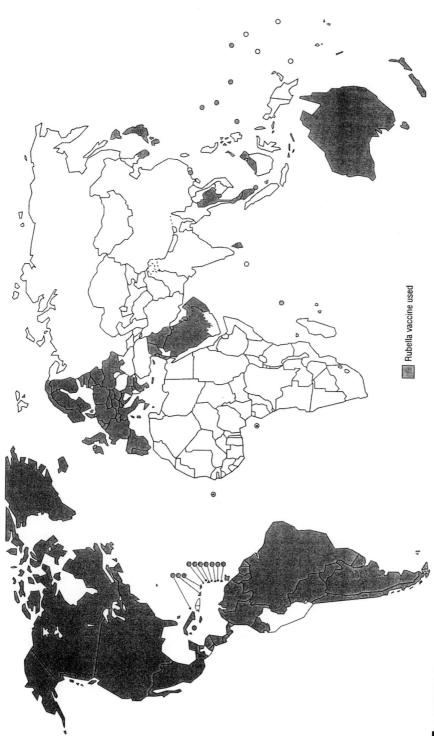

FIGURE 5 Countries–areas with rubella vaccine in their national immunization programs, April 2000 (reproduced with permission from the World Health Organization. The designations employed and the presentation of material on this map do not imply the expression of any opinion whatsoever on the part of the World Health Organization concerning the legal status of any country, territory, city or area or of its authorities, or concerning the delimitation of its frontiers or boundaries. Dotted lines represent approximate border lines for which there may not yet be full agreement.).

Rubella vaccine used

and establish surveillance for rubella and congenital rubella [32]. WHO also published guidelines for the surveillance of congenital rubella syndrome and rubella [33] and a position paper on rubella vaccines [34].

It has been suggested that, if rubella were combined with measles vaccine in all campaigns for the eradication of measles [35], rubella could also be eradicated for little extra cost. However, care must be taken to include the vaccination of susceptible adolescents and adult women, as they will initially remain at risk [32]. As of April 2000, 214 countries–areas reported use of rubella vaccine in their national immunization programs, but it is used in few countries in Africa and Asia (Fig. 5) [32]. A combination of childhood vaccination and mass vaccination, with MR or MMR vaccines, has been employed by some countries in South and Central America and the Caribbean [36, 37]. Such initiatives show that it may be possible to eradicate rubella. Without further progress, congenital rubella will continue to be a burden in developing countries, with occasional imported cases presenting in developed countries.

REFERENCES

1. Bosma, T. J., Etherington, J., O'Shea, S., *et al.* (1998). Rubella virus and chronic joint disease: Is there an association? *J. Clin. Microbiol.* **36**:3524–3526.
2. Best, J. M., and O'Shea, S. (1995). Rubella virus. *In* "Diagnostic Procedures for Viral, Rickettsial and Chlamydial Infections," 7th ed., E. H. Lennette, D. A. Lennette, and E. T. Lennette, Eds., pp. 583–600. Washington, DC: American Public Health Association.
3. Best, J. M., and Banatvala, J. E. (2000). Rubella. *In* "Principles and Practice of Clinical Virology," 4th ed., A. J. Zuckerman, J. E. Banatvala, and J. R. Pattison, Eds., pp. 387–418. Chichester, UK: John Wiley & Sons.
4. Gershon, A. A. (2000). Rubella virus (German measles). *In* "Principles & Practice of Infectious Diseases," 5th ed., G. L. Mandell, J. E. Bennett, and R. Dolin, Eds., pp. 1708–1714. Philadelphia: Churchill Livingstone.
5. Cooper, L. Z., Zirling, P. R., *et al.* (1969). Rubella: Clinical manifestations and management. *Am. J. Dis. Child* **118**:18–29.
6. Enders, G., Nickerl-Pacher, U., Miller, E., and Cradock-Watson, J. E. (1988). Outcome of confirmed periconceptional maternal rubella. *Lancet* **1**:1445–1447.
7. Cooper, L. Z. (1975). Congenital rubella in the United States. *In* "Progress in Clinical and Biological Research," S. Krugman and A. A. Gershon, Eds., pp. 1–22. New York: A. R. Liss.
8. Miller, C. L. (1991). Rubella in the developing world. *Epidemiol. Infect.* **107**:63–68.
9. Cutts, F. T., Robertson, S. E., Diaz-Ortega, J.-L., and Samuel, R. (1997). Control of rubella and congenital rubella syndrome (CRS) in developing countries, part 1: Burden of disease from CRS. *Bull. WHO* **75**:55–68.
10. Lindegren, M., Fehrs, L. J. *et al.* (1991). Update: Rubella and congenital rubella syndrome, 1980–1990. *Epidemiol. Rev.* **13**:341–348.
11. Tookey, P. A., and Peckham, C. S. (1999). Surveillance of congenital rubella in Great Britain, 1971–96. *Br. Med. J.* **318**:769–770.
12. Peltola, H., Heinonen, O. P., Valle, M. *et al.* (1994). The elimination of indigenous measles, mumps, and rubella from Finland by a 12-year, two-dose vaccination program. *New England J. Medicine* **331**:1397–1402.
13. Robertson, S. E., Cutts, R. F., Samuel, R., and Diaz-Ortega, J.-L. (1997). Control of rubella and congenital rubella syndrome (CRS) in developing countries, part 2: Vaccination against rubella. *Bull. WHO* **75**:69–80.
14. Pebody, R. G., Edmunds, W. J., Conyn-van Spaendonck, M., *et al.* (2000). The seroepidemiology of rubella in western Europe. *Epidemiol. Infect.* **125**:347–357.
15. Banatvala, J. E., and Best, J. M. (1989). Rubella vaccines. *In* "Recent Developments in Prophylactic Immunization," A. J. Zuckerman, Ed., pp. 155–180. Lancaster, UK: Kluwer Academic Publishers.

16. Perkins, F. T. C. (1985). Licensed vaccines. *Rev. Infect. Dis.* 7(Suppl. 1):S73–78.

17. Shishido, A., and Ohtawara, M. (1976). Development of attenuated rubella virus vaccines in Japan. *Japan J. Med. Sci. Biol.* **29**:227–253.

18. Plotkin, S. A. (1999). Rubella vaccine. *In* "Vaccines," 3rd ed., S. A. Plotkin and W. A. Orenstein, Eds., pp. 409–439. Philadelphia: W. B. Saunders Co.

19. Best, J. M. (1991). Rubella vaccines: Past, present and future. *Epidemiol. Infect.* **107**:17–30.

20. Wakefield, A. J., Murch, S. H., Anthony, A., *et al.* (1998). Ileal–lymphoid nodular hyperplasia, non-specific colitis, and pervasive developmental disorder in children. *Lancet* **351**:1327–1328.

21. Chen, R. T., and DeStefano, F. (1998). Vaccine adverse events: Causal or co-incidental? Commentary. *Lancet* **351**:611–612.

22. Peltola, H., and Heinonen, O. P. (1986). Frequency of true adverse reactions to measles–mumps–rubella vaccine. A double-blind placebo-controlled trial in twins. *Lancet* **1**:939–942.

23. Peltola, H., Patja, A., Leinikki, P., *et al.* (1998). No evidence for measles, mumps and rubella vaccine associated inflammatory bowel disease or autism in a 14-year prospective study. *Lancet* **351**:1327–1328.

24. Patja, A., Davidkin, I., Kurki T., *et al.* (2000). Serious adverse events after measles–mumps–rubella vaccination during a fourteen-year follow-up. *Ped. Infect. Dis. J.* **19**:1127–1134.

25. Chief Medical Officer, Chief Nursing Officer, and the Chief Pharmaceutical Officer (2001). "Current Vaccine and Immunisation Issues," pp. 1–12. London: Department of Health.

26. Elliman, D. A. C., and Bedford, H. E. (2001). MMR vaccine—Worries are not justified. *Arch. Dis. Child* **85**:271–273.

27. Halsey, N. A., Hyman, S. L., and Conference Writing Panel (2001). Measles–mumps–rubella vaccine and autistic spectrum disorder: Report from the New Challenges in Childhood Immunizations Conference convened in Oak Brook, IL, June 12–13, 2000. *Pediatrics* **107**:1–23.

28. Miller, E. (2001). MMR vaccine—Worries are not justified. Commentary. *Arch. Dis. Child* **85**:273–274.

29. World Health Organization (1998). Expanded Programme on Immunization (EPI). Association between measles infection and the occurrence of chronic inflammatory bowel disease. *Weekly Epidemiol. Record* **73**:33–40.

30. Reef, S. E., Plotkin, S., Cordero, J. F., *et al.* (2000). Preparing for elimination of congenital rubella syndrome (CRS): Summary of a Workshop on CRS elimination in the United States. *Clin. Infect. Dis.* **31**:85–95.

31. Miller, E., Waight, P., Gay, N., *et al.* (1997). The epidemiology of rubella in England and Wales before and after the 1994 measles and rubella vaccination campaign: Fourth joint report from the PHLS and the National Congenital Rubella Surveillance Programme. *Commun. Dis. Rep.* **7**:R26–32.

32. World Health Organization (2000). Preventing congenital rubella syndrome. *Weekly Epidemiol. Record* **75**:290–295.

33. World Health Organization (1999). Guidelines for surveillance of congenital rubella syndrome and rubella. Field test version, May (1999). WHO/V&B/99.22. WHO, Geneva.

34. World Health Organization (2000). Rubella vaccines. WHO position paper. *Weekly Epidemiol. Record* **75**:161–172.

35. World Health Organization (1999). Manual for the laboratory diagnosis of measles viral infection, December 1999. WHO/V&B/00.16. WHO, Geneva.

36. Centers for Disease Control (2001). Nationwide campaign for vaccination of adults against rubella and measles—Costa Rica, 2001. *Morbid. Mortal. Weekly Rep.* **50**:976–979.

37. Irons, B., Lewis, M, J., Dahl-Regis, M., *et al.* (2001). Strategies to eradicate rubella in the English-speaking Caribbean. *Am. J. Pub. Health* **90**:1545–1549.

6 DISEASE STATES AND VACCINES: SELECTED CASES

PART D. Pertussis

STEFANIA SALMASO

Reparto Malattie Infettive, Laboratorio di Epidemiologia e Biostatistica, Istituto Superiore di Sanità, 00161 Rome, Italy

I.   INTRODUCTION

Pertussis is a bacterial disease due to *Bordetella pertussis*, which causes some 20–40 million cases worldwide with 200,000–400,000 fatalities each year. Ninety percent of these cases occur in developing countries. In the United States about 7000 cases occur annually, and in western Europe more than 400,000 cases were reported in 1998. Humans are the only reservoir for infections with *B. pertussis*, and therefore control of the disease theoretically could lead to its elimination or eradication. Vaccination is universally recommended. Vaccines currently available are made of the entire bacterial cell (whole cell vaccines) or with selected purified and inactivated virulence factors (acellular vaccines). Whole cell vaccines were associated with adverse events, and the lower reactogenicity of the acellular products led to high acceptance of pertussis vaccination in most western countries. However, the persistence of immunity either naturally acquired or induced by vaccination is not lifelong, and to date no biological marker of protective immunity has been identified. Despite the great amount of research performed on various aspects of controlling and eliminating pertussis, many issues remain to be clarified, and a resurgence of the disease has been reported in countries with long records of good vaccination coverage in infancy. The priorities for pertussis, therefore, are to improve surveillance to better identify the burden of the disease, to reduce and eliminate pertussis as a cause of death and serious illness, to improve coverage of childhood immunization, and to identify the need for additional vaccination in adolescence and adult life.

II. PERTUSSIS

Pertussis is a bacterial infection of the respiratory mucosa caused by the bacterium *Bordetella pertussis*. The genus *Bordetella* includes seven species with pathogenic potential for different host organisms. Members of the genus are aerobic, gram-negative, minute coccobacilli. *B. pertussis*, *B. parapertussis*, and *B. bronchiseptica* have been studied since the early twentieth century and have been found to be genetically very similar [1]. *B. pertussis* can undergo genetic changes known as "phase variation" and phenotypic "modulation" in response to environmental changes. Most of the phenotypic differences among the *Bordetella* species can be explained by adaptation to different mammalian hosts, leading to different expression of genes common to the three species. Expression of the most important virulence factors for human hosts are regulated by a *bvg* gene, which is responsive to temperature and allows the expression of genes coding virulence factors only at body temperature and in temporal sequence. The *bvg* gene system originally may have been required for *B. pertussis* to become infectious to humans [2].

At least two *Bordetella* species have been identified to cause disease in humans: *B. pertussis* and *B. parapertussis*. Humans are the only identified host for *B. pertussis*, whereas *B. parapertussis* may also infect sheep. However, the strains of ovine origin have been shown to be genetically different from strains isolated from humans, and humans are therefore the only reservoir for human infections with *B. pertussis* and *B. parapertussis*. For this reason, no perfect animal model is available for the human disease. Many laboratory assays are based on experimental infection of mice, even though they do not develop clinical pertussis [3].

A. Virulence Factors

Virulence factors of *B. pertussis* are numerous and can be classified in two groups according to their biological effect: (1) adhesins (FHA, PRN, and FIM), important for the development of the infection, and (2) toxins (PT, ACT, TCT, and DNT), important for the development of clinical symptoms [4].

1. Adhesins

a. Filamentous Hemagglutinin (FHA)

FHA is a protein found on the surface of *B. pertussis* and secreted by the bacterium. The protein is also produced by *B. parapertussis*, and antibodies cross-reacting with FHA can also be elicited by nontypable *Haemophilus influenzae* [5]. In pertussis, FHA is involved in bacterial adhesion to ciliated epithelium, and in an animal model, secreted FHA appears to be used by other pathogenic bacteria as a heterologous adhesin. This may explain in part the occurrence of secondary respiratory infections following pertussis. Strains of *B. pertussis* lacking FHA are still virulent in mice. Studies have shown that FHA from *B. pertussis* influences the cell-mediated immunity response to other concurrent antigens by interacting with receptors on macrophages and interfering with the production of various interleukins [6].

b. Pertactin (PRN)

PRN is a 69-kD a protein found in the outer bacterial membrane. It too is involved in bacterial adhesion, and, as is the case for FHA, mutant strains lacking PRN are still virulent in mice. Purified PRN induces protection in animal models, and two independent studies on children have shown that high levels of anti-PRN antibodies, together with antibodies against PT and FIM, are associated with a low probability of developing the disease [7–9].

c. Fimbriae (FIM)

FIM are also known as agglutinogens 2 and 3 because they induce antibodies able to agglutinate the bacterium (agglutinins). It is generally accepted that fimbriae are also involved in the adhesion of B. pertussis to the ciliated cells, although their precise role in the development of infection has not been clarified. Studies on the efficacy of the whole cell vaccines indicated that the acquisition of agglutinins was associated with protection from pertussis [10]. Two studies examining serological markers for clinical protection have suggested that antibodies against fimbriae, associated with antibodies against PRN, correlate with clinical protection and, therefore, suggest that these two antigens may play a relevant role in the development of infection [7–9].

2. Toxins

a. Pertussis Toxin (PT)

PT is a toxin with a two-component A–B structure similar to that secreted by a number of other bacteria (e.g., Escherichia coli, Vibrio cholerae) [11]. Part A is an enzymatically active subunit (S1) that has ADP ribosyltransferase activity on certain members of the G-protein family located on mammalian cells that are involved in transduction of receptor signals on cell membranes. The membrane receptors affected by the toxin are involved in signals of diverse nature, including neurotransmitters, hormones, and growth factors. Biologic effects of PT include hypoglycemia and hyperinsulinemia, attenuated hyperinsulinemia in response to adrenaline, and lymphocytosis. The B part of PT is an oligomer made of four subunits (S2–S5) necessary for efficient binding to receptors of target cells. Part B is, by itself, able to induce some PT-specific effects, such as T-lymphocyte mitogenesis and platelet activation. PT has been shown in experimental conditions on animals to modulate immune response and to enhance Th1 and Th2 responses to co-injected antigens [12]. Because of the biologically important effects of PT, pertussis has been considered as a toxin-mediated disease comparable to diphtheria [13].

b. Adenylate Cyclase (ACT)

ACT is an extracytoplasmatic enzyme capable of entering mammalian cells and impairing leukocyte functions. Mutant strains defective in ACT are avirulent in some animal models.

c. Tracheal Cytotoxin (TCT)

TCT is a protein located on the bacterial membrane causing ciliostasis, inhibiting DNA synthesis, and killing tracheal epithelial cells in vitro. It seems to play an important role in preventing the regeneration of ciliated cells damaged by the infection.

d. Dermonecrotic Toxin (DNT)

DNT causes vascular smooth muscle contractions, resulting in ischemic necrosis in animal models. Its role in the pathogenesis of pertussis is not well-understood [1].

B.   Infection and Disease Progression

The infection and its progression to disease are characterized by three stages: (1) attachment and colonization of B. pertussis in the respiratory tract of the host; (2) production of local damage and onset of aspecific symptoms; and (3) systemic disease and typical cough. The incubation period before the onset of symptoms ranges from 1 to 3 weeks, during which time the bacteria replicate and produce a sufficient amount of toxins to cause clinical symptoms. Onset of pertussis is insidious with coryza and mild cough; fever occurs infrequently in the first catarrhal phase. These initial symptoms result from damage to the ciliated epithelium.

After approximately 1 week, the cough in unvaccinated children becomes paroxysmal and is characterized by the sudden onset of protracted spasms of cough followed by an inspiratory whoop [14]. The probability of recovering B. pertussis by culturing the mucous from the nasopharyngeal tract is greatest within 15 days from the onset of cough. Symptoms in this phase are mainly due to PT, whose effects persist after the natural clearance of the bacterium from the respiratory tract. In many individuals, the coughing spasms may result in cyanosis and posttussive vomiting. Small infants, who often lack the posttussive whoop, may develop apnea after coughing episodes. In some children, the high intrathoracic pressure that develops during coughing spasms may result in subconjunctival hemorrhages, pneumothorax, or rectal prolapse. In a trial conducted in 1992–1994 in Italy, among 95 unvaccinated children between 7 and 23 months of age who had laboratory-confirmed B. pertussis infection, 70% had whoop, 93% spamodic cough, 78% posttussive vomiting, 68% posttussive apnea, and 47% cyanosis. One child experienced seizures without sequelae [15, 16]. This so-called paroxysmal phase lasts several weeks, during which the spasms become increasingly frequent. Between spasms, the child may have no respiratory signs or symptoms. In the children studied in Italy, the cough lasted for a median time of 61 days, and spasms occurred for a median of 37 consecutive days. A few children had a cough lasting up to 9 months.

The most severe complications are superimposed infections causing pneumonia pertussis. Surveillance in the United States reported that, in the period 1989–1991, 11% of pertussis patients of all ages developed pneumonia, 1.6% developed seizures, and 0.1% developed encephalopathy [17]. The latter two appear to be attributable to hypoxia caused by cough, directly related to the systemic effects of PT, or attributable to intracranial hemorrhages. The case–fatality rate in the North American studies was found to be 1.3% for infants in their first month of life and 0.3% for those between the ages of 2 and 11 months.

The convalescent phase, which follows the paroxysmal phase, is protracted. Over time, paroxysms become less frequent and less intense, although infections with common respiratory pathogens can precipitate the return of spasmodic cough. In this stage, laboratory confirmation of the infection by *B. pertussis* is obtained by demonstrating of a significant increase in the circulating antibodies of IgG or IgA against PT and FHA.

In vaccinated children, adolescents, and adults, the clinical picture of pertussis is less severe and the disease may be more difficult to diagnose. Culture and serologic confirmation are less sensitive at detecting infection in vaccinated children and adults. Because pertussis in these individuals may go unrecognized, and because naturally acquired immunity wanes over time, adults and adolescents may serve as the reservoir of *B. pertussis* in populations with high infant vaccination coverage [18, 19].

C. Frequency of Pertussis

Worldwide, pertussis was estimated by the World Health Organization (WHO) to cause some 20–40 million cases in 1999 with 200,000–400,000 fatalities each year [20]. Ninety percent of these cases occur in developing countries. In the United States about 7000 cases occur annually [21], and in western Europe more than 400,000 cases were reported in 1998, representing a substantial disease burden. The infection is airborne, and during the catarrhal phase it is highly contagious in close household settings, with attack rates as high as 90–100% of susceptible subjects [22]. Up to 50% of fully immunized or naturally immune individuals may develop subclinical infection after intense exposure [23] and may be able to transmit to other close contacts. Chronic asymptomatic carriage appears to be rare and of limited epidemiologic importance.

In the prevaccine era or in countries where vaccine coverage is low, the vast majority of cases occur in children between 1 and 5 years of age. Serological studies conducted in conditions of low vaccine coverage show that, by the age of 18 years, almost all of the subjects in the general population have immunological markers (either humoral or cellular) of exposure to *B. pertussis* [24]. In areas with high vaccination coverage in childhood, a considerable reduction in the incidence of pertussis has been recorded. In these settings, most cases occur in infants before the age at which the first dose of vaccine is administered and in adolescents and adults who escaped vaccination at younger ages or who have lost their immunity [25]. In situations where pertussis is endemic, natural or vaccine-induced immunity is likely to be maintained by frequent reexposure to *B. pertussis*. However, when childhood vaccination succeeds in reducing the number of cases, the persistence of acquired immunity may be shorter. In such settings, if transmission of infection is not interrupted, cases continue to occur in the large pool of adults whose immunity has waned. This mechanism is believed to be the reason for the increase in incidence of pertussis observed in areas that have had high pertussis vaccination coverage for decades, such as Finland, the United States, and France, but where the proportion of cases among the population older than 10 years is increasing [26, 27]. In the Netherlands, selective pressure on the circulating strains of

B. pertussis by extensive vaccination programs appears to have affected the incidence of the disease [28], as the bacterial strain included in the vaccine and those strains currently isolated from patients with pertussis show a wider genetic divergence than in the past. It has been suggested that such genetic changes in circulating strains of *B. pertussis* have caused the large epidemic of pertussis reported there since 1996.

D. Vaccination

Because pertussis is a worldwide problem, vaccination is universally recommended and has always been an integral part of the Expanded Program Immunisation (EPI). WHO estimates that approximately 80% of the world's children are vaccinated against pertussis and about 750,000 deaths are prevented annually in developing countries [20].

Primary immunization is usually provided by three doses of vaccine injected intramuscularly within the first year of life. However, schedules of vaccine administration vary between countries according to local epidemiology and logistics and the type of vaccine employed. In the United Kingdom, for example, an accelerated schedule of whole cell vaccine is used with doses at 2, 3, and 4 months of age, whereas in many other European countries, doses are administered at 2, 5, and 12 months [29]. In the United States they are given at 2, 4, and 6 months [14]. The first dose is not recommended before the 6th week of life (in order to avoid interference of maternal antibodies observed with the whole cell vaccine), and each subsequent dose should be administered at least 4 weeks apart. In addition, the policy of administering subsequent doses varies across countries: for example, no additional doses are recommended in the United Kingdom, a fourth ("reinforcing" or "booster") dose is recommended in the United States at 15–18 months of age, and a fifth dose is administered in many other countries before 6 years of age. Because the traditional whole cell vaccine was too reactogenic to be administered after the age of 6 years, immunization schedules prior to the mid-1990s did not include additional pertussis immunization after that age. However, the availability of the acellular vaccines has opened the possibility of further vaccination in adolescents and adults, and the issue is currently under debate.

III. VACCINES

A. Whole Cell Vaccines

1. Content

Whole cell vaccines have been available for more than four decades. They are prepared from a *B. pertussis* culture, which is concentrated and suspended in saline solution. Bacteria are killed with heat and treated with formalin in order to denature the toxic protein components [30]. In infant vaccines, pertussis antigens are combined with tetanus and diphtheria toxoids and adsorbed on aluminum hydroxide or phosphate as adjuvant. These vaccines are referred to as DTPw. There are a variety of commercial preparations of DTPw available in the world, and several countries, including many developing countries, produce the vaccine for their national needs. In addition to the various DTPw preparations, commercial vaccines now are available in many countries that combine DTPw with IPV (injectable inactivated polio vaccine) and Hib (*Haemophilus influenzae* type B).

2. Adverse Events

Whole cell vaccines have always been associated with a certain rate of adverse events, which have been considered acceptable in the face of danger posed by acquiring pertussis, especially in early infancy. Most of the events are not clinically serious, although about 40% of vaccinees develop a fever over 38°C (100.4°F) within 3 days following the administration of each dose, and about 30% develop local redness and swelling [14]. The occurrence of persistent crying, seizures, and hypotonic–hyporesponsive episodes has also been described. Systemic and local reactions appear to be attributable to the amount of different antigens included in the whole bacterial cell and to the presence of small amounts of active PT [13].

In the 1970s, concerns raised about the safety of the whole cell vaccines in Japan and the United Kingdom reduced public confidence in pertussis vaccination [31]. Despite the absence of sound scientific evidence, pertussis immunization was nearly abandoned in some countries (Japan, Sweden), with the effect of a resurgence of the disease. Whole cell vaccine was reintroduced in the United Kingdom, reaching a coverage close to 90% in 1990, but in other countries the uptake of pertussis vaccine remained low. The concerns stimulated research on new vaccines containing limited numbers of antigens from *B. pertussis* rather than extracts from the entire bacterial cell. These new vaccines are known as "acellular" vaccines.

B. Acellular Pertussis Vaccines

1. Content

Acellular pertussis vaccines are prepared by purifying and inactivating selected antigens among the identified virulence factors of *B. pertussis* [30]. To reduce reactogenicity to the lowest possible levels, researchers have attempted to minimize the number of antigens and maximize protection from the disease.

All acellular vaccines include PT because it is recognized to play a primary role in the development of disease. PT used in vaccines can be chemically or genetically inactive. Chemical inactivation is performed with formaldehyde, glutaraldehyde, or hydrogen peroxide treatment. Genetically inactived PT is obtained by a mutant strain whose enzymatic activity is impaired by the substitution of two amino acids in subunit S1 [32]. In addition to PT, most acellular vaccines include surface antigens such as FHA, PRN, and FIM, resulting in acellular vaccines with one to five different components (the two types of FIM are counted separately). A variant of the acellular vaccines known as Takeda vaccines also exists. This type of vaccine is produced according to a technique developed in Japan involving the copurification of *B. pertussis* antigens. They differ from other acellular vaccines in that the exact amount of each antigen cannot be predefined. A list of acellular vaccines and their antigenic content is shown in Table 1.

TABLE 1 Antigenic Composition per Dose of Acellular Pertussis Vaccines Studied

Manufacturer	PT (μg)	FHA (μg)	PRN (μg)	FIM (μg)
North American Vaccine	40			
Connaught (Biken)	23.4	23.4		
Glaxo SmithKline Beecham GSB	25	25		
Aventis Pasteur Merieux	25	25		
Chiron[a]	5	2.5	2.5	
Glaxo SmithKline Beecham	25	25	8	
Wyeth-Lederle (Takeda)	3.2	34.4	1.6	0.8
Aventis Connaught	20	20	3	5

[a]Genetically inactive PT.

Similar to whole cell vaccines, the acellular vaccines designed for use in infancy are combined with diphtheria and tetanus toxoids (indicated as DTPa). Additional combinations with Hib, IPV, and HBV have also been developed and licensed in the European Union and some Asian countries.

2. Adverse Events

As expected, the reactogenicity of the acellular vaccines was found in various studies to be much lower than that of the whole cell vaccine. Table 2 reports the frequency of adverse events following three doses at 2, 4, and 6 months of age for two, three-component acellular vaccines and one whole cell vaccine studied in the Italian Pertussis Trial [14]. The reported reactogenicity has been similar for all acellular vaccines studied to date. This has led to a rapid increase in acceptance of the acellular pertussis vaccination in many

TABLE 2 Percentages of Observed Events Out of the Total Number of Doses Administered in the Italian Pertussis Trial

Event	Vaccine received (total doses)			
	DTPa Chiron (13,713)	DTPa GSB (13,761)	DTPw Connaught (13,520)	DT only (4540)
Fever ≥ 38°C	5.3	8.0	41.0	4.2
Somnolence	21.9	21.9	37.4	21.9
Loss of appetite	13.9	14.0	24.5	13.2
Irritability	30.3	30.0	49.9	29.8
Vomiting	4.2	4.4	5.4	4.5
Local redness	8.4	9.7	26.4	8.3
Local swelling	7.3	9.3	26.1	6.5
Local pain	4.6	4.6	29.7	4.5

countries. Administration of booster doses with acellular vaccines at preschool age has been found to be associated with a higher frequency of common adverse events (especially at the site of injection) in children primed with the same vaccine [33].

IV. EFFICACY

Evidence of the efficacy of pertussis vaccines is reliant on clinical trials because animal models show poor correlation with human response, and the correlation between antibody response and clinical protection in humans is poor. Studies including various whole cell vaccines used in clinical trials demonstrate that they vary widely in their ability to induce antibodies against the major virulence factors (e.g., PT or FHA) and to prevent pertussis in children, even though each had passed the standard potency test in which mice are injected peritoneally with pertussis vaccine and challenged intracerebrally 2 weeks later with pertussis. These findings underline the inability of the experimental tests on mice to predict clinical efficacy as well as the large differences that exist among various commercial products and even among production lots. Moreover, the potency test is not suitable for assessing the efficacy of cellular vaccines [13].

The lack of antibody markers predicting efficacy is perhaps best illustrated by studies examining PT antibodies, which are correlated with protection in the mouse model. In recipients of whole cell vaccines, low PT antibody titers were induced, whereas levels of PT antibodies induced by acellular vaccines were high despite the fact that both vaccines protect against disease. Furthermore, the preexposure quantity of antibodies against PT poorly correlates with the probability of acquiring clinical pertussis [8–10] and clinical protection also persists when the level of vaccine-induced circulating antibodies decreases [34]. These findings suggest that protection is provided by a more complex synergy between humoral and cellular mediated immunity [35, 36].

Given the lack of identified immunologic markers (related to humoral or cellular immune response) that can serve as predictors of clinical protection from pertussis, the assessment of efficacy of each of the developed vaccines has required clinical studies designed to assess the fraction of pertussis cases prevented. In Japan, the acellular vaccines were introduced starting in 1981 on the basis of evidence from nonrandomized field trials, with administration starting at 2 years of age [37]. Since that time, however, proof of efficacy has been obtained from epidemiological studies using different study designs [14, 38–44]. In particular, in the early 1990s various randomized clinical trials (RCTs) involving one or more acellular vaccines were conducted in Europe and Africa to identify candidate vaccines for licensure. In these RCTs, which are considered superior in design to retrospective studies, the disease incidence in children vaccinated with acellular vaccines was compared to that of a placebo group of unvaccinated children [40], that of a group of children vaccinated with the whole cell vaccine, [41–43], or to both types of control groups [14, 39]. In these studies, vaccinated children were followed to detect clinical events compatible with pertussis, which were then evaluated extensively.

Others were so-called household studies in which the vaccinated and unvaccinated subjects were observed after an ascertained household exposure to a pertussis case [44].

In all of the studies conducted after 1991, a common definition for pertussis was included according to a WHO recommendation [45]. The definition of pertussis was a case with at least 21 days of spasmodic cough, in whom *B. pertussis* infection was confirmed in the laboratory by a positive culture of nasopharyngeal mucous or a significant increase in antibody titer against PT or FHA. The figures of vaccine efficacy commonly reported for each product have been computed by using this definition, although valuable information on the clinical performance of each product is also provided by computing the efficacy in preventing a wider spectrum of clinical events and in altering the length and type of cough. The estimated efficacy, according to case definitions including variable severity, obtained in clinical studies of various vaccines are reported in Table 3. The studies, conducted independently in different settings, provided consistent results. Most of the vaccines tested prevented at least 70% of the pertussis cases (defined according to the WHO recommendation) occurring in unvaccinated individuals. The multicomponent acellular pertussis vaccines were found to be more effective and prevented a higher proportion of milder pertussis. Acellular vaccines were also found to be more efficacious than one of the available whole cell pertussis vaccines, but less effective than two other whole cell vaccines.

V. OPEN ISSUES

As a result of the preceding studies, acellular pertussis vaccines are currently used in much of the developed world during infant primary immunizations, although most of the pertussis immunizations in the developing world are still performed with the less expensive whole cell vaccines. Although pertussis is an old disease, the advancements in knowledge of its natural history, pathogenesis, and interaction with the human immune system are relatively recent, and many points remain unclear. Humans are the only reservoir for pertussis, and therefore in theory the disease can be eradicated. However, the absence of knowledge about the persistence of immunity provided by vaccination and by natural infection, as well as the lack of serological markers for distinguishing immune from susceptible individuals, of standardized laboratory methods for confirming the diagnosis in common practice, and of standardized techniques for characterization of strains isolated in different contexts (e.g., latitude, vaccination coverage, type of vaccine used), makes it difficult to establish whether a target for the elimination of the disease can be set. To date, the immunization programs have been aimed at a reduction in pertussis morbidity and mortality without setting precise short-term or long-term operative objectives. Evidence on the persistence of the protection provided by the new acellular vaccines is limited [46–48], and whether it will significantly change over prolonged follow-up remains to be assessed.

To date, the priorities for pertussis, therefore, are to improve surveillance to better identify the burden of the disease in different geographical contexts

TABLE 3 Proportion of Prevented Cases of Pertussis of Variable Severity Estimated in Prospective Studies

Vaccine	Components	Design,[a] country, year	Duration of cough (days) Any type of cough >7 (%)	>14 (%)	>21 (%)	Spasmodic cough >7 (%)	>14 (%)	>21 (%)
North American Vaccine	PT	RCT, Sweden, 1992	54					71
Glaxo SmithKline Beecham	PT + FHA	RCT, Sweden,1992	44		54		57	59
Aventis Pasteur Merieux	PT + FHA	RT, Senegal, 1990			31			74
Glaxo SmithKline Beecham	PT + FHA + PRN	RCT, Italy, 1992	71	73	79		78	84
		HH, Germany, 1992						83
Chiron	PT + FHA + PRN	RCT, Italy, 1992	71	74	77		81	84
Connaught	PT + FHA + PRN + FIM	RCT, Sweden, 1992	78		81		82	85
Wyeth-Lederle	Takeda	RT, Germany, 1991	62					78

[a]RT, prospective randomized trial with nonrandomized controls; RTC, prospective randomized trial with randomized controls; HH, household study.

and age groups, to reduce and eliminate pertussis as a cause of death and serious illness, to improve coverage of childhood immunization, and to identify the need for additional vaccination in adolescence and adult life.

REFERENCES

1. Hewlett, E. L. (2000). *Bordetella* species. *In* Mandell GL, Bennet JE, Dolin R, ed. "Mandell, Douglas, and Bennet's Principles and Practice of Infectious Disease," 5th ed., G. L. Mandell, J. E. Bennet, and R. Dolin Eds., pp. 2414–2422. Churchill Livingston, Philadelphia.
2. Gerlach, G., von Wintzingerode, F., Middendorf, B. *et al.* (2001). Evolutionary trends in the genus *Bordetella. Microbes Infect.* 3:61–72.
3. Kendrick, P. L., Eldering, G., and Dixon, M. M. (1947). Mouse protection tests in the study of pertussis vaccine: A comparative series using the intracerebral route for challenge. *Am. J. Pub. Health* 37:803–810.
4. Granstrom, M., Blennow, M., and Winberry L. (1991). Pertussis vaccine. *In* "Vaccines and Immunotherapy," S. J. Cryz, Ed., pp. 20–35. McGraw-Hill, Inc.
5. Barenkamp, S. J., and Leininger, E. (1992). Cloning, expression and DNA sequence anaysis of genes encoding nontypeable *Haemophilus influenzae* high-molecular-weight surface-exposed proteins related to filamentous hemagglutinin of *Bordetella pertussis. Infect. Immun.* 60:1302–1313.
6. McGuirk, P., Johnson, P. A., Ryan, E. J. *et al.* (2000). Filamentous hemagglutinin and pertussis toxin from *Bordetella pertussis* modulate immune responses to unrelated antigens. *J. Infect. Dis.* 182:1286–1288.
7. Cherry, J. D., Gombein, J., Heinger, U. *et al.* (1998). A search for serological correlates of immunity to *Bordetella pertussis* cough illnesses. *Vaccine* 16:1901–1906.
8. Storsaeter, J., Hallander, H. O., Gustafsson, L. *et al.* (1988). Levels of antibodies related to protection after household exposure to *Bordetella pertussis. Vaccine* 16:1907–1916.
9. Hewlett, E. L., Halperin, S. A. (1998). Serologic correlates of immunity to *Bordetella pertussis. Vaccine* 16:1899–1900
10. Miller, J. J. Jr., Silverberg, R., Saito, T. M. *et al.* (1943). An agglutinative reaction for *Haemophilus pertussis* II. Its relation to clinical immunity. J. Pediatr. 22:644–651.
11. Tamura, M. K., Nogimori, S., Murai, S. *et al.* (1982). Subunit structure of islet-activating protein, pertussis toxin, in conformity with the A–B model. *Biochemistry* 21:5516–5522.
12. Ryan, M., McCarthy, L., Mahon, B. *et al.* (1998). Pertussis toxin potentiates Th1 and Th2 responses to co-injected antigen: Adjuvant action is associated with enhanced regulatory cytokine production and expression of the co-stimulatory molecules B7-1, B7-2 and CD28. *Int. Immunol.* 10:651–662.
13. Robbins, J. B., Pittman, M., Trollfors, B. *et al.* (1993). Primum non nocere: A pharmacologically inert pertussis toxoid alone should be the next pertussis vaccine. *Pediatr. Infect. Dis. J.* 12:795–807.
14. American Academy of Pediatrics (2000). *In* "2000 Red Book: Report of the Committee on Infectious Disease," 25th ed., L. K. Pickering, Ed., pp. 435–448. Elk Grove Village, IL: American Academy of Pediatrics.
15. Greco, D., Salmaso, S., Mastrantonio, P. *et al.* (1996). A controlled trial of two acellular vaccines and one whole-cell vaccine against pertussis. *New England J. Medicine* 334:341–348.
16. Salmaso, S., Moiraghi, A., Barale, A. *et al.* (1987). Case definitions. *In* "Pertussis Vaccine Trials," F. Brown , D. Greco, P. Mastrantonio, S. Salmaso, and S. Wassilak, Eds., pp. 135–142. Basel, Switzerland: Karger.
17. Centers for Disease Control and Prevention (1993). Pertussis surveillance: United States 1989–91. *J. Am. Med. Assoc.* 269:1489–1498.
18. Wright, W. S., Edwards, K. M., Decker, M. D., and Zeldin, M. H. (1995). Pertussis infection in adults with persistent cough. *J. Am. Med. Assoc.* 273:1044–1046.
19. Cromer, B. A., Goydos, J., Hackell, J. *et al.* (1993). Unrecognized pertussis infection in adolescents. *Am. J. Dis. Child* 147:575–577.
20. World Health Organization (1999). Pertussis vaccines. WHO position paper. *Weekly Epidemial. Record* 74:137–144.

21. Centers for Disease Control and Prevention (1997). Pertussis vaccination: Use of acellular pertussis vaccines among infants and young children. Recommendations of the Advisory Committee on Immunization Practices (ACIP). *Morbidity and Mortality Weekly Report* **46**(No. RR-7):1–25

22. Lambert, H. J. (1965). Epidemiology of a small pertussis outbreak in Kent County, Michigan. *Pub. Health Rep.* **80**:365–369.

23. Long, S.S. (1996). Pertussis. *In* "Textbook of Pediatrics," W. E. Nelson, Ed., pp. 779–784. Philadelphia: Saunders.

24. Rota, M. C., Ausiello, C. M., D'Amelio, R. *et al.* (1998). Prevalence of markers of exposure to *Bordetella pertussis* among Italian young adults. *Clin. Infect. Dis.* **26**:297–302.

25. Miller, E., and Gay, N. J. (1997). Epidemiological determinants of pertussis. *In* "Pertussis Vaccine Trials," F. Brown, D. Greco, P. Mastrantonio, S. Salmaso, and S. Wassilak Eds., pp. 15–23 Basel, Switzerland: Karger.

26. Baron, S., Njamkepo, E., Grimpel, E. *et al.* (1998). Epidemiology of pertussis in French hospitals in 1993 and 1994: Thirty years after a routine use of vaccination. *Pediatr. Infect. Dis. J.* **17**:412–418.

27. Centers for Disease Control and Prevention (1993). Resurgence of pertussis—United States 1993. *Morbidity and Mortality Weekly Report* **42**:952–953, 959–960.

28. de Melker, H. E., Schellekens, J. F. P., Nepplenbroek, S. E. *et al.* (2000). Reemergence of pertussis in the highly vaccinated population of the Netherlands: Observations on surveillance data. *Emerg. Infect. Dis.* **6**:348–357.

29. Therre, H., and Baron, S. (2000). Pertussis immunisation in Europe—The situation in late 1999. *Eurosurveillance* **5**:6–10.

30. Edwards, K., Dekker, M. D., Mortimer, E. A. Jr. (1999). Pertussis vaccine. *In* "Vaccines," 3rd ed., S. Plotkin, and W. Orenstein, Eds., pp. 293–344. Philadelphia: Sauders, Hartcourt Brace and Co.

31. Gangarosa, E. J., Galazka, A.M., Wolfe, C. R. *et al.* (1998). Impact of anti-vaccine movements on pertussis control: The untold story. *Lancet* **351**:356–361.

32. Nencioni, L., Volpini, G., Peppoloni, S. *et al.* (1990). Properties of the pertussis toxin mutant PT-9K/129G after formaldehyde treatment. *Infect. Immun.* **59**:625–630.

33. Tozzi, A. E., Anemona, A., Stefanelli, P. *et al.* (2001). Reactogenicity and immunogenicity at pre-school age of a booster dose of two three-component DTaP vaccines in children primed in infancy with acellular vaccines. *Pediatrics* **107**:E2.

34. Giuliano, M., Mastrantonio, P., Giammanco, A. *et al.* (1998). Antibody responses and persistence in the two years after immunization with two acellular vaccines and one whole-cell vaccine against pertussis. *J. Pediatr.* **132**:983–988.

35. Cassone, A., Ausiello, C. M., Urbani, F. *et al.* (1997). Cell-mediated and antibody responses to *Bordetella pertussis* antigens in children vaccinated with acellular or whole-cell pertussis vaccines. *Arch. Pediatr. Adolesc. Med.* **151**:283–289.

36. Ausiello, C., Lande, R., Urbani, F. *et al.* (1999). Cell-mediated immune responses in four year old children after primary immunization with acellular pertussis vaccines. *Infect. Immun.* **67**:4064–4071.

37. Kimura, M., and Kuno-Sakai, H. (1990). Developments in pertussis immunization in Japan. *Lancet* **336**:30–32.

38. Liese, J. G., Meschievitz, C. K., Harzer, E. *et al.* (1997). Efficacy of a two-component acellular pertussis vaccine in infants. *Pediatr. Infect. Dis. J.* **16**:1038–1044.

39. Gustafsson, L., Hallander, H. O., Olin, P. *et al.* (1996). A controlled trial of a two-component acellular, a five component acellular, and a whole-cell pertussis vaccine. *New England J. Medicine* **334**:349–355.

40. Trollfors, B., Taranger, J., Lagergård, T. *et al.* (1995). A placebo-controlled trial of a pertussis-toxoid vaccine. *New England J. Medicine* **333**:1045–1050.

41. Simondon, F., Preziosi, M. P., Yam, A. *et al.* (1997). A randomized double-blind trial comparing a two component acellular to a whole-cell pertussis vaccine in Senegal. *Vaccine* **15**:1606–1612.

42. Stehr, K., Cherry, J. D., Heininger, U. *et al.* (1998). A comparative efficacy trial in Germany in infants who received either the Lederle/Takeda acellular pertussis component DTAP (DTaP) vaccine, the Lederle whole-cell component DTAP vaccine, or DT vaccine. *Pediatrics* **101**:1–11.

43. Olin, P., Rasmussen, F., Gustafsson, L. *et al.* (1997). Randomised controlled trial of two-component, three-component, and five-component acellular pertussis vaccines compared with whole-cell pertussis vaccine. *Lancet* **350**:1569–1577.

44. Schmitt, H.-J., Wirsing von Konig, C. H., Neiss, A. *et al.* (1996). Efficacy of acellular pertussis vaccine in early childhood after household exposure. *J. Am. Med. Assoc.* **275**:37–41.

45. World Health Organization (1991). WHO Report from the meeting on case definition pertussis, Geneva, January 10–11, 1991. MIM/EPI/PERT/91.1.

46. Taranger, J., Trollfors, B., Lagerard, T. *et al.* (1997). Unchanged efficacy of a pertussis toxoid vaccine throughout the two years after the third vaccination of infants. *Pediatr. Infect. Dis. J.* **16**:180–184.

47. Salmaso, S., Mastrantonio, P., Wassilak, S. G. *et al.* (1998). Persistence of protection through 33 months of age provided by immunization in infancy with two three-component acellular pertussis vaccines. *Vaccine* **16**:1270–1275.

48. Salmaso, S., Mastrantonio, P., Tozzi, A. E. *et al.* (2001). Sustained efficacy during the first six years of life of three-component acellular pertussis vaccines administered in infancy: The Italian Experience. *Pediatrics* **108**:E81.

6 DISEASE STATES AND VACCINES: SELECTED CASES

PART E. Rotavirus

JOSEPH S. BRESEE, ROGER I. GLASS, UMESH PARASHAR, AND JON GENTSCH

Viral Gastroenteritis Section, Respiratory and Enteric Virus Branch, National Center for Infectious Diseases, Centers for Disease Control and Prevention, Atlanta, Georgia 30333

I. INTRODUCTION

Diarrheal diseases remain a leading cause of morbidity and mortality among young children worldwide, with estimates of 2.3–2.9 million deaths each year among children under 5 years of age [36]. Rotavirus is the most common cause of acute, severe gastroenteritis, and, unlike many other diarrheal pathogens, affects children in both developed and developing countries [8, 28]. Although virtually all children in the world will be infected and become ill with rotavirus during their first five years of life, the vast majority of deaths caused by rotavirus occur among children in low-income countries. Vaccines against rotavirus have been promoted as the most promising method to reduce the burden of rotavirus disease worldwide, and development and testing of candidate vaccines began in the early 1980s. This part of Chapter 6 will review the features and burden of rotavirus disease, the principles and history of rotavirus vaccine development, and future directions for rotavirus research.

II. ROTAVIRUS DISEASE AND PATHOPHYSIOLOGY

Rotavirus infections cause a syndrome of acute diarrhea and vomiting, often associated with fever. The vomiting and fever are prominent early in the illness and last for less than 48 hr, whereas watery, nonbloody diarrhea may be profuse (>10 stools/day) and commonly persists for 3–8 days. Other associated symptoms include malaise, anorexia, irritability, and abdominal pain. The primary complication of rotavirus gastroenteritis is dehydration, requiring oral or intravenous fluid replacement. Although the clinical manifestations of rotavirus gastroenteritis are usually indistinguishable from those associated with other viral agents of gastroenteritis, rotavirus diarrhea is more likely to be severe and to lead to dehydration and hospitalization [7]. Finally, severe disease and prolonged shedding of virus are more common among immunocompromised persons.

Rotaviruses are transmitted by the fecal–oral route, often by person-to-person spread through close contact. Other mechanisms of transmission occur less commonly, such as through fomites (most common in daycare centers and hospitals) and by contaminated water and food. Although airborne transmission has been postulated, direct evidence for this is lacking. Finally, animal-to-human transmission may be possible, although uncommon, as RNA–RNA hybridization studies have identified strains isolated from humans that are reassortants of human and animal parent strains [37].

Once ingested, rotavirus infects the epithelial cells in the small intestine, particularly the mature cells on the villus tips, and infection probably begins in the proximal small intestine and proceeds distally. The onset of symptoms occurs 1–3 days following ingestion of the virus. Diarrhea associated with infection is likely caused by several different mechanisms. Rotavirus first elaborates an enterotoxin, NSP4, that may induce secretory diarrhea through disruption of Cl$^-$ channels. Then, rotavirus infection can lead to lysis and sloughing of mature apical cells in the intestinal villus, which results in a loss of absorptive capacity for water, sodium, and glucose and a decrease in lactase and sucrase activity. Replacement of mature epithelial cells with immature crypt cells leads to further losses of water and salt, and the increased intraluminal osmotic load resulting from the loss of brush border enzymes results in further fluid loss. Shedding of virus in the stool persists throughout the symptomatic period, and children may shed prior to and for 1–2 days following resolution of symptoms. A role for the enteric nervous system has been implicated as an additional mechanism of diarrhea [34].

Children generally become most severely ill following their first infection, with less severe illness following subsequent exposures. Among a cohort of children followed for 2 years in Mexico, an initial infection with rotavirus conferred 77% protection against rotavirus diarrhea of any severity and 87% protection against moderate-to-severe disease [53]. Protection increased with increasing number of infections and was greater against severe disease. Although natural or vaccine-induced infection clearly protects against subsequent disease, the immune correlates of protection are incompletely understood. Protection has been associated with immunoglobin A (IgA) coproantibodies and serum IgA [38], but neither IgA nor serotype-specific serum-neutralizing antibodies have been a reliable indicator of protection in vaccine studies [56]. The lack of clear immune correlates of protection has made vaccine development problematic, because large trials are necessary to examine the efficacy of each candidate vaccine.

III. EPIDEMIOLOGY AND DISEASE BURDEN

Rotavirus gastroenteritis occurs most commonly among children between 6 and 24 months of age, with virtually all children infected at least once during their first 3–5 years [24, 53]. Most children living in developing countries become infected earlier in life, usually by 12 months of age, and more than 50% of rotavirus-associated hospitalizations occur in this age group [39] (Table 1). Infants under 3 months of age are less likely to develop symptoms when infected, perhaps due to protection conferred by the presence of maternal antibodies, breast-feeding, or an immature intestinal epithelium. Furthermore, because each natural infection confers greater protection against subsequent disease [5, 53], symptomatic illness is uncommon among children older than 5 years. In the United States and most temperate countries, rotavirus activity peaks in the cold months, and few cases occur in the summer [50]. In tropical countries, illness can occur year-round, although peak seasons can still be identified in some countries.

Worldwide, 450,000–650,000 children die each year from rotavirus infection, with 85% of the deaths occurring in low-income countries [8] (Fig. 1). Rotavirus has been estimated to account for 20% of all childhood diarrheal deaths and 5% of all deaths among children less than 5 years. Rates of rotavirus-associated deaths are difficult to estimate due the limitations of establishing cause of death in settings in which most deaths occur at home. Nonetheless, estimates from Bangladesh indicate that approximately 1 in every 111–203 children will die of rotavirus by 5 years of age [52], and in India similar rates of 1 death in 200 children have been estimated [26]. Deaths are comparatively rare in developed countries; in the United States, approximately 1 death occurs for every 200,000 infants born for an annual total of 20–40 deaths [23]. However, rotavirus is a leading cause of hospitalizations and clinic visits in developed countries, where between 1 child in 72 and 1 child in 19 is hospitalized in the first 5 years of life [19, 47].

IV. VIROLOGY

Human rotaviruses were first detected by immune electron microscopy by Bishop and colleagues in 1972 as wheel-shaped particles in the bowel contents of infants with severe gastroenteritis [6]. Rotaviruses are 100 nm in diameter and structured as triple-layered particles, with an outer capsid, an inner capsid, and a core [20], and they belong to the family *Reoviridae*. The genome consists of 11 segments of double-stranded RNA, which code for 6 structural proteins (VP1–VP4, VP6, VP7) and 6 nonstructural proteins (NSP1–NSP6). VP6, the most abundant protein, defines the antigenic group of rotavirus and is the protein to which common immune diagnostics are directed. The two proteins that comprise the outer capsid, VP7 (the glycoprotein or G protein) and VP4 (the protease-cleaved protein or P protein), are major antigens to which neutralizing antibodies are directed, and so they are potentially important targets for vaccines.

Current methods to classify rotavirus strains are based on their G and P types [22]. Because the genes encoding these two proteins segregate independently, a typing system that accounts for VP7 (G) and VP4 (P) specificities has been adopted. G types are commonly determined by enzyme immunoassay (EIA) using type-specific monoclonal antibodies for predominant serotypes (usually G1–G4)

TABLE 1 Differences in the Epidemiology of Rotavirus between Developed and Developing Countries: Implications for Vaccine Strategies

	Developed country	Developing country	Impact on vaccine evaluation or program
Age at first infection			
Infected by 12 mos.	40%	80%	Vaccine must be given earlier in developing countries; failure to immunize on time could decrease vaccine effectiveness and underestimate efficacy.
Median age	12–18 months	6–9 months	Consider alternative vaccine schedules, e.g., neonatal dose
Neonatal infection	Uncommon	More common	May result in inability to estimate efficacy in trials if many newborns are already infected and immune.
Seasonality	Winter	Year-round	In developed countries, ensure vaccination prior to seasonal peak; in developing countries, encourage vaccination at target age.
Presence of other enteric pathogens	Uncommon	Common (10–30%)	May limit vaccine take in developing countries and necessitate increased dosage or number of doses
Case fatality	Low	High	Outcomes of vaccine trials–programs will be different between settings.
Mixed infections	Uncommon	Common	May limit estimates of vaccine efficacy in developing countries or suggest different modes of transmission.
No. of serotypes	4–5 common types	More diverse types	Formulation of vaccines for developing countries may require additional serotype representation.
Inoculum of challenge	Low	High?	Possible differences related to alternative modes of transmission; could require higher doses of vaccine in developing countries.

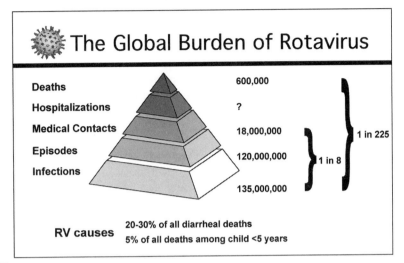

FIGURE 1   Distribution of rotavirus infection and outcomes.

and possibly for other strains depending on previous local data [49]. Alternatively, reverse transcriptase polymerase chain reaction (RT-PCR) and hybridization are increasingly used to genotype rotavirus strains [22] because they are sensitive and specific and can be used when reagents for type-specific immunoassays are not available. P types are determined by RT-PCR and hybridization or sequencing, because immunoassays incorporating monoclonal antibodies are not widely available [21].

V.   ROTAVIRUS STRAIN PREVALENCE

Although 10 G types and 9 P types have been described, only 5 G types (serotypes 1–4 and 9) and 2 P types (genotypes P1A[8] and P1B[4]) account for the majority of strains in circulation [22]. In fact, of the 90 possible G–P combinations, a review of G and P types from >2700 specimens found that 96% of strains contained one of four common combinations: P1A[8],G1; P1B[4],G3; P1B[4],G2; and P1A[8],G4. Consequently, vaccine development has generally focused on eliciting protection against these key strains.

In the past several years, greater availability and use of typing methods have resulted in the recognition that rotavirus strains may be more diverse than originally expected. For instance, although G9 strains had been described sporadically in the 1980s, they have more recently been found to have a wide geographic distribution and appear to represent the fifth globally important strain [44]. In surveys in the United States, Australia, France, Bangladesh, and the United Kingdom, G9 strains account for between 0.4 and 13% of strains. Other serotypes including G5, G8, and G10, though globally uncommon, may predominate within a given country and are found in combination with a variety of P types. Finally, in developing countries, strain patterns may be quite different, and common G types may be combined with uncommon P types. In India, the four common types represented only 33% of the total, whereas P[6] strains with common G types accounted for 43% of typeable strains detected from the stool of children with diarrhea [43]. One-third of all single infections identified in one study in Brazil were serotypes uncommon elsewhere, including P[6],G1, P[6],G3, P[6],G4, and P[3],G1, and P[8],G5 was the second most common strain detected in another study. Of note, children in developing countries are more likely to have mixed rotavirus infections which may promote the emergence of reassortant strains.

VI. RATIONALE FOR IMMUNIZATION AGAINST ROTAVIRUS

The decision to develop and promote the use of vaccines as the primary method to prevent rotavirus disease was based on several considerations. First, the epidemiology of rotavirus suggested that natural infection led to protection [5, 53]. The rationale for vaccination is predicated on the ability of a vaccine to elicit the same protective immunity that is conferred through natural infection. Second, the magnitude of the disease burden made the development of an effective prevention method a high global priority [8]. Third, despite the use and advocacy of oral rehydration solutions worldwide and general improvements in sanitation and hygiene in many places, rotavirus remains a major cause of childhood morbidity and mortality. In developed countries such as the United States, the estimated rate of rotavirus hospitalizations has declined minimally in the past 15 years despite efforts to promote the use of oral rehydration therapies and the presence of relatively good access to early treatment. Globally, estimates indicate that a vaccine with features of those in current development could prevent as many as 326,000 of the remaining rotavirus deaths each year in developing countries. In a study of the expected impact of a vaccine in the United States, a rotavirus immunization program would be expected to prevent 95,000 of the 160,000 emergency room and clinic visits, 33,600 of the 50,000 hospitalizations each year, and 13–26 of the 20–40 deaths that occur each year and result in overall cost savings [51].

VII. ROTAVIRUS VACCINES

Rotavirus vaccine development began in the late 1970s, with the first trials commencing in the early 1980s, barely 10 years after the discovery of rotavirus. Since that time, a variety of strategies for vaccine development have been pursued [8]. Most of the candidate vaccines have been based on the use of attenuated live strains, including animal and human rotaviruses, and the creation of human–animal reassortants. Even so, novel approaches, including DNA vaccines, inactivated parenterally administered vaccines, vaccine-like particles, and others, are in active development as well. Each strategy, the history of the vaccine candidates, and the current status of vaccine development are reviewed next.

A. Nonhuman Strain Vaccines: "Jennerian" Approach

The first generation of vaccines were based on the "Jennerian" approach, in which a related animal strain of an agent is given to humans with the expectation that it will be naturally attenuated in humans and will confer heterotypic protection against natural challenge with human strains. These vaccine candidates were generally of bovine, rhesus, or ovine origin and were further attenuated by multiple passages in cell culture. All of the vaccines were administered orally, generally with buffer to ensure safe passage through the stomach.

1. Bovine Strain RIT4237

The first such vaccine to be tested was the bovine strain RIT4237 (P6[1],G6). The vaccine induced seroconversions in 61–86% of 4- to 12-month-old children given a dose of either $10^{7.2}$ or $10^{8.3}$ $TCID_{50}$ (tissue culture infectious dose), and efficacy trials were conducted using one of these doses. In Finland, a one- or two-dose schedule yielded an efficacy of 55–62% against any rotavirus diarrhea and 80–88% efficacy against more severe disease [14, 54] (Table 2). One subsequent Finnish trial among neonates revealed little efficacy against mild diarrhea, but similarly high rates of protection against more severe illness. With these encouraging results, three trials were conducted in populations in developing countries, the Gambia, Peru, and Rwanda, as well as in an underserved population in the United States. These trials uniformly demonstrated little or no efficacy against the outcomes measured, and despite the promising result in Finland, further efforts to develop the vaccine were not pursued. Nonetheless, the principle that live oral vaccines could work had been established, and efforts were subsequently directed at producing more consistent results.

2. Bovine Strain WC3

One hypothesis to explain the poor performance of RIT4237 was that the strain was overattenuated by extensive passaging. Consequently, a second bovine strain vaccine candidate, WC3 (P7[5],G6), was produced using fewer cell culture passages than RIT4237 in the hope that a less attenuated strain would be more immunogenic and efficacious [15]. Indeed, 95% of 5- to 11-month-olds developed neutralizing antibodies to the vaccine strain in the first trial in the United States. Like RIT4237, efficacy trials were conducted in both developed and developing countries, and the results were again inconsistent (Table 2). In two trials conducted in similar populations in the United States, when G1 strains were the predominant serotype, a single dose of vaccine was 76–100% efficacious in one trial but 0% in the second. In Shanghai, a single dose provided 50% efficacy against all rotavirus disease, but in the Central African Republic, two doses yielded no efficacy. Despite excellent immune responses to the vaccine strain (61–97%) in all studies, the variable efficacy was striking and difficult to explain.

3. Rhesus Rotavirus Strain (RRV)

The rhesus strain vaccine candidate, MMU18006 (P5B[3],G3), was developed following serial passage in cell culture. The presence of a G3 VP7 similar to those on common human strains made the strain an attractive candidate. Early studies using doses of 10^5 or 10^6 PFU demonstrated good immunogenicity, but some mild adverse effects including low-grade fever and loose stools [29]. At lower doses (10^4), the vaccine remained immunogenic and adverse effects (fever and loose stools) remained higher among recipients of

TABLE 2 Live Oral Vaccine Candidates for Which Development Has Been Discontinued

Vaccine	First tested	Developer	Vaccine			Comments
			Strain type	Strategy	Experience	
RIT4237	1983		Bovine	Monovalent	8 efficacy trials	Variable efficacy (low efficacy in LDCs)
RIT4256	1986		Bovine	Monovalent, fewer passages than RIT4237	1 phase 2 trial	Immunogenicity not superior to RIT4237
WC3	1984	Wistar	Bovine	Monovalent	4 efficacy trials	Variable efficacy—evolved to multivalent reassortant vaccine
RRV	1985	NIH	Rhesus	Monovalent	8 efficacy trials	Variable efficacy—evolved to tetravalent reassortant vaccine
M37	1989	NIH	Human neonatal	Monovalent	1 efficacy trial	No efficacy in small trial

vaccine versus placebo. When tested in neonates in Venezuela, however, no adverse events were observed, possibly due to the effects of high levels of maternally acquired neutralizing antibodies. Like trials using the bovine strains, the results of the studies with RRV were inconsistent (Table 1). Efficacy against all rotavirus diarrhea was highly variable, ranging from 0 to 29%–66%. Similarly, although efficacy against severe disease was higher (67%–80%) in two trials, little efficacy was observed in most others [29]. A small trial in Venezuela provided an explanation for the highly variable results of these monovalent vaccines based on animal strains [41]. In this trial, excellent efficacy against all disease (64%) and severe disease (85–90%) suggested first that live oral rotavirus vaccines could work in a poor, developing country population. Second, the finding that the vaccine protected best against G3 strains, the same serotype as the vaccine strain, provided evidence that serotype-specific immunity might be important for rotavirus vaccines.

4. Lanzhou Lamb Rotavirus (LLR)

Despite the disappointing history of these vaccines, a single nonhuman strain has been developed and licensed as a vaccine, the ovine strain LLR (for Lanzhou lamb rotavirus) produced at the Lanzhou Institute in China (Table 3). The strain was first isolated in 1985 and is a P[12],G10 rotavirus that was attenuated by 42 passages in cell culture. Several early trials indicated that 6- to 24-month-old children administered a single dose of vaccine mounted good immune responses and no adverse events [57]. A large trial suggested some efficacy, and the vaccine was licensed in China in 2000 for use in children. Further postlicensure evaluations are planned.

B. Human–Animal Reassortant Vaccines

The variable success of monovalent, nonhuman strain vaccines led to three strategic improvements. First, reassortant strains were created from the parent animal strains that contained the neutralizing outer capsid genes for VP7 and VP4 from common human rotaviruses. These reassortants induced immune responses to the human capsid proteins while maintaining the attenuation properties of the parent strain. Second, multiple reassortants were combined in the same vaccine to provide strain-specific protection against the common rotavirus serotypes. Finally, vaccines were administered in multiple doses to ensure optimal immune responses in vaccinees. Several reassortant vaccines have been produced and tested, based on either bovine (WC3 or UK) or simian (RRV) strains.

1. WC3-Based Bovine–Human Reassortants

A monovalent human–bovine reassortant, WI79-9, that incorporated genes from the WC3 parent strain along with a human rotavirus gene for a serotype 1 VP7 was produced initially and was found to be adequately immunogenic and efficacious in limited trials [15]. This was quickly followed by a quadrivalent human–bovine reassortant vaccine that was developed that included genes encoding the VP7 for three major serotypes of human rotavirus with bovine genes from WC3 [WI79-9 (G1), SC2-9 (G2), WI78-8 (G3)] and WI79-4,

TABLE 3 Vaccines Currently Licensed or in Clinical Trials

Product	Company/developer	Concept	Status of vaccine
Licensed vaccines			
Rotashield	Wyeth Ayerst (U.S.A)[a]/A.Z. Kapikian, NIH (U.S.A)	RRV-TV-based quadrivalent human–rhesus reassortant	Licensed in U.S.A. (1998), withdrawn following intussusception (1999), status pending
LLR	Lanzhou Institute of Biological Products (China)/Z.-S. Bai	Monovalent lamb strain (P[12],G10)	Licensed in China (2000)
In development			
Rotateq	Merck (U.S.A.)/H.F. Clark, P. Offit (U.S.A.)	WC-3-based multivalent human–bovine reassortant	Phase 3
Rotarix (89-12)	Glaxo SmithKline (Belgium)/ R. Ward, D. Bernstein	Monovalent human strain (P1A[8],G1)	Phase 3
RV3	Univ. of Melbourne (Australia)/R. Bishop	Neonatal strain (P2A[6],G3)	Phase 2
UK- reassortant strain	Wyeth Ayerst[a]/A.Z. Kapikian, NIH (U.S.A.)	Multivalent bovine–human reassortant	Phase 2
116E	Bharat BioTech, Ltd. (India)/ M.K. Bhan, R. Glass	Neonatal strain (P8[11],G9)	Phase 1
I132	Bharat BioTech, Ltd. (India)/ C.D. Rao, H. Greenberg	Neonatal strain (P8[11],G10)	Phase 1

[a]Changed to Wyeth Lederle Vaccines and Pediatrics and then to American Home Products.

a human VP4 reassortant with P1A[8] specificity. The addition of the most common human VP4 would help to ensure protection against challenges with the three common serotypes sharing P1A[8] specificity. Three doses of this vaccine, called WC-QV, conferred 69% efficacy against mild and severe rotaviral disease [16]. Based on these results, the development of pentavalent and hexavalent vaccines with additional reassortants is proceeding (Table 2).

2. RRV-Based Rhesus–Human Reassortants and RRV-TV

Individual single-gene reassortants were also produced based on RRV and were found to be acceptably safe, immunogenic, and efficacious [29]. In order to broaden the coverage against circulating serotypes, a quadrivalent vaccine was produced based on RRV that included RRV (with G3 specificity) with three reassortants, including D × RRV (G1), DS-1 × RRV (G2), and ST3 × RRV (G4). Seven large efficacy trials were conducted: three with a "low-dose" formulation (4×10^4 PFU/dose) and four with a "high-dose" formulation (4×10^5 PFU/dose) [10] (Table 2). The results of these trials were notable for the consistency of the estimates of efficacy, particularly in light of the variability observed in trials with previous vaccines. The vaccine conferred 50–60% protection against all cases of rotavirus diarrhea and 70–100% against severe rotavirus disease, such as dehydrating diarrhea and hospitalizations [2, 27, 42, 45]. The trials also demonstrated that the vaccine conferred protection against more than one serotype and that efficacy persisted over at least three rotavirus seasons.

Like the parent RRV strain, RRV-TV was associated with mild, short-lived fevers 3–5 days following the first vaccination; 21% of vaccinee versus 6% of placebo recipients had temperatures >38°C, whereas 2% versus 1%, respectively, had temperatures >39°C. The fevers generally occurred in the 3–5 days following the first dose of vaccine and were short-lived and mild [29]. No other adverse events were commonly associated with vaccine in prelicensure studies.

Because of the demonstrated efficacy along with an acceptable safety profile, this vaccine was the first rotavirus vaccine in the world to be licensed and recommended for routine use when it was recommended for inclusion in the U.S. childhood vaccine schedule in 1998 [10]. Between September 1998 and July 1999, about 1 million children received the vaccine in the United States (CDC, unpublished data). However, the recommendations for use of the vaccine were suspended in 1999 following reports of intussusception among vaccine recipients [9, 40]. Following an investigation that demonstrated a significant association between receipt of the vaccine and the occurrence of intussusception, the vaccine was withdrawn and is no longer produced. As a result, future use of this vaccine is unlikely in the United States; however, its use in developing countries remains possible.

3. UK-Based Bovine–Human Reassortant Vaccine

A second bovine-based reassortant vaccine has been developed and tested based on the UK rotavirus, a P7[5],G6 strain (Table 3). Reassortants representing each of the four common human VP7 serotypes have been produced: D × UK (serotype 1), DS-1 × UK (serotype 2), P × UK (serotype 3), and ST3 × UK (serotype 4). In a U.S. trial, the vaccine demonstrated satisfactory levels of attenuation, safety, infectivity, and immunogenicity of each monovalent reassortant in infants [17].

C. Human Rotavirus Strain Vaccine

A vaccine derived from a live attenuated human rotavirus strain (89-12), a P1A[8],G1 strain that has been attenuated by multiple passages in cell culture, has also been field-tested [3] (Table 3). The rationale for use of a human strain is that, because of its antigenic similarity to natural challenge strains, it may induce better protection. In a U.S. field trial, the vaccine conferred 89% efficacy against all rotavirus diarrhea, and none of the vaccinees required medical attention compared to 10 of the placebo recipients. In this trial, mild fever was reported in a significantly higher proportion of the group receiving the vaccine versus the placebo (19% versus 5%). Virtually all rotavirus gastroenteritis in the community in which the trial was conducted was caused by serotype 1 strains. Larger trials will be needed to confirm the efficacy of this vaccine, especially in developing countries and against other rotavirus serotypes.

D. Neonatal Human Strain Vaccines

Early observations indicated that neonates infected with rotavirus in hospital nurseries were usually asymptomatic [4, 5]. Follow-up studies demonstrated that these neonates asymptomatically infected in hospital nurseries with "neonatal" strains were later protected against severe disease in early childhood. Because of these findings, four rotavirus strains isolated from asymptomatic neonates have been developed as vaccine candidates.

1. M37 Strain

The first neonatal strain to be tested was M37, a P2A[6],G1 strain isolated from a neonate in Venezuela in 1982. Early trials demonstrated acceptable safety and immunogenicity. In a small trial in Finland, a single dose of 10^4 PFU conferred no protection against disease [55], and development was halted.

2. RV3 Strain

RV3, a P2A[6],G3 strain, was isolated by Bishop and colleagues from newborns at the Children's Hospital in Melbourne [5] (Table 3). Neonates that were infected in the hospital were protected from rotavirus disease during 2 years of follow-up. The strain was attenuated by multiple passages in cell culture. Initial safety trials that used a single dose of vaccine at 6×10^5 FFU/ml demonstrated no significant side effects, but serum immune responses were poor [1]. A trial of three doses of vaccine (6×10^5 IU/dose) induced immune responses in half of the infants, and those vaccinees were protected from rotavirus-associated disease. No efficacy was observed in the nonresponders, so that further efforts to develop this strain are underway and additional testing is expected.

3. Indian Neonatal Strains

Two Indian strains isolated from newborns are also being prepared in India as vaccine candidates (Table 3). Strain 116E isolated from a newborn in New Delhi is a P8[11],G9, natural, single-gene reassortant between a human parent strain and the VP4 gene of bovine origin. The strain has been found to be common in neonatal nurseries in India [4]. Children infected asymptomatically with this strain were found to be protected (46%) against rotavirus diarrhea during 1 year of follow-up. A second neonatal strain, I321 (P8[11],G10), isolated from a newborn in Bangalore is also a bovine–human reassortant strain in which most genes are from bovine origin [18]. Both strains are in preclinical development and human trials are planned.

E. Other Approaches to Rotavirus Vaccines

Although live virus, orally administered vaccines represent the primary approach to rotavirus vaccines, other types of vaccine and routes of administration are being evaluated. Work on virus-like particles (13), cold-adapted strains, inactivated strains (8), and DNA vaccines (25) has been undertaken. These approaches may hold some distinct advantages in the future if they could improve the variable immune response to oral vaccines, could be combined with other parenterally administered vaccines, or might avoid the association with intussusception.

VIII. CHALLENGES TO ROTAVIRUS VACCINE DEVELOPMENT AND INTRODUCTION

A number of challenges lie ahead for rotavirus vaccines. Some of these challenges are generic obstacles to getting any new vaccine incorporated into a country's immunization program. These include the high cost of any new vaccine compared to existing vaccines, programmatic issues regarding addition of a new vaccine to EPI programs, ability to produce enough for the expected demand, and obtaining data to assess the need and demand in countries interested in using the vaccine. In addition to these, other challenges exist that are more specific to rotavirus vaccines, and addressing these will be a priority.

A. Ensuring Vaccine Safety: Intussusception

Currently, the most pressing challenge facing efforts to develop and introduce rotavirus vaccines is the need to ensure vaccine safety with respect to intussusception. Intussusception is a relatively common cause of bowel obstruction among infants worldwide. It occurs when a portion of the intestine invaginates into an adjacent portion of bowel, resulting in venous congestion, vascular compromise, and tissue necrosis if not emergently corrected. Intussusception was rarely noted in recipients of a rhesus-rotavirus-based vaccine prior to licensure of RRV-TV [46]; 5 intussusception cases occurred among 10,054 vaccine recipients, a rate that was not significantly different from that among placebo recipients. Even so, intussusception was included as a possible adverse event in the package insert and in U.S. recommendations [10].

Following licensure, 15 cases of intussusception were reported to the Centres for Disease Control (CDC) among vaccinees between September 1998 and July 1999 [9]. The cases were clustered in the first week following vaccination and following the first dose of vaccine. Because of these and other data, recommendations for use of the vaccine were suspended so that definitive studies could be conducted to assess the risk. Since this time, several studies have been completed. A large case–control study along with a related case series analysis demonstrated approximately 2-fold increased risk of intussusception among vaccinees [35]. A retrospective cohort study found a similar risk in a cohort of children enrolled in several large health maintenance organizations [30]. The risk in both studies was limited to a 3- to 14-day period following the first two doses of vaccine. From these studies, it has been estimated that approximately 1 child in 4600–11,000 vaccinees would develop intussusception, and these studies led to the withdrawal of the vaccine from the U.S. vaccination schedule [11]. Notably, two more recent studies have produced lower attributable risk estimates of 1 case for every 18,000–302,000 vaccinees [12, 48]. The future of this vaccine is uncertain.

The withdrawal of the first licensed rotavirus vaccine because of the association with a serious adverse event has led to considerable concern among vaccine advocates and the public health community regarding future prospects for any rotavirus vaccine. Several issues remain to be resolved that will clarify the path for future vaccine candidates. Foremost among these is whether other live oral rotavirus vaccines will also lead to intussusception. No direct data are available to address this, and the answer will ultimately depend on adequate postlicensure surveillance once the next vaccine is introduced into an immunization program. Even so, studies to better understand the pathogenesis of intussusception in general, and RRV-TV-associated intussusception in particular, will be helpful, as they might lead to animal models to predict intussusception risk or to correlates of risk that might be used to select the best vaccine candidates. Data to determine whether naturally acquired rotavirus or other enteric pathogens are associated with intussusception would help to address the question of whether the condition is likely to be specific to rhesus strain infections or a more general reaction to a broader array of gut infections. Finally, because the risks and benefits of rotavirus vaccines will vary considerably between developed and developing countries, additional data that might better predict the risk–benefit equations in the various settings should be collected. These data might include baseline estimates of intussusception rates and epidemiology in countries considering use of a rotavirus vaccine, as well as better data on rotavirus-attributable disease and cost burden. Even if the attributable risk of intussusception following vaccination is as high in developing countries as in the United States, the benefits of vaccination against rotavirus may far exceed the risks. A WHO report recommended that surveillance for intussusception be included in all planned vaccine trials with any rotavirus vaccines [57]. It is unclear whether prelicensure trials must include sufficient subjects to ensure a defined level of safety with respect to this rare adverse event.

B. Differences between Rotavirus in Developed and Developing Country Settings and Impact on Vaccine Development

Although rotavirus infection is virtually universal early in childhood, the epidemiology of the disease is quite different in developed and developing countries (Table 1). Differences in seasonality, strain prevalence, age distribution, and outcomes between settings may affect decisions about vaccine composition, schedule, dose, and priority. For instance, in developing countries, where the age of first infection and severe disease is lower than in developed countries, vaccine schedules that include a neonatal dose may be more effective than provision of the first dose of vaccine at 2 months of age. Higher doses of vaccine or additional doses may be needed in these settings to overcome the inhibitory effects of competing gut flora, concomitant use of OPV, and high levels of maternal antibodies against rotavirus. Finally, because of differences in strain prevalence between countries, vaccines that protect against strains common in the United States might perform poorly in developing country settings. Studies to better define the epidemiology of rotavirus disease and trials of new vaccines in developing countries are clearly needed.

C. Poor Past Performance of Rotavirus Vaccines When Tested in Developing Countries

Generally, rotavirus vaccine trials conducted in developing countries have yielded poor efficacy results. Each of the first-generation, Jennerian vaccines was tested in trials in developing countries, either in Africa or in South America. RIT4237 and WC3 vaccines failed to provide any efficacy when tested in Africa [15], and variable, but low levels of protection were observed when one, two, or three doses of RIT4237 were given to Peruvian children. Although the RRV trial in Venezuela yielded good efficacy [41], trials of this vaccine in Peru were disappointing [31]. More recently, in the three trials of RRV-TV carried out in South America, results were similarly mixed. Again, efficacy estimates from a trial in Venezuela were excellent and comparable to those from U.S. and Finnish trials (42), whereas trials in Peru and Brazil found little efficacy [31, 32]. The observation that several different vaccines have fared poorly in developing countries has led to concerns about the potential effectiveness of live oral rotavirus vaccines in vaccine programs in these settings.

Clearly, the differences in host factors, virologic characteristics, and disease epidemiology discussed previously may have played a role in the poor results in trials conducted in developing countries. Most early trials used a single- or two-dose schedule. Additional doses of vaccine or larger dosages might be required in settings where vaccine take may be poor in the face of the variety of factors mentioned previously. Indeed, this might be one explanation for the improved efficacy of RRV-TV when tested in a 10-fold higher dose in Venezuela compared to the low-dose studies in Brazil and Peru. However, the data are conflicting; no differences were observed between one- and three-dose schedules in the Peru study, and even if an increase in the virus challenge dose might enhance the immune response, no clear correlation exists between serologic response to vaccine and protection.

Methodologic limitations in some studies may have led to an underestimation of efficacy, including limited follow-up periods, use of suboptimal outcome measures [33], and enrollment of older children who had likely already been infected with rotavirus. Attention to appropriate study design will be crucial in the next generation of trials of new vaccine candidates in developing countries.

IX. FUTURE DIRECTIONS

Rotavirus diarrhea represents a large disease burden in all areas of the world. The previous two decades have proved that rotavirus disease is preventable through vaccination and that these vaccines should be one of the most cost-effective strategies to reduce diarrheal disease morbidity and mortality worldwide. The next decade should produce the vaccines that will fulfill this promise. Several challenges and opportunities remain, not the least of which is to develop vaccines that will be effective and safe in multiple types of settings. In addition to this, significant attention must be paid to advocacy for new vaccines, overcoming obstacles in the expansion of vaccines available to children in developed countries to those living in developing countries, and vaccine safety.

REFERENCES

1. Barnes, G. L., Lund, J. S., Adams, L., Mora, A., Mitchell, S. V., Caples, A., and Bishop, R. F. (1997). Phase 1 trial of a candidate rotavirus vaccine (RV3) derived from a human neonate. *J. Pediatr. Child Health* **33**:300–304.
2. Bernstein, D. I., Glass, R. I., Rodgers, G., Davidson, B. L., and Sack, D. A. (1995). Evaluation of rhesus rotavirus monovalent and tetravalent reassortant vaccines in US children. *J. Am. Med. Assoc.* **273**:1191–1196.
3. Bernstein, D. I., Sack, D. A., Rothstein, E., Reisinger, K., Smith, V. E., O'Sullivan, D., Spriggs, D. R., and Ward, R. L. (1999). Efficacy of live, attenuated, human rotavirus vaccine 89-12 in infants: A randomised placebo-controlled trial. *Lancet* **354**:287–290.
4. Bhan, M. K., Lew, J. F., Sazawal, S., Das, B. K., Gentsch, J. R., and Glass, R. I. (1993). Protection conferred by neonatal rotavirus infection against subsequent diarrhea. *J. Infect. Dis.* **168**:282–287.
5. Bishop, R. F., Barnes, G. L., Cipriani, E., and Lund, J. S. (1983). Clinical immunity after neonatal rotavirus infection: A prospective longitudinal study in young children. *New England J. Medicine* **309**:72–76.
6. Bishop, R. F., Davidson, G. P., Holmes, I. H., and Ruck, B. J. (1974). Detection of a new virus by electron microscopy of fecal extracts from children with acute gastroenteritis. *Lancet* **1**:149–151.
7. Brandt, C. D., Kim, H. W., Rodriguez, J. O., Arrobio, W. J., Jeffries, B. C., Stallings, E. P., Lewis, C., Miles, A. J., Chanock, R. M., Kapikian, A. Z., and Parrott, R. H. (1983). Pediatric viral gastroenteritis during eight years of study. *J. Clin. Microbiol.* **18**:71–78.
8. Bresee, J., Glass, R. I., Ivanoff, B., and Gentsch, J. (1999). Current status and future priorities for rotavirus vaccine development, evaluation, and implementation in developing countries. *Vaccine* **17**:2207–2222.
9. Centers for Disease Control and Prevention (1999). Intussusception among recipients of rotavirus vaccine, United States, 1998–1999. *Morbidity and Mortality Weekly Report* **48**(27):577–581.
10. Centers for Disease Control and Prevention (1999). Rotavirus vaccine for the prevention of rotavirus gastroenteritis among children. Recommendations of the Advisory Committee on Immunization Practices (ACIP). *Morbidity and Mortality Weekly Report* **48**(No. RR-2):1–20.

11. Centers for Disease Control and Prevention (1999). Withdrawal of rotavirus vaccine recommendation. *Morbidity and Mortality Weekly Report* **48**(43):1007.

12. Chang, H. G., Smith, P. F., Ackelsberg, J., Morse, D. L., and Glass, R. I. (2001). Intussusception, rotavirus diarrhea, and rotavirus vaccine use among children in New York State. *Pediatrics* **108**:54–60.

13. Ciarlet, M., Crawford, S. E., Barone, C., Bertolotti-Ciarlet, A., Ramikg, R. F., Estes, M. K., and Conner, M. E. (1998). Subunit rotavirus vaccine administered parenterally to rabbits induces active protective immunity. *J. Virol.* **72**:9233–9246.

14. Clark, H. F., Glass, R. I., and Offit, P. A. (1999). Rotavirus vaccines. *In* "Vaccines," 3rd ed., S.A. Plotkin and W.A. Orenstein, Eds., pp.987–1005. Philadelphia: W.B. Saunders Co.

15. Clark, H. F., Offit, P. A., Ellis, R. W., Eiden, J. J., Krah, D., Shaw, A. R., Pichichero, M., Treanor, J. J., Borian, F. E., Bell, L. M., and Plotkin, S. A. (1996). The development of multivalent bovine rotavirus (strain WC3) reassortant vaccine for infants. *J. Infect. Dis.* **174**(Suppl. 1):S73–80.

16. Clark, H. F., White, C. J., Offit, P. A. (1995). Preliminary evaluation of safety and efficacy of quadrivalent human–bovine rotavirus vaccine. *Pediatr. Res.* **37**:172A.

17. Clements-Mann, M. L., Makhene, M. K., Mrukowicz, J., Wright, P. F., Hoshino, Y., Midthun, K., Sperber, E., Karron, R., and Kapikian, A. Z. (1999). Safety and immunogenicity of live attenuated human–bovine (UK) reassortant rotavirus vaccines with VP7-specificity for serotypes 1, 2, 3, or 4 in adults, children and infants. *Vaccine* **17**:2715–2725.

18. Das, B. K., Gentsch, J. R., Hoshino, Y., Ishida, S.-I., Nakagomi, O., Bhan, M. K., Kumar, R., and Glass, R. I. (1993). Characterization of the G serotype and genogroup of New Delhi newborn rotavirus strain 116E. *Virology* **197**:99–107.

19. de Wit, M. A. S., Koopmans, M. P. G., van der Blig, J. F., and van Duynhoven, Y. T. H. P. (2000). Hospital admissions for rotavirus infection in the Netherlands. *Clin. Infect. Dis.* **31**:698–704.

20. Estes, M. K. (2001). Rotaviruses and their replication. *In* "Fields Virology," 4th ed., Vol. 2, Knipe, D. M., and P. M. Howley Eds., pp. 1747–1785. Philadelphia: Lippicot-Raven.

21. Gentsch, J. R., Glass, R. I., Woods, P., Gouvea, V., Gorziglia, M., Flores, J., Das, B. K., and Bhan, M. K. (1992). Identification of group A rotavirus gene 4 types by polymerase chain reaction. *J. Clin. Microbiol.* **30**:1365–1373.

22. Gentsch, J. R., Woods, P. A., Ramachandran, M., Das, B. K., Leite, J. P., Alfieri, A., Kumar, R., Bhan, M. K., and Glass, R. I. (1996). Review of G and P typing results from a global collection of strains: Implications for vaccine development. *J. Infect. Dis.* **174**(Suppl. 1):S30–S36.

23. Glass, R. I., Kilgore, P. E., Holman, R. C., Jin, S., Smith, J. C., Woods, P. A., Clarke, M. J., Ho, M. S., and Gentsch, J. R. (1996). The epidemiology of rotavirus diarrhea in the United States: Surveillance and estimates of disease burden. *J. Infect. Dis.* **174**(Suppl. 1):S5–S11.

24. Gurwith, M., Wenman, W., Gurwith, D., Brunton, J., Feltham, S., and Greenberg, H. (1983). Diarrhea among infants and young children in Canada: A longitudinal study in three northern communities. *J. Infect. Dis.* **147**:685–692.

25. Herrmann, J. E., Chen, S. C., Fynan, E. F., Santoro, J. C., Greenberg, H. B., Wang, S., and Robinson, H. L. (1996). Protection against rotavirus infections by DNA vaccination. *J. Infect. Dis.* **174**(Suppl.):S93–S97.

26. Jain, V., Parashar, U. D., Glass, R. I., and Bhan, M. K. (2000). Epidemiology of rotavirus in India: A review to assess the disease burden and potential benefits of a national immunization program. *Indian J. Pediatr*, in press.

27. Joensuu, J., Koskenniemi, E., Pang, X.-L., and Vesikari, T. (1997). Randomised, double-blind, placebo-controlled trial of rhesus–human reassortant rotavirus vaccine for prevention of severe rotavirus gastroenteritis. *Lancet* **350**:1205–1209.

28. Kapikian, A. Z., Hoshino, Y., and Chanock, R. M. (2001). Rotaviruses. *In* "Fields Virology," 4th ed., Vol. 1, D. M., Knipe, and P. M. Howley, Eds., pp. 1787–1834. Philadelphia: Lipincott Williams and Wilkins.

29. Kapikian, A. Z., Hoshino, Y., Chanock, R. M., and Perez-Schael, I. (1996). Efficacy of a quadrivalent rhesus rotavirus-based human rotavirus vaccine aimed at preventing severe rotavirus diarrhea in infants and young children. *J. Infect. Dis.* **174**(Suppl. 1):S65–S72.

30. Kramarz, P., France, E. K., Destefano, F., Black, S. B., Shinefield, H., Ward, J. I., Chang, E. J., Chen, R. T., Shatin, D., Hill, J., Lieu, T., and Ogren, J. M. (2001). Population-based study of rotavirus vaccination and intussusception. *Pediatr. Infect. Dis. J.* **20**:410–416.

31. Lanata, C. F., Black, R. E., Flores, J., Lazo, F., Butron, B., Linares, A., Huapaya, A., Ventura, G., Gil, A., and Kapikian, A. Z. (1996). Immunogenicity, safety and protective efficacy of one dose of the rhesus rotavirus vaccine and serotype 1 and 2 human–rhesus rotavirus reassortants in children from Lima, Peru. *Vaccine* **14**:237–243.

32. Linhares, A. C., Gabbay, Y. B., Mascarenhas, J. D. P., de Freitas, R. B., Olivera, C. S., Bellesi, N., Monteiro, T. A. F., Lins-Lainson, Z., Ramos, F. P., and Valente, S. A. (1996). Immunogenicity, safety and efficacy of tetravalent rhesus–human, reassortant rotavirus vaccine Belem, Brazil. *Bull. WHO* **74**:491–500.

33. Linhares, A. C., Lanata, C. F., Hausdorff, W. P., Gabbay, Y. B., and Black, R. E. (1999). Reappraisal of the Peruvian and Brazilian lower titer tetravalent rhesus–human reassortant rotavirus vaccine efficacy trials: Analysis by severity of diarrhea. *Pediatr. Infect. Dis. J.* **18**:1001–1006.

34. Lundgren, O., Peregrin, A. T., Persson, K., Kordasti, S., Uhnoo, I., and Svensson, L. (2000). Role of the enteric nervous system in the fluid and electrolyte secretion of rotavirus diarrhea. *Science* **287**:491–495. (www.sciencemag.org).

35. Murphy, T. V., Gargiullo, P. M., Massoudi, M. S., Nelson, D. B., Jumaan, A. O., Okoro, C. A., Zanardi, L. R., Setia, S., Fair, E. L., LeBaron, C. W. Schwartz, B., Wharton, M., and Livingood, J. R. (2001). Intussusception among infants given an oral rotavirus vaccine. *New England J. Med.* **344**:564–572.

36. Murray, C. J., and Lopez, A. D. (1997). Global mortality, disability, and the contribution of risk factors: Global burden of disease study. *Lancet* **349**:1436–1442.

37. Nakagomi, O., and Nakagomi, T. (1991). Genetic diversity and similarity among mammalian rotaviruses in relation to interspecies transmission of rotavirus. *Arch. Virol.* **120**:43–55.

38. Offit, P. A. (1996). Host factors associated with protection against rotavirus disease: The skies are clearing. *J. Infect. Dis.* **174** (Suppl. 1):S59–S64.

39. Parashar, U. D., Holman, R. C., Clarke, M. J., Bresee, J. S., and Glass, R. I. (1997). Hospitalizations associated with rotavirus diarrhea in the United States, 1993 through 1995: Surveillance based on the new ICD-9-CM rotavirus-specific diagnostic code. *J. Infect. Dis.* **177**:13–17.

40. Pavia, A. T., Long, E. G., Ryder, R. W., Nsa, W., Puhr, N. D., Wells, J. G., Martin, P., Tauxe, R. V., and Griffin, P. M. (1992). Diarrhea among African children born to human immunodeficiency virus 1-infected mothers: Clinical, microbiologic and epidemiologic features. *Pediatr. Infect. Dis. J.* **11**:996–1003.

41. Perez-Schael, I., Garcia, D., Gonzalez, M., Gonzalez, R., Daoud, N., Perez, M., Cunto, W., Kapikian, A. Z., and Flores, J. (1990). Prospective study of diarrheal diseases in Venezuelan children to evaluate the efficacy of rhesus rotavirus vaccine. *J. Med. Virol.* **30**:219–229.

42. Perez-Schael, I., Guntinas, M. J., Perez, M., Pagone, V., Rojas, A. M., Gonzalez, R., Cunto, W., Hoshino, Y., and Kapikian, A. Z. (1997). Efficacy of the rhesus rotavirus-based quadrivalent vaccine in infants and young children in Venezuela. *New Engl. J. Medicine* **337**:1181–1209.

43. Ramachandran, M., Das, B. K., Vij, A., Kumar, R., Bhambal, S. S., Kesari, N., Rawat, H., Bahl, L., Thakur, S., Woods, P. A., Glass, R. I., Bhan, M. K., and Gentsch, J. R. (1996). Unusual diversity of human rotavirus G and P genotypes in India. *J. Clin. Microbiol.* **34**:436–439.

44. Ramachandran, M., Gentsch, J. R., Parashar, U. D., Jin, S., Woods, P. A., Holmes, J. L., Kirkwood, C. D., Bishop, R. F., Greenberg, H. B., Urasawa, S., Gerna, G., Coulson, B. S., Taniguchi, K., Bresee, J. S., Glass, R. I., and The National Rotavirus Strain Surveillance System Collaborating Laboratories (1998). Detection and characterization of novel rotavirus strains in the United States. *J. Clin. Microbiol.* **36**:3223–3229.

45. Rennels, M. B., Glass, R. I., Dennehy, P. H., Bernstein, D. I., Pichichero, M. E., and Zito, E. T. (1996). Safety and efficacy of high-dose rhesus–human reassortant rotavirus vaccines: Report of the national multicenter trial. *Pediatrics* **97**:7–13.

46. Rennels, M. B., Parashar, U. D., Holman, R. C., Le, C. T., Chang, H.-G., and Glass, R. I. (1998). Lack of an apparent association between intussusception and wild or vaccine rotavirus infection (brief report). *Pediatr. Infect. Dis. J.* **17**:924–925.

47. Ryan, M. J., Ramsay, M., Brown, D., Gay, N. J., Farrington, C. P., and Wall, P. G. (1996). Hospital admissions attributable to rotavirus infection in England and Wales. *J. Infect. Dis.* **174** (Suppl. 1):S12–S18.

48. Simonsen, L., Morens, D. M., Elixhauser, A., Gerber, M., Van Raden, M., and Blackwelder, W. (2001). Effect of rotavirus vaccination programme on trends in admission of infants to hospital for intussusception. *Lancet* **358**:1224–1229.

49. Taniguchi, K., Urasawa, T., Morita, Y., Greenberg, H. B., and Urasawa, S. (1987). Direct serotyping of human rotavirus in stools using serotype 1-, 2-, 3-, and 4-specific monoclonal antibodies to VP7. *J. Infect. Dis.* **155**:1159–1166.

50. Torok, T. J., Kilgore, P. E., Clarke, M. J., Holman, R. C., Bresee, J. S., and Glass, R. I. (1997). Visualizing geographic and temporal trends in rotavirus activity in the United States, 1991 to 1996. *Pediatr. Infect. Dis. J.* **16**:941–946.

51. Tucker, A. W., Haddix, A. C., Bresee, J. S., Holman, R. C., Parashar, U. D., and Glass, R. I. (1998). Cost-effectiveness analysis of a rotavirus immunization program for the United States. *J. Am. Med, Assoc.* **279**:1371–1376.

52. Unicomb, L. E., Kilgore, P. E., Faruque, A. S. G., Hamadani, J. D., Fuchs, G. J., Albert, M. J., and Glass, R. I. (1997). Anticipating rotavirus vaccines: Hospital-based surveillance for rotavirus diarrhea and estimates of disease burden in Bangladesh. *Pediatr. Infect. Dis. J.* **16**:947–951.

53. Velazquez, F. R., Matson, D. O., Calva, J. J., Guerrero, M. L., Morrow, A. L., Carter-Campbell, S., Glass, R. I., Estes, M. K., Pickering, L. K., and Ruiz-Palacios, G. M. (1996). Rotavirus infection in infants as protection against subsequent infections. *New England J. Medicine* **335**:1022–1028.

54. Vesikari, T., Isolauri, E., and D'Hondt, E. (1984). Protection of infants against rotavirus diarrhea by RIT4237 attenuated bovine rotavirus strain vaccine. *Lancet* **1**:977–981.

55. Vesikari, T., Ruuska, T., Koivu, H., Green, K., Flores, J., and Kapikian, A. (1991). Evaluation of the M37 human rotavirus vaccine in 2- to 6-month-old infants. *Pediatr. Infect. Dis. J.* **10**(December):912–917.

56. Ward, R. L., and Bernstein, D. I. (1995). Lack of correlation between serum rotavirus antibody titers and protection following vaccination with reassortant RRV vaccines. *Vaccine* **13**:1226–1232.

57. World Health Organization (2000). Report of the meeting on future directions for rotavirus vaccine research in developing countries, Geneva, February 9–11, 2000, WHO/V&B 00.23.

6 DISEASE STATES AND VACCINES: SELECTED CASES

PART F. AIDS Vaccines: Challenges and Prospects

ELISA I. CHOI AND NORMAN L. LETVIN

Harvard Medical School, Beth Israel Deaconess Medical Center, Boston, Massachusetts 02215

I. INTRODUCTION

The development of a successful AIDS (acquired immunodeficiency syndrome) vaccine presents a unique challenge to the scientific community. HIV (human immunodeficiency virus) infections and AIDS-related deaths are increasing at a staggering rate, particularly in the developing world. This is occurring despite significant advances in drug therapies that have provided increased longevity and quality of life for those HIV-infected individuals with access to medications. However, for the vast majority of the world's HIV-infected population, particularly in sub-Saharan Africa and southeast Asia, long-term treatment with multiple drugs is not feasible because of limited healthcare resources and the absence of a healthcare infrastructure. The development of an HIV vaccine offers the best hope to curb the spreading AIDS epidemic.

However, the biology of HIV and the immune response to this virus present considerable obstacles to the development of a successful AIDS vaccine. Unique aspects of HIV have rendered traditional vaccine approaches ineffective in preventing HIV infection. Thus, almost two decades after the discovery of HIV as the cause of AIDS, we are still awaiting the development of an effective HIV vaccine.

II. THE BURDEN OF HIV–AIDS

HIV–AIDS is the largest cause of death in the world from any infectious disease. According to the Joint United Nations Programme on HIV–AIDS [27],

245

almost 19 million people worldwide have died from AIDS, with more than 34 million people infected with HIV. In 1999, 5.4 million individuals were newly infected with HIV [28]. This disease has had a tremendous impact on the rates of infant, childhood, and maternal mortality, as well as on overall life expectancy and economic growth. More than 13 million children have been orphaned as a result of AIDS [28]. In at least 16 countries, more than one-tenth of the adult population, aged 15–49 years, is infected with HIV, and in several countries, the rates of HIV infection are substantially higher (Botswana, 35.8% of adults with HIV infection; South Africa, 19.9% of adults with HIV infection) [27, 28].

III. BIOLOGY OF HIV

HIV represents a genetically diverse population of viruses [1]. The HIVs responsible for causing AIDS in most of the world are referred to as HIV-1. Within this HIV-1 grouping are clusters of related viruses, classified by their DNA similarities into distinct subtypes or clades. Particular clades of HIV-1 are found in different geographic regions of the world. These HIV-1 clades vary sufficiently from one another in genetic sequence that it may prove necessary to develop distinct HIV vaccines for each of these groups of viruses. It should be noted, however, that no evidence exists to support a relationship between this genetic taxonomy and any biologic characteristics of the viruses.

Another distinct grouping of HIVs, known as HIV-2, causes AIDS in western Africa. The HIV-1 and HIV-2 groups of viruses differ so much in their genetic sequence that portions of these viruses do not even exhibit immunologic cross-reactivity. HIV vaccines therefore must provide protection against not only viruses that are members of distinct HIV-1 clades but also viruses of this dramatically different grouping of HIV-2 viruses.

The propensity of HIV to mutate poses another obstacle for the successful development of an AIDS vaccine. In the process of ongoing HIV replication in an infected individual, genetically distinct viruses are generated continuously. This process results in the creation of a population of viruses in a single individual that is so heterogeneous that an antibody that binds to one virus and blocks its ability to infect a cell may not be able to bind to another of these viruses.

The various routes of HIV transmission, and how the virus is seen by the immune system, also have important implications for AIDS vaccine development. HIV can be transmitted by sexual contact and by contaminated blood products. Therefore, a successful HIV vaccine must generate effective immunity in the anatomic regions that first come into contact with the virus during sexual exposure, as well as effective immunity in the bloodstream. In addition, HIV can be transmitted either by cells that are infected with the virus or by cell-free virus. The types of immune responses needed to control these forms of the virus are likely to be very different. Virus within a cell can be recognized and eliminated by cytotoxic or killer T lymphocytes, whereas free virus can only be controlled by antibody. The types of vaccines that can elicit these different immune responses are quite distinct. For an HIV vaccine to be successful, it would appear that both humoral and cellular immunity will need to be generated against the virus.

However, even if a vaccine elicits both cytotoxic T-lymphocyte and antibody responses, successful containment of HIV may still not be achieved. Infection with HIV is associated with the production of large amounts of virus. This virus production persists despite the generation of strong cytotoxic T-lymphocyte and antibody responses in the infected person. The amount of virus production is directly related to the rapidity of disease progression in the infected individual, with high levels of HIV production associated with the rapid development of clinical disease. The persistence of virus production in the face of strong immune responses directed against HIV raises serious questions about the potential for any HIV vaccine to generate immune responses that can contain or eliminate HIV from the infected individual.

IV. IMMUNE CONTROL OF HIV

In most viral infections, antibodies produced in response to the virus facilitate successful clearance of that virus. These antibodies play critical roles in blocking viral attachment to and subsequent infection of cells. However, accumulating evidence suggests that antibodies may not play an important role in preventing the spread of HIV. A substantial anti-HIV antibody response develops in infected individuals and is the basis for the most common diagnostic tests for infection. However, such antibodies demonstrate only weak biologic activity against HIV. In addition, although HIV production is partially controlled within the first few weeks after infection in an infected individual, this control occurs before the development of specific antibodies with activity against HIV.

Nevertheless, the ability of an AIDS vaccine to elicit antibodies may still be important in the generation of an immune response that might prevent an infection. Antibodies that bind to HIV will be needed for the control of cell-free circulating virus. Several specific antibodies have been shown to be capable of binding to a variety of HIV-1 viruses, suggesting that it should be possible to induce antibodies through vaccination that recognize many of the existing HIV-1 viruses. Studies have shown that AIDS virus infections can be prevented by antibodies injected into monkeys, if sufficient quantities of these antibodies are given [4]. Therefore, if a vaccine could elicit large quantities of selected antibodies that block infection, the transmission of HIV-1 may be successfully prevented. However, creation of a vaccine that induces such an antibody response has proven to be an enormous challenge.

Many other viruses are contained solely by antibodies. Therefore, vaccines to prevent infections with those viruses merely need to elicit such antibodies. However, antibody responses appear to play a relatively minor role in the normal containment of HIV. $CD8^+$ cytotoxic T-lymphocyte immunity appears to play an unusually critical role in controlling HIV. $CD8^+$ T lymphocytes can inhibit the replication of HIV in $CD4^+$ T lymphocytes *in vitro* [5]. Cytotoxic T-lymphocyte populations have been implicated in controlling the early spread of HIV in recently infected individuals [6]. Further evidence supporting the importance of these cytotoxic T lymphocytes includes the demonstration that there are large numbers of these cells in the blood of infected individuals, and those with the highest number of circulating cytotoxic T lymphocytes have the best preserved immunologic function [7, 8]. Collectively, these observations would seem to suggest that an effective HIV vaccine should elicit a CD8+ cytotoxic T-lymphocyte response.

The importance of CD4+ T lymphocytes in providing appropriate immunologic help to support the functioning of CD8+ cytotoxic T lymphocytes also appears to be validated in human and mouse experiments. For the cytotoxic T lymphocytes to function optimally, there must be an adequate number of normally functioning CD4+ T helper cells. An association has been demonstrated between high levels of CD4+ T lymphocyte and both successful control of HIV spread and stable clinical status [10]. On the basis of such evidence, it would appear that a successful HIV vaccine must elicit CD4+ T-lymphocyte help, as well as CD8+ cytotoxic T-lymphocyte responses.

V. EVIDENCE FROM ANIMAL EXPERIMENTS

Much of what has been learned about the biology of HIV infections has come from the use of animal models for AIDS. The most powerful of these models involves the use of nonhuman primates. HIV-1 and HIV-2 are members of a closely related family of lentiviruses that includes a diversity of viruses that infect the nonhuman primate species of Africa. These viruses are known as simian immunodeficiency viruses (SIVs). These SIVs do not cause disease in their natural host species. However, some of these viruses can cause AIDS in Asian monkeys [2].

Scientists have also engineered viruses that express the HIV-1 envelope glycoproteins on a viral backbone of SIV. These chimeric viruses, known as simian human immunodeficiency viruses (SHIVs), can also cause a rapidly progressive AIDS-like illness in Asian monkeys [3]. Both SIV-infected and SHIV-infected Asian monkeys have served as useful models to elucidate the mechanisms by which HIV is controlled by the immune system and to evaluate the efficacy of various HIV vaccine strategies.

The SIV-infected monkey model has been particularly useful in demonstrating the importance of CD8+ cytotoxic T lymphocytes in controlling HIV infections. When monkeys were depleted of CD8+ lymphocytes by treatment with a monoclonal antibody and then infected with SIV, they were unable to control early viral replication and died from a rapidly progressive AIDS-like disease [9]. Observations such as this in SIV-infected monkeys have helped to establish the importance of CD8+ cytotoxic T lymphocytes in the control and containment of HIV replication, a finding that has influenced HIV vaccine strategies.

VI. HIV VACCINE DESIGNS

A. Problems with Traditional Approaches

An understanding of the unique biology of HIV infections has provided insights into some of the obstacles that have slowed the development of an effective HIV vaccine. Many of the traditional vaccine designs that proved successful in preventing infections by various other viruses are not likely to be effective in preventing HIV infections. Studies in nonhuman primates and early phase clinical trials in humans suggest that live attenuated or weakened virus, inactivated virus, and protein vaccine strategies will not prove viable as AIDS vaccines.

The live attenuated virus vaccine approach is likely to be limited in its utility by the inability to create an AIDS virus that can induce immunity but not cause disease. Viruses grown in tissue culture accumulate small numbers of mutations that can limit the potential of the viruses to cause disease. Viruses altered genetically in this fashion are being used successfully worldwide for vaccination to prevent measles, polio, and chicken pox.

In studies using the SIV–rhesus monkey model, it was demonstrated that SIV could be altered by deletion of small amounts of genetic material, such that the variant viruses remained infectious but were attenuated in their ability to cause disease. Further experiments showed that prior infection with these attenuated SIV variants could prevent subsequent infection with unaltered disease-causing virus [11]. The early successes of these preliminary studies raised the possibility that a live virus vaccine approach for HIV might be feasible. However, this optimism was dampened by the subsequent demonstration that rhesus monkeys chronically infected with these genetically altered SIVs eventually developed AIDS and died, albeit more slowly than those infected with unaltered virus [12]. Further evidence from humans also raised similar concerns. A cluster of individuals was found that was infected through the transfusion of blood products with a single HIV having a genetic alteration that, on the basis of the results of earlier monkey studies, would have been predicted to eliminate the ability of that virus to cause disease. Nevertheless, these individuals all eventually developed AIDS, although a delay was seen from the time of initial infection to the onset of AIDS. This finding reinforced the notion that it may not be possible to create a fully attenuated live AIDS virus that will be able to elicit a protective immune response but not cause disease.

Another traditional vaccine approach, the use of whole inactivated viruses, has been successfully employed in the prevention of influenza and polio. However, evidence from the SIV–monkey model has indicated significant limitations in this strategy for an HIV vaccine [13, 14]. Vaccination with inactivated virus has not elicited the CD8+ cytotoxic T lymphocytes needed to control viral spread [15]. Moreover, these vaccines to date have generated antibodies that can block the infection of only a very limited range of AIDS viruses that is too limited a range to be useful.

A third traditional vaccine strategy, the use of protein vaccination, has also been explored for making an HIV vaccine. This vaccine modality has been successfully used for preventing hepatitis B infection in humans. However, in much the same manner as that seen in the other traditional vaccine approaches, studies in nonhuman primates have suggested that vaccination with available viral proteins will not confer protection against HIV infection. HIV envelope proteins (gp160 or gp120) have been assessed as vaccine candidates in chimpanzee studies. Protection against virus challenge was demonstrated, but only when the challenge virus and the vaccine virus had identical envelope sequences [16]. Such a situation would almost certainly not occur for HIV in vaccinated human populations.

Despite the limitations of protein vaccination that were demonstrated in these nonhuman primate studies, early phase human trials were pursued. These trials demonstrated the elicitation of only modest antibody responses, and the biologic activity of these antibodies was apparent against only a limited spectrum of HIV isolates. As with the inactivated virus vaccine approach, envelope protein vaccinations elicited no cytotoxic T-lymphocyte responses. Although these studies suggested little likelihood that this vaccination strategy could be successful, phase 3 clinical efficacy trials of a gp120 vaccine have been initiated, both in the United States and in Southeast Asia, supported by private sector resources. However, there is not much optimism in the scientific community that this strategy will result in an effective HIV vaccine.

B. New Approaches

The apparent limitations of these traditional vaccine design approaches for the development of an HIV vaccine have stimulated the investigation of novel vaccine design strategies. The use of live recombinant vectors and naked DNA immunization appear to be the most promising of these strategies.

The use of live recombinant vectors is currently being explored as a tool for eliciting immune responses against HIV. By using recombinant molecular biology techniques, genes of HIV can be inserted into live microorganisms or "vectors." The biologically engineered microorganisms then can serve as "carriers" of the HIV genes of interest. Subsequently, infection by the recombinant microorganisms will generate immune responses against both the vectors and the products of the HIV genes that the vectors carry. The use of such live recombinant vectors has proven to be a very efficient means of inducing cytotoxic T-lymphocyte responses.

A variety of microorganisms can be used as vaccine vectors. One of the most thoroughly explored of these is the pox family of viruses. The prototype of these poxviruses is vaccinia, the primary vaccine virus that was used in the worldwide smallpox eradication program. Experiments in nonhuman primates have demonstrated that vaccinia virus vectors are capable of eliciting powerful CD8+ cytotoxic T-lymphocyte responses against HIV and SIV proteins [19]. However, there are significant concerns about the potential for the vaccinia virus vector itself to cause severe adverse effects or even disease. Vaccinia virus can cause an infection that spreads throughout the body in people with diminished immune function, occasionally causing fatal brain inflammation [20]. The areas of the world in which there is a high prevalence of HIV infection are the areas in which large-scale HIV immunization programs are most urgently needed. There is a concern that many who might receive this vaccine in these areas are already HIV-infected. Thus, these individuals are immunosuppressed, and they might develop vaccinia disease.

As a result of this concern regarding the safety of vaccinia virus, other members of the pox family of viruses have been evaluated as possible HIV vectors. One of the most interesting of these vectors is modified vaccinia Ankara (MVA), a virus derived by long-term growth of vaccinia virus in cells. During this period of growth and passage in tissue culture cells, multiple genes of vaccinia virus were deleted, leaving it attenuated in its ability to cause disease. Nevertheless, this attenuated virus remains infectious and capable of eliciting potent immune responses. Studies in nonhuman primates have demonstrated that MVA-carrying genes of an AIDS virus can elicit potent immune responses, both when used alone or when used in combination with other vaccine modalities. Another similarly derived vaccinia virus variant, NYVAC, has also shown promise as a vaccine vector in nonhuman primates. These two pox vectors will soon be evaluated in human studies.

Other members of this family of viruses have also been evaluated as potential HIV vaccine vectors. The most important of these are the avian poxviruses. Avian poxviruses begin to make viral protein when they infect human cells, but they do not successfully make progeny viruses. This level of protein expression is sufficient to elicit an immune response to the poxvirus and to the products of the genes it is carrying. Both canary- and fowlpox vectors have been evaluated extensively in human studies [21]. They have been shown to be immunogenic and safe, eliciting both antibody responses in approximately 70% and cytotoxic T-lymphocyte responses in approximately 30% of vaccine recipients. Canarypox-vector-based vaccines are currently being considered for large-scale human efficacy trials. However, there is ongoing discussion regarding whether the potency of these vaccines that has been demonstrated to date in human studies warrants the evaluation of this live recombinant vector in large, costly trials.

Studies suggest that adenovirus is a particularly promising vector for HIV vaccination. The adenovirus family of viruses causes the common cold in humans. These viruses have received a great deal of attention as potential vectors for gene therapy, for carrying new genes into cells. Interestingly, adenoviruses that have been crippled in their ability to grow through the deletion of selected genes have been shown to elicit substantial antibody and cytotoxic T-lymphocyte responses in a variety of experimental animals [22]. Limited human studies of HIV vaccine prototypes are underway using this gene-deleted adenovirus vector vaccine approach. There is, however, a concern that this approach may not prove as effective in humans as it has in monkeys. Antibodies in the circulation of individuals previously infected with adenovirus may blunt the expression of protein production by genetically engineered adenovirus vectors, decreasing the ability of these vaccines to elicit optimal immune responses.

One solution to this potential problem may be to combine an adenovirus vector with a second vaccine approach. Alternatively, similar vaccines could be engineered using particular types of adenoviruses that only rarely infect humans. Additionally, vaccines could also be constructed using adenoviruses obtained from nonhuman primate species. These alternatives would decrease the likelihood that preexisting circulating antibodies to adenovirus could interfere with the ability of the vaccine to generate immune responses.

Other microorganisms that are being evaluated as potential vectors for HIV vaccines include the single-strand RNA alphaviruses (Venezuelan equine encephalitis, Semliki Forest virus) and the parvovirus adeno-associated virus. Weakened bacteria (i.e., *Salmonella* and *Mycobacterium*) vector systems also will likely be tested in the near future in human trials.

Finally, considerable effort has been devoted to exploring "naked DNA" as a vaccine strategy. Over a decade ago it was demonstrated that the direct inoculation into animals of large quantities of DNA encoding a viral protein could elicit immune responses against that protein. Vaccines based on this approach have proven efficient in generating cytotoxic T-lymphocyte responses in experimental animals [17]. Studies have also supported the use of naked DNA as an effective initial vaccine in the two-modality strategy of immunization [18]. Although naked DNA vaccines provide a safe approach for eliciting cytotoxic T-lymphocyte responses, some preliminary studies have raised the possibility that extremely large quantities of DNA may be necessary to stimulate immune responses in humans. A number of human trials are planned at this time to evaluate the potential utility of naked DNA as an HIV vaccine.

One of the main obstacles facing investigators in the field of HIV vaccine research has been the difficulty in constructing a single protein that is able to elicit an antibody response with activity against a diversity of HIV viruses. A variety of strategies for achieving this goal are currently being pursued. One approach to this challenge has focused on developing proteins that "unshield" structurally conserved regions of the HIV envelope by constructing biochemically and structurally altered versions of HIV envelope. Another approach to this problem that is being taken is the attempt to create an envelope protein that folds into the three-dimensional conformation of native, virus-associated HIV envelope glycoproteins. Some have suggested that the best way to elicit anti-envelope antibodies with biologic activity against a wide array of HIV isolates is through the use of mixtures of different HIV envelope proteins. Others are assessing the possibility of generating a protein that mimics the conformational form of HIV envelope seen as the virus fuses with the target cell membrane. All of this investigative work underscores the need for novel HIV vaccine designs that take into account the unique biology of HIV.

Although the ultimate configuration of an effective HIV vaccine remains unclear, it is becoming increasingly evident that such a product will include more than a single vaccine modality. Several of the novel vaccine approaches that have been evaluated appear to be effective in eliciting either an antibody or a cytotoxic T-lymphocyte response. Thus, a protein will likely be needed to elicit biologically effective anti-envelope antibodies, and a naked DNA and/or live recombinant vector vaccine will be required for the induction of a CD8+ cytotoxic T-lymphocyte response. However, an effective HIV vaccine will likely need to elicit both an antibody and a cytotoxic T-lymphocyte response. Moreover, experimentation suggests that the combination of two complementing vaccine strategies will likely generate a more potent cytotoxic T-lymphocyte response than any single vaccine modality. The ultimate HIV vaccine, therefore, will likely make use of a combination of strategies, an approach that radically departs from any vaccines that have previously been developed.

VII. VACCINE-ELICITED CLINICAL PROTECTION

A growing awareness of the biologic obstacles that exist for creating an effective HIV vaccine had long ago created a sense of pessimism among many HIV researchers. However, observations made in nonhuman primate studies have suggested that currently available HIV vaccine strategies may confer clinical benefit, if not complete sterilizing immunity, to vaccine recipients who are subsequently exposed to HIV [18, 22–25]. In these studies, monkeys that had vaccine-elicited cytotoxic T-lymphocyte responses and were subsequently challenged with AIDS virus were not protected from infection. However, these monkeys showed a dramatic control of virus spread and an associated slowing of progression to clinical AIDS. These results suggest that, even if a vaccine cannot prevent HIV infection, it may still induce an immune response that can slow the progression of disease in individuals who become infected. Such a therapeutic benefit may be extremely important in regions of the world in which HIV is endemic and financial resources for healthcare are scarce. Such an HIV vaccine might impact morbidity and even mortality, even if it cannot prevent HIV infection.

In any infectious disease, the most important public health priority is to block transmission of the infection. If vaccine-elicited immune protection can be achieved against disease progression, it may also alter the rate of HIV transmission in a population. It has previously been shown that the likelihood of HIV transmission from an infected to an uninfected individual is determined by the amount of virus in that infected person's blood and secretions [26]. A vaccinated individual who subsequently becomes infected with HIV may have a very low viral burden. Given this low viral burden, such an individual may be less likely than an unvaccinated, infected individual to transmit HIV to an uninfected individual. Such a process may ultimately slow the spread of HIV infection within a population. Thus, although a vaccine that prevents HIV infection may not be within reach at this time, an HIV vaccine that can both slow the progression to AIDS when infection occurs and slow the transmission of virus spread in a population may be a readily attainable goal.

Vaccines to prevent infections by other infectious agents have been evaluated in populations in the developed world and then used worldwide to eradicate the microbe. The testing of HIV vaccine candidates poses an unprecedented problem in this regard. The populations at greatest risk for HIV infection, and therefore the populations in which an HIV vaccine can be most readily assessed for efficacy, are in the developing world. If an HIV vaccine prevents overt disease but does not prevent infection from occurring, a highly sophisticated evaluation of vaccinees will be required in any large-scale studies of vaccine efficacy. However, the infrastructure that will be needed to monitor such vaccine efficacy does not exist in these geographic regions. Therefore, the testing of HIV vaccine approaches in at-risk human populations presents a challenge to the medical and scientific community.

VIII. CONCLUSIONS

Although the historically successful approaches to vaccine design have proven to be of limited utility in the search for an effective HIV vaccine, a number of novel vaccine strategies that address the unique biology of this virus appear to be promising. Live vectors, naked DNA, and multimodality vaccine approaches will likely play a significant role in the final design of a successful HIV vaccine. Observations in nonhuman primate studies have provided hope that the immunity generated by an HIV vaccine may be capable of attenuating clinical disease and slowing HIV transmission, even if infection by HIV is not prevented.

REFERENCES

1. Malim, M. H., and Emerman, M. (2001). HIV-1 sequence variation: Drift, shift, and attenuation. *Cell* **104**:469–472.
2. Hirsch, V. M., and Johnson, P. R. (1994). Pathogenic diversity of simian immunodeficiency viruses. *Virus Res.* **32**:183–203.
3. Reimann, K. A., Li, J. T., Veazey, R., Halloran, M., Park, I. -W., Karlsson, G. B., Sodroski, J., and Letvin, N. L. (1996). A chimeric simian/human immunodeficiency virus expressing a primary patient human immunodeficiency virus type 1 isolate *env* causes an AIDS-like disease after *in vivo* passage in rhesus monkeys. *J. Virol.* **70**:6922–6928.
4. Mascola, J. R., Stiegler, G., VanCott, T. C., Katinger, H., Carpenter, C. B., Hanson, C. E., Beary, H., Hayes, D., Frankel, S. S., Birx, D. L., and Lewis, M. G. (2000). Protection of macaques against vaginal transmission of a pathogenic HIV-1/SIV chimeric virus by passive infusion of neutralizing antibodies. *Nat. Med.* **6**:207–210.
5. Walker, C. M., Moody, D. J., Stites, D. P., and Levy, J. A. (1986). CD8+ lymphocytes can control HIV infection *in vitro* by suppressing virus replication. *Science* **234**:1563–1566.
6. Koup, R. A., Safrit, J. T., Cao, Y., Andrews, C. A., McLeod, G., Borkowsky, W., Farthing, C., and Ho, D. D. (1994). Temporal association of cellular immune responses with the initial control of viremia in primary human immunodeficiency virus type 1 syndrome. *J. Virol.* **68**:4650–4655.
7. Ogg, G. S., Jin, X., Bonhoeffer, S., Dunbar, P. R., Nowak, M. A., Monard, S., Segal, J. P., Cao, Y., Rowland-Jones, S. L., Cerundolo, V., Hurley, A., Markowitz, M., Ho, D. D., Nixon, D. F., and McMichael, A. J. (1998). Quantitation of HIV-1-specific cytotoxic T lymphocytes and plasma load of viral RNA. *Science* **279**:2103–2106.
8. Altman, J. D., Moss, P. A. H., Goulder, P. J. R., Barouch, D. H., McHeyzer-Williams, M. G., Bell, J. I., McMichael, A. J., and Davis, M. M. (1996). Phenotypic analysis of antigen-specific T lymphocytes. *Science* **274**:94–96.
9. Schmitz, J. E., Kuroda, M. J., Santra, S., Sasseville, V. G., Simon, M. A., Lifton, M. A., Racz, P., Tenner-Racz, K., Dalesandro, M., Scallon, B. J., Ghrayeb, J., Forman, M. A., Montefiori, D. C., Rieber, E. P., Letvin, N. L., and Reimann, K. A. (1999). Control of viremia in simian immunodeficiency virus infection by CD8+ lymphocytes. *Science* **283**:857–860.
10. Rosenberg, E. S., Billingsley, J. M., Caliendo, A. M., Boswell, S. L., Sax, P. E., Kalams, S. A., and Walker, B. D. (1997). Vigorous HIV-1-specific CD4+ T cell responses associated with control of viremia. *Science* **278**:1447–1450.
11. Daniel, M. D., Kirchhoff, F., Czajak, S. C., Sehgal, P. K., and Desrosiers, R. C. (1992). Protective effects of a live attenuated SIV vaccine with a deletion in the *nef* gene. *Science* **258**:1938–1941.
12. Baba, T. W., Jeong, Y. S., Penninck, D., Bronson, R., Greene, M. F., and Ruprecht, R. M. (1995). Pathogenicity of live, attenuated SIV after mucosal infection of neonatal macaques. *Science* **267**:1820–1825.
13. Murphey-Corb, M., Martin, L. N., Davison-Fairburn, B., Montelaro, R. C., Miller, M., West, M., Ohkawa, S., Baskin, G. B., Zhang, J. -Y., Putney, S. D., Allison, A. C., and Eppstein, D. A. (1989). A formalin-inactivated whole SIV vaccine confers protection in macaques. *Science* **246**:1293–1297.
14. Stott, E. J. (1991). Anti-cell antibody in macaques. *Nature* **353**:393.
15. Levine, A. M., Groshen, S., Allen, J., Munson, K. M., Carlo, D. J., Daigle, A. E., Ferre, F., Jensen, F. C., Richieri, S. P., Trauger, R. J., Parker, J. W., Salk, P. L., and Salk, J. (1996). Initial studies on active immunization of HIV-infected subjects using a gp120-depleted HIV-1 immunogen: Long-term follow-up. *J. Acquir. Immune Defic. Syndr.* **11**:351–364.

16. Berman, P. W., Gregory, T. J., Riddle, L., Nakamura, G. R., Champe, M. A., Porter, J. P., Wurm, F. M., Hershberg, R. D., Cobb, E. K., and Eichberg, J. W. (1990). Protection of chimpanzees from infection by HIV-1 after vaccination with recombinant glycoprotein gp120 but not gp160. *Nature* **345**:622–625.

17. Egan, M. A., Charini, W. A., Kuroda, M. J., Voss, G., Schmitz, J. E., Racz, P., Tenner-Racz, K., Manson, K., Wyand, M., Lifton, M. A., Nickerson, C. E., Fu, T.-M., Shiver, J. W., and Letvin, N. L. (2000). Simian immunodeficiency virus (SIV) *gag* DNA-vaccinated rhesus monkeys develop secondary cytotoxic T-lymphocyte responses and control viral replication after pathogenic SIV infection. *J. Virol.* **74**:7485–7495.

18. Amara, R. R., Villinger, F., Altman, J. D., Lydy, S. L., O'Neil, S. P., Staprans, S. I., Montefiori, D. C., Xu, Y., Herndon, J. G., Wyatt, L. S., Candido, M. A., Kozyr, N. L., Earl, P. L., Smith, J. M., Ma, H.-K., Grimm, B. D., Hulsey, M. L., McClure, H. M., McNicholl, J. M., Moss, B., and Robinson, H. L. (2001). Control of a mucosal challenge and prevention of clinical AIDS in rhesus monkeys by a multiprotein DNA/MVA vaccine. *Science* **292**:69–74.

19. Shen, L., Chen, Z. W., Miller, M. D., Stallard, V., Mazzara, G. P., Panicali, D. L., and Letvin, N. L. (1991). Recombinant virus vaccine-induced SIV-specific CD8+ cytotoxic T lymphocytes. *Science* **252**:440–443.

20. Redfield, R. R., Wright, D. C., James, W. D., Jones, T. S., Brown, C., and Burke, D. S. (1987). Disseminated vaccinia in a military recruit with human immunodeficiency virus (HIV) disease. *New England J. Medicine* **316**:673–676.

21. Evans, T. G., Keefer, M. C., Weinhold, K. J., Wolff, M., Montefiori, D., Gorse, G. J., Graham, B. S., McElrath, M. J., Clements-Mann, M. L., Mulligan, M. J., Fast, P., Walker, M. C., Excler, J. L., Duliege, A. M., and Tartaglia, J. (1999). A canarypox vaccine expressing multiple human immunodeficiency virus type 1 genes given alone or with rgp120 elicits broad and durable CD8+ cytotoxic T lymphocyte responses in seronegative volunteers. *J. Infect. Dis.* **180**:290–298.

22. Shiver, J. W., Fu, T.-M., Chen, L., Casimiro, D. R., Davies, M.-E., Evans, R. K., Zhang, Z.-Q., Simon, A. J., Trigona, W. L., Dubey, S. A., Huang, L., Harris, V. A., Long, R. S., Liang, X., Handt, L., Schleif, W. A., Zhu, L., Freed, D. C., Persaud, N. V., Guan, L., Punt, K. S., Tang, A., Chen, M., Wilson, K. A., Collins, K. B., Heidecker, G. J., Fernandez, V. R., Perry, H. C., Joyce, J. G., Grimm, K. M., Cook, J. C., Keller, P. M., Kresock, D. S., Mach, H., Troutman, R. D., Isopi, L. A., Williams, D. M., Xu, Z., Bohannon, K. E., Volkin, D. B., Montefiori, D. C., Miura, A., Krivulka, G. R., Lifton, M. A., Kuroda, M. J., Schmitz, J. E., Letvin, N. L., Caulfield, M. J., Bett, A. J., Youil, R., Kaslow, D. C., and Emini, E. A. (2002). AIDS vaccines: Good news, bad news. *Nature* **415**:331–335.

23. Ourmanov, I., Brown, C. R., Moss, B., Carroll, M., Wyatt, L., Pletneva, L., Goldstein, S., Venzon, D., and Hirsch, V. M. (2000). Comparative efficacy of recombinant modified vaccinia virus Ankara expressing simian immunodeficiency virus (SIV) Gag-Pol and/or Env in macaques challenged with pathogenic SIV. *J. Virol.* **74**:2740–2751.

24. Barouch, D. H., Santra, S., Schmitz, J. E., Kuroda, M. J., Fu, T.-M., Wagner, W., Bilska, M., Craiu, A., Zheng, X. X., Krivulka, G. R., Beaudry, K., Lifton, M. A., Nickerson, C. E., Trigona, W. L., Punt, K., Freed, D. C., Guan, L., Dubey, S., Casimiro, D., Simon, A., Davies, M.-E., Chastain, M., Strom, T. B., Gelman, R. S., Montefiori, D. C., Lewis, M. G., Emini, E. A., Shiver, J. W., and Letvin, N. L. (2000). Control of viremia and prevention of clinical AIDS in rhesus monkeys by cytokine-augmented DNA vaccination. *Science* **290**:486–492.

25. Barouch, D. H., Santra, S., Kuroda, M. J., Schmitz, J. E., Plishka, R., Buckler-White, A., Gaitan, A. E., Zin, R., Nam, J.-H., Wyatt, L. S., Lifton, M. A., Nickerson, C. E., Moss, B., Montefiori, D. C., Hirsch, V. M., and Letvin, N. L. (2001). Reduction of simian–human immunodeficiency virus 89.6P viremia in rhesus monkeys by recombinant modified vaccinia virus Ankara vaccination. *J. Virol.* **75**:5151–5158.

26. Quinn, T. C., Wawer, M. J., Sewankambo, N., Serwadda, D., Li, C., Wabwire-Mangen, F., Meehan, M. O., Lutalo, T., and Gray, R. H. (2000). Viral load and heterosexual transmission of human immunodeficiency virus type 1. *New England J. Medicine* **342**:921–929.

27. Joint United Nations Programme on HIV/AIDS. (2000). Report on the Global HIV–AIDS Epidemic. Geneva, Switzerland.

28. Letvin, N. L., Bloom, B. R., and Hoffman, S. L. (2001). Prospects for vaccines to protect against AIDS, tuberculosis, and malaria. *J. Am. Med. Assoc.* **285**:606–611.

6 DISEASE STATES AND VACCINES: SELECTED CASES

PART G. Pneumococcus, Pneumococcal Disease, and Prevention

JILL G. HACKELL

Wyeth Vaccines Research, Pearl River, New York 10965

I. INTRODUCTION

Streptococcus pneumoniae, also known as the pneumococcus, colonizes the human respiratory tract and is an important cause of invasive infections such as bacteremia and meningitis, lower respiratory infections (pneumonia, empyema), and upper respiratory infections, including sinusitis and otitis media. The incidence of serious pneumococcal disease is greatest at the extremes of life, that is, in children under the age of 2 years and in the elderly. Additionally, a number of conditions, such as functional or anatomic asplenia, immunosuppressive conditions, chronic cardiovascular, pulmonary, liver disease, and diabetes, are associated with an increased risk of pneumococcal infection [1]. Day care attendance has been associated with a higher risk of invasive pneumococcal disease in children under 2 years of age [2, 3].

Pneumococcal infection causes 1–2 million of the approximately 11.7 million deaths per year worldwide among children under 5 years of age [4]. Annual incidence rates of bacteremic pneumococcal disease in the United States are estimated to be 10–30 cases per 100,000 persons overall, 160 cases per 100,000 persons 2 years of age or under, 50–83 cases per 100,000 persons 65 years of age or over, and as high as 940 cases per 100,000 persons with acquired immunodeficiency syndrome (Fig. 1) [1]. The pneumococcus is the most common cause of bacterial meningitis in the United States (incidence: 1 case per 100,000 persons overall), and pneumococcal meningitis has the highest mortality rate [5].

In the past decade, the proportion of *S. pneumoniae* isolates resistant to antibiotics has been increasing in the United States and worldwide [3]. The

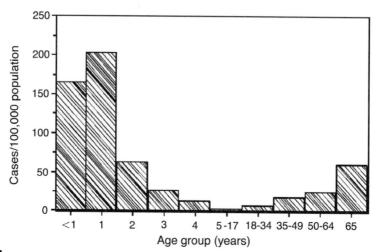

FIGURE 1   Rates of invasive pneumococcal disease by age group—United States, 1998. Source: CDC, Active Bacterial Core Surveillance (ABCs)/Emerging Infectious Program (EIP) Network, 2000. Available at <http://www.cdc.gov/ncidod/dbmd/abcs/survreports/spneu98.pdf>. Accessed August 22, 2000.

burden of pneumococcal disease and the emergence of resistant organisms have led to a high priority for the development of new pneumococcal vaccines.

II. ORGANISM AND NOMENCLATURE

S. pneumoniae is a lancet-shaped, gram-positive coccus, usually seen in pairs (diplococci), with a polysaccharide capsule (Fig. 2). Designation of serotype is based on the immunological response of the capsule to antiserum (Quelling reaction). Based on this immunological classification, over 90 serotypes have been described. Serotypes are classified by two systems. The Danish system classifies similar, potentially cross-reacting serotypes into serogroups (e.g., 6A and 6B within serogroup 6), whereas the U.S. system numbers serotypes in order of their discovery. (In this system, types 6A and 6B are designated 6 and 26, respectively.) This part of Chapter 6 will follow the Danish system, which has been adopted for worldwide use [6].

III. EPIDEMIOLOGY

A. Disease

The 23 pneumococcal serotypes included in the currently licensed polysaccharide vaccines cover approximately 90% of invasive disease due to pneumococcus in the target population of high-risk individuals over 2 years of age and the elderly. Newer generation pneumococcal conjugate vaccines will, at least initially, contain fewer capsular types, because of the complexity of conjugating each polysaccharide serotype to a protein carrier and because of concerns regarding limitation in the amount of carrier that can be administered at

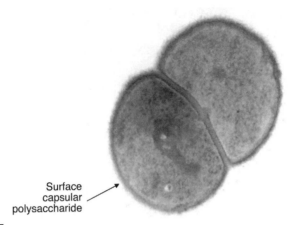

Surface
capsular
polysaccharide

FIGURE 2   Electron micrograph of pneumococcus.

a time. Selection of vaccine types must be based on a firm knowledge of the epidemiology of pneumococcal disease. Factors to be considered are (1) population, including both age and geography, (2) target illness (e.g., invasive disease or otitis media), (3) potential for cross-protection across serotypes within a serogroup, and (4) serotypes associated with antibiotic resistance.

On the basis of a review of worldwide published epidemiological studies of clinical isolates of *S. pneumoniae,* a relatively small number of serogroups accounts for most of the invasive pneumococcal disease in each geographic region, and fewer serogroups account for disease in young children than for older children and adults. Thus, 5–8 serogroups in young children and 10–11 serogroups in older individuals are responsible for at least 75% of invasive disease isolates in each geographic region [7].

Serotypes in the currently available 7-valent pneumococcal conjugate vaccine (4, 6B, 9V, 14, 18C, 19F, 23F) account for approximately 80% of invasive pneumococcal disease in young children in the United States and Canada and at least 50% in all other regions, except Asia (30%). If protection extends to potentially cross-reactive serotypes (especially 6A and 19A), coverage rises by approximately 8–15% in each region. Addition of serotypes 1 and 5 to a future generation of conjugate vaccine would increase coverage by 15–20% in Latin America, Asia, and Africa, 8–12% in Europe, but only about 1% in the United States. Finally, the addition of types 3 and 7, as proposed for a future generation 11-valent vaccine, would increase coverage most significantly in Asia (about 9%) [7]. Figure 3a shows the anticipated coverages of 7-, 9-, and 11-valent pneumococcal conjugate vaccines in different regions of the world. Figure 3b shows the incidence of individual serotypes in North American children and the cumulative percentage of invasive pneumococcal disease caused by these serotypes.

Some serotypes may be associated more often with a particular disease manifestation. On the basis of a meta-analysis across multiple published studies, Hausdorff *et al.* [8] found that serogroups 23, 14, and 4 were isolated more frequently from blood than from cerebrospinal fluid (CSF), and that serogroups 1, 6, and 10 were isolated more frequently from CSF than from

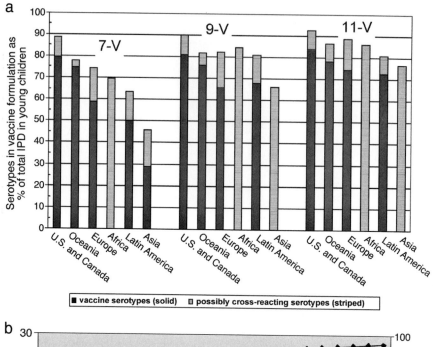

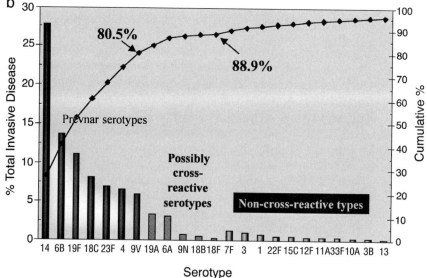

FIGURE 3 Vaccine serotype coverage (young children) for IPD (a) by Region and (b) in North America. Adapted with permission from Hausdorff *et al.* [7].

blood. Serogroups 3, 19, and 23 were more commonly isolated from middle ear fluid than from either blood or CSF in younger children.

Methodological differences across studies and in different geographic regions may play a role in the observed differences between regions in serogroup distribution. For example, serogroups 4, 6, 9, 14, 18, 19, and 23 are among the most common serogroups causing disease in young children in both the United States and Europe. However, the reported incidence of disease due to these serogroups is higher in the United States than in Europe, whereas the reported incidence of type 1 and 5 diseases is similar in the two regions. By one hypothesis, types 1 and 5 may be more virulent than the other pediatric serotypes; hence, types 4, 6, 9, 14, 18, 19, and 23 might be less likely to be isolated in regions (e.g., Europe) in which the threshold for blood culturing is higher than in the United States [9].

It is also revealing to examine those serogroups that are associated with antibiotic resistance. In a study of over 850 middle ear isolates in Israel, five serotypes (6B, 9V, 14, 19F, and 23F) were predominantly associated with antibiotic resistance. Whereas these serotypes accounted for 53% of all middle ear isolates, they accounted for 74% of penicillin-resistant isolates, 69% of isolates resistant to one or more drugs, and 87% of multiresistant (≥ 3) isolates. Thus, although the coverage of the pneumococcal conjugate vaccines against middle ear isolates was only 56, 60, and 65% for the 7-, 9-, and 11-valent vaccines, respectively, coverage of these vaccines against penicillin-resistant isolates (MIC > 1.0) was 92, 93, and 93%. Children with frequent previous otitis media episodes or the recent receipt of antibiotics were most likely to have resistant isolates and, therefore, high anticipated coverage by pneumococcal conjugate vaccine [10].

Similar findings have been described in other regions of the world and in association with isolates from invasive pneumococcal disease [11–14], where at least 80% of antibiotic-resistant serotypes are represented in the 7-valent vaccine formulation [7].

B. Nasopharyngeal Colonization

Nasopharyngeal colonization is an important marker for describing the epidemiology of *S. pneumoniae*, the development of resistance, and the role of interventions such as the use of pneumococcal vaccines.

The *amount* of carriage differs in different regions of the world and by different ages of the host. Carriage increases with age in the first months to years of life [15], is more common in young children than in adults, and varies by geographic region [16, 17]. Genetic and immunological differences in the host and socioeconomic conditions, such as crowding, sanitation, family size, and day care contact, can play a role. In addition, carriage is a very dynamic situation, so that frequency of culture may substantially increase the percentage of children noted to carry at least one strain of pneumococcus over time [17]. Carriage of pneumococci increases in winter, presumably related to closer interperson contact, viral infections, and the use of antibiotics [15, 16]. Parental smoking has been identified as a risk factor for increased pneumococcal carriage in young children [16, 18a].

The *duration* of carriage depends on the age of the host and the serotype of the strain [18b]. The younger the infant at the time of acquisition, the longer a strain is likely to be carried. In addition, the first strain acquired is likely to be carried longer than subsequent strains. Serotypes 6, 14, 19, and 23 may be carried significantly longer than other serotypes [15].

The serotype distribution of isolates in the nasopharynx usually predicts the isolates associated with disease in a population, although the rank order may be different. Positive predictive value in an individual is low, although a negative nasopharyngeal culture for a given serotype or resistance profile makes that strain an unlikely cause of otitis media for that individual [16]. Approximately 15% of new acquisitions in young children result in clinical infection (usually otitis media), with most infections (23/31, 74%) caused by types carried a month or less after acquisition [15].

As demonstrated by Gray and others, the common pediatric serotypes are those most likely to be carried in the nasopharynges of young children for a long time. They are the most likely to be present, associated with acute otitis media, and exposed to antibiotics, and therefore they are most likely to develop resistance. In turn, these now resistant serotypes will spread to others, in day care centers or to younger siblings. Thus, resistant serotypes are most likely to be found in children with previous otitis, recent antibiotic exposure, or an older sibling [10, 19, 20].

IV. PATHOGENESIS

As with other infectious diseases, the pathogenesis of pneumococcal disease is due to a complex balance of the virulence factors of the bacterium and the immune response of the host [21].

The capsule is the major virulence factor of S. pneumoniae. Studies in mice have demonstrated that encapsulated strains were at least 10^5 times more virulent than strains lacking the capsule [22]. Both the chemical structure of the capsular polysaccharide and, to a lesser extent, the quantity produced by the organism determine the ability of the pneumococcus to survive in the bloodstream [6, 23]. The major role of the capsule is to provide resistance to phagocytosis. Opsonization is initiated in the presence of anticapsular antibodies, but is complement-dependent. Optimal clearance of pneumococci requires both the classical and the alternate pathways [21]. Differences between serotypes in the degree of activation of the alternate pathway of complement, deposition and degradation of complement on the capsule, resistance to phagocytosis, and ability to induce antibody may account for differences in virulence among capsular serotypes [23].

Changes in capsule production, expressed phenotypically as opaque versus transparent colonies, have been associated with colonization versus invasion of the host. Organisms can spontaneously alternate between the two phases. Transparent variants produce increased amounts of cell-wall-associated teichoic acid (C-polysaccharide). They are better able to adhere to human buccal and lung epithelial cells and vascular endothelial cells and have been shown to colonize the nasopharynx in the infant rat more easily than opaque strains [24]. In addition, they are 3–5 times more likely to invade cerebral microvascular endothelium than opaque phenotypes, and so they may be associated with the development of meningitis. The opaque phenotype is associated with increased production of capsule and pneumococcal surface protein A (PspA). Opaque organisms are more likely to invade the bloodstream [21].

Also important in the pathogenesis of pneumococcal disease are factors that contribute to the host inflammatory response. These include pneumolysin, C-polysaccharide, autolysin, and PspA.

Pneumolysin is not secreted by the pneumococcus, but it can be released upon lysis of pneumococci under the influence of autolysis [23]. Pneumolysin is a cytoplasmic toxin that binds to target cell membrane, causing a colloid–osmotic lysis of the cell. By this mechanism, pneumolysis can damage a wide range of cells [21]. It interferes with the beating of cilia and disrupts the respiratory tract epithelium, thereby facilitating access of pneumococci into the bloodstream [25]. In addition, it can be neurotoxic upon invasion of the central nervous system in meningitis [21]. Pneumolysin is a general inflammatory agonist, stimulating the production of inflammatory cytokines such as tumor necrosis factor-α (TNFα) and interleukin-1β (IL-1β) by human monocytes and also activating complement by the classical pathway. Pneumolysin decreases the migration and bactericidal activity of neutrophils, as well as inhibits lymphocyte proliferation and antibody synthesis [23].

Cell wall polysaccharide (C-polysaccharide) is also a potent inflammatory agent. Injection of purified C-polysaccharide into animals can mimic the clinical findings of otitis media, meningitis, and pneumonia. C-polysaccharide stimulates IL-12 by monocytes more effectively than endotoxin, activates complement by the alternative pathway, and may play an important role in the adhesion of pneumococci to human cells [21, 23].

Autolysin is responsible for cell wall turnover in the pneumococcus. Its role in virulence is mediated by the generation of inflammatory cell-wall-breakdown products and the release of pneumolysin [23].

Pneumococcal surface protein A appears to facilitate pneumococcal survival and host invasion by interference with complement activation [21, 23].

V. IMMUNOLOGY

The immunological response of humans to the pneumococcus is mounted against the capsular antigen. Antibody is produced, complement is activated, and bacteria are ingested by phagocytes. Clearance of organisms occurs in the reticuloendothelial system of the liver and the spleen [25]. Thus, patients with liver disease or functional/anatomical asplenia are at increased risk of invasive pneumococcal disease.

Antibody to the capsule is the critical protective factor, although a precise protective level has not been established. Some infants may be protected at birth from antibody obtained passively across the placenta during gestation, but this protection wanes and is essentially gone by 6 months of age. This fact, combined with the inability of young children to mount a response to polysaccharide antigen, explains the propensity to pneumococcal disease in this age group.

In older children and adults, the kinetics of the immune response to polysaccharide is such that antibodies reach peak titers at 2–8 weeks after exposure and decline gradually after the ensuing 6–12 months. Antibodies may persist thereafter for as long as 8 years at one-third to one-half of peak values, presumably because capsular polysaccharides are not biodegradable by human enzymes and therefore persist in the reticuloendothelial system for protracted periods [6]. Typically, there is little response to a subsequent immunization.

Titers may return to previous levels, but there is no evidence of a booster response.

Why is this so? Responses to polysaccharide antigen are T-cell-independent. Antigens interact directly with B cells and do not require antigen-presenting cells (APCs) or T cells. A T-cell-independent immune response is characterized by a poor response in young children, a limited response at all ages dominated by immunoglobulin M (IgM) and IgG_2 subtypes [26], the absence of immunological memory and consequent lack of a booster response, and a restricted immune repertoire at the molecular level manifested by a lack of affinity maturation [27a].

In contrast, the immune response to protein antigens has an absolute requirement for T-cell help. Response to these antigens occurs in very young children. IgG_1 and IgG_3 are the predominant antibody subtypes. Immunity is long-lived and boostable, due to the evocation of immunological memory, and affinity maturation can be demonstrated [26].

The poor response to polysaccharide in infants and young children is a result of the delayed ontogeny of the B-cell response to polysaccharide, compounded by no previous exposure to naturally conjugated bacterial polysaccharides, which may play a role in enhanced responses in older individuals. Interestingly, the maturation of the immune response varies by serotype. For type 3, responsiveness occurs by 6 months of age [6]. For serotypes that are responsible for much of the infections in early childhood (6A, 6B, 14, 19F, 19A, 23 F), responsiveness may not appear until as late as 4–8 years of age [27b]. These are the very serotypes associated with prolonged nasopharyngeal carriage and the development of resistance in young children [10, 15].

Not only the antibody quantity but also its quality may be influenced by interactions with T cells and the development of antigenic memory, as exemplified by the concept of affinity maturation. The original response to an antibody is dependent upon antibody genes initially formed from unmodified germline gene segments. Because there are a limited number of these segments in the genome, there is a constraint on the number of different antibody structures in the repertoire of the naive host. Thus, the first antibodies produced by the host are likely to have an imperfect fit with the antigens, resulting in lower avidity of binding. In T-dependent immune responses, in which memory is present, continued maturation occurs in the lymph nodes. Through this process, under the combined action of somatic mutation and selection, the combining site geometries in the descendants of the originally activated B cells are altered, so that interactions with antigen are enhanced [28]. This leads to greater affinity in binding antibody and, hence, affinity maturation.

VI. PNEUMOCOCCAL CONJUGATE VACCINE IMMUNOLOGY

The development of conjugate vaccine technology allows us to improve upon nature by converting immunological responses to the polysaccharide from a T-independent to a T-dependent response. The desired response following vaccination with the new vaccines can be determined by using analyses that measure these immunological properties.

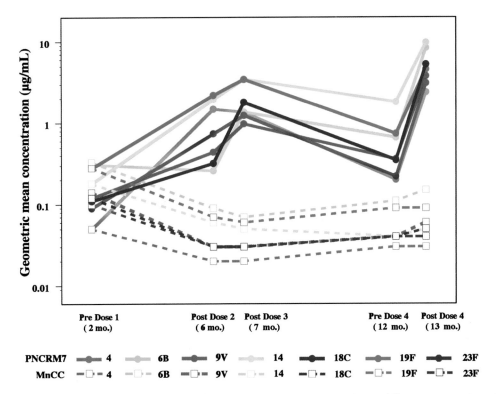

FIGURE 4 Geometric mean concentrations of antibody to 7 pneumococcal types following immunization with pneumococcal 7-valent conjugate vaccine conjugated to diphtheria CRM_{197} protein, compared to a control group. Adapted from Rennels *et al.* [29].

A. Is There a Response in Children in Whom There Is No Response to Polysaccharide?

The ability to raise an antibody response with conjugate vaccines in children under 2 years of age has been demonstrated in response to pneumococcal serotypes conjugated to various proteins, including diphtheria CRM_{197} protein [29–35], diphtheria or tetanus toxoid [36, 37], and a meningococcal outer membrane protein complex [38]. Interestingly, the kinetics of the antibody response varies with serotype. For type 4, antibody response is close to maximal following two doses of vaccine in young infants. For type 6B, the greatest rise in antibody level is seen after the third dose. The remaining types show an intermediate pattern (Fig. 4, see color insert) [29, 38].

B. What Is the Isotype Profile Developed in Vaccinees?

For infants, the primary response to conjugate vaccine is predominantly IgG_1 [39, 40]. Adults immunized with pneumococcal polysaccharide have a response dominated by IgG_2. However, if they are administered pneumococcal conjugate vaccine, the response is both IgG_1 and IgG_2 [26, 41, 42]. IgG_1 is more efficient as an opsonin and activator of the classical pathway of complement, and it is indicative of a T-dependent response.

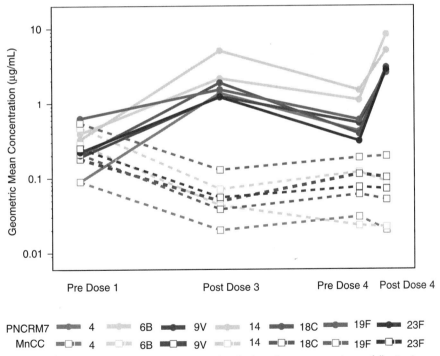

FIGURE 4 Geometric mean concentrations of antibody to 7 pneumococcal types following immunization with pneumococcal 7-valent conjugate vaccine conjugated to diphtheria CRM_{197} protein, compared to a control group. Adapted from Rennels et al. [29]. (See color insert.)

C. Is There Evidence of Priming?

Evidence of priming can be demonstrated by boosting with a polysaccharide after the primary series. If the age at boost is lower than the age at which polysaccharide (PS) responsiveness is usually seen (and a control group can be used to demonstrate this clearly), a good PS response will provide clear evidence of priming [36, 38, 40, 43]. For example, Obaro et al. [43] administered polysaccharide vaccine to 2-year-old Gambian children initially randomized to receive 5-valent pneumococcus conjugated to CRM_{197} or Hib vaccine as a control. Responses were 5- to 100-fold higher, depending on serotype, in the group that had been primed as compared to controls.

D. Is There Evidence of Affinity Maturation?

The use of conjugate vaccine is anticipated to evoke memory and with it the ability to continually mature the antibody response toward greater avidity of binding with antigen. As an example, Goldblatt et al. [44], measured avidity immediately following a primary series of Haemophilus influenzae type b (Hib) conjugate vaccine and then 6 months later, just prior to boost. In children with good evidence of priming (as seen by good antibody titers following the original series), the affinity index increased during a time when antibodies were declining. The majority of the increase in avidity was seen in the months after primary immunization, and boosting only marginally increased the avidity.

In the subgroup of children with suboptimal primary ELISA responses to Hib conjugate vaccine, the avidity after boosting was low and was the same as that seen after primary immunization. Hence, these children had not been adequately primed, and this response to the booster was akin to a primary response. The postboost titer achieved was lower in these children and more similar to the response in children after a single dose of conjugate vaccine (at this age, a primary response).

A similar role for avidity has been seen with pneumococcal conjugate vaccines [45]. Antilla et al. [46a], found an increase in avidity following the primary series for four different pneumococcal conjugate vaccines. Similar results were seen by Goldblatt et al. following a single dose of pneumococcal CRM_{197} conjugate in toddlers [46b].

This information suggests that the measurement of maturation of the affinity of antibody to bind with antigen may be a marker for memory. However, methodology is not yet standardized, and the measurement of affinity is often difficult in prebooster samples, where antibody levels are low. In addition, further understanding of the relationship between primary antibody responses and long-term protection is clearly needed.

E. Is There a Functional Response?

For the pneumococcus, where host response is mediated mainly by opsonin-dependent phagocytosis, a measure of the opsonic activity of antibodies to pneumococcal capsular polysaccharides is believed to measure their functional activity and to be a better surrogate of in vivo protection than antibody concentration [39]. Good opsonophagocytosis (OPA) response is a functional end point to which a variety of qualitative and quantitative elements contribute, including antibody concentration, isotype, and avidity [39, 47].

F. Other Considerations

The quality and quantity of the immune response to pneumococcal conjugate vaccines are dependent on the size of the capsular polysaccharide, the type of carrier protein, and the nature of the interaction between the capsular polysaccharide and the carrier protein , including the presence or absence of a linker molecule. For example, for the pneumococcal CRM_{197} conjugate vaccine, the polysaccharide conjugated to CRM_{197} without a linker was found to be superior to vaccine formulated with oligosaccharide or using a six-carbon linker molecule [30, 48].

In addition, interactions among polysaccharide vaccines that share the same carrier protein may result in either enhancement of or interference with the immune response to the polysaccharide. Several studies have shown a decreased antibody response to Hib capsular polysaccharide when pneumococcus conjugated to tetanus toxoid carrier was administered simultaneously with Hib–tetanus toxoid conjugate [49, 50]. On the contrary, the response to Hib polysaccharide conjugated to CRM_{197} was noted to be significantly enhanced when administered simultaneously with pneumococcal CRM_{197} conjugate vaccine [48; T. Kilpi, personal communication]. Neither enhancement nor interference was observed when the Hib conjugate vaccine carrier was different from that of the simultaneously administered pneumococcal conjugate vaccine [T. Kilpi, personal communication]. The complex immunological interactions that may provide an explanation for these effects have been reviewed by Insel [51].

VII. PNEUMOCOCCAL CONJUGATE VACCINE EFFICACY

The ultimate proof of success of the pneumococcal conjugate vaccine is its ability to protect infants against pneumococcal disease. Several large-scale field studies have been completed to date demonstrating the efficacy of pneumococcal conjugate vaccine against invasive pneumococcal disease and otitis media due to vaccine serotypes.

In a study conducted at Northern California Kaiser Permanente, approximately 38,000 children were randomized to receive either the 7-valent pneumococcal CRM_{197} conjugate vaccine (now licensed as Prevnar®) or a meningococcal CRM_{197} conjugate vaccine as a control. Children were immunized at 2, 4, 6, and 12–15 months of age. Vaccine efficacy against invasive pneumococcal disease, defined as a positive culture for vaccine-type *S. pneumoniae* from a normally sterile body fluid, was 97.4% (95% CI 82.7, 99.9) in children who had received at least three doses, 93.9% (95% CI 79.6, 98.5) in children who had received at least one dose, and 89.1% (95% CI 73.7, 95.8) when all cases of invasive *S. pneumoniae*, regardless of serotype, were included. Statistically significant serotype-specific vaccine efficacy was demonstrated for types 19F, 14, 18C, 23F, and 6B. For the remaining serotypes, no vaccine failures occurred, but there were too few cases overall to demonstrate efficacy [32].

Efficacy against invasive disease was also seen in a field study of the 7-valent pneumococcal CRM_{197} conjugate in Native Americans, a population with a very high risk of pneumococcal disease. In this community-randomized study, 8292 infants in 42 study communities were randomized to receive 7-valent pneumococcal CRM_{197} conjugate vaccine or a control vaccine. After controlling for community randomization, the efficacy of 7VPnC vaccine (intent-to-treat analysis) was 86.4% (95% CI 40.3, 96.9) [52].

The Kaiser study also found that the pneumococcal conjugate vaccine reduced the incidence of otitis media episodes regardless of cause by 7% (95% CI 4.1, 9.1) and the placement of ear tubes by 20.1% (95% CI 1.5, 35.2). Bacteriology was only obtained on draining ears, for which a reduction of approximately 65% was seen, and all vaccine failures were due to 19F [32].

In the Finnish otitis media study [35], 1662 infants were randomized to receive the 7-valent pneumococcal CRM_{197} conjugate vaccine or a hepatitis B vaccine control, also at 2, 4, 6, and 12 months of age. Children with otitis media underwent myringotomy in order to obtain a bacterial diagnosis. The vaccine reduced the number of episodes of otitis media due to vaccine serotypes by 57% (95% CI 44, 67) and those due to any culture-confirmed pneumococcal infection by 34% (95% CI 21, 45). The incidence of otitis media of any cause was reduced by 6% (−4, 16), a result similar to that obtained in the Kaiser study.

The occurrence of episodes due to serotypes cross-reactive with those in the vaccine was reduced 51% (95% CI 27, 67). and statistically significant serotype-specific efficacy was demonstrated for types 4, 6B, 14, and 23 F and for the cross-reactive serotype 6A. Some replacement with nonvaccine serotypes was seen.

A number of studies have shown a decrease in nasopharyngeal colonization with vaccine type pneumococcus when pneumococcal conjugate vaccine is administered [53–58]. All studies show a decrease in the colonization rate with vaccine serotypes in vaccinees as compared to controls. In addition, a significant reduction in antibiotic-resistant pneumococci was also seen, as would be expected, because the majority of such strains are vaccine types. Dagan *et al.* [59] demonstrated herd immunity in the family setting; siblings of recipients of a 9-valent pneumococcal CRM_{197} conjugate vaccine were significantly less likely to carry antibiotic-resistant prenumococci compared with siblings of control-vaccine recipients. Finally, some, but not all, of these studies show an increase in nasopharyngeal colonization with nonvaccine strains [53, 54, 57]. Whether this is true replacement or an unmasking of serotypes that were co-colonizing the nasopharynx is unknown.

The mechanism of protection against colonization remains uncertain. Both serum-derived IgG and mucosally derived IgA can be found in the saliva of vaccinated children [60–61b]. Furthermore, the level of serum antibody that may correlate with protection against invasive disease or against otitis media also has not been determined. Estimates from infant mouse and rat models have estimated type-specific anticapsular concentrations required for protection ranging from less than 0.05 to 2 µg/ml depending on serotype and model [62, 63]. Vaccine efficacy studies can use population methods for estimating serological correlates of protection by comparing antibody levels in protected and nonprotected populations [64, 65]. By using data obtained in the Kaiser efficacy study,

Black *et al.* [32] have estimated that the minimum antibody titer associated with long-term protection is in the range of 0.15–0.5 µg/ml. Although no estimates of protection have been reported in the Finnish otitis media study, it is noted that there was little correlation between the height of the geometric mean titer achieved and the degree of protection for a given serotype [35].

VIII. EPIDEMIOLOGY OF *S. PNEUMONIAE* IN THE CONJUGATE VACCINE ERA

With the introduction of *Haemophilus* b conjugate vaccines, the epidemiology of the disease changed significantly. In fact, the impact on disease rates in the youngest children preceded vaccine introduction into those age groups, suggesting an important role for herd immunity [66]. Furthermore, concerns about the replacement of *Haemophilus influenzae* type b with other serotypes failed to materialize.

The pneumococcus is a different organism than Hib, with higher rates of carriage in the population and more numerous strains. Data from Northern California Kaiser Permanente on the incidence of disease in the first year following vaccine licensure show a reduction in invasive disease that is higher than the percent of children vaccinated, suggesting that herd immunity is operating. In children less than 1 year of age, the incidence of invasive disease due to vaccine serotypes fell from 51.5–98.2 cases per 100,000 person years in the 4 years prior to licensure to 9.4 cases per 100,000 person years in the year following vaccine introduction, an 87.3% reduction. Vaccine coverage with one or more doses during this period averaged 57.8%, and only 16.2% of children in this age group were fully vaccinated. To date, no increase in invasive disease risk due to nonvaccine serotypes has been observed (Fig. 5) [67]. Ongoing surveillance will be necessary to better understand the effects of

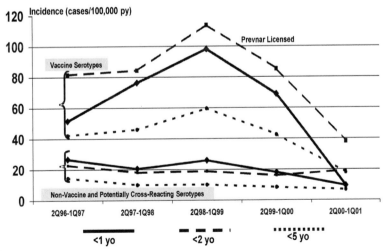

FIGURE 5 Age-specific impact of introduction of pneumococcal conjugate vaccine on invasive disease incidence for vaccine serotypes and nonvaccine serotypes. From Black *et al.* [67], with permission from Lippincott Williams & Wilkins, Copyright 2001.

broad vaccine usage on the microenvironment of the nasopharynx, antibiotic resistance patterns, and clinical disease. Ultimately, broad use of pneumococcal conjugate vaccine may change practice standards for the evaluation of high fever and the use of antibiotics in young children.

IX. PROTECTION OF OTHER POPULATIONS

Some data are available demonstrating that pneumococcal conjugate vaccine is immunogenic in certain populations at high risk for pneumococcal disease. These include older children who are nonresponsive to polysaccharide vaccine [68, 69], children with sickle cell disease, [34, 70], HIV-positive children [71], and patients undergoing bone marrow transplantation [72]. Additional data are needed in these and other populations.

Although many children less than 6 months of age are naturally protected by maternal immunity, there is nonetheless a significant disease burden in these youngest children, particularly in developing countries. The herd immunity afforded by immunization of older siblings may be sufficient to protect these young infants. However, other approaches, including neonatal immunization and maternal immunization, should be explored. Conjugate vaccines may have an advantage over polysaccharide vaccine in maternal immunization, as they stimulate a higher proportion of IgG_1 antibodies; this subtype is transferred more efficiently across the placenta than IgG_2 [42, 73].

Adults, too, might benefit from a vaccine that converts a T-independent response to one that is T-dependent and associated with the induction of immunologic memory. In the absence of memory, antibody levels to many serotypes decline to insignificant levels in 3–10 years [74–79]. In addition, some older individuals may respond poorly to many serotypes in the polysaccharide vaccine. Studies of the conjugate vaccine in this population have not shown a dramatic improvement, although some increase in antibody response to some serotypes was seen [80–82]. However, little is known about the optimum schedule or dosage for conjugate vaccine in adults, a population that differs from infants in its prior exposure to both the polysaccharide and the carrier protein components of most conjugate vaccines. In addition, little is known about the decline in immune function that must play a role in the increase in pneumococcal disease in the elderly. Romero-Steiner *et al.* [83] have suggested that functional immune response to pneumococcal polysaccharide vaccine, as measured by opsonophagocytosis and avidity, is significantly reduced for all serotypes in the elderly, compared with young subjects. Evaluation of new vaccines in the elderly should include measurement of these parameters.

X. FUTURE DIRECTIONS

New approaches to the prevention of pneumococcal disease are being examined. Vaccines directed against common proteins of the pneumococcus have the potential to protect against a wide variety of pneumococcal serotypes, without the need to include each one in the vaccine. The major candidates for such a strategy include pneumolysin and PspA.

Antibodies to pneumolysin increase during convalescence in patients with bacteremic pneumococcal pneumonia. Musher *et al.* [84] further found that mice who were passively protected by anti-pneumolysin IgG were protected from challenge with *S. pneumoniae* types 1 and 4. Alexander *et al.* [85] immunized mice with a pneumolysin toxoid and found protection against at least nine pneumococcal serotypes.

Virolainen *et al.* [86] measured antibodies to PspA in adults, children, and infants, which were cross-reactive against a variety of pneumococcal strains. Furthermore, children with pneumococcal disease had significantly lower titers of PspA in their acute sera than children with infection caused by other bacteria, suggesting that low levels of PspA at least correlate with susceptibility to disease.

A PspA-recombinant vaccine has been studied in a phase 1 study in adults. Responses were seen as early as 7 days postimmunization, suggesting that adults are primed for response to this protein [87]. Antibody demonstrated significant cross-reactivity against pneumococci containing different clades of PspA antibody, and passive immunization of mice with this antibody was protective against fatal infection with *S. pneumoniae*, including strains of different PspA clades and different capsular types (3, 6A, and 6B were tested) [88].

Another approach under consideration is mucosal immunization. Hvalbye *et al.* [89] administered heat-killed *S. pneumoniae* type 4 intranasally to mice. Antibodies against type-specific capsular polysaccharide were induced in serum and in mucosal secretions. Furthermore, immunized mice were protected against systemic infection and death after intraperitoneal challenge with type 4 organisms.

Finally, one group of investigators has suggested a completely different approach to protection against *S. pneumoniae* using xylitol. Xylitol is known to inhibit the growth of *S. pneumoniae* and to inhibit adhesion of both *S. pneumoniae* and *H. influenzae* in the nasopharyx. Uhari *et al.* [90] hypothesized that it might, therefore, represent a new modality for the prevention of acute otitis media due to these organisms. In fact, these authors found a reduction in otitis media in children receiving xylitol syrup (30% decrease, 95% CI 4.3, 55.4) or xylitol chewing gum (40% decrease, 95% CI 10, 71.1) compared with controls.

XI. CONCLUSION

S. pneumoniae has been an important cause of morbidity and mortality globally. The development of polysaccharide vaccines and the use of antibiotics have provided effective weapons against this disease. However, our success has been limited by an increase in the resistance of the organism to antibiotics and by the lack of effectiveness of the polysaccharide vaccine in young children, for whom the risk of disease is highest.

Improved understanding of immunology, together with advances in conjugate technology, has led to the development of new vaccines. This technology, first proven effective in virtually eliminating invasive diseases due to *H. influenzae* type b in young children, has now been applied to produce multivalent pneumococcal conjugate vaccines containing the serotypes of greatest importance in pediatric disease. The 7-valent pneumococcal vaccine, conjugated to a diphtheria CRM_{197} protein carrier, has proven to be highly effective in the prevention of disease due to vaccine serotypes and is now licensed for routine use in infants in the United States and other countries around the world. Additional studies of this vaccine are anticipated to demonstrate a substantial impact on pneumococcal disease burden, as well as to extend our understanding of pneumococcal epidemiology and the nature of the immune response in the elderly and other high-risk populations. Further knowledge about the organism and pathogenesis of disease has opened up additional lines of research that may provide an even broader defense against the pneumococcus in the future.

REFERENCES

1. Advisory Committee on Immunization Practices (1997). Prevention of Pneumococcal Disease: Recommendations of the ACIP. *Mortality Morbidity Weekly Report* **46**(RR-08):1–24.
2. Levine, O., Farley, M., Harrison, L. H., *et al.* (1999). Risk Factors for Invasive Pneumococcal Disease in Children: A Population-Based Case–Control Study in North America. *Pediatrics* **103**:1–5.
3. Kaplan, S. L., Mason, E. O., Barson, W. J., *et al.* (1998). Three-Year Multicenter Surveillance of Systemic Pneumococcal Infections in Children. *Pediatrics* **102**:538–544.
4. Eskola, J., and Anttila, M. (1999) Pneumococcal Conjugate Vaccines. *Pediatr. Infect. Dis.* **18**:543–551.
5. Schuchat, A., Robinson, K., Wenger, J. D., *et al.* (1997). Bacterial Meningitis in the United States in 1995. *New England J. Medicine* **337**:970–976.
6. Austrian, R. (1984). Pneumococcal Infections. *In* "Bacterial Vaccines," R. Germanier, Ed., Chapter 9, pp. 257–285. Orlando: Academic Press.
7. Hausdorff, W. P., Bryant, J., Paradiso, P., and Siber, G. R. (2000). Which Pneumococcal Serogroups Cause the Most Invasive Disease: Implications for Conjugate Vaccine Formulation and Use, Part 1. *Clin. Infect. Dis.* **30**:100–121.
8. Hausdorff, W. P., Bryant, J., Kloek, C., Paradiso, P., and Siber, G.R. (2000). The Contribution of Specific Pneumococcal Serogroups to Different Disease Manifestations: Implications for Conjugate Vaccine Formulation and Use, Part II. *Clin. Infect. Dis.* **30**:122–140.
9. Hausdorff, W. P., Siber, G., and Paradiso, P. R. (2001). Geographical Differences in Invasive Pneumococcal Disease Rates and Serotype Frequency in Young Children. *Lancet* **357**:950–952.
10. Dagan, R., Givon-Lavi, N., Shkolnik, L., Yagupsky, P., and Fraser, D. (2000). Acute Otitis Media Caused by Antibiotic-Resistant *Streptococcus pneumoniae* in Southern Israel: Implication for Immunizing with Conjugate Vaccines. *J. Infect. Dis.* **181**:1322–1329.
11. Friedland, I. R., and Klugman, K. P. (1992). Antibiotic-Resistant Pneumococcal Disease in South African Children. *Am. J. Dis. Child.* **146**:920–923.
12. Nava, J. M., Bella, F., Garau, J., *et al.* (1994). Predictive Factors for Invasive Disease Due to Penicillin-Resistant *Streptococcus pneumoniae*: A Population-Based Study. *Clin. Infect. Dis.* **19**:884–890.
13. Inostroza, J., Trucco, O., Prado, V., *et al.* (1998). Capsular Serotype and Antibiotic Resistance of *Streptococcus pneumoniae* Isolated in Two Chilean Cities. *Clin. Diagn. Lab. Immunol.* **5**:176–180.
14. Koh, T. H., and Lin, R. V. T. P. (1997). Increasing Antimicrobial Resistance in Clinical Isolates of *Streptococcus pneumoniae*. *Ann. Acad. Med. Singapore* **26**:604–608.
15. Gray, B. M., Converse, G. M., III, and Dillion, H. C., Jr. (1980). Epidemiologic Studies of *Streptococcus pneumoniae* in Infants: Acquisition, Carriage, and Infection during the First 24 Months of Life. *J. Infect. Dis.* **142**:923–933.

16. Ghaffar, F., Friedland, I. R., and McCracken, G. H., Jr. (1999). Dynamics of Nasopharyngeal Colonization by *Streptococcus pneumoniae*. *Pediatr. Infect. Dis. J.* **18**:638–645.

17. Vives, M., Garcia, M. E., Saenz, P., *et al*. (1997). Nasopharyngeal Colonization in Costa Rican Children during the First Year of Life. *Pediatr. Infect. Dis. J.* **16**:852–858.

18a. Sung, R. Y. T, Ling, J. M., and Fung, S. M. (1995). Carriage of Haemophilus influenzae and *Streptococcus pneumoniae* in Healthy Chinese and Vietnamese Children in Hong Kong. *Acta Paediatr.* **84**:1262–1267.

18b. Schrag, S. J., Beall, B., and Dowell, S. F. (2000). Limiting the Spread of Resistant Pneumococci: Biological and Epidemiologic Evidence for the Effectiveness of Alternative Interventions. *Clin. Microbiol. Rev.* **13**:588–601

19. Dagan, R., Melamed, R., Muallem, M., Piglansky, L., and Yagupsky, P. (1996). Nasopharyngeal Colonization in Southern Israel with Antibiotic-Resistant Pneumococci during the First Two Years of Life: Relation to Serotypes Likely To Be Included in Pneumococcal Conjugate Vaccines. *J. Infect. Dis.* **174**:1352–1355.

20. Dagan, R., and Fraser, D. (2000). Conjugate Pneumococcal Vaccine and Antibiotic-Resistant *Streptococcus pneumoniae*: Herd Immunity and Reduction of Otitis Morbidity. *Pediatr. Infect. Dis.* **19**:S79–88.

21. Gillespie, S. H., and Balakrishnan, I. (2000). Pathogenesis of Pneumococcal Infection. *J. Med. Microbiol.* **49**:1057–1067.

22. Avery, O. T., and Dubos, R. (1931). The Protective Action of a Specific Enzyme against Type III Pneumococcus Infection in Mice. *J. Exp. Med.* **54**:73–89.

23. Alonso deVelasco, E., Verheul, A. F. M., Verhoef, J., and Snippe, H. (1995). *Streptococcus pneumoniae*: Virulence Factors, Pathogenesis, and Vaccines. *Microbiol. Rev.* **59**:591–603.

24. Cundell, D. R., Weiser, J. N., Shen, J., Young, A., and Tuomanen, E. I. (1995). Relationship between Colonial Morphology and Adherence of *Streptococcus pneumoniae*. *Infect. Immun.* **63**:757–761.

25. Bruyn, G. A. W., Zegers, B. J. M., and van Fruth, R. (1992). Mechanisms of Host Defense against Infection with *Streptococcus pneumoniae*. *Clin. Infect. Dis.* **14**:251–262.

26. Goldblatt, D. (2000). Conjugate Vaccines. *Clin. Exp. Immunol.* **119**:1–3.

27a. Baxendale, H. E., Davis, Z., White, H. N., Spellerberg, M. B., Stevenson, F. K., and Goldblatt, D. (2000). Immunogenetic Analysis of the Immune Response to Pneumococcal Polysaccharide. *Eur. J. Immunol.* **30**:1214–1223.

27b. Douglas, R. M., Paton, J. C., Duncan, S. J., and Hansman, D. J. (1983). Antibody Response to Pneumococcal Vaccination in Children Younger Than Five Years of Age. *J. Infect. Dis.* **148**:131–137.

28. Foote, J., and Eisen, H. N. (1995). Kinetic and Affinity Limits on Antibodies Produced during Immune Responses. *Proc. Natl. Acad. Sci. U.S.A.* **92**:1254–1256.

29. Rennels, M. B., Edwards, K. M., Keyserling, H. L., Reisinger, K. S., Hogerman, D. A., Madore, D. V., Chang, I., Paradiso, P. R., Malinoski, F. J., and Kimura, A. (1998). Safety and Immunogenicity of Heptavalent Pneumococcal Vaccine Conjugated to CRM_{197} in United States Infants. *Pediatrics* **101**:604–611.

30. Steinhoff, M. C., Edwards, K., Keyserling, H., Thoms, M. L., Johnson, D. C., Madore, D., and Hogerman, D. (1994). A Randomized Comparison of Three Bivalent *Streptococcus pneumoniae* Glycoprotein Conjugate Vaccines in Young Children: Effect of Polysaccharide Size and Linkage Characteristics. *Pediatr. Infect. Dis. J.* **13**:368–372.

31. Shinefield, H. R., Black, S., Ray, P., Chang, I., Lewis, N., Fireman, B., Hackell, J., Paradiso, P. R., Siber, G., Kohberger, R., Madore, D. V., Malinoski, F. J., Kimura, A., Le, C., Landaw, I., Aguilar, J., and Hansen, J. (1999). Safety and Immunogenicity of Heptavalent Pneumococcal CRM_{197} Conjugate Vaccine in Infants and Toddlers. *Pediatr. Infect. Dis.* **18**:757–763.

32. Black, S., Shinefield, H., Fireman, B., Lewis, E. Ray, P., Hansen, J. R., Elvin, L., Ensor, K. M., Hackell, J., Siber, G., Malinoski, F., Madore, D., Chang, I., Kohberger, R., Watson, W., Austrian, R., Edwards, K., and The Northern California Kaiser Permanente Vaccine Study Center Group (2000). Efficacy, Safety and Immunogenicity of Heptavalent Pneumococcal Conjugate Vaccine in Children. *Pediatr. Infect. Dis. J.* **19**:187–195.

33. Choo, S., Seymour, L., Rhonwen, M., Quataert, S., Lockhart, S., Cartwright, K., and Finn, A. (2000). Immunogenicity and Reactogenicity of a Pneumococcal Conjugate Vaccine Administered Combined with a *Haemophilus influenzae* Type B Conjugate Vaccine in United Kingdom Infants. *Pediatr. Infect. Dis. J.* **19**:854–862.

34. O'Brien, K. L., Swift, A. J., Winkelstein, J. A., Santosham, M., Stover, B., Luddy, R., Gootenberg, J. E., Nold, J. T., Eskenazi, A., Snader, S. J., and Lederman, H. M. (2000). Safety and Immunogenicity of Heptavalent Pneumococcal Vaccine Conjugated to CRM_{197} among Infants with Sickle Cell Disease. *Pediatrics* 106:965–972.

35. Eskola, J., Kilpi, T., Palmu, A., Jokinen, J., Haapakoski, J., Herva, E., Takala, A. Kayhty, H., Karma, P., Kohberger, R., Siber, G., and Makela, H. (2001). Efficacy of Pneumococcal Conjugate Vaccine against Acute Otitis Media. *New Engl. J. Medicine* 344:403–409.

36. Dagan, R., Melamed, R., Zamir, O., and Leroy, O. (1997). Safety and Immunogenicity of Tetravalent Pneumococcal Vaccines Containing 6B, 14, 19F and 23F Polysaccharides Conjugated to Either Tetanus Toxoid or Diphtheria Toxoid in Young Infants and Their Boosterability by Native Polysaccharide Antigens. *Pediatr. Infect. Dis. J.* 16:1053–1059.

37. Wuorimaa, T., Kayhty, H., Leroy, O., and Eskola, J. (2001). Tolerability and Immunogenicity of an 11-Valent Pneumococcal Conjugate Vaccine in Adults. *Vaccine* 19:1863–1869.

38. Anderson, E. L., Kennedy, D. J., Geldmacher, K. M., Donnelly, J., and Mendelman, P. M. (1996). Immunogenicity of Heptavalent Pneumococcal Conjugate Vaccine in Infants. *J. Pediatr.* 128:649–653.

39. Anttila, M., Voutilainen, M., Jantti, V., Eskola, J., and Kayhty, H. (1999). Contribution of Serotype-Specific IgG Concentration, IgG Subclasses and Relative Antibody Avidity to Opsonophagocytic Activity against *Streptococcus pneumoniae*. *Clin. Exp. Immunol.* 118:402–407.

40. O'Brien, K. L., Steinhoff, M. C., Edwards, K., Keyserling, H., Thoms, M. L., and Madore, D. (1996). Immunologic Priming of Young Children by Pneumococcal Glycoprotein Conjugate, but Not Polysaccharide Vaccines. *Pediatr. Infect. Dis.* 15:425–430.

41. Vella, P. P., Marburg, S., Staub, J. M., *et al.* (1992). Immunogenicity of Conjugate Vaccines Consisting of Pneumococcal Capsular Polysaccharide Types 6B, 14, 19F, and 23F and a Meningococcal Outer Membrane Protein Complex. *Infect. Immunol.* 60:4977–4983.

42. Soininen, A., Seppala, I., Nieminen, T., Eskola, J., and Kayhty, H. (1999). IgG Subclass Distribution of Antibodies after Vaccination of Adults with Pneumococcal Conjugate Vaccines. *Vaccine* 17:1889–1897.

43. Obaro, S. K., Huo, Z., Banya, W. A. S., Henderson, D. C., Monteil, M. A., Leach, A., and Greenwood, B. M. (1997). A Glycoprotein Pneumococcal Conjugate Vaccine Primes for Antibody Responses to a Pneumococcal Polysaccharide Vaccine in Gambian Children. *Pediatr. Infect. Dis. J.* 16:1135–1140.

44. Goldblatt, D., Pinto Vaz, A. R. J. P. M., and Miller, E. (1998). Antibody Avidity as a Surrogate Marker of Successful Priming by *Haemophilus influenzae* Type b Conjugate Vaccines following Infant Immunization. *J. Infect. Dis.* 177:1112–1115.

45. Eskola, J. (2000). Immunogenicity of Pneumococcal Conjugate Vaccines. *Pediatr. Infect. Dis. J.* 19:388–393.

46a. Anttila, M., Eskola, J., Ahman, H., and Kayhty, H. (2000). Differences in the Avidity of Antibodies Evoked by Four Different Pneumococcal Conjugate Vaccines in Early Childhood. *Vaccine* 17:1970–1977.

46b. Goldblatt, D., Akoto, O. Y., Ashton, L., Asafo-Adjei, E., Brainsby, K., Twumai, P., and Baffoe-Bonnie, B. (2000). Does One Dose of Pneumococcal Conjugate Vaccine Induce Immunological Memory in Toddlers? 40th Interscience Conference on Antimicrobial Agents and Chemotherapy, Presentation 42.

47. Usinger, W. R., and Lucas, A. H. (1999). Avidity as a Determinant of the Protective Efficacy of Human Antibodies to Pneumococcal Capsular Polysaccharides. *Infect. Immun.* 67:2366–2370.

48. Daum, R. S., Hogerman, D., Rennels, M. B., Bewley, K., Malinoski, F., Rothstein, E., Reisinger, K., Block, S., Keyserling, H., and Steinhoff, M. (1997). Infant Immunization with Pneumococcal CRM_{197} Vaccines: Effect of Saccharide Size on Immunogenicity and Interactions with Simultaneously Administered Vaccines. *J. Infect. Dis.* 176:445–455.

49. Dagan, R., Eskola, J., Leclerc, C., and Leroy, O. (1998). Reduced Response to Multiple Vaccines Sharing Common Protein Epitopes That Are Administered Simultaneously to Infants. *Infect. Immun.* 66:2093–2098.

50. Ahman, H., Kayhty, H., Vuorela, A., Leroy, O., and Eskola, J. (1999). Dose Dependency of Antibody Response in Infants and Children to Pneumococcal Polysaccharides Conjugated to Tetanus Toxoid. *Vaccine* 17:2726–2732

51. Insel, R. A. (1995). Potential Alterations in Immunogenicity by Combining or Simultaneously Administering Vaccine Components. *In* "Combined Vaccines and Simultaneous Administration: Current Issues and Perspectives," Vol. 754, pp. 35–47. New York Academy of Sciences, New York.

52. O'Brien, K. L., Moulton, L., Reid, R., Weatherholtz, R., Hackell, J., Kohberger, R., Chang, I., Siber, G., and Santosham, M. (2001). "Invasive Disease Efficacy of a 7-Valent Pneumococcal Conjugate Vaccine among Navajo and White Mountain Apache (N/WMA) Children". Presented at 19th Annual Meeting of the European Society for Paediatric Infectious Diseases (ESPID), Istanbul, Turkey.

53. Obaro, S., Adegbola, R. A., Banya W., and Greenwood, B. (1996). Carriage of Pneumococci after Pneumococcal Vaccination. *Lancet* **348**:271–272.

54. Mbelle, N., Huebner, R. E., Wasas, A. D., Kimura, A., Chang, I., and Klugman, K. P. (1999). Immunogenicity and Impact on Nasopharyngeal Carriage of a Nonavalent Pneumococcal Conjugate Vaccine. *J. Infect. Dis.* **180**:1171–1176.

55. Dagan, R., Melamed, R., Muallem, M., Piglansky, L., Greenberg, D., Abramson, O., Mendelman, P. M., Bohidar, N., and Yagupsky, P. (1996). Reduction of Nasopharyngeal Carriage of Pneumococci during the Second Year of Life by a Heptavalent Conjugate Pneumococcal Vaccine. *J. Infect. Dis.* **174**:1271–1278.

56. Dagan, R., Muallem, M., Melamed, R., Leroy, O., and Yagupsky, P. (1997). Reduction of Pneumococcal Nasopharyngeal Carriage in Early Infancy after Immunization with Tetravalent Pneumococcal Vaccines Conjugated to Either Tetanus Toxoid or Diphtheria Toxoid. *Pediatr. Infect. Dis. J.* **16**:1060–1064.

57. O'Brien, K. L., Bronsdon, M. A., Carlone, G. M., Facklam, R. R., Schwartz, B., Reid, R. R., and Santosham, M. (2001). "Effect of a 7-Valent Pneumococcal Conjugate Vaccine (7vPnC) on Nasopharyngeal Carriage (NP) among Navajo and White Mountain Apache (N/WMA) Children." Presented at 19th Annual Meeting of the European Society for Paediatric Infectious Diseases (ESPID), Istanbul, Turkey.

58. Dagan, R., Fraser, D., Sikuler-Cohen, L., Guy, N., Givon-Lavi, N., and Janco, J. (2000). "Reduction of Nasopharyngeal (NP) Carriage in Day Care Center (DCC) Attendees after Vaccination with a 9-Valent CRM_{197} Conjugate Pneumococcal Vaccine (PncCRM9)— Protection against Individual Serotypes." Presented at 40th Interscience Conference on Antimicrobial Agents and Chemotherapy, Abstract 686.

59. Dagan, R., Givon-Lavi, N., Porat, N., Sikuler-Cohen, L., and Fraser, D. (2000). "Immunization of Toddlers Attending Day Care Centers (DCCs) with a 9-Valent Conjugate Pneumococcal Vaccine (PncCRM9) Reduced Transmission of *Streptococcus pneumoniae* (Pnc) and Antibiotic Resistant *S. pneumoniae* (R-Pnc) to Their Young Siblings (YS). Presented at 40th Interscience Conference on Antimicrobial Agents and Chemotherapy, Abstract 687.

60. Nieminen, T., Kayhty, H., Leroy, O., and Eskola, J. (1999). Pneumococcal Conjugate Vaccination in Toddlers: Mucosal Antibody Response Measured as Circulation Antibody-Secreting Cells and as Salivary Antibodies. *Pediatr. Infect. Dis. J.* **18**:764–772.

61a. Korkeila, M., Lehtonen, H., Ahman, H., Leroy, O., Eskola, J., and Kayhty, H. (2000). Salivary Anti-Capsular Antibodies in Infants and Children Immunised with *Streptococcus pneumoniae* Capsular Polysaccharides Conjugated to Diphtheria or Tetanus Toxoid. *Vaccine* **18**:1218–1226.

61b. Choo, S., Zhang, Q., Seymour, L., Akhtar, S., and Finn, A. (2000). Primary and Booster Salivary Antibody Responses to a 7-Valent Pneumococcal Conjugate Vaccine in Infants. *J. Infect. Dis.* **182**:1260–1263.

62. Johnson, S. E., Rubin, L., Romero-Steiner, S., Dykes, J. K., Pais, L. B., Rizvi, A., Ades, E., and Carlone, G. M. (1999). Correlation of Opsonophagocytosis and Passive Protection Assays Using Human Anticapsular Antibodies in an Infant Mouse Model of Bacteremia for *Streptococcus pneumoniae*. *J. Infect. Dis.* **180**:133–140.

63. Stack, A. M., Malley, R., Thompson, C. M., Kobzik, L., Siber, G. R., and Saladino, R. A. (1998). Minimum Protective Serum Concentrations of Pneumococcal Anti-Capsular Antibodies in Infant Rats. *J. Infect. Dis.* **177**:968–990.

64. Siber, G. R. (1997). Methods for Estimating Serological Correlates of Protection. *In* "Pertussis Vaccine Trials," F. Brown, D. Greco, P. Mastrantonio, S. Salmaso, and S. Wassilak, Eds., pp. 283–296. Basel, Switzerland: Karger.

65. Klugman, K. P., Koornhof, H. J., Robbins, J. B., and Le Cam, N. N. (1996). Immunogenicity, Efficacy and Serological Correlate of Protection of *Salmonella typhi* Vi Capsular Polysaccharide Vaccine Three Years after Immunization. *Vaccine* 14:435–438.

66. Adams, W. G., Deaver, K. A., Cochi, S. L., Pikaytis, B. D., Zell, E. R., Broome C. V., and Wenger, J. D. (1993) Decline of Childhood *Haemophilus influenzae* type b (Hib) Disease in the Hib Vaccine Era. *J. Am. Med. Assoc.* 269:221–226.

67. Black, S., Shinefield, H., Hansen, J., Elvin, L., Laufer, D. and Malinoski, F. (2001). Post Licensure Evaluation of the Effectiveness of Seven Valent Pneumococcal Conjugate Vaccine. *Pediatr. Infect. Dis. J.* 20(12):1105–1107.

68. Zielen, S., Buhring, I., Strnad, N., Reichenback, J., and Hoffman, D. (2000). Immunogenicity and Tolerance of a 7-Valent Pneumococcal Conjugate Vaccine in Nonresponders to the 23-Valent Pneumococcal Vaccine. *Infect. Immun.* 68(3):1435–1440.

69. Sorensen, R. U., Leiva, L. E., Giangrosso, P. A., Butler, B., Javier, F. C. III, Sacerdote, D. M., Bradford, N., and Moore, C. (1998). Response to a Heptavalent Conjugate *Streptococcus Pneumoniae* Vaccine in Children with Recurrent Infections Who Are Unresponsive to the Polysaccharide Vaccine. *Pediatr. Infect. Dis. J.* 17:685–691.

70. Vernacchio, L., Neufeld, E. J., MacDonald, K., *et al.* (1998). Combined Schedule of 7-Valent Pneumococcal Conjugate Vaccine Followed by 23-Valent Pneumococcal Vaccine in Children and Young Adults with Sickle Cell Disease. *J. Pediatr.* 103:275–278.

71. King, J. C., Vink, P. E., Farley, J. J., Parks, M., Smilie, M., Madore, D., Lichenstein, R., and Malinoski, F. (1996). Comparison of the Safety and Immunogenicity of a Pneumococcal Conjugate with a Licensed Polysaccharide Vaccine in Human Immunodeficiency Virus and Non-Human Immunodeficiency Virus-Infected Children. *Pediatr. Infect. Dis. J.* 15:192–196.

72. Molrine, D., Antin, J., Guinan, E., Soiffer, R., Ambrosino, D., Maldonado, J., and Wilson, M. (2001). Pneumococcal Conjugate Vaccine (PCV) Elicits Protective Responses in Allogeneic Bone Marrow Transplant (BMT) Recipients. 41st Interscience Conference on Antimicrobial Agents and Chemotherapy, Abstract 2035.

73. Glezen, W. P., and Alpers, M. (1999). Maternal Immunization. *Clin. Infect. Dis.* 28:219–224.

74. Fedson, D. S. (2000). Pneumococcal Conjugate Vaccination for Adults: Why It's Important for Children. *Pediatr. Infect. Dis. J.* 19:183–186.

75. Sankilampi, U., Honkanen, P. O., Bloigu, A., and Leinonen, M. (1997). Persistence of Antibodies to Pneumococcal Capsular Polysaccharide Vaccine in the Elderly. *J. Infect. Dis.* 176:1100–1104.

76. Mufson, M. A., Krause, H. E., and Schiffman, G. (1983). Long-Term Persistence of Antibody following Immunization with Pneumococcal Polysaccharide Vaccine. *Proc. Soc. Exp. Biol. Med.* 173:270–275.

77. Mufson, M. A., Krause, H. E., Schiffman, G., and Hughley, D. F. (1987). Pneumococcal Antibody Levels One Decade after Immunization of Healthy Adults. *Am. J. Med. Sci.* 293:279–289.

78. Hilleman, M. R., Carlson, A. J., McLean, A. A., Vella, P. P., Weibel, R. E., and Woodhour, A. F. (1981). *Streptococcus Pneumoniae* Polysaccharide Vaccine: Age and Dose Responses, Safety, Persistence of Antibody, Revaccination, and Simultaneous Administration of Pneumococcal and Influenza Vaccines. *Rev. Infect. Dis.* 3(Suppl.):S31–S42.

79. Kraus, C., Fischer, S. Ansorg, R., and Hottemann, U. (1985). Pneumococcal Antibodies (IgG, IgM) in Patients with Chronic Obstructive Lung Disease 3 Years after Pneumococcal Vaccination. *Med. Microbiol. Immunol.* 174:51–58.

80. Shelly, M. A., Jacoby, H., Riley, G. J., Graves, B. T., Pichichero, M., and Treanor, J. J. (1997). Comparison of Pneumococcal Polysaccharide and CRM_{197} Conjugated Pneumococcal Oligosaccharide Vaccines in Young and Elderly Adults. *Infect. Immun.* 65:242–247.

81. Powers, D. C., Anderson, E. L., Lottenbach, K., and Mink, C. M., (1996). Reactogenicity and Immunogenicity of a Protein-Conjugated Pneumococcal Oligosaccharide Vaccine in Older Adults. *J. of Infect. Dis.* 173:1014–1018.

82. Musher, D. M., Groover, J. E., Watson, D. A., Rodriguez-Barradas, M. C., and Baughn, R. E. (1998). IgG Responses to Protein-Conjugated Pneumococcal Capsular Polysaccharides in Persons Who Are Genetically Incapable of Responding to Unconjugated Polysaccharides. *Clin. Infect. Dis.* 27:1487–1490.

83. Romero-Steiner, S., Musher, D. M., Cetron, M.., Pais, L. B., Groover, J. E., Fiore, A. E., Plikaytis, B. D., and Carlone, G. M. (1999). Reduction in Functional Antibody Activity against *Streptococcus pneumoniae* in Vaccinated Elderly Individuals Highly Correlates with Deceased IgG Antibody Avidity. *Clin. Infect. Dis.* **29**:281–288.

84. Musher, D. M., Hoang, M. P., and Baughn, R. E. (2001). Protection against Bacteremic Pneumococcal Infection by Antibody to Pneumolysin. *J. Infect. Dis.* **183**:827–830.

85. Alexander, J. E., Lock, R., Peeters, C. C., *et al.* (1994). Immunization of Mice with Pneumolysin Toxoid Confers a Significant Degree of Protection against at Least Nine Serotypes of *Streptococcus pneumoniae. Infect. Immun.* **62**:5683–5688.

86. Virolainen, A., Russell, W., Crain, M. J., Rapola, S., Kayhty, H., and Briles, D. E. (2000). Human Antibodies to Pneumococcal Surface Protein A in Health and Disease. *Pediatr. Infect. Dis. J.* **19**:134–138.

87. Nabors, G. S., Braun, P. A., Herrmann, D. J., Heise, M. L., Pyle, D. J., Gravenstein, S., Schilling, M., Ferguson, L. M., Hollingshead, S. K., Briles, D. E., and Becker, R. S. (2000). Immunization of Healthy Adults with a Single Recombinant Pneumococcal Surface Protein A (PspA) Variant Stimulates Broadly Cross-Reactive Antibodies to Heterologous PspA Molecules. *Vaccine* **18**:1743–1754.

88. Briles, D. E., Hollingshead, S. K., King, J., Swift, A., Braun, P. A., Park, M. K., Ferguson, L. M., Nahm, M. H., and Nabors, G. S. (2000). Immunization of Humans with Recombinant Pneumococcal Surface Protein A (rPspA) Elicits Antibodies That Passively Protect Mice from Fatal Infection with *Streptococcus pneumoniae* Bearing Heterologous PspA. *J. Infect. Dis.* **182**:1694–1701.

89. Hvalbye, B. K. R., Aaberge, I. S., Lovik, M., and Haneberg, B. (1999). Intranasal Immunization with Heat-Inactivated Streptococcus Pneumoniae Protects Mice against Systemic Pneumococcal Infection. *Infect. Immun.* **67**:4320–4325.

90. Uhari, M., Kontiokari, T., and Niemela, M. (1998). A Novel Use of Xylitol Sugar in Preventing Acute Otitis Media. *Pediatrics* **102**:879–884.

6 DISEASE STATES AND VACCINES: SELECTED CASES

PART H. Tuberculosis

DOUGLAS B. YOUNG

Centre for Molecular Microbiology and Infection, Imperial College of Science, Technology and Medicine, London SW7 2AZ, United Kingdom

I. INTRODUCTION

Tuberculosis (TB) has been a scourge of humankind throughout recorded history, leaving its indelible imprint on Egyptian mummies, Victorian romance, and a current annual toll of 2 million deaths. Aerosol emissions from the lungs of an estimated 8 million incident cases provide an efficient means for spreading *Mycobacterium tuberculosis*, with one-third of the global population having immunological signs of infection [1]. Strategies to control tuberculosis are based on efforts to reduce the number of people exposed to infection—by prompt diagnosis and effective treatment of patients—together with efforts to prevent progression to active disease among the exposed population. Efficient delivery of drugs currently available provides the mainstay of control programs, but advances on both fronts will be required to reverse the current global epidemic [2].

Infection with *M. tuberculosis* results in disease in only a minority of individuals. For the most part, the presence of inhaled bacilli in the lungs stimulates a cell-mediated immune response that controls the infection. Evidence of this encounter is seen in the form of a delayed type hypersensitivity response to skin testing with mycobacterial antigens and, perhaps, the presence of an apparently benign lesion on X-ray examination. Between 5 and 10% of infected individuals will go on to develop clinical tuberculosis at some stage in their lives. Around one-half of these cases will occur during the first few years after infection. In young children, this is generally in the form of disseminated disease, such as tuberculous meningitis, and is frequently fatal; in adults, the highly infectious pulmonary tuberculosis predominates. The remaining cases of tuberculosis develop years or decades after initial infection. In most instances, it is not possible to determine whether this is the result of reactivation of the initial infection or is triggered by a subsequent reinfection [3]. It is likely that both processes are important. Onset of disease in individuals many years after any exposure to infection, together with detection of bacteria during autopsy in the absence of evidence of disease [4], suggests that control—rather than elimination—of the initial infection is probably a common outcome.

The immune response is central to the complex dynamics of tuberculosis infection and disease. The consequences of impaired immunity are dramatically demonstrated by the 5–10% *annual* risk of tuberculosis in individuals co-infected with HIV and *M. tuberculosis* [5]; the goal of a tuberculosis vaccine is to achieve the reverse effect by an enhancement of the immune response.

II. THE OLD GENERATION OF TB VACCINES

Within a decade of his description of the tubercle bacillus, Robert Koch made the dramatic announcement at the 10th International Congress of Medicine in Berlin that he had developed a vaccine that would both protect against the development of tuberculosis in guinea pigs and effect a cure in patients with disease. This involved injection of an *M. tuberculosis* culture filtrate preparation, "old tuberculin," which was subsequently partially refined to produce the current skin-test reagent, "purified protein derivative" (PPD). His claims were quickly disproven. Vaccination clearly elicited an immune response, but it was ineffective or even detrimental to the course of the disease in most patients.

Subsequent studies by researchers such as Edward Trudeau showed that exposure to live mycobacteria engendered a more effective immune response than exposure to killed organisms, leading Albert Calmette and Camille Guérin to follow the strategy of Louis Pasteur in developing a live attenuated vaccine. The resulting bacillus of Calmette and Guérin (BCG) was prepared by serial *in vitro* passage of an isolate of *Mycobacterium bovis*—the closely related agent of bovine tuberculosis—and was introduced as a human vaccine in 1923. With the tools of modern genomics, it is possible to reconstruct the genetic events that occurred during this attenuation process [6]. These included a series of gene deletions, the first of which (RD1) resulted in the loss of two of the prominent antigens of *M. tuberculosis* (ESAT6 and CFP10, see later discussion) [7]. The vaccine strain continued to change during subsequent passage, generating a series of closely related, but genetically distinct substrains of BCG [8, 9]. By taking these deletions together with the original genetic differences between *M. bovis* and *M. tuberculosis*, something in the region of 100–150 of the 4000 genes present in *M. tuberculosis* clinical isolates are absent from BCG. Current research focuses on analysis of the potential contribution of the deletion genes to virulence and immunogenicity and the search for additional less obvious genetic alterations in BCG.

BCG has a distinguished record of efficacy in animal models of tuberculosis. Administration of BCG prior to challenge with *M. tuberculosis* results in an accelerated immune response that reduces the initial growth of the virulent challenge organisms and prevents dissemination in the bloodstream. This provides a restriction of the disease, which is evident from a reduced number of bacteria in infected organs, reduced pathological damage, and prolonged survival. The mechanism is that of a classical prophylactic vaccine, priming the natural immune response by mimicking the pathogenic challenge. This approach shares the limitations inherent in the natural immune response to tuberculosis. In marked contrast to a disease like smallpox, those who survive their initial encounter with M. *tuberculosis* are by no means guaranteed immunity to a subsequent challenge. Patients treated for tuberculosis are susceptible to reinfection as well as to relapse [10], and, as noted earlier, most cases of tuberculosis arise in individuals who mounted an apparently protective response to initial infection.

The protective effect of BCG has been demonstrated in mice [11], guinea pigs [12], rabbits [13], possum [14], cattle [15], deer [16], and nonhuman primates [17]. It would be unlucky if humans were an exception to this zoological diaspora, and similar efficacy has indeed been demonstrated in human trials that have focused on the effect of BCG vaccination on tuberculosis in children. Neonatal BCG consistently confers around 70% protection against the fatal forms of childhood tuberculosis [18]. The story is less straightforward when it comes to adult tuberculosis. In trials initiated in the United Kingdom in the 1950s and 1960s, BCG was found to confer almost 80% protection when administered to 12- to 13-year-old schoolchildren. However, this activity was not observed in other settings; an extensive trial of BCG in south India demonstrated no overall protection against adult tuberculosis [19, 20]. These results, together with further disparate findings in a series of randomized and case–control trials, have puzzled tuberculosis researchers and seeded uncertainty and a degree of pessimism at the heart of the vaccine development efforts.

A reasonable resolution of the seeming contrariness of BCG is that it provides an excellent means of establishing the fundamental basis of antimycobacterial immunity in the naive host, but it achieves little in the face of a preexisting response. In addition to immune responses engendered by tuberculosis infection itself or by prior BCG vaccination, exposure to environmental mycobacteria may be sufficient to abrogate the protective efficacy associated with BCG. This could occur if the protection induced by environmental exposure is equivalent to that conferred by BCG [21] or if the preexisting immune response inhibits BCG replication and immunogenicity [22]. Although these two explanations would look similar in the context of a BCG trial, they have quite distinct implications for future vaccines. In the latter case, a new vaccine capable of inducing the *same* response as BCG but in a refractory host would be an important goal. If the former explanation proves true, however, a useful new vaccine would have to surpass the protection conferred by BCG.

In addition to the question of variation between target populations, it would seem that BCG resembles natural immunity in offering no guarantee of lifelong protection and in failing to prevent reactivation disease [23]. An attractive goal for future vaccination strategies would be to modulate the immune response in the 2 billion people who have already weathered their first encounter with M. *tuberculosis*, with the aim of eliminating persistent organisms or at least reducing the risk of breakdown to active disease. This may require modifying rather than simply reproducing the natural immune response.

III.   THE NEW GENERATION OF TB VACCINES

There are myriad opportunities for the enthusiastic researcher in the field of tuberculosis vaccines. Our knowledge of the nuances of the immune response has blossomed since the days of Calmette and Guerin [24]; we can now direct and control the genetic events underlying their empirical approach to attenuation [25, 26], and from the *M. tuberculosis* genome we have at hand the sequence of every antigen of the bacterium [27]. It just remains to assemble these opportunities in some propitious organization.

A.   Immunity

Extensive evidence from the murine infection model identifies macrophages activated by interferon-γ (IFNγ) as major effector cells in the control of *M. tuberculosis*. This process is regulated by interleukin-12 (IL-12) and tumor necrosis factor-α (TNFα), with CD4+ T cells as the dominant source of IFNγ [24]. Evidence in human infection is less direct, but the enhanced incidence of tuberculosis in HIV-infected individuals and hypersusceptibility to mycobacterial disease associated with congenital defects in IL-12–IFNγ signaling [28, 29] are consistent with an analogous role for this pathway as a basic effector mechanism in humans. BCG vaccination is very effective in establishing this CD4-mediated Th1 response. Further evidence from the mouse model identifies a role for CD8+ T cells [30]. This function is expressed at a later stage in the infection [31] and may involve the production of additional IFNγ [32], lysis of infected macrophages that are refractory to activation, and perhaps also direct mycobacterial killing by means of microbicidal effector molecules [33]. CD8+ cells are not required for control of the attenuated BCG vaccine strain, and BCG may be less effective in stimulating this class of T cells [34]. Two other types of T cell are activated by mycobacterial infection. T cells carrying a γδ receptor proliferate in response to mycobacterial infection or exposure to low-molecular-weight phosphorylated ligands produced by the bacilli [35], though their precise effector function remains to be determined. Finally, T cells restricted by CD1 molecules on antigen-presenting cells respond to nonprotein ligands from mycobacteria, including the copious cell-wall-associated mycolic adids and glycolipids [36]. Again, the functional contribution of these cells to protection and pathology is the subject of ongoing research.

The picture emerging from these studies is of a highly complex immune response, with multiple cell types influencing different stages of the infection. On top of this are local spatial considerations relating to cell recruitment and formation of the organized granuloma that serves to isolate the infectious focus [37]. Although we have been able to identify the key players in this response, we have a very limited grasp of the quantitative rules that govern their performance (if some IFNγ is good, is more necessarily better?) or the ways in which their contributions change over time. In addition, because it is hard to distinguish the Th1 inflammatory response implicated in protective immunity from that responsible for the destruction of lung tissue during active disease, immunomodulation carries potential risks as well as benefits. In the context of vaccine development, one might elect either to try to orchestrate this complex response in its entirety—with a live attenuated vaccine, for example—or to specifically alter one or another particular element of the response by using a subunit vaccine.

B. Live Attenuated Vaccines

New tools for genetic manipulation of mycobacteria provide an opportunity to refine the live vaccine strategy of Calmette and Guérin by the addition of genes to BCG or by the targeted removal of genes from initially virulent *M. tuberculosis* [38]. Modified strains of BCG have been engineered to express recombinant mammalian cytokines designed to augment immunogenicity, for example [39], and the addition of a hemolysin gene has been shown to enhance the ability of BCG to induce a CD8$^+$ T-cell response [40]. BCG has also been used as a vehicle for the expression of recombinant antigens from a range of infectious agents. In the case of mycobacterial antigens, promising results have been generated by overexpression of the dominant antigen 85 protein carried on a multicopy plasmid in recombinant BCG [41]. Deletion of regulatory genes provides an alternative strategy for stable overexpression of selected proteins and has been exploited in the case of the hsp70 antigen [42]. Genes encoding *M. tuberculosis* RD1 antigens that are absent from existing BCG strains represent potentially interesting candidates for the construction of improved BCG vaccines.

Several laboratories have generated novel attenuated strains by the deletion of genes from *M. tuberculosis*. Three strategies have been pursued. The first involves introduction of auxotrophic mutations that render the bacteria dependent on exogenous growth factors that are in limited supply within the infected host. Successful examples include attenuated strains dependent on leucine and tryptophan [43, 44]. A second approach is to inactivate *M. tuberculosis* genes analogous to those involved in virulence in other bacterial pathogens. An interesting example is the construction of a *phoP* attenuated mutant, based on knowledge of the key role of this regulator in pathogenic *Salmonella* [45]. The third strategy for the identification of attenuated strains exploits the technique of signature-tagged transposon mutagenesis (STM). This involves screening of a library of transposon mutants to identify clones selectively lost during murine infection. The STM approach has identified attenuated mutants with defects in the biosynthesis of cell wall lipids [46, 47].

Although the technology of mycobacterial attenuation is now well-established, screening for vaccine candidates remains empirical. Aside from the overriding need for safety, the properties required for an improved live vaccine strain are unknown; interactions with antigen-presenting cells, the precise mode of death, and the duration of persistence within the host are all likely to influence vaccine efficacy. Information from the current round of trial-and-error screening of novel attenuated strains, together with progress in understanding of the cellular microbiology of mycobacterial infection, may ultimately allow the design of rational selection procedures for improved live vaccines.

C. Subunit Vaccines

Subunit vaccines offer advantages over live mycobacteria in terms of safety, ease of manufacture, and precise immune targeting. This approach requires identification of relevant antigens and selection of appropriate delivery systems.

Two criteria have been used to identify *M. tuberculosis* antigens for incorporation into programs for subunit vaccine development. The first is the ability to elicit a strong recall response in humans or experimental animals exposed to *M. tuberculosis* infection. The response is assessed primarily in terms of the release of IFNγ from antigen-specific CD4+ T cells. Although this represents a realistic strategy to identify antigens available for T-cell recognition, it will (like BCG) tend to reproduce rather than augment the natural immune response. Amplification of naturally subdominant responses represents a potentially interesting alternative approach [48], particularly in the context of postexposure vaccination (see later discussion). A second common criterion for antigen selection is identification in the supernatant of *in vitro* cultures. This is based on the rationale that the presence of secreted proteins may be a critical factor in determining the superior protective efficacy of live as compared to killed mycobacteria and is supported by experiments from several laboratories demonstrating successful vaccination with culture filtrate preparations [49–51]. An extensive body of literature indicates that, using appropriate delivery systems, a wide range of secreted and nonsecreted mycobacterial proteins have the potential to confer some degree of protection against subsequent challenge with *M. tuberculosis*.

Three antigens are currently at the forefront of subunit vaccine development. The first is antigen 85, a set of three closely related proteins that function as mycolyl transferases and are the most abundant protein components in culture filtrate preparations [52, 53]. The second is ESAT6, a low-molecular-weight protein antigen present in culture filtrates of *M. tuberculosis*, but absent from BCG [54, 55]. ESAT6 is a member of an *M. tuberculosis* gene family comprising several prominent immunogens [56, 57] including CFP10, which is encoded by the adjacent gene and is also deleted from BCG [58]. A third candidate is 72f, generated by fusing prominent protein antigens [59, 60]. Because any single protein may lack epitopes suitable for universal immune recognition, it is probable that the fusion approach will be an important aspect of future subunit vaccine development.

The choice of delivery system is at least as important as antigen selection in the development of subunit vaccines. Vaccination with purified proteins in adjuvant generates a strong CD4 immune response and has been successfully exploited in several experimental models [51, 61, 62]. Adjuvants currently used in human vaccines are generally optimized for the induction of antibody rather than T-cell responses, but novel preparations, such as the SBAS2 adjuvant [63], may be useful in moving mycobacterial candidates into clinical trials. DNA vaccination has also proved effective in small animal models [53]. In addition to prophylactic use, DNA vaccination has been shown to enhance the immune response when delivered after infection [64]. This study focused on the mycobacterial hsp65 protein, a prominent antigen but with the potential to induce autoimmune reactions in some protocols [65]. Viral delivery systems have also been used for mycobacterial antigens [66, 67]. As with DNA vaccination, this approach is able to induce strong CD8 responses. A combination of different vaccination strategies in "prime–boost" protocols would seem to offer an attractive approach to exploit the potential of the various delivery systems.

IV. MOVING NEW VACCINES INTO HUMANS

The past decade has witnessed a dramatic transformation in the status of tuberculosis vaccine development in research laboratories. Hundreds of preparations have been evaluated in experimental animals, with several matching and some perhaps surpassing BCG in protective efficacy [68]. To translate this progress into clinical application, it is important that we have a clear view of what we might hope that new vaccines will actually achieve in humans.

The ideal vaccine would be delivered prior to exposure to *M. tuberculosis* and confer lifelong protection against disease. An obvious target population for evaluation of such a vaccine would be infants in high-prevalence areas. It would be hard to justify withdrawing neonatal BCG from this population, however, and therefore difficult to evaluate any new vaccine. This problem could be avoided by carrying out trials in countries where BCG is not used— the United States, for example—but the low incidence of disease would necessitate large and prolonged trials. A third target group would be young adults living in high-prevalence countries and screened for the absence of immunological markers of previous infection. Of particular relevance would be groups in which BCG had previously been proven ineffective.

Given the practical limitations inherent in evaluating a conventional preexposure vaccine, it is attractive to consider vaccination strategies that could be implemented after initial exposure to *M. tuberculosis* or as an adjunct to BCG. It is not clear why the protective response to initial infection should prove ineffective in later life. Resistance to the adult form of disease may depend on some immune mechanism distinct from that required for childhood tuberculosis, or cells required for protection may be lost in later life. In either case, priming of extra T-cell clones by some novel vaccination procedure may be of benefit. Induction of responses to antigens that are neglected by the natural or BCG-induced response or presentation of familiar antigens by way of novel delivery systems provides potentially relevant approaches. In contrast to preexposure vaccination, there is only limited experience in the evaluation of postexposure or booster vaccines in experimental models [69]. The postexposure paradigm is attractive in directly addressing the one-third of the global population already exposed to tuberculosis and also in the fact that it is amenable to evaluation in relatively short-term trials in high-risk young adult cohorts.

Although the fundamental conceptual aspects of tuberculosis vaccination continue to present formidable challenges best addressed in experimental models, analysis of the safety and immunogenicity of new vaccine candidates in early human trials will provide an important element in setting future research directions.

REFERENCES

1. Dye, C., Scheele, S. *et al.* (1999). Consensus statement. Global burden of tuberculosis: estimated incidence, prevalence, and mortality by country. WHO Global Surveillance and Monitoring Project. *J. Am. Med. Assoc.* **282**:677–686.
2. Murray, C. J., and Salomon, J. A. (1998). Modeling the impact of global tuberculosis control strategies. *Proc. Natl. Acad. Sci. U.S.A.* **95**:13881–13886.
3. Fine, P. E., and Small, P. M. (1999). Exogenous reinfection in tuberculosis. *New England J. Medicine* **341**:1226–1227.
4. Hernandez-Pando, R., Jeyanathan, M. *et al.* (2000). Persistence of DNA from *Mycobacterium tuberculosis* in superficially normal lung tissue during latent infection. *Lancet* **356**:2133–2138.
5. Hopewell, P. C. (1992). Impact of human immunodeficiency virus infection on the epidemiology, clinical features, management, and control of tuberculosis. *Clin. Infect. Dis.* **15**:540–547.
6. Behr, M. A., and Small, P. M. (1999). A historical and molecular phylogeny of BCG strains. *Vaccine* **17**:915–922.
7. Mahairas, G. G., Sabo, P. J. *et al.* (1996). Molecular analysis of genetic differences between *Mycobacterium bovis* BCG and virulent *M. bovis*. *J. Bacteriol.* **178**:1274–1282.
8. Behr, M. A., Wilson, M. A. *et al.* (1999). Comparative genomics of BCG vaccines by whole-genome DNA microarray. *Science* **284**:1520–1523.
9. Gordon, S. V., Brosch, R. *et al.* (1999). Identification of variable regions in the genomes of tubercle bacilli using bacterial artificial chromosome arrays. *Mol. Microbiol.* **32**:643–655.
10. van Rie, A., Warren, R. *et al.* (1999). Exogenous reinfection as a cause of recurrent tuberculosis after curative treatment. *New England J. Medicine* **341**:1174–1179.
11. Orme, I. M., and Collins, F. M. (1994). Mouse model of tuberculosis. *In "Tuberculosis: Protection, Pathogenesis, and Control,"* B. R. Bloom, Ed., pp. 113–134. Washington, DC: American Society for Microbiology.
12. McMurray, D. N. (1994). Guinea pig model of tuberculosis. *In "Tuberculosis: Protection, Pathogenesis, and Control,"* B. R. Bloom, Ed., pp. 135–148. Washington, DC: American Society for Microbiology.
13. Dannenberg, A. M., Bishai, W. R. *et al.* (2000). Efficacies of BCG and vole bacillus (*Mycobacterium microti*) vaccines in preventing clinically apparent pulmonary tuberculosis in rabbits: A preliminary report. *Vaccine* **19**:796–800.
14. Corner, L. A., Buddle, B. M. *et al.* (2001). Aerosol vaccination of the brushtail possum (*Trichosurus vulpecula*) with bacille Calmette–Guerin: The duration of protection. *Vet. Microbiol.* **81**:181–191.
15. Buddle, B. M. (2001). Vaccination of cattle against *Mycobacterium bovis*. *Tuberculosis (Edinburgh)* **81**:125–132.
16. Griffin, J. F., Chinn, D. N. *et al.* (2001). Optimal models to evaluate the protective efficacy of tuberculosis vaccines. *Tuberculosis (Edinburgh)* **81**:133–139.
17. Langermans, J. A., Andersen, P. *et al.* (2001). Divergent effect of bacillus Calmette–Guerin (BCG) vaccination on *Mycobacterium tuberculosis* infection in highly related macaque species: Implications for primate models in tuberculosis vaccine research. *Proc. Natl. Acad. Sci. U.S.A.* **98**:11497–11502.
18. Colditz, G. A., Berkey, C. S. *et al.* (1995). The efficacy of bacillus Calmette–Guerin vaccination of newborns and infants in the prevention of tuberculosis: Meta-analyses of the published literature. *Pediatrics* **96**:29–35.

19. Bloom, B. R., and Fine, P. E. M. (1994). The BCG experience: Implications for future vaccines against tuberculosis. *In* "Tuberculosis: Protection, Pathogenesis, and Control," B. R. Bloom, Ed., pp. 531–557. Washington, DC: American Society for Microbiology.

20. Fine, P. E. (1995). Variation in protection by BCG: Implications of and for heterologous immunity. *Lancet* **346**:1339–1345.

21. Palmer, C. E., and Long, M. W. (1966). Effects of infection with atypical mycobacteria on BCG vaccination and tuberculosis. *Am. Rev. Respir. Dis.* **94**:553–568.

22. Fine, P. E., and Vynnycky, E. (1998). The effect of heterologous immunity upon the apparent efficacy of (e.g., BCG) vaccines. *Vaccine* **16**:1923–1928.

23. Fine, P. E. (2001). BCG: The challenge continues. *Scand. J. Infect. Dis.* **33**:243–245.

24. Flynn, J. L., and Chan, J. (2001). Immunology of tuberculosis. *Annu. Rev. Immunol.* **19**:93–129.

25. Bardarov, S., Kriakov, J. *et al.* (1997). Conditionally replicating mycobacteriophages: A system for transposon delivery to *Mycobacterium tuberculosis*. *Proc. Natl. Acad. Sci. U.S.A.* **94**:10961–10966.

26. Pelicic, V., Jackson, M. *et al.* (1997). Efficient allelic exchange and transposon mutagenesis in *Mycobacterium tuberculosis*. *Proc. Natl. Acad. Sci. U.S.A.* **94**:10955–10960.

27. Cole, S. T., Brosch, R. *et al.* (1998). Deciphering the biology of *Mycobacterium tuberculosis* from the complete genome sequence. *Nature* **393**:537–544.

28. Newport, M. J., Huxley, C. M. *et al.* (1996). A mutation in the interferon-γ-receptor gene and susceptibility to mycobacterial infection. *New England J. Medicine* **335**:1941–1949.

29. Altare, F., Durandy, A. *et al.* (1998). Impairment of mycobacterial immunity in human interleukin-12 receptor deficiency. *Science* **280**:1432–1435.

30. Sousa, A. O., Mazzaccaro, R. J. *et al.* (2000). Relative contributions of distinct MHC class I-dependent cell populations in protection to tuberculosis infection in mice. *Proc. Natl. Acad. Sci. U.S.A.* **97**:4204–4208.

31. van Pinxteren, L. A., Cassidy, J. P. *et al.* (2000). Control of latent *Mycobacterium tuberculosis* infection is dependent on CD8 T cells. *Eur. J. Immunol.* **30**:3689–3698.

32. Tascon, R. E., Stavropoulos, E. *et al.* (1998). Protection against *Mycobacterium tuberculosis* infection by CD8+ T cells requires the production of γ interferon." *Infect. Immun.* **66**:830–834.

33. Stenger, S., Hanson, D. A. *et al.* (1998). An antimicrobial activity of cytolytic T cells mediated by granulysin. *Science* **282**:121–125.

34. Flynn, J. L., Goldstein, M. M. *et al.* (1992). Major histocompatibility complex class I-restricted T cells are required for resistance to *Mycobacterium tuberculosis* infection. *Proc. Natl. Acad. Sci. U.S.A.* **89**:12013–12017.

35. Constant, P., Davodeau, F. *et al.* (1994). Stimulation of human γδ T cells by nonpeptidic mycobacterial ligands. *Science* **264**:267–270.

36. Ulrichs, T., and Porcelli, S. A. (2000). CD1 proteins: Targets of T cell recognition in innate and adaptive immunity. *Rev. Immunogenet.* **2**:416–432.

37. Gonzalez-Juarrero, M., Turner, O. C. *et al.* (2001). Temporal and spatial arrangement of lymphocytes within lung granulomas induced by aerosol infection with *Mycobacterium tuberculosis*. *Infect. Immun.* **69**:1722–1728.

38. Glickman, M. S. and Jacobs, W. R., Jr. (2001). Microbial pathogenesis of *Mycobacterium tuberculosis*: Dawn of a discipline." *Cell* **104**:477–485.

39. Murray, P. J., Aldovini, A. *et al.* (1996). Manipulation and potentiation of antimycobacterial immunity using recombinant bacille Calmette–Guerin strains that secrete cytokines. *Proc. Natl. Acad. Sci. U.S.A.* **93**:934–939.

40. Hess, J., Miko, D. *et al.* (1998). *Mycobacterium bovis* bacille Calmette–Guerin strains secreting listeriolysin of *Listeria monocytogenes*. *Proc. Natl. Acad. Sci. U.S.A.* **95**:5299–5304.

41. Horwitz, M. A., Harth, G. *et al.* (2000). Recombinant bacillus Calmette–Guerin (BCG) vaccines expressing the *Mycobacterium tuberculosis* 30-kDa major secretory protein induce greater protective immunity against tuberculosis than conventional BCG vaccines in a highly susceptible animal model. *Proc. Natl. Acad. Sci. U.S.A.* **97**:13853–13858.

42. Stewart, G. R., Snewin, V. A. *et al.* (2001). Overexpression of heat-shock proteins reduces survival of *Mycobacterium tuberculosis* in the chronic phase of infection. *Nat. Med.* **7**:732–737.

43. Hondalus, M. K., Bardarov, S. *et al.* (2000). Attenuation of and protection induced by a leucine auxotroph of *Mycobacterium tuberculosis*. *Infect. Immun.* **68**:2888–2898.

44. Smith, D. A., Parish, T. *et al.* (2001). Characterization of auxotrophic mutants of *Mycobacterium tuberculosis* and their potential as vaccine candidates. *Infect. Immun.* **69**: 1142–1150.

45. Perez, E., Samper, S. *et al.* (2001). An essential role for *phoP* in *Mycobacterium tuberculosis* virulence. *Mol. Microbiol.* **41**:179–187.

46. Camacho, L. R., Ensergueix, D. *et al.* (1999). Identification of a virulence gene cluster of *Mycobacterium tuberculosis* by signature-tagged transposon mutagenesis. *Mol. Microbiol.* **34**:257–267.

47. Cox, J. S., Chen, B. *et al.* (1999). Complex lipid determines tissue-specific replication of *Mycobacterium tuberculosis* in mice. *Nature* **402**:79–83.

48. Olsen, A. W., Hansen, P. R. *et al.* (2000). Efficient protection against *Mycobacterium tuberculosis* by vaccination with a single subdominant epitope from the ESAT-6 antigen. *Eur. J. Immunol.* **30**:1724–1732.

49. Andersen, P. (1994). Effective vaccination of mice against *Mycobacterium tuberculosis* infection with a soluble mixture of secreted mycobacterial proteins. *Infect. Immun.* **62**:2536–2544.

50. Horwitz, M. A., Lee, B. W. *et al.* (1995). Protective immunity against tuberculosis induced by vaccination with major extracellular proteins of *Mycobacterium tuberculosis*. *Proc. Natl. Acad. Sci. U.S.A.* **92**:1530–1534.

51. Baldwin, S. L., D'Souza, C. *et al.* (1998). Evaluation of new vaccines in the mouse and guinea pig model of tuberculosis. *Infect. Immun.* **66**:2951–2959.

52. Content, J., de la Cuvellerie, A. *et al.* (1991). The genes coding for the antigen 85 complexes of *Mycobacterium tuberculosis* and *Mycobacterium bovis* BCG are members of a gene family: Cloning, sequence determination, and genomic organization of the gene coding for antigen 85-C of *M. tuberculosis*. *Infect. Immun.* **59**:3205–3212.

53. Huygen, K., Content, J. *et al.* (1996). Immunogenicity and protective efficacy of a tuberculosis DNA vaccine. *Nat. Med.* **2**:893–898.

54. Sorensen, A. L., Nagai, S. *et al.* (1995). Purification and characterization of a low-molecular-mass T-cell antigen secreted by *Mycobacterium tuberculosis*. *Infect. Immun.* **63**:1710–1717.

55. Brandt, L., Elhay, M. *et al.* (2000). ESAT-6 subunit vaccination against *Mycobacterium tuberculosis*. *Infect. Immun.* **68**:791–795.

56. Alderson, M. R., Bement, T. *et al.* (2000). Expression cloning of an immunodominant family of *Mycobacterium tuberculosis* antigens using human CD4(+) T cells. *J. Exp. Med.* **191**:551–560.

57. Louise, R., Skjot, V. *et al.* (2001). Antigen discovery and tuberculosis vaccine development in the post-genomic era. *Scand. J. Infect. Dis.* **33**:643–647.

58. Berthet, F. X., Rasmussen, P. B. *et al.* (1998). A *Mycobacterium tuberculosis* operon encoding ESAT-6 and a novel low-molecular-mass culture filtrate protein (CFP-10). *Microbiology* **144**(Pt. 11):3195–3203.

59. Dillon, D. C., Alderson, M. R. *et al.* (1999). Molecular characterization and human T-cell responses to a member of a novel *Mycobacterium tuberculosis* mtb39 gene family. *Infect. Immun.* **67**:2941–2950.

60. Skeiky, Y. A., Lodes, M. J. *et al.* (1999). Cloning, expression, and immunological evaluation of two putative secreted serine protease antigens of *Mycobacterium tuberculosis*. *Infect. Immun.* **67**:3998–4007.

61. Coler, R. N., Campos-Neto, A. *et al.* (2001). Vaccination with the T cell antigen Mtb 8.4 protects against challenge with *Mycobacterium tuberculosis*. *J. Immunol.* **166**:6227–6235.

62. Weinrich Olsen, A., van Pinxteren, L. A. *et al.* (2001). Protection of mice with a tuberculosis subunit vaccine based on a fusion protein of antigen 85b and esat-6. *Infect. Immun.* **69**:2773–2778.

63. Lalvani, A., Moris, P. *et al.* (1999). Potent induction of focused Th1-type cellular and humoral immune responses by RTS,S/SBAS2, a recombinant *Plasmodium falciparum* malaria vaccine. *J. Infect. Dis.* **180**:1656–1664.

64. Lowrie, D. B., Tascon, R. E. *et al.* (1999). Therapy of tuberculosis in mice by DNA vaccination. *Nature* **400**:269–271.

65. Turner, O. C., Roberts, A. D. *et al.* (2000). Lack of protection in mice and necrotizing bronchointerstitial pneumonia with bronchiolitis in guinea pigs immunized with vaccines directed against the hsp60 molecule of *Mycobacterium tuberculosis*. *Infect. Immun.* **68**:3674–3679.

66. Zhu, X., Venkataprasad, N. *et al.* (1997). Vaccination with recombinant vaccinia viruses protects mice against *Mycobacterium tuberculosis* infection. *Immunology* **92**:6–9.

67. McShane, H., Brookes, R. *et al.* (2001). Enhanced immunogenicity of CD4(+) T-cell responses and protective efficacy of a DNA-modified vaccinia virus Ankara prime-boost vaccination regimen for murine tuberculosis. *Infect. Immun.* **69**:681–686.

68. Orme, I. M., McMurray, D. N. *et al.* (2001). Tuberculosis vaccine development: Recent progress. *Trends Microbiol.* **9**:115–118.

69. Brooks, J. V., Frank, A. A. *et al.* (2001). Boosting vaccine for tuberculosis. *Infect. Immun.* **69**:2714–2717.

6 DISEASE STATES AND VACCINES: SELECTED CASES

PART I. Malaria

STEPHEN L. HOFFMAN[*,1] **AND THOMAS L. RICHIE**[†]
*Celera Genomics, Rockville, Maryland 20850
†Naval Medical Research Center, Silver Spring, Maryland 20910

I. NEED FOR A MALARIA VACCINE

Malaria, caused by *Plasmodium falciparum, Plasmodium vivax, Plasmodium malariae*, or *Plasmodium ovale*, is a major health problem in most countries lying between the Tropics of Cancer and Capricorn, with over 2 billion people living in countries where malaria is transmitted. For years it has been estimated that there may be 300–500 million new infections and 1–3 million deaths annually caused by malaria and that the majority of cases and more than 90% of the deaths occur in sub-Saharan Africa, where in many places malaria is the leading cause of death among children less than 5 years of age. Analyses suggest that the medical impact of malaria may actually have been significantly underestimated [9] and that the enormous economic impact of malaria has never been adequately considered [26]. It has been estimated that malaria reduces annual gross domestic product (GDP) in affected countries in sub-Saharan Africa by greater than 1.3% [26]. As we begin the twenty-first century, it is inconceivable that a treatable infectious disease has such an "intolerable" impact, yet the increasing resistance of the *Plasmodium* sp. parasite to chemoprophylactic and chemotherapeutic agents, the resistance of the *Anopheles* sp. mosquito vector to insecticides, including the pyrethroids used in insecticide-impregnated bednets, and the inability of the most affected countries to mobilize and sustain the resources required for malaria control, as evidenced by the resurgence of malaria in areas formerly free of the disease, highlight the urgency for developing an effective malaria vaccine. A vaccine would dramatically improve the chances of optimally controlling and eventually eradicating malaria.

[1]Current address: Sanaria, Gaithersburg, Maryland 20878

II. INDICATIONS FOR A MALARIA VACCINE

Many malariologists believe that different types of malaria vaccines may be necessary for different populations. The primary requirement is to reduce the incidence of severe malaria and malaria-associated mortality in infants and children with heavy exposure to *P. falciparum*, such as those living in sub-Saharan Africa (type 1 vaccine). At the other extreme is the requirement to prevent all clinical manifestations of malaria in individuals from areas with no exposure who travel to regions where malaria is endemic, primarily malaria caused by *P. falciparum* and *P. vivax* (type 2 vaccine). This "extremes" approach to malaria vaccine development does not take into account specifically populations affected by malaria that fall between these extremes, such as individuals in endemic regions at high risk of *P. vivax* infections [49]. In fact, as type 1 and type 2 vaccines are developed, they will need to be assessed in many different populations.

For the type 1 vaccine designed to reduce mortality among children in sub-Saharan Africa, the problem is complicated by the varied epidemiology of the disease. In areas with extremely intense transmission, such as northern Ghana, it is primarily infants who are dying of malaria, with severe anemia as a major cause of death. In areas with less intense transmission, such as The Gambia, it appears to be 2- to 5-year-olds who are most at risk of dying, with cerebral malaria as the major cause of death [71]. It is possible that different vaccines, tailored to the predominant pathophysiology of a particular region, or different vaccination strategies, targeting the age groups at greatest risk, may be the most effective response to this heterogeneous epidemiology.

For the type 2 vaccine designed to prevent all clinical manifestations of malaria in nonimmune travelers to malaria-endemic areas, the target populations are also varied. One generally thinks that the major recipients of such a vaccine would be travelers from North America, Europe, Japan, Australia, and other highly developed malaria-free areas of the world. However, there are hundreds of millions of people living in nonmalarious areas of malaria-endemic countries, who travel to malarious areas of their own country. For example, it is not infrequent for someone born in western Kenya, where transmission is high, to attend university in Nairobi, where there is no malaria transmission, and then settle in Nairobi. The children of these Nairobi residents are nonimmune to malaria, and the parents themselves may lose their immunity if sufficient time passes. When they, their children, or other Kenyans born in Nairobi visit western Kenya on holidays, they are at high risk of contracting malaria and rapidly developing severe disease. There is little mention in the malaria literature of the increasing numbers of nonimmune individuals living in countries with endemic malaria, who must receive short-term protection against malaria by a vaccine. Because of their susceptibility to rapidly developing severe disease and because of their brief exposure to transmission, we think that these in-country travelers, including their children, would require a vaccine with the same protective profile as a vaccine for travelers from North America or Europe.

There is a third type of vaccine being developed that is a transmission-blocking vaccine. This vaccine is not designed to protect the immunized individual but the entire community by reducing transmission intensity. We believe that such a vaccine will eventually be combined with the type 1 and 2 vaccines described earlier. The target populations for such a vaccine are not yet clearly defined. Such a vaccine unquestionably would be of great value, perhaps even on its own, on islands with malaria and in areas with only modest transmission. It might also be useful during prolonged epidemics. It is not clear how useful such a vaccine will be in areas like sub-Saharan Africa, where transmission is intense. However, an article has suggested that such a vaccine could be quite useful, particularly when combined with vaccines targeting other parasite stages, for preventing a selective process favoring the emergence of rapid-growing and therefore virulent parasites [27]. Even if not able to impact overall transmission levels, a transmission-blocking component included in a type 1 or 2 vaccine could block the escape of resistant mutants, thus preserving vaccine efficacy in the community and preventing the emergence of virulent parasite strains.

III. EXPOSURE OF HUMANS TO WHOLE PARASITES INDICATES TYPE 1 AND 2 MALARIA VACCINES ARE POSSIBLE

In areas of sub-Saharan Africa with the most intense malaria transmission and highest rates of malaria-associated morbidity and mortality, children who survive to the age of 7–10 years rarely develop life-threatening *P. falciparum* infections and rarely die of malaria. This acquired immunity occurs more rapidly, the more intense the transmission of *P. falciparum*. Older children and adults become infected frequently, but it is thought that their immune systems limit the infections, thereby preventing severe disease. A vaccine that duplicated naturally acquired immunity and "turned infants and children into 15-year-olds" in regard to protective immunity would essentially eliminate severe disease and death without eliminating infection or mild-to-moderate disease. Such a vaccine would fulfill the requirements of a type 1 vaccine.

In the late 1960s, it was reported that mice immunized with radiation-attenuated rodent malaria sporozoites were protected against challenge [53]. Based on these rodent studies, human trials were conducted in the early 1970s. It was reported that exposure of humans to the bite of irradiated mosquitoes carrying *P. falciparum* sporozoites in their salivary glands protected these volunteers [12–14, 59, 60]. These studies have since been repeated more systematically [21, 22, 32, 36]. The results are striking [36]. Among 14 volunteers exposed to greater than 1000 infected, irradiated mosquitoes and first challenged 2–10 weeks after last exposure, 13 of 14 were completely protected against infection; there is apparently little or no genetic restriction of the protective immunity elicited by this immunization approach. Six volunteers were challenged 23–42 weeks after last boost, and 5 of 6 were protected. A total of seven challenges have been done with strains of *P. falciparum* different from the one used for immunization, and there was protection in all seven challenges; individuals immunized with parasites from Africa and challenged with parasites from South America were protected. To combine all studies, among volunteers exposed to at least 1000 infective bites and challenged within 1 year after last exposure, 33 of 35 challenges were associated with complete protection. Thus, the irradiated sporozoite vaccine protects greater than 90% of recipients against experimental challenge for at least 10 months, and the protection is not strain-specific. This is as good a record of protection as has been achieved with any vaccine for any disease. Such a vaccine would fulfill the indications for a type 2 vaccine, and perhaps a type 1 vaccine.

IV. WHY THERE IS NOT A MALARIA VACCINE

Vaccines have eliminated the natural transmission of smallpox from the world, nearly eliminated polio, and eliminated *Haemophilus influenzae* as a cause of severe disease in industrialized countries. We know that it is possible for the immune system to control malaria, yet despite much effort, there is currently no malaria vaccine on the market. It is not possible to consider immunizing people by exposure to live parasites, which is what occurs in naturally acquired immunity, as the individuals would become ill. It has not been practical to consider immunization by the bite of irradiated, infected mosquitoes or by intravenous injection of live radiation-attenuated sporozoites to duplicate the immunity elicited by the irradiated sporozoite vaccine. The approach has been to try to determine the immune responses responsible for the protective immunity seen in these two models, and the parasite life cycle stages and proteins against which this protective immunity is directed, and then to develop vaccine delivery systems that would induce the required responses against the identified targets. Current and future advances in understanding human immunology and the biology of malaria parasites, many of which will be dependent on data from the human genome sequencing projects [44, 75] and the malaria genome projects [8, 28, 28a], should allow the identification of key antigens associated with protection and the formulation of vaccines effective in all recipients, regardless of their genetic background. Publication of the entire *P. falciparum* genome (estimated 5000–6000 genes on 14 chromosomes) is anticipated in late 2002, providing a quantum leap in the list of candidate antigens and an unprecedented opportunity for solving the malaria problem.

A. Targets of Protective Immune Responses

The parasites that cause malaria are much more complex than the viruses and bacteria that heretofore have been controlled by vaccination, and this complexity is reflected in their multistage life cycle. When sporozoites are inoculated into humans by *Anopheles* mosquitoes, they circulate extracellularly in the bloodstream for less than 30 min before entering the liver. Within the hepatocyte, a uninucleate sporozoite develops into a schizont with an estimated 10,000–40,000 uninucleate merozoites. These merozoites rupture from the hepatocyte and each can invade an erythrocyte, initiating the cycle of intraerythrocytic stage development, rupture, and reinvasion that leads to a 10- to 20-fold increase in the number of parasites in the bloodstream every 48 hr. These asexual erythrocytic stage parasites are responsible for the clinical manifestations and pathology of malaria. Erythrocytic stage sexual parasites, called gametocytes, are ingested by mosquitoes, in which they develop to sporozoites over 14–21 days.

Frequently, the proteins expressed by each of these various stages are antigenically distinct. Thus, if a vaccine elicits high levels of antisporozoite antibodies, those antibodies generally will not recognize the asexual erthrocytic stages that follow. Furthermore, for many of these genes–proteins, allelic or antigenic variation has been demonstrated. A single individual can be infected simultaneously with at least eight different strains [23], which may vary at critical T- and B-cell epitopes. This allelic heterogeneity is expanded by extensive antigenic variation present within a single strain. For example, PfEMP1, a protein expressed on the surface of erythrocytes, is encoded by 50–100 different genes, each with some variation in its sequence [20]. The parasite probably expresses only one at a time, with each new variant escaping from antibodies induced by the previous variant and therefore capable of generating a new wave of parasitemia. PfEMP1 is thought to mediate cytoadherence of infected erythrocytes to endothelial cells in the microcirculation during maturation of the parasite, thereby preventing removal of the infected erythrocyte in the spleen during this vulnerable phase of its life cycle. Cytoadherence is thought to be responsible for the microcirculatory obstruction important to the pathogenesis of severe disease. In summary, stage-specific expression of proteins, the presence of multiple antigenically distinct strains in nature, and within-strain antigenic variation are critical to the parasite's survival, are unfavorable for the host, and greatly complicate the challenge for vaccine developers.

During the past 20 years, considerable work has been done to develop subunit vaccines that provide protective immunity comparable to that of the human models of naturally acquired immunity and radiation-attenuated sporozoite immunization. However, no such vaccine has provided comparable protection. There are numerous potential explanations for the lack of current success. One is that exposure to the whole parasite elicits a more potent, protective immune response than the subunit vaccines tested thus far. However, in the case of people protected by the irradiated sporozoite vaccine or by naturally acquired immunity, T-cell and antibody responses against our current candidate antigens [5, 10, 34, 38, 40, 70] or, in the case of antibody responses, against whole parasites are generally modest at best and in many instances lower than those achieved by subunit vaccination. It is more likely that immunization with only a few parasite proteins cannot duplicate the immunity elicited by exposure to a parasite that has thousands of proteins; modest immune response against tens, hundreds, or thousands of parasite proteins may be additive or synergistic and may underlie the protection observed in these human models. In the case of naturally acquired immunity, this "breadth" would be further expanded by exposure to many polymorphic strains of *P. falciparum*. In summary, whole-parasite-induced immunity could be directed at many of the 5000–6000 malaria parasite proteins. The malaria genome project and the single-nucleotide polymorphism (SNP) projects currently nearing completion may provide knowledge of all these potential targets and their variability at the epitope level, thereby laying the foundation for duplicating whole-organism immunity with subunit vaccines.

B. Immune Responses Responsible for Protection

Vaccines against infectious agents are effective by destroying the infectious agent or the host cells in which they reside or by inhibiting a function of the infectious agent critical for its survival. As described previously, there is little doubt that immune responses against *P. falciparum* can protect against malaria, probably through both mechanisms.

In naturally acquired immunity, all arms of the immune system are probably activated against all stages of the parasite life cycle. However, three responses are thought to be primary: antibodies directed against parasite proteins expressed on the surface of erythrocytes that prevent sequestration in the microcirculation (function-inhibiting activity) [20], antibodies directed against parasite proteins expressed on the surface of merozoites that prevent invasion of erythrocytes (function-inhibiting activity) [69], and antibodies expressed against either type of parasite protein that are capable of mediating antibody dependent cellular inhibition (ADCI) [7], whereby biologically active molecules, including cytokines, nitric oxide, and free oxygen intermediates, are released from reticuloendothelial or other cells after activation through the Fc component of the bound antibody molecule (function-independent killing). In addition, biologically active molecules such as cytokines, free oxygen radicals, and nitric oxide molecules released from CD4 T cells after an antigen-specific interaction may also contribute to this immunity. The pathogenesis of the disease may be mediated by these same host-derived biologically active molecules, perhaps elicited by putative malarial toxins released from the infected erythrocytes [11, 56, 64a]. Antibodies against these toxins may contribute to naturally acquired immunity. It seems likely that immune responses against sporozoites or infected hepatoctyes that limit the number of parasites that emerge from the liver into the bloodstream also play an important role in the resistance to clinical disease demonstrated by individuals with naturally acquired immunity.

The most persuasive human data regarding the importance of antibodies against asexual erythrocytic stage antigens come from passive transfer studies in humans [15, 48, 62]. A number of experiments have shown that passive transfer of immunoglobulin (I_g) from west African adults with life-long exposure to malaria can be used to treat children with malaria. The most striking experiment took place in Thai children with drug-resistant *P. falciparum*, in which administration of immune IgG from Africa led to a 99% reduction in parasitemia, proving that naturally acquired antibodies can have a profound effect on asexual erythrocytic stage *P. falciparum* parasites in humans.

For irradiated sporozoite-induced immunity, the protection is thought to be mediated primarily by T cells directed against peptides from parasite proteins expressed in infected hepatocytes (function-independent killing via elimination of infected host cells) (Fig. 1), although antibodies that reduce sporozoite invasion of hepatocytes (function-inhibiting activity) probably play a role [34]. Humans immunized with radiation-attenuated sporozoites [47, 82, 83] and also humans naturally exposed to malaria [1, 17, 24, 43, 66] have been shown to have CD8[+] cytotoxic T lymphocytes and CD8[+] T cell-derived interferon-γ (IFN-γ) responses against preerythrocytic stage proteins. However, data regarding

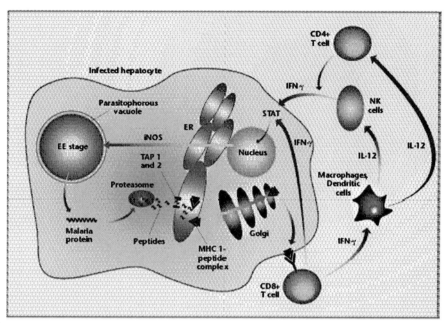

FIGURE 1 Proposed mechanism of protective immunity directed against the *Plasmodium*-infected hepatocyte. Protective immunity directed against preerythrocytic stage (EE stage) malaria is primarily mediated by antigen-specific CD8$^+$ T cells that recognize parasite-derived peptides presented in association with class 1 MHC molecules on the surface of infected hepatocytes. Short peptides are derived from the cytoplasmic malaria protein by the proteolytic action of proteosomes. The peptides are imported into the endoplasmic reticulum (ER) via the transporters associated with antigen processing, TAP1 and TAP2. In the ER, the peptides associated with the MHC class 1 molecule and the peptide–MHC complex pass through the Golgi apparatus to the cell surface where they can be recognized by the T-cell receptor on the surface of CD8$^+$ T-cells. IFN-γ is produced as a direct consequence of the CD8$^+$ T-cell activation, and subsequent production of IFN-γ may be up-regulated by a positive feedback loop involving IL-12 (produced by dendritic cells, macrophages or other cells), NK cells, and/or CD4$^+$ T-cells, depending on the host. IFN-γ, via signal transducers associated with transcription (STAT), activates nitric oxide synthase (iNOS) and induces the L-arginine-dependent NO pathway to eliminate the infected hepatocyte or the intrahepatic schizonts. Reproduced with permission from Hoffman, S. L., and Doolan, D. L. (2000). *Nature Medicine*, **6**(11):1218–1219.

the importance of T cells in the immunity elicited by radiation-attenuated sporozoites come virtually entirely from rodent experiments [18, 19, 42, 61, 65, 79, 80], in which T-cell subsets were depleted or adoptively transferred. These data indicate that in rodent models protection is always dependent on CD8$^+$ T cells and IFN-γ, and often dependent on interleukin-12 (IL-12) and nitric oxide (Fig. 1) [18, 19]. Nonetheless, adoptive transfer of antigen-specific CD4$^+$ T cells can also protect against sporozoite challenge [57], and in some strains of mice depletion of CD4$^+$ T cells also eliminates protection [19, 76].

In the case of understanding and duplicating naturally acquired immunity, the complexity of the immune response to infection may add to the difficulty of developing a vaccine. Human host genetics, transmission dynamics of the parasite, and probably even the age of the host contribute to this complexity. Although it is clear that sickle-cell trait protects against severe disease and that other genetic traits may influence outcome, our understanding of the relationship between host genetics and the response to infection is very limited. The elucidation of the sequence of the human genome and the development of scientific tools to use these data should lead to a better understanding of the role of host factors in determining the severity of disease associated with infection.

The human immune response is also dependent on transmission dynamics. In areas in which transmission is most intense, infants are at the highest risk of developing severe and fatal malaria, whereas, in areas with less intense transmission, older children have a higher incidence of severe and fatal disease than infants. The age of the individual when first exposed to parasites (or a vaccine) thus may play a role in determining the immune response. A number of reports have suggested that, among nonimmune children and nonimmune adults, adults are actually more susceptible than children to developing severe disease after their first infection [4]. However, adults acquire immunity faster than children [3]. Only a few adults have been immunized with the irradiated sporozoite vaccine; thus, there are no such data regarding the effect of age on irradiated sporozoite-induced immunity. However, it is quite likely that such factors will have to be taken into consideration for a vaccine that duplicates this type of immunity.

C. Vaccine Delivery Systems

For a vaccine to be optimally effective, it must elicit the appropriate protective immune responses and sustain those immune responses over time, either due to vaccine administration or due to boosting by exposure to parasites. As has been discussed, both protective antibody and T-cell responses against potentially multiple antigens are needed. Experimental recombinant protein and synthetic peptide malaria vaccines have been tested in humans with multiple adjuvants including aluminum hydroxide, monophosphoryl lipid A (MPL), MPL and cell wall skeleton of mycobacteria [33], liposomes with MPL [25], Montanide [29], different combinations of MPL, QS-21, and oil-in-water emulsion [72], recombinant viruses [55], and *Salmonella typhi* [84]. More recently, DNA vaccines have also been tested, anticipating that their simplicity of design, ease of production, capacity to elicit CD8+ T-cell responses [45, 77, 78], and suitability for combination will be ideal for delivering the multiple antigens likely required to induce immune responses comparable to those elicited by natural exposure to parasites or by the irradiated sporozoite vaccine. Much progress has been made, but no vaccine delivery system has been shown to be optimal or adequate.

V. CURRENT APPROACHES TO MALARIA VACCINE DEVELOPMENT

A. Strategies for Vaccine Design

Currently, three general approaches to malaria vaccine development are being pursued. The most work has been done and progress achieved in attempts to maximize the magnitude and quality of immune responses to a single or a few key antigens, such as the circumsporozoite protein (CSP, dominant antigen on the surface of sporozoites) and merozoite surface protein 1 (MSP-1), by immunizing with synthetic peptides or recombinant proteins in an adjuvant. These vaccines are being designed primarily to induce antibody and CD4 T-cell responses, but there is also interest in eliciting CD8 T-cell responses. The second approach is to simultaneously induce good or optimum immune responses against all of the approximately 15–20 identified potential target proteins by immunizing with DNA vaccines or recombinant viruses and boosting with DNA vaccines, recombinant viruses or bacteria, or recombinant proteins in adjuvant. The goal is to elicit antibody and CD8 and CD4 T-cell responses. The third approach is to try to duplicate the whole-organism immunity induced by immunization with radiation-attenuated sporozoites and natural exposure to malaria. Success in this area will be dependent on the sequencing of the genome and developing methods for exploiting this genomic sequence data. It remains to be established how vaccines of the third type will be constructed.

VI. CURRENT STATUS OF CLINICAL TRIALS

A. Maximizing the Magnitude and Quality of Immune Responses to a Few Key Antigens

1. Preerythrocytic Stages

A number of clinical trials with preerythrocytic stage *P. falciparum* vaccines have already been completed, and numerous others are planned or in progress. The Walter Reed Army Institute of Research (WRAIR) and GlaxoSmithKline Biologicals (GSK) have done extensive work on a CSP recombinant vaccine, RTS,S/ASO2. The vaccine is formulated with an oil-in-water emulsion plus the immunostimulants monophosphoryl lipid A (MPL) and the saponin derivative QS21. It consistently protects 40–50% of volunteers experimentally challenged 2–3 weeks after their last immunization [41, 72, 73]. When five protected volunteers were rechallenged 6 months after the last immunization, only one of the five was protected [73]. The first field trials with this vaccine have been conducted in The Gambia by the Medical Research Council (UK). Vaccine efficacy (end point defined as "time to first infection") was 71% (95% CI 46% to 85%) during the first 9 weeks of surveillance, but subsequently declined to 0% (95% CI −52% to 34%) in the last 6 weeks [6]. The first trial of RTS,S/ASO2 in a pediatric population (6- to 11-year-olds) began in The Gambia in May 2001. Concomitantly, efforts aimed at improving the vaccine are ongoing and include the evaluation of alternative adjuvants as well as the combination of RTS,S with other preerythrocytic and/or erythrocytic stage antigens.

A synthetic multiple antigen peptide (MAP) malaria vaccine developed at New York University, containing minimal *P. falciparum* CSP repeat epitopes, elicited high levels of parasite-specific antibodies and CD4+ T cells in volunteers with specific human leukocyte antigen (HLA) genotypes [50]. More recently, this group, in collaboration with the University of Geneva, has demonstrated the immunogenicity of a synthetic triepitope polyoxime malaria vaccine containing B-cell epitopes and a universal T-cell epitope of PfCSP [51]. In an open-labeled phase 1 study, volunteers of diverse HLA types developed antirepeat antibodies and T cells specific for the universal T-cell epitope [51]. A carboxy-terminal synthetic peptide from the PfCSP has been developed by the University of Lausanne. Based on a phase 1 trial demonstrating induction of CSP-specific antibodies as well as CD4 and CD8 T-cell responses [46], the European Malaria Vaccine Initiative (EMVI), in collaboration with researchers in Nijmegen, The Netherlands, plans to bring this vaccine into phase 2a testing in 2002. Clinical trials with several *P. falciparum* liver stage antigen-3 (PfLSA-3) constructs, including an LSA-3 long synthetic peptide, are planned to take place in Nijmegen in early 2002. Collaborators from Apovia, Inc. New York University, Malaria Vaccine Initiative, NIH, and University of Maryland are initiating phase 1 studies of a preerythrocytic stage vaccine based on a virus-like particle consisting of a modified hepatitis B core particle and circumsporozoite protein-specific T- and B-cell epitopes.

2. Asexual Erythrocytic Stages

A number of human trials are planned or in progress of erythrocytic stage *P. falciparum* vaccines. The most work has been done with SPf66, which was developed at the Institute of Immunology in Bogota, Colombia. SPf66 is a polymer containing an 11-amino acid sequence from the N-terminus of PfMSP1, the 4-amino acid repeat of PfCSP, and 2 other short amino acid sequences thought to be from *P. falciparum* but not yet proved to be so. Initial trials in Colombia and Tanzania demonstrated moderate efficacy in adults and in children aged 1–5 years [2, 74]. More recently, trials in the Gambia (6- to 12-month-olds), Thailand (2- to 15-year-olds), Tanzania (infants), and Brazil (7- to 70-year-olds) have shown no significant efficacy [31]. In Australia and Papua New Guinea, trials have been conducted with vaccines containing purified recombinant proteins based on three blood stage *P. falciparum* proteins [a fragment of MSP-1, MSP-2, and a portion of the ring-infected erythrocyte surface antigen (RESA)] formulated in the adjuvant Montanide ISA 720, demonstrating induction of both T-cell and antibody responses. A field trial in 5- to 9-year-olds has shown promising results [30]. The first phase 1 trial with the *P. falciparum* apical membrane antigen-1 (PfAMA-1) was initiated by this group. Another group, the National Institutes of Allergy and Infectious Diseases (NIAID) Malaria Vaccine Development Unit (MVDU), is anticipating the initiation of phase 1 clinical trials with a PfAMA-1 recombinant vaccine in 2002.

A phase 1 dose-escalation trial of a purified recombinant PfMSP-1 has been conducted by WRAIR, with phase 2a testing ongoing. The EMVI, together with investigators at the Institut Pasteur in Paris and the University of Lausanne, has initiated a phase 1 trial in Lausanne with an MSP-3 synthetic peptide, with phase 2a testing tentatively planned for early in 2002. EMVI, in collaboration with Statens Serum Institut in Denmark, began a clinical trial in the Netherlands in September 2001 with a glutamate-rich protein (GLURP) synthetic peptide vaccine.

3. Sexual Stages

The NIAID MVDU is currently developing candidate transmission-blocking vaccines based on sexual stage antigens of *P. falciparum* and *P. vivax*, Pfs25 and Pvs25. It is anticipated that phase 1 clinical trials will begin in 2002.

B. Induction of Immune Responses to Multiple Antigens from Different Stages

1. Preerythrocytic Stages

a. Single-Gene Vaccines

In order to establish the safety of DNA vaccines and to determine the immunogenicity of such vaccines, three different phase 1 trials of a PfCSP DNA vaccine have been conducted at the Naval Medical Research Center (NMRC). There have been two phase 1 safety and immunogenicity clinical trials of a PfCSP DNA vaccine and a study assessing the capacity of a PfCSP recombinant protein to boost the immune responses elicited by DNA vaccination. In the first two studies, the vaccine appeared to be safe and well-tolerated, and most of the volunteers developed antigen-specific, genetically restricted CD8 cytotoxic lymphocytes (Fig. 2) [45, 77]. In the second trial, which compared needle and needle-free Biojector® jet injection (Fig. 3) [78], antigen-specific CD8 T-cell-dependent IFN-γ responses were detected in all 14 volunteers, with the Biojector® intramuscular route appearing to be most immunogenic. The results of the third trial are under analysis.

b. Multigene Vaccines

Most recently, the PfCSP plasmid has been combined with four additional plasmids encoding preerythrocytic stage antigens (MuStDO 5 vaccine, see later discussion). Thirty-one volunteers received three monthly IM jet injections of this plasmid cocktail consisting of 500 μg each of plasmids encoding PfCSP, PfSSP2/TRAP, PfEXP1, N'C-terminus PfLSA1, and PfLSA3, combined with

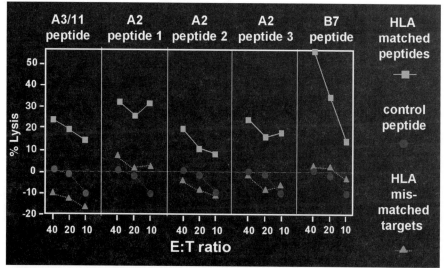

FIGURE 2 DNA-induced CTL responses are restricted by multiple HLA alleles. Fresh PBMCs from volunteer 36 (500-μg-dosage group), who expressed the alleles HLA-A2, A3, and B7, were assayed for antigen-specific, genetically restricted CTLs. The assay was repeated with coded frozen PBMCs that were collected before and after vaccination; the results confirmed that the peptide-specific (the same five peptides) genetically restricted CTLs were induced by vaccination with plasmid DNA. Reprinted with permission from Wang, R., Doolan, D., Le, T., Hedstrom, R., Coonan, K., Charoenvit, Y., Jones, T., Hobart, P., Margalith, M., Ng, J., Weiss, W., Sedegah, M., de Taisne, C., Norman, J., Hoffman, S. L. (1998). *Science* **282**:476–480, American Association for the Advancement of Science.

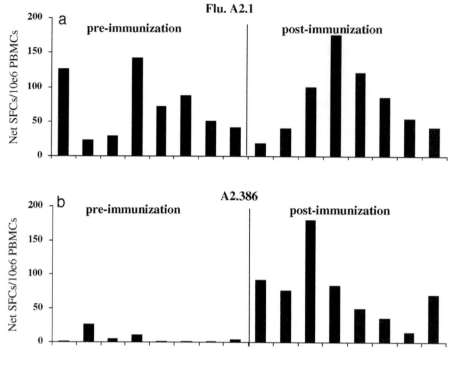

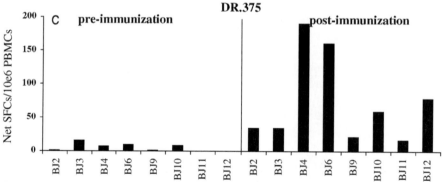

FIGURE 3   Antigen-specific IFN-γ responses induced by DNA vaccination. IFN-γ responses against: (a) HLA-A2.1-restricted positive control peptide from influenza matrix protein (Flu. A2.1); (b) HLA-A2.1-restricted peptide from PfCSP (A2.386); or (c) HLA-DR-restricted peptide from PfCSP (DR.375), with coded and double-blinded frozen PBMCs collected from eight volunteers pre- and postimmunization with PfCSP DNA administered by Biojector intramuscularly (BJ2, -33, and –4), Biojector intramuscularly and intradermally (BJ6, -9, and –10), and needle intramuscularly (BJ11 and –1). From Wang, R., Epstein, J., Baraceros, F., Gorak, E., Charoenvit, Y., Carucci, D., Hedstrom, R., Rahadjo, N., Gay, T., Hobart, P., Stout, R., Jones, T., Richie, T., Parker, S., Doolan, D., Norman, J., Hoffman, S. L. (2001). *Proc. Natl. Acad. Sci. U.S.A.* **98**(19):10817–10822.

escalating doses of a sixth plasmid encoding human GM-CSF. T-cell responses, as measured by IFN-γ production by ELIspot assay, were induced against all five antigens, although there was no evidence of enhancement associated with the addition of hGM-CSF. As with the single-gene PfCSP vaccine [45], no antibodies were detected by ELISA or IFA. Volunteers were challenged with Pf sporozoites 18 days after the third immunization, and none were protected. However, vaccinees developed stronger T-cell and antibody responses following the challenge than unvaccinated infectivity controls, indicating the DNA immunization primed for boosting of both humoral and cellular immune responses elicited by exposure to parasites. This finding suggests that DNA-based vaccines may be effective as type 1 vaccines for use in endemic areas by priming for boosting by natural exposure and thereby accelerating the transition from malaria-naive to malaria-immune. The MuStDO 5 plasmids have now been remanufactured following optimization of their codon frequencies for mammalian expression and will be tested in combination with several blood stage antigens, alone and in prime boost approaches (see later discussion).

At Oxford University, phase 1–2a trials were conducted with a *P. falciparum* preerythrocytic stage vaccine expressing multiple malaria CD8+ T-cell epitopes and the entire TRAP/SSP2 antigen. Recipients received the priming doses of a DNA vaccine, delivered either by intramuscular needle or by Powderject gene gun, and booster doses of a recombinant attenuated vaccinia virus (MVA) expressing the same MultiEpitope-TRAP insert delivered intradermally. This group has begun to test the DNA prime–MVA boost strategy in a field trial in the Gambia, and has also conducted promising clinical trials using a recombinant fowlpox prime and MVA boost (Long and Hoffman, 2002).

2. Combined Preerythrocytic and Erythrocytic Stage Vaccines

As mentioned previously, the NMRC (in collaboration with multiple partners, including USAID, the Noguchi Memorial Institute of Medical Research in Accra, Ghana, and the Navrongo Health Research Centre in Navrongo, Ghana) has been moving forward with a vaccine program known as the Multi-Stage Malaria DNA-Based Vaccine Operation (MuStDO). The cloning of the genes and the construction of plasmids have been a collaboration between NMRC, Monash University (Melbourne, Australia), and Entremed Inc. (Maryland, U.S.A.), with manufacturing at Vical Inc. (California, U.S.A.). The program is based on the concept that an effective malaria vaccine will require a two-tiered immune response, the first tier attacking preerythrocytic stage parasites and the second blood stage parasites that break through the preerythrocytic defense. Thus, the final product would be a vaccine that incorporates both the type 1 and type 2 strategies. The preerythrocytic component of the vaccine will be designed to induce antibodies that bind to the sporozoite surface before hepatocyte invasion and, most importantly, CD8 and CD4 T cells that recognize parasite-derived peptides complexed with class I and class II HLA molecules, respectively, on the surface of infected hepatocytes. The hypothesis is that the preerythrocytic stage component will drastically reduce the number of parasites emerging from the liver. The erythrocytic component of the vaccine will be designed to prime the immune system to antigens from parasite proteins expressed on infected erythrocytes. These immune responses should limit replication of the parasite, thereby preventing severe disease and death in those experiencing breakthrough parasitemia. Recent studies in rhesus monkeys strongly support this approach (Rogers *et al.*, 2002).

NMRC is focusing on DNA vaccines as the core technology for this effort because of their demonstrated ability to preferentially induce CD8 T-cell and Th1-type immune responses, which has been difficult with the more traditional vaccines, and because of their simplicity of design and the ease with which they can be modified. As described previously, the current MuStDO 5 antigens will be supplemented with erythrocytic stage antigens. Eventually, the testing of DNA vaccines in healthy nonimmune adults in the United States will be followed by testing in semi-immune adults in the region of Accra, Ghana, in collaboration with the Noguchi Memorial Institute of Medical Research, and finally in younger children and infants in highly endemic regions of northern Ghana in collaboration with the Navrongo Health Research Centre.

As these first-generation vaccines are transitioned to the field for initial safety testing, new technologies to improve antigen expression, immunogenicity, and protective efficacy of DNA vaccines will be assessed in animal models and then clinical studies conducted in the United States. These technologies include replacement of native genes with synthetic genes utilizing codon frequencies optimized for host expression [52], absorbing plasmids onto microparticles to increase antigen uptake by dendritic cells [54], particle-mediated gene transfer (gene gun) to enhance plasmid delivery [16], and further assessment of costimulation with plasmids encoding cytokines such as GM-CSF [63, 68, 81] or other immunostimulatory molecules such as CpG motifs [39]. However, by far the most effective way to enhance immunogenicity and protection afforded by DNA vaccines is to combine them with a heterologous vaccine containing the same antigens (prime–boost approach) [64, 67, 68]. Trials will soon be underway evaluating priming with DNA plasmids such as those in the MuStDO 5 vaccine followed by boosting with the poxvirus or adenovirus constructs encoding the same antigen and/or the homologous recombinant proteins.

3. Duplicate Whole-Organism-Elicited Immunity

This approach is based on the idea of duplicating immunity induced by exposure to the whole parasite (irradiated sporozoites or natural exposure) by down-selecting a large number of proteins or epitopes from the sequence data obtained from the malaria genome project (6% currently published, the remainder anticipated in late 2002) [28a]. Thousands of new proteins identified in this way will be assessed by applied genomics (gene expression, proteomics, and other techniques) for potential inclusion in vaccines. This will constitute one of the first attempts to integrate microbial and human genomics with molecular and cell biology, immunology, and epidemiology in a concerted effort to overcome an entrenched infectious disease [35, 37].

VII. CHALLENGE OF DESIGNING AND EXECUTING FIELD TRIALS OF MALARIA VACCINES

It will be critical to consider which outcome variables to measure in field trials of malaria vaccines and which populations to study. A primary goal is to reduce mortality and severe disease. There is a potential problem that a vaccine may be discarded as a result of initial studies because the proper outcome variable(s) was (were) not measured. It will be difficult to use severe disease and death as the primary outcome variables in initial studies because of the very large sample sizes required. It is important to identify groups at highest risk so that sample sizes can be reduced. Current investigations seek to identify surrogates of severe disease and death: parasitological, hematological, biochemical, or clinical manifestations that are predictive of severe outcome [58].

Several other areas of field research could provide data to aid vaccine development. The most important is the identification of target groups for vaccines in different areas (see earlier discussion) and the exclusion of groups, like those with sickle-cell trait, who are at decreased risk and do not need to be immunized. It is important to determine whether there are measurable outcome variables that have a high predictive value for severe disease and malaria-associated mortality. The impact of bednets and other interventions on epidemiology and the age-specific attributable reduction in mortality must be assessed. Better assays are needed for predicting protective immunity, involving more detailed characterization of the proteins and epitopes on these proteins involved in protective immunity.

VIII. PROSPECTS FOR MALARIA VACCINES

The human models (irradiated sporozoite and naturally acquired immunity) indicate that the development of a malaria vaccine is feasible. Genomics, proteomics, molecular biology, molecular immunology, vaccinology, population genetics, population biology, and quantitative epidemiology have created great expectations for the development, licensing, and deployment of effective malaria vaccines. It will be a formidable task to determine which antigens–epitopes from which stages of the life cycle are required for sustainable protection, how to measure immune responses that predict protection, which vaccine delivery systems are optimal, who and when in life to immunize, and how to establish surveillance systems able to assess the true impact of a malaria vaccine on public health. However, we believe that the next 10–25 years will see the development of effective malaria vaccines and that these will be used to control the effects of the disease worldwide and, when combined with other interventions, will be able to eradicate malaria from many areas.

The importance and difficulty of this task must not be underestimated. We are still far from controlling the enormous suffering and loss of life caused by malaria. We believe that the development of malaria vaccines will be crucial to successfully achieving these goals.

ACKNOWLEDGMENT

This work was supported in part (T.L.R.) by funds from the Naval Medical Research Center, work units 62787A.870.F.A0010,60000.000.000.A0062. The assertions here are private ones by the authors and are not to be construed as official or reflecting the views of the United States Navy or the Naval service.

REFERENCES

1. Aidoo, M., Lalvani, A., Allsopp, C. E., Plebanski, M., Meisner, S. J., Krausa, P. *et al.* (1995). Identification of conserved antigenic components for a cytotoxic T lymphocyte-inducing vaccine against malaria. *Lancet* **345**:1003–1007.

2. Alonso, P. L., Smith, T., Armstrong Schellenberg, J. R., Masanja, H., Mwankusye, S., Urassa, H. *et al.* (1994). Randomised trial of efficacy of SPf66 vaccine against *Plasmodium falciparum* malaria in children in southern Tanzania. *Lancet* **344**:1175–1181.

3. Baird, J. K. (1995). Host age as a determinant of naturally acquired immunity to *Plasmodium falciparum*. *Parasitol. Today* **11**:105–111.

4. Baird, J. K., Masbar, S., Basri, H., Tirtokusumo, S., Subianto, B., and Hoffman, S. L. (1998). Age-dependent susceptibility to severe disease with primary exposure to *Plasmodium falciparum*. *J. Infect. Dis.* **178**:592–595.

5. Berzins, K., and Perlmann, P. (1996). Malaria vaccines: Attacking infected erythrocytes. *In* "Malaria Vaccine Development: A Multi-Immune Response Approach," S. L. Hoffman, Ed., pp. 105–144. Washington, DC: ASM Press.

6. Bojang, K. A., Obaro, S. K., and D'Alessandro, U. (2001). Randomized, double-blind controlled trial of efficacy of RTS,S/ASO2 malaria vaccine against *Plasmodium falciparum* infection in semi-immune adult men in the Gambia: A randomised trial. *Lancet* **358**:1927–1934.

7. Bouharoun Tayoun, H., Oeuvray, C., Lunel, F., and Druilhe, P. (1995). Mechanisms underlying the monocyte-mediated antibody-dependent killing of *Plasmodium falciparum* asexual blood stages. *J. Exp. Med.* **182**:409–418.

8. Bowman, S., Lawson, D., Basham, D., Brown, D., Chillingworth, T., Churcher, C. M., *et al.* (1999). The complete nucleotide sequence of chromosome 3 of *Plasmodium falciparum*. *Nature* **400**(6744):532–538.

9. Breman, J. G., (2001). The ears of the hippopotamus: manifestations, determinants, and estimates of the malaria burden. *Am. J. Trop. Med. Hyg.* **64**(1–2 Suppl):1–11.

10. Brown, G. V., and Rogerson, S. J. (1996). Preventing cytoadherence of infected erythrocytes to endothelial cells and noninfected erythrocytes. *In* "Malaria Vaccine Development: A Multi-Immune Response Approach," S. L. Hoffman, Ed., pp. 145–166. Washington, DC: ASM Press.

11. Clark, I. A., and Schofield, L. (2000). Pathogenesis of malaria. *Parasitol. Today* **16**(10):451–454.

12. Clyde, D. F., McCarthy, V. C., Miller, R. M., and Hornick, R. B. (1973). Specificity of protection of man immunized against sporozoite-induced falciparum malaria. *Am. J. Med. Sci.* **266**:398–401.

13. Clyde, D. F., Most, H., McCarthy, V. C., and Vanderberg, J. P. (1973). Immunization of man against sporozoite-induced falciparum malaria. *Am. J. Med. Sci.* **266**:169–177.

14. Clyde, D. F., McCarthy, V. C., Miller, R. M., and Woodward, W. E. (1975). Immunization of man against falciparum and vivax malaria by use of attenuated sporozoites. *Am. J. Trop. Med. Hyg.* **24**:397–401.

15. Cohen, S., McGregor, I. A., and Carrington, S. (1961). Gamma-globulin and acquired immunity to human malaria. *Nature* **192**:733–737.

16. Degano, P., Schneider, J., Hannan, C. M., Gilbert, S. C., and Hill, A. V. (1999). Gene gun intradermal DNA immunization followed by boosting with modified vaccinia virus Ankara: Enhanced CD8+ T cell immunogenicity and protective efficacy in the influenza and malaria models. *Vaccine* **18**(7–8):623–632.

17. Doolan, D. L., Hoffman, S. L., Southwood, S., Wentworth, P. A., Sidney, J., Chestnut, R. W., *et al.* (1997). Degenerate cytotoxic T cell epitopes from *P. falciparum* restricted by HLA-A and HLA-B supertypes alleles. *Immunity* **7**:97–112.

18. Doolan, D. L., and Hoffman, S. L. (1999). IL-12 and NK cells are required for antigen-specific adaptive immunity against malaria initiated by CD8+ T cells in the *Plasmodium yoelii* model. *J. Immunol.* **163**:884–892.

19. Doolan, D. L., and Hoffman, S. L. (2000). The complexity of protective immunity against liver-stage malaria. *J. Immunol.* **165**:1453–1462.

20. Duffy, P. E., Craig, A. G., and Baruch, D. I. (2001). Variant proteins on the surface of malaria-infected erythrocytes—Developing vaccines. *Trends Parasitol.* **17**(8):354–356.

21. Edelman, R., Hoffman, S. L., Davis, J. R., Beier, M., Sztein, M. B., Losonsky, G. *et al.* (1993). Long-term persistence of sterile immunity in a volunteer immunized with X-irradiated *Plasmodium falciparum* sporozoites. *J. Infect. Dis.* **168**:1066–1070.

22. Egan, J. E., Hoffman, S. L., Haynes, J. D., Sadoff, J. C., Schneider, I., Grau, G. E. *et al.* (1993). Humoral immune responses in volunteers immunized with irradiated *Plasmodium falciparum* sporozoites. *Am. J. Trop. Med. Hyg.* **49**:166–173.

23. Felger, I., Irion, A., Steiger, S., and Beck, H. P. (1999). The epidemiology of multiple *Plasmodium falciparum* infections. 2. Genotypes of merozoite surface protein 2 of *Plasmodium falciparum* in Tanzania. *Trans. R. Soc. Trop. Med. Hyg.* **93**(Suppl 1):3–9.

24. Flanagan, K. L., Lee, E. A., Gravenor, M. B., Reece, W. H., Urban, B. C., Doherty, T. *et al.* (2001). Unique T cell effector functions elicited by *Plasmodium falciparum* epitopes in malaria-exposed Africans tested by three T cell assays. *J. Immunol.* **167**(8): 4729–4737.

25. Fries, L. F., Gordon, D. M., Richards, R. L., Egan, J. E., Hollingdale, M. R., Gross, M. et al. (1992). Liposomal malaria vaccine in humans: A safe and potent adjuvant strategy. *Proc. Natl. Acad. Sci. U.S.A.* **89**:358–362.

26. Gallup, J. L., and Sachs, J. D. (2001). The economic burden of malaria. *Am. J. Trop. Med. Hyg.* **64**(1–2 Suppl):85–96.

27. Gandon, S., Mackinnon, M. J., Nee, S., and Read, A. F. (2001). Imperfect vaccines and the evolution of pathogen virulence. *Nature* **414**(6865):751–756.

28. Gardner, M. J., Tettelin, H., Carucci, D. J., Cummings, L. M., Aravind, L., Koonin, E. V. *et al.* (1998). Chromosome 2 sequence of the human malaria parasite *Plasmodium falciparum*. *Science* **282**:1126–1132. [Published erratum appears in *Science* (1998) **282**:1827.]

28a. Gardner, M. J., Hall, N., Fung, E., White, O., Berriman, M., Hyman, R. W., Carlton, J. M., Pain, A., Nelson, K. E., Bowman, S., Paulsen, I. T., James, K., Eisen, J. A., Rutherford, K., Salzberg, S. L., Craig, A., Kyes, S., Chan, M.-S., Nene, V., Shallom, S. J., Suh, B., Peterson, J., Angiuoli, S., Pertea, M., Allen, J., Selengut, J., Haft, D., Mather, M. W., Vaidya, A. B., Martin, D., Fairlamb, A. H., Fraunholz, M. J., Roos, D. S., Ralph, S., McFadden, G. I., Cummings, L. M., Subramanian, G. M., Mungall, C., Venter, J. C., Carucci, D. J., Hoffman, S. L., Newbold, C., Davis, R. W., Fraser, C. M., Barrell, B. The genome sequence of the human malaria parasite Plasmodium falciparum. *Science.* In press, will be published on October 2, 2002.

29. Genton, B., al Yaman, F., Anders, R., Saul, A., Brown, G., Pye, D. *et al.* (2000). Safety and immunogenicity of a three-component blood-stage malaria vaccine in adults living in an endemic area of Papua New Guinea. *Vaccine* **18**(23):2504–2511.

30. Genton, B., Betuela, I., Felger, I. *et al.* (2002). A recombinant blood-stage malaria vaccine reduces *Plasmodium falciparum* density and exerts selective pressure on parasite populations in a Phase I/IIb trial in Papua New Guinea. *J. Infect. Dis.* **185**(6):820–827.

31. Graves, P., and Gelbrand H. (2001). Vaccines for preventing malaria. Cochrane Review Issue 3. Update Software, Oxford. Electronic Citation.

32. Herrington, D., Davis, J., Nardin, E., Beier, M., Cortese, J., Eddy, H. *et al.* (1991). Successful immunization of humans with irradiated sporozoites: Humoral and cellular responses of the protected individuals. *Am. J. Trop. Med. Hyg.* **45**:539–547.

33. Hoffman, S. L., Edelman, R., Bryan, J., Schneider, I., Davis, J., Sedegah, M. *et al.* (1994). Safety, immunogenicity, and efficacy of a malaria sporozoite vaccine administered with monophosphoryl lipid A, cell wall skeleton of mycobacteria and squalene as adjuvant. *Am. J. Trop. Med. Hyg.* **51**:603–612.

34. Hoffman, S. L., Franke, E. D., Hollingdale, M. R., and Druilhe, P. (1996). Attacking the infected hepatocyte. *In* "Malaria Vaccine Development: A Multi-Immune Response Approach," S. L. Hoffman, Ed., pp. 35–75. Washington, DC: ASM Press.

35. Hoffman, S. L., Rogers, W. O., Carucci, D. J., and Venter, J. C. (1998). From genomics to vaccines: Malaria as a model system. *Nat. Med.* **4**:1351–1353.

36. Hoffman, S. L., Goh, L. M., Luke, T. C., Schneider, I., Le, T. P., Doolan, D. L. *et al.* (2002). Protection of humans against malaria by immunization with radiation-attenuated *Plasmodium falciparum* sporozoites. *J. Infect. Dis.* **185**(8):1155–1164.

37. Hoffman, S. L., Subramanian, G. M., Collins, F. H., and Venter, J. C. (2002). *Plasmodium*, human, and Anopheles genomics and malaria. *Nature* **415**(6872):702–709.

38. Holder, A. A. (1996). Preventing merozoite invasion of erythrocytes. *In* "Malaria Vaccine Development: A Multi-Immune Response Approach," S. L. Hoffman, Ed., pp. 77–104. Washington, DC: ASM Press.

39. Jones, T. R., Obaldia, N., Gramzinski, R. A., Charoenvit, Y., Kolodny, N., Kitov, S. *et al.* (1999). Synthetic oligodeoxynucleotides containing CpG motifs enhance immunogenicity of a peptide malaria vaccine in *Aotus* monkeys. *Vaccine* 17:3065–3071.

40. Kaslow, D. C. (1996). Transmission-blocking vaccines. *In* "Malaria Vaccine Development: A Multi-Immune Response Approach," S. L. Hoffman, Ed., pp. 181–227. Washington, DC: ASM Press.

41. Kester, K. E., McKinney, D. A., Tornieporth, N., Ockenhouse, C. F., Heppner, D. G., Hall, T. *et al.* (2001). Efficacy of recombinant circumsporozoite protein vaccine regimens against experimental *Plasmodium falciparum* malaria. *J. Infect. Dis.* 183(4):640–647.

42. Khusmith, S., Sedegah, M., and Hoffman, S. L. (1994). Complete protection against *Plasmodium yoelii* by adoptive transfer of a CD8+ cytotoxic T cell clone recognizing sporozoite surface protein 2. *Infect. Immun.* 62:2979–2983.

43. Lalvani, A., Hurt, N., Aidoo, M., Kibatala, P., Tanner, M., and Hill, A. V. (1996). Cytotoxic T lymphocytes to *Plasmodium falciparum* epitopes in an area of intense and perennial transmission in Tanzania. *Eur. J. Immunol.* 26(4):773–779.

44. Lander, E. S., Linton, L. M., Birren, B., Nusbaum, C., Zody, M. C., Baldwin, J. *et al.* (2001). Initial sequencing and analysis of the human genome. *Nature* 409(6822):860–921.

45. Le, T. P., Coonan, K. M., Hedstrom, R. C., Charoenvit, Y., Sedegah, M., Epstein, J. E. *et al.* (2000). Safety, tolerability and humoral immune responses after intramuscular administration of a malaria DNA vaccine to healthy adult volunteers. *Vaccine* 18:1893–1901.

45a. Long, C. A. and Hoffman, S. L. (2002). Malaria from infants to genomics to vaccines. *Science* 297:345–347.

46. Lopez, J. A., Weilenman, C., Audran, R., Roggero, M. A., Bonelo, A., Tiercy, J. M. *et al.* (2001). A synthetic malaria vaccine elicits a potent CD8(+) and CD4(+) T lymphocyte immune response in humans. Implications for vaccination strategies. *Eur. J. Immunol.* 31(7): 1989–1998.

47. Malik, A., Egan, J. E., Houghten, R. A., Sadoff, J. C., and Hoffman, S. L. (1991). Human cytotoxic T lymphocytes against the *Plasmodium falciparum* circumsporozoite protein. *Proc. Natl. Acad. Sci. U.S.A.* 88:3300–3304.

48. McGregor, I. A., and Carrington, S. P. (1963). Treatment of East African *P. falciparum* malaria with West African human γ-globulin. *Trans. R. Soc. Trop. Med. Hyg.* 57(3):170–175.

49. Mendis, K., Sina, B. J., Marchesini, P., and Carter R. (2001). The neglected burden of *Plasmodium vivax* malaria. *Am. J. Trop. Med. Hyg.* 64(1–2 Suppl.):97–106.

50. Nardin, E. H., Oliveira, G. A., Calvo-Calle, J. M., Castro, Z. R., Nussenzweig, R. S., Schmeckpeper, B. *et al.* (2000). Synthetic malaria peptide vaccine elicits high levels of antibodies in vaccinees of defined HLA genotypes. *J. Infect. Dis.* 182(5):1486–1496.

51. Nardin, E. H., Calvo-Calle, J. M., Oliveira, G. A., Nussenzweig, R. S., Schneider, M., Tiercy, J. M. *et al.* (2001). A totally synthetic polyoxime malaria vaccine containing *Plasmodium falciparum* B cell and universal T cell epitopes elicits immune responses in volunteers of diverse HLA types. *J. Immunol.* 166(1):481–489.

52. Narum, D. L., Kumar, S., Rogers, W. O., Fuhrmann, S. R., Liang, H., Oakley, M. *et al.* (2001). Codon optimization of gene fragments encoding *Plasmodium falciparum* merozoite proteins enhances DNA vaccine protein expression and immunogenicity in mice. *Infect. Immun.* 69(12):7250–7253.

53. Nussenzweig, R. S., Vanderberg, J., Most, H., and Orton, C. (1967). Protective immunity produced by the injection of X-irradiated sporozoites of *Plasmodium berghei*. *Nature* 216:160–162.

54. O'Hagan, D., Singh, M., Ugozzoli, M., Wild, C., Barnett, S., Chen, M. *et al.* (2001). Induction of potent immune responses by cationic microparticles with adsorbed human immunodeficiency virus DNA vaccines. *J. Virol.* 75(19):9037–9043.

55. Ockenhouse, C. F., Sun, P. F., Lanar, D. E., Wellde, B. T., Hall, B. T., Kester, K. *et al.* (1998). Phase I/IIa safety, immunogenicity, and efficacy of NYVAC-Pf7, a pox-vectored, multiantigen, multistage vaccine candidate for *Plasmodium falciparum* malaria. *J. Infect. Dis.* 177:1664–1673.

56. Playfair, J. H. L. (1996). An antitoxic vaccine for malaria? *In* "Malaria Vaccine Development: A Multi-Immune Response Approach," S. L. Hoffman, Ed., pp. 167–180. Washington, DC: ASM Press.

57. Renia, L., Grillot, D., Marussig, M., Corradin, G., Miltgen, F., Lambert, P. H. *et al.* (1993). Effector functions of circumsporozoite peptide-primed CD4+ T cell clones against *Plasmodium yoelii* liver stages. *J. Immunol.* 150:1471–1478.

58. Richie, T. L., and Saul A. (2002). Progress and challenges for malaria vaccines. *Nature* 415(6872):694–701.

59. Rieckmann, K. H., Carson, P. E., Beaudoin, R. L., Cassells, J. S., and Sell, K. W. (1974). Sporozoite induced immunity in man against an Ethiopian strain of *Plasmodium falciparum*. *Trans. R. Soc. Trop. Med. Hyg.* 68:258–259.

60. Rieckmann, K. H., Beaudoin, R. L., Cassells, J. S., and Sell, D. W. (1979). Use of attenuated sporozoites in the immunization of human volunteers against falciparum malaria. *Bull. World. Health. Organ.* 57(Suppl. 1):261–265.

60a. Rogers, W. O., Weiss, W. R., Kumar, A., Aguiar, J. C., Tine, J. A., Gwadz, R., Harre, J. G., Gowda, K., Rathore, D., Kumar, S., and Hoffman, S. L. (2002). Protection of rhesus macaques against lethal Plasmodium knowlesi malaria by a heterologous DNA priming and poxvirus boosting immunization regimen. *Infect. Immun.* 70(8):4329–4335.

61. Romero, P., Maryanski, J. L., Corradin, G., Nussenzweig, R. S., Nussenzweig, V., and Zavala F. (1989). Cloned cytotoxic T cells recognize an epitope in the circumsporozoite protein and protect against malaria. *Nature* 341:323–325.

62. Sabchareon, A., Burnouf, T., Ouattara, D., Attanath, P., Bouharoun-Tayoun, H., Chantavanich, P. *et al.* (1991). Parasitologic and clinical human response to immunoglobulin administration in falciparum malaria. *Am. J. Trop. Med. Hyg.* 45(3):297–308.

63. Sakai, T., Horii, T., Hisaeda, H., Zhang, M., Ishii, K., Nakano, Y. *et al.* (1999). DNA immunization with *Plasmodium falciparum* serine repeat antigen: Regulation of humoral immune response by coinoculation of cytokine expression plasmid. *Parasitol. Int.* 48(1):27–33.

64. Schneider, J., Gilbert, S. C., Blanchard, T. J., Hanke, T., Robson, K. J., Hannan, C. M. *et al.* (1998). Enhanced immunogenicity for CD8+ T cell induction and complete protective efficacy of malaria DNA vaccination by boosting with modified vaccinia virus Ankara. *Nat. Med.* 4:397–402.

64a. Schofield, L., Hewitt, M. C., Evans, K., Siomos, M. A., and Seeberger, P. H. (2002). Synthetic GPI as a candidate anti-toxic vaccine in a model of malaria. *Nature* 418(6899):785–789.

65. Schofield, L., Villaquiran, J., Ferreira, A., Schellekens, H., Nussenzweig, R. S., and Nussenzweig, V. (1987). Gamma-interferon, CD8+ T cells and antibodies required for immunity to malaria sporozoites. *Nature* 330:664–666.

66. Sedegah, M., Sim, B. K. L., Mason, C., Nutman, T., Malik, A., Roberts, C. *et al.* (1992). Naturally acquired CD8+ cytotoxic T lymphocytes against the *Plasmodium falciparum* circumsporozoite protein. *J. Immunol.* 149:966–971.

67. Sedegah, M., Jones, T. R., Kaur, M., Hedstrom, R. C., Hobart, P., Tine, J. A. *et al.* (1998). Boosting with recombinant vaccinia increases immunogenicity and protective efficacy of malaria DNA vaccine. *Proc. Natl. Acad. Sci. U.S.A.* 95:7648–7653.

68. Sedegah, M., Weiss, W., Sacci, J. B., Jr., Charoenvit, Y., Hedstrom, R., Gowda, K. *et al.* (2000). Improving protective immunity induced by DNA-based immunization: Priming with antigen and GM-CSF encoding plasmid DNA and boosting with antigen expressing recombinant poxvirus. *J. Immunol.* 164:5905–5912.

69. Sim, B. K., Narum, D. L., Liang, H., Fuhrmann, S. R., Obaldia, N., III, Gramzinski, R. *et al.* (2001). Induction of biologically active antibodies in mice, rabbits, and monkeys by *Plasmodium falciparum* EBA-175 region II DNA vaccine. *Mol. Med.* 7(4):247–254.

70. Sinnis, P., and Nussenzweig, V. (1996). Preventing sporozoite invasion of hepatocytes. *In* "Malaria Vaccine Development: A Multi-Immune Response Approach," S. L. Hoffman, Ed., pp. 15–33. Washington, DC: ASM Press.

71. Snow, R. W., Omumbo, J. A., Lowe, B., Molyneux, C. S., Obiero, J. O., Palmer, A. *et al.* (1997). Relation between severe malaria morbidity in children and level of *Plasmodium falciparum* transmission in Africa. *Lancet* 349:1650–1654.

72. Stoute, J. A., Slaoui, M., Heppner, D. G., Momin, P., Kester, K. E., Desmons, P. *et al.* (1997). A preliminary evaluation of a recombinant circumsporozoite protein vaccine against *Plasmodium falciparum* malaria. *New England J. Medicine* 336:86–91.

73. Stoute, J. A., Kester, K. E., Krzych, U., Wellde, B. T., Hall, T., White, K. *et al.* (1998). Long-term efficacy and immune responses following immunization with the RTS,S malaria vaccine. *J. Infect. Dis.* **178**:1139–1144.

74. Valero, M. V., Amador, L. R., Galindo, C., Figueroa, J., Bello, M. S., Murillo, L. A. *et al.* (1993). Vaccination with SPf66, a chemically synthesised vaccine, against *Plasmodium falciparum* malaria in Colombia. *Lancet* **341**:705–710.

75. Venter, J. C., Adams, M. D., Myers, E. W., Li, P. W., Mural, R. J., Sutton, G. G. *et al.* (2001). The sequence of the human genome. *Science* **291**(5507):1304–1351.

76. Wang, R., Charoenvit, Y., Corradin, G., De la Vega, P., Franke, E. D., Hoffman, S. L. (1996). Protection against malaria by *Plasmodium yoelii* sporozoite surface protein 2 linear peptide induction of CD4+ T cell- and IFN-gamma-dependent elimination of infected hepatocytes. *J. Immunol.* **157**:4061–4067.

77. Wang, R., Doolan, D. L., Le, T. P., Hedstrom, R. C., Coonan, K. M., Charoenvit, Y. *et al.* (1998). Induction of antigen-specific cytotoxic T lymphocytes in humans by a malaria DNA vaccine. *Science* **282**:476–480.

78. Wang, R., Epstein, J. E., Baraceros, F. M., Gorak, E. J., Charoenvit, Y., Carucci, D. J. *et al.* (2001). Induction of CD4+ T cell dependent CD8+ Type 1 responses in humans by a malaria DNA vaccine. *Proc. Natl. Acad. Sci. U.S.A.* **98**(19):10817–10822.

79. Weiss, W. R., Sedegah, M., Beaudoin, R. L., Miller, L. H., and Good, M. F. (1988). CD8+ T cells (cytotoxic/suppressors) are required for protection in mice immunized with malaria sporozoites. *Proc. Natl. Acad. Sci. U.S.A.* **85**:573–576.

80. Weiss, W. R., Berzovsky, J. A., Houghten, R. A., Sedegah, M., Hollingdale, M., and Hoffman, S. L. (1992). A T cell clone directed at the circumsporozoite protein which protects mice against both *Plasmodium yoelii* and *Plasmodium berghei*. *J. Immunol.* **149**:2103–2109.

81. Weiss, W. R., Ishii, K. J., Hedstrom, R. C., Sedegah, M., Ichino, M., Barnhart, K. *et al.* (1998). A plasmid encoding murine granulocyte-macrophage colony-stimulating factor increases protection conferred by a malaria DNA vaccine. *J. Immunol.* **161**:2325–2332.

82. Wizel, B., Houghten, R., Church, P., Tine, J. A., Lanar, D. E., Gordon, D. M. *et al.* (1995). HLA-A2-restricted cytotoxic T lymphocyte responses to multiple *Plasmodium falciparum* sporozoite surface protein 2 epitopes in sporozoite-immunized volunteers. *J. Immunol.* **155**:766–775.

83. Wizel, B., Houghten, R. A., Parker, K., Coligan, J. E., Church, P., Gordon, D. M. *et al.* (1995). Irradiated sporozoite vaccine induces HLA-B8-restricted cytotoxic T lymphocyte responses against two overlapping epitopes of the *Plasmodium falciparum* surface sporozoite protein 2. *J. Exp. Med.* **182**:1435–1445.

84. Wu, S., Beier, M., Sztein, M. B., Galen, J., Pickett, T., Holder, A. A. *et al.* (2000). Construction and immunogenicity in mice of attenuated *Salmonella typhi* expressing *Plasmodium falciparum* merozoite surface protein 1 (MSP-1) fused to tetanus toxin fragment C. *J. Biotechnol.* **83**(1–2):125–135.

6 DISEASE STATES AND VACCINES: SELECTED CASES

PART J. Human Papillomaviruses[1]

LUTZ GISSMANN

Deutsches Krebsforschungszentrum, Im Neuenheimer Feld 242, 69120 Heidelberg, Germany

I. DESCRIPTION OF THE VIRUS

Human papillomaviruses (HPV) can induce different diseases: cutaneous warts (papillomas), condylomata acuminata (genital warts), laryngeal papillomas, and intraepithelial lesions at the skin or mucosa, such as cervical intraepithelial neoplasia. These diseases are primarily self-limited, benign epithelial proliferations. Only some of them can, under rare circumstances, develop into malignant tumors. The most relevant clinical entity is cancer of the uterine cervix arising from intraepithelial neoplasias (CIN).

The papillomavirus family consists of more than 100 members, of which 86 are specific for humans and 16 for animal species (e.g., cattle, dogs, or rabbits) (for a review, see ref 1). The different human papillomaviruses (HPV types) have been numbered according to their first descriptions as HPV 1, 2, etc. Classification is based upon the nucleotide sequences of the about 8000-base-pair-sized, double-stranded, circular DNA genome. Papillomaviruses share remarkable similarities in virus morphology, genome structure (6–8 early and 2 late genes), and biological properties (growth exclusively in differentiating epithelia). Yet the lesions that are arising as a consequence of an infection by a particular HPV type can differ in location, clinical appearance (exophytic warts, flat lesions), and natural history (self-limited and benign or exhibiting a potential toward malignant progression).

---

[1]The references cited in this article are mainly review articles. For original literature the reader should consult a source such as medline.

311

Human papillomaviruses are classified either according to their preferred tropism in skin or mucosa as cutaneous (e.g., HPV 1–4) or mucosal types (e.g., HPV 6, 16, 18) or according to their transforming potential as low-risk or high-risk types. Because the causal association between cervical cancer and infection by particular HPV types (16, 18, 31, 33, 45 and others) [2] has been firmly established through both experimental studies and epidemiological surveys (for reviews see refs 3 and 4), the use of virus-specific diagnostics and antiviral strategies in medical practice is now under investigation. There are attempts to develop virus-specific small molecules, such as antisense oligonucleotides, but the most straightforward approach for intervention is the development of virus-specific vaccines.

Other cancers are also associated with infection by high-risk HPV types, although the link is less consistent (e.g., squamous cancer of the head and neck) or they are less important in terms of total numbers (e.g., cancer of the penis and vulva). Obviously the demand for an HPV-specific vaccine varies according to the severity of the disease and clinical importance. Today the major focus regarding an HPV-specific vaccine(s) is toward the prevention and therapy of cervical cancer and its precursors.

II. DESCRIPTION AND EPIDEMIOLOGY OF THE DISEASES

Papillomavirus infections are transmitted through direct contact with the infected epithelium (skin, genital mucosa) but also through shared contaminated objects, such as bath mats in public swimming pools. It is well-established (and best documented in case of genital HPV infections) that only a fraction of infected individuals will develop a clinically manifest disease. On the other hand, infections are known to persist, in some instances, for life.

A. Cutaneous Warts

On the basis of clinical features, histology, and location, cutaneous warts can be classified as common warts, plantar warts, and flat warts. These lesions can be broadly correlated with infection by particular HPV types, such as HPV 2, HPV 1, or HPV 3, respectively (for a review, see ref 3). Deep plantar warts may cause considerable pain, yet treatment of warts is mostly for cosmetic reasons. Cutaneous warts are common (estimated incidence is 7–10%), especially in children, but they usually regress spontaneously in non-immunosuppressed patients. Standard procedures include chemical treatment (e.g., with salicylic acid), cryotherapy, or surgical intervention.

B. Laryngeal Papillomas

The vocal fold is the usual site of a laryngeal papillomavirus infection, resulting in hoarseness as the cardinal symptom. Because of the clinical characteristics and resistance to therapy, the disease is now referred to as recurrent respiratory papillomatosis (RRP), which can be of juvenile or adult onset. Either HPV type 11 or 6 is responsible for RRP, which can be life-threatening due to an obstructive growth within the airway. In affected children operations may be necessary every 7–10 days, and there are patients who have more than 100 endoscopic excisions [5]. Long-term administration of α-interferon appears to have some supporting effect toward remission in about 40% of patients, although interferon withdrawal may result in a rebound of papilloma growth.

Transmission of HPV 6 or HPV 11 infection to the upper respiratory tract resulting in RRP can occur during vaginal delivery from an infected birth canal (juvenile-onset RRP) or through sexual contact (adult-onset RRP). The incidence in children between 0 and 14 years of age was shown to vary more than 10-fold in different countries (0.1–2.8 per 100,000) (for a review, see ref 3).

C. Genital Warts

Genital warts (condylomata acuminata) are located in the genital and perianal regions, frequently at the mucocutaneous junction, and have a characteristic papillomatous appearance. They are normally visible with the naked eye, and often grow in clusters appearing as cauliflower-like lesions [6]. They are caused by the closely related HPV types 6 and 11, with a higher preponderance of HPV 6. Genital warts rarely develop into still benign "giant condylomas" first described by Buschke and Löwenstein in 1925. Anogenital warts are difficult to treat and recur with great frequency (38–81% of cases; reviewed in ref 7). Standard treatments include cytotoxic agents, such as podophyllin or 5-fluorouracil, mechanical intervention (surgery, cryotherapy, laser), or medical treatment using immune–modulatory substances (for reviews, see refs 7and 8). Genital warts are sexually transmitted and, consequently, have been on the rise during recent years (data cited in ref 9). The annual incidence per 100,000 inhabitants in the United States was calculated as 107, mounting to a lifetime risk of 10%.

D. Cervical Neoplasias

Precursor lesions to cervical cancer have been known to clinicians and pathologists since the beginning of the twentieth century. More recently, it was recognized that there is a continuum of cervical lesions with increasing severity, culminating in a malignant tumor. This observation triggered the introduction of the term of cervical intraepithelial neoplasia (CIN). It is assumed that there is a transition from low-grade to high-grade CIN (formerly called CIN I, II, III), although there is evidence that high-grade CIN may be the immediate result of infection with "high-risk" HPV types, such as HPV 16 or 18 (for a review, see ref 3). CIN is often detected by cytology ("Pap smear") performed during the screening programs that exist in most industrialized countries. In the case of cellular abnormalities suggesting a high-grade CIN, colposcopy is performed and eventually the final diagnosis will be made by histology of a colposcopically guided punch biopsy. Low-risk as well as high-risk HPV types are found in low-grade CIN, but there is a strong increase in the prevalence of high-risk types with increasing severity.

Because of the relatively poor sensitivity of the Pap test [10] and the close association between HPV and the development of high-grade CIN, detection of virus infection is discussed as a complementary or even alternative method for cervical cancer screening. HPV infection is diagnosed on the basis of detection of the viral DNA. Cervical swabs are tested for the presence of genomes of high-risk HPV types (e.g., HPV 16, 18, 31, 45) by nucleic acid hybridization with or without previous amplification by PCR.

Intraepithelial neoplasias similar to CIN occur also within the vagina, at the vulva, penis, or perianal region (called VAIN, VIN, PIN, or AIN, respectively), but the natural history is best studied for cervical lesions. There are approximately 1.2 million new cases of low-grade CIN and 300,000 cases of high-grade CIN each year in the United States. In contrast, the HPV infection rate is at least 30 times higher, as can be estimated from the number of abnormal smears [11].

Cervical cancer arises in about 90% of cases from squamous epithelial cells, though in some instances it is of glandular origin. Both tumor types are related to HPV infection, but the prevalence of particular types varies between the two variants. Altogether cervical cancer represents about 10% of cases of invasive tumors in women worldwide (estimated for 1990), accounting for approximately 370,000 new cases per year (estimated annual death rate is 210,000). There is more than a threefold higher incidence of this disease in developing countries [12]. This difference is mostly due to the lack of Pap screening programs in resource-poor settings.

III. MECHANISM OF PROTECTION AND SURROGATE END POINTS

A. Immune Biology of Papillomavirus Infections

Papillomaviruses are not particularly immunogenic during the natural course of infection, most likely because replication of theses viruses is confined to an immunological privileged site, i.e., the terminally differentiated keratinocytes. Antibodies to virus particles develop in only a fraction of cases of proven incident HPV infection and in women more likely than in men. On the other hand, a humoral immune response correlates with persistent infection and clinically visible disease (for a review, see ref 13). T-cell responses are not readily detected after HPV infection (reviewed in ref 14), yet there are several lines of evidence strongly suggesting that a Th1-type immune response is controlling the outcome of an HPV infection, i.e., the development of a clinically apparent disease [15]. (1) A local immune response, i.e., the influx of mononuclear cells and the appearance of Th1 cytokine patterns, correlates with regressing lesions. Immune modifiers leading to regression of genital warts in a proportion of patients were shown to induce these immunologic events, whereas early proteins of some HPV types seemed to negatively interfere with them. Systemic, HPV-specific, T-cell responses (cytotoxic T cells, delayed type hypersensitivity reaction measured by skin test) were also detected in patients with regressing lesions. It is unknown which event(s) triggers such immune responses during the natural course of infection. (2) HPV infection is more likely to persist and less likely to respond to treatment in individuals with immune suppression. (3) The fact that the individual HPV types originally identified on the basis of their genomic sequence (for a review, see ref 1) seem to represent antigenically distinct serotypes suggests that the immune system has been actively driving the diversity of this virus family.

B. Animal Experiments

Due to the species specificity of papillomaviruses, there is no animal model for HPV infections. Yet the striking similarities across members of the papillomavirus family justify that evidence by experiments with animal papillomaviruses in their natural hosts can be transferred to the situation in humans.

The strongest evidence that papillomavirus infections can be controlled by the immune system stems from experiments with animal papillomaviruses in their natural hosts, i.e., the cottontail rabbit papillomavirus (CRPV), bovine papillomavirus (BPV), and canine oral papillomavirus (COPV). Upon experimental inoculation, the animals develop papillomas with a time course and histopathologic features similar to these developed during natural infection. Efficient protection against tumor development or an enhancement of regression was obtained with inactivated virions, recombinant structural proteins, or, to some degree, early viral proteins (for a review, see ref 16). Protection was shown to be conferred by neutralizing antibodies directed against conformational epitopes. The most striking results have been obtained when virus-like particles (VLP) were used for immunization. VLPs assemble spontaneously from the major structural protein L1 together with, but also in the absence of, the minor capsid protein L2 upon expression in recombinant systems, such as yeast or vaccinia- or baculovirus-infected cells, as first described in 1991 (reviewed in ref 17). In numerous preclinical studies, VLPs of different animal and human papillomaviruses were demonstrated to induce L1-specific neutralizing antibodies with high efficiency when inoculated into small laboratory animals and also to induce protective immunity in their natural hosts (reviewed in ref 16). Regarding the need of a local immune response to prevent mucosal HPV infections, L1-specific serum antibodies were found to transude into vaginal secretions following systemic vaccination. Alternatively, a strong mucosal, mostly IgA-specific, antibody response was induced by intranasal immunization with VLPs. L1-specific T-cell responses have also been demonstrated, but quantitative data are just being generated as the appropriate reagents, such as L1-positive cell lines, become available.

These data demonstrate that virion-specific neutralizing antibodies are protecting against papillomavirus infection or at least against their clinical manifestation. There is also evidence from experiments with animal papillomaviruses in their natural host that a cell-mediated immune response directed against early viral proteins is critical for controlling the course of papillomavirus infections (for a review, see ref 16).

In the analysis of HPV-specific tumor rejection antigens, investigation has focused on the oncoproteins E6 and E7 of high-risk types for cervical cancer, i.e., HPV 16 and 18. These proteins are prime candidates as targets for immune therapy because their constitutive expression in tumors is indispensable for cell proliferation (for a review, see ref 4). Numerous studies in rodents have demonstrated that a protective or sometimes even therapeutic effect against the growth of HPV E6- and/or E7-positive syngeneic tumor cells can be induced by immunization with either one of these antigens. Vaccination proved to be successful with the purified proteins plus the appropriate adjuvants, recombinant viral or bacterial vectors, protein-derived peptides, or chimeric virus-like particles (CVLP; for a review, see refs 18 and 19). CVLPs have been generated by the expression of heterologous sequences fused with either the L1 or the L2 protein (for a review, see ref 19). HPV 16 E7-containing CVLPs were shown to induce (L1-specific) neutralizing antibodies, as well as E7-specific cytotoxic T cells; hence, they are considered a suitable vaccine in a scenario in which both prophylactic and therapeutic aspects are required (see later discussion and Fig. 1).

C. Vaccine Scenarios

Because benign skin warts usually regress spontaneously, vaccination against them is not considered a pressing medical problem. A different situation exists in immunosuppressed individuals, who often suffer from extended warts on the face and extremities. In organ transplant recipients waiting for their grafts, preventive vaccination would actually be an option prior to immunosuppression, provided that a small number of HPV types that are characteristically present in these patients can be identified. Immune therapy is not likely to be successful in such patients. Recurrent laryngeal papillomatosis (RRP) is extremely rare; thus, despite its severity, development of a prophylactic vaccine at present is not under consideration. As the same HPV types (i.e., HPVs 6 and 11) responsible for this disease are also causing genital warts, protection against RRP may be gained as a side effect in the case that vaccination against genital warts is introduced (see later discussion).

Cervical cancer is the most relevant HPV-induced disease. Hence, the development of vaccines against this tumor comprising the most prevalent cancer-related HPV types 16 and 18 is most advanced and will be the major

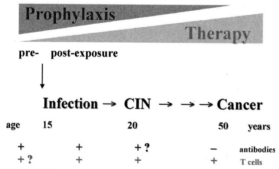

FIGURE 1 Vaccination strategies for prevention and therapy of cervical cancer, which develops after several years following infection by high-risk HPV types via low- and high-grade cervical intraepithelial neoplasia (here simplified as CIN). High-risk HPV is mostly transmitted by sexual intercourse; thus, prevention of infection has to occur in children or teenagers. Postexposure intervention also aims at the prevention of cervical neoplasia; treatment of CIN can also be considered as prophylaxis of cancer. Prophylactic (therapeutic) components of vaccination become less (more) important during the natural history of HPV infection, here also indicated by the need for neutralizing antibodies and a T-cell (Th1) response, respectively.

focus of this part of chapter 6. Based on the natural history of cervical cancer, several scenarios for vaccination can be conceived, i.e., pre- and postexposure prophylaxis and therapy (Fig. 1).

D. Prophylaxis

Genital HPV infections are, in most instances, transmitted venereally; hence, vaccination should occur prior to first intercourse and, as with any sexually transmitted infection, must include males and females in order to reduce the virus load within the population. Among women, there seems to be broad acceptance for a prophylactic HPV-specific vaccine aimed at the prevention of cervical cancer [20]. It remains to be explored whether there is a similar willingness among men. It may indeed be helpful to incorporate a component against genital warts (i.e., an HPV-6- and HPV-11-specific vaccine) to further increase the male incentive, as suggested by some researchers [20].

On the Basis of animal experiments (see earlier discussion), it is expected that neutralizing antibodies will prevent the infection; thus, VLPs are prime candidates for prophylactic vaccination to be given before exposure. It is expected that intramuscular injection of purified particles will lead to a systemic immune response and that there will be IgG transuding through mucosal surfaces, thus preventing cervical infection, although levels may vary during the menstrual cycle. Alternatively, immunization at mucosal sites (e.g., intranasally) is also being considered, possibly leading to a response of secretory IgA that, however, is not as long-lasting. It is tempting to speculate that mucosal delivery of the antigen can be achieved via genetic immunization using mucosotropic bacteria (e.g., apathogenic Salmonellae) or viruses as carriers for the HPV genes. This option is of particular relevance for application in developing countries, where the use of needles should be avoided and where there is the need for a stable vaccine in the absence of facilities for refrigerated storage. Ultimately, production of VLPs in plants also can be considered, allowing economical manufacture and eventually even the application as an edible vaccine. In fact, oral immunization of mice with purified VLPs has been successfully demonstrated. From the animal experiments mentioned earlier, no conclusions about the duration of a protective response against natural exposure can be drawn; hence, the immunization scheme still needs to be worked out following the results of the initial human trials.

According to the animal experiments, different HPV types represent serotypes, i.e., most likely there is no or only very limited cross-protection across types. Despite the plurality of different HPV types that were found to be responsible for the development of cervical cancer, the four most prevalent ones (HPVs 16, 18, 31, and 45) account for approximately 80% of the cancer cases worldwide, although certain geographic variations do exist [2]. It has been discussed that a "cervical cancer vaccine" consisting of four different types would be a reasonable compromise between sufficient protection and the expense that can realistically be incurred in clinical development. This concept undoubtedly applies to countries without implemented programs for the early detection of cervical cancer and its precursors. It is unclear, however, to what extent such screening programs existing in more developed countries can be modified if the four most frequent HPV types have been eliminated, yet 20% of women (those being infected with high-risk types other than 16, 18, 31, or 45) are still at risk for this disease. This consideration does not even include a possible interdependence between the individual HPV types as the dynamics of the HPV infections as a whole is difficult to predict if the distribution of the individual player has changed after successful vaccination [14]. Besides the use of multiple VLPs, there are attempts to develop a cross-protecting vaccine by engineering heterologous epitopes into a virus-like particle and generating intertypic VLPs. Alternatively, it has been shown that the minor structural protein L2 induces cross-neutralizing antibodies, although the titers are much lower than after L1-specific immunization.

Apart from the strategy for the primary prevention of HPV infection, there is also the option of a postexposure vaccination aiming to prevent cervical cancer (Fig. 1). The rationale behind this scenario is the fact that persistent infection with high-risk types over several years is the prerequisite for developing a high-grade cervical dysplastic lesion. In fact, the risk of women with persistent HPV infection for a cervical abnormality within the next 3–6 years is reported to be between 18 and 35%, depending on the population under study [21]. Whereas the virus structural proteins are produced as late events during vegetative virus replication, the early viral genes are expressed during persistent infection. Hence, it appears to be appropriate to induce a Th1-biased immune response directed against one or several of the early antigens. Based on the data from animal experiments, the E7 protein is the prime target in a number of vaccine projects (see later discussion). Other candidates include the E6 and E2 proteins, although an efficient immune response against the latter may actually select for E2-negative escape mutants. Silencing of E2 expression is thought to be one of the steps toward malignancy as E2 is often deleted from tumor cells, resulting in a constitutive expression of the HPV oncoproteins E6 and E7.

Given the young age of women with persistent infection that would be eligible for postexposure immunization, they are still at risk for a new infection. Thus, a vaccine eliciting both neutralizing antibodies and T cells directed against early viral proteins is the most logical approach in this situation. Some investigators favor a combined prophylactic and therapeutic vaccine in the preexposure scenario too. There may actually be some individuals who have already been infected prior to onset of their sexual life. Second and more importantly, sterilizing immunity may be difficult to obtain especially for an extended period of time following vaccination. Therefore, it appears to be advisable also to induce a cellular immune response directed against persistent infection as a "safety net." Treatment of CIN can be considered to prevent cervical cancer, but it will be discussed in this article from a mechanistic point of view and thus be included in the aspect of therapy.

Due to the long period between HPV infection and the development of cervical cancer, reduction of its incidence cannot be used as a clinical end point in human trials. Consequently, surrogate parameters are, besides the induction of neutralizing antibodies, the prevention of either CIN or even a new HPV infection.

E. Therapy

The ultimate goal of HPV-specific vaccination is undoubtedly the reduction of the incidence of cervical cancer. On the other hand, a significant worldwide effect of prophylactic vaccination would be visible only three decades after the launch of such programs [22]. Hence, it is mandatory that attempts toward immune therapy also are made that obviously will have a much earlier benefit. The studies on animal papillomaviruses in their natural host also provide arguments in favor of a successful immune therapy against papillomavirus-induced tumors in humans, even if the evidence is less conclusive than in the situation of prophylactic vaccination (for a review, see ref 16). The early proteins E6 and E7 are considered the most promising targets, and the development of first-generation vaccines has focused on these antigens. Aspects of therapeutic vaccination have been discussed elsewhere in more detail [18, 19].

Because of the high regression rate of low-grade CIN, therapy of intraepithelial neoplasia will most likely be restricted to high-grade lesions. It is generally believed that full-blown invasive cancer will be extremely difficult to treat by immune therapy only. It appears unlikely that the immune system is able to cope with the high burden of tumor cells because they can exert immunosuppressive functions and also have a complete or heterogeneous loss of HLA class I expression. On the other hand, one can assume that HPV-specific therapy is helpful as an adjunct to standard treatment aiming to reduce the risk of relapse and also to improve the quality of life. HPV-specific immune therapy of CIN appears more promising because expression of class I molecules is not compromised within such lesions.

Because a Th1-biased immune response is necessary for the induction of "antitumor" immunity, soluble proteins per se are insufficient and the addition of suitable adjuvants is mandatory (for a more detailed discussion, see ref 18). Therefore, investigators have looked for alternative means of antigen delivery. Actually, immunization of HPV early proteins fused to peptides that direct the molecules into the MHC class I pathway (e.g., the hsp65 heat shock protein) was shown to induce a CD8$^+$ immune response in mice. Immunization with fusions of E7 (or E6 plus E7) and the L2 minor protein also induce good CTL and T helper responses, especially when applied in a prime–boost scheme with recombinant vaccinia [23]. Use of long overlapping peptides or ex vivo loading of dendritic cells with proteins are other promising approaches that have been successfully explored in preclinical studies (for a review, see ref 19). Delivery of HPV DNA through recombinant viral vectors is also under investigation (for a review, see ref 19). HPV E6/E7-positive vaccinia virus has already been tested in clinical trials (see later discussion). If a vaccination strategy is being promoted that depends upon the viral oncogenes E6 and E7, appropriate modifications of the DNA must be included. It can be questioned, however, whether point mutations at biologically important sites of the protein (e.g., the Rb binding site) provide a sufficient level of safety. One should consider building in additional safety features, such as the use of minigenes or shuffled sequences.

In human trials the response to a therapeutic vaccination can easily be measured by cytology or colposcopy and by the disappearance of detectable HPV DNA. The analysis of immunological parameters (HPV-specific cytotoxic

T cells and T helper cells) will provide additional information about the duration of a response and about possible cross-reactivity and cross-protection between different HPV types.

IV. CURRENT STATE OF PLAY

As yet there is no HPV-specific vaccine on the market. However, several clinical trials evaluating the safety and immunogenicity of HPV 6, 11, and 16 vaccines are ongoing or have already been completed [24–32] (for reviews, see refs 18, 33, and 34; preliminary data are presented in refs 35 and 36).

Therapeutic trials were conducted with late-stage cervical cancer patients or with patients suffering from anogenital intraepithelial neoplasias or benign warts. Preliminary information on the efficacy has been published for some of the therapeutic trials, including patients with premalignant or benign HPV-induced lesions (for reviews, see refs 18, 33, and 34; preliminary data are presented in refs 35 and 36). Clinical response in some of the vaccine recipients was reported (1) in HLA-A2-positive patients with either HPV-16-positive high-grade cervical or vulval intraepithelial neoplasia treated with an HPV 16 E7-derived peptide vaccine [29]; (2) in patients with high-grade anal intraepithelial neoplasia after treatment with an hsp–HPV 16 E7 fusion protein; (3) in patients with genital warts, laryngeal papillomas, or cutaneous warts treated with an HPV 6 L2–E7 fusion protein; or (4) in genital warts patients immunized with HPV 6 L1 VLPs [30]. These data are clearly promising, but they also pose some questions as to the unexpected cross-type protection between HPV types or the mode of action of an HPV 6 L1-specific therapeutic vaccine [30]. Generally, all therapeutic vaccines were reasonably well-tolerated and induced an immune response in some of the patients. For more detailed information about the early therapeutic trials, see references 18 and 34.

Development of prophylactic vaccines is being pursued independently by the U.S. National Cancer Institute (NCI) and pharmaceutical companies in collaboration with researchers from academic institutions. Some details of the study designs and preliminary results have been disclosed [31, 32] (for preliminary data, see refs 35 and 36). The existing data can be summarized as follows: (1) Application of VLPs into humans is well-tolerated. (2) VLPs are highly immunogenic at relatively low doses (3×10 to 50 µg) and even in the absence of adjuvants. (3) Induction of high-titer neutralizing antibodies and of cell-mediated (Th1 and Th2 -type) immune responses was observed.

Based upon results of a dose-escalation trial in young healthy individuals using HPV 16 VLPs [32] and upon data from a cohort of women from Guanacaste, Costa Rica [37], the NCI is preparing a large population-based, double-blind, placebo-controlled prophylactic trial in Costa Rica (including 21,000 women, age 18–25 years) to prevent persistent HPV infection and CIN [38]. At a later stage, as a third arm of the trial, chimeric VLPs will be included that are composed of HPV 16 L1 and HPV 16 L2 fused to mutated forms of the HPV 16 E2 and E7 genes. In the prophylactic scenario, data on efficacy will only become available if an appropriate follow-up period (4 years in case of the NCI trial) has been completed.

REFERENCES

1. de Villiers, E. M. (2001). Taxonomic Classification of Papillomaviruses. *Papillomavirus Rep.* **12**:57–63.
2. Bosch, F. X., Manos, M. M., Munoz, N., Sherman, M., Jansen, A. M., Peto, J., Schiffman, M. H., Moreno, V., Kurman, R., and Shah, K. V. (1995). Prevalence of human papillomavirus in cervical cancer: A worldwide perspective. International Biological Study on Cervical Cancer (IBSCC) Study Group [see comments]. *J. Natl. Cancer Inst.* **87**:796–802.
3. IARC (1995). Human Papillomaviruses. Lyon, France: IARC.
4. zur Hausen, H. (1999). Papillomaviruses in human cancers. *Proc. Assoc. Am. Physicians* **111**:581–587.
5. Kashima, H., and Mounts, P. (1987). Tumors of the Head and Neck, Larynx, Lung and Esophagus and Their Possible Relation to HPV. *In* "Papillomaviruses and Human Disease," K. Syrjänen, L. Gissmann, and L. Koss, Eds., pp. 138–157. Berlin: Springer Verlag.
6. Sonnex, C. (1995). The clinical features of genital and perigenital human papillomavirus infection. *In* "Genital Warts: Human Papillomavirus Infection," A. Mindel, Ed., pp. 82–104. London: Edward Arnold.
7. Gross, G., von Krogh, G., and Barrasso, R. (1997). Genitoanal Lesions. *In* "Human Papillomavirus Infections in Dermatovenerology," G. Gross and G. von Krogh, Eds., pp. 389–416. Boca Raton, FL: CRC Press.
8. Edwards, L. (2000). Imiquimod in clinical practice. *J. Am. Acad. Dermatol.* **43**:S12–17.
9. Franco, E. L. (1996). Epidemiology of anogenital warts and cancer. *Obstet. Gynecol. Clin. North Am.* **23**: 597–623.
10. Naryshkin, S. (1997). The false-negative fraction for Papanicolaou smears: How often are "abnormal" smears not detected by a "standard" screening cytologist? *Arch. Pathol. Lab. Med.* **121**:270–272.
11. National Cancer Institute (2001). Cancer Facts. http://cis.nci.nih.gov/fact/5_16.htm.
12. Parkin, D. M., Pisani, P., and Ferlay, J. (1999). Estimates of the worldwide incidence of 25 major cancers in 1990. *Int. J. Cancer* **80**:827–841.
13. Dillner, J. (1999). The serological response to papillomaviruses. *Semin. Cancer Biol.* **9**:423–430.
14. Konya, J., and Dillner, J. (2001). Immunity to oncogenic human papillomaviruses. *Adv. Cancer Res.* **82**:205–238.
15. Stern, P. L., Brown, M., Stacey, S. N., Kitchener, H. C., Hampson, I., Abdel-Hady, E. S., and Moore, J. V. (2000). Natural HPV immunity and vaccination strategies. *J. Clin. Virol.* **19**:57–66.
16. Breitburd, F., and Coursaget, P. (1999). Human papillomavirus vaccines. *Semin. Cancer Biol.* **9**:431–444.
17. Schiller, J. T., and Roden, R. B. S. (1995). Papillomavirus-like particles. *Papillomavirus Rep.* **6**:121–128.
18. Gissmann, L., Osen, W., Müller, M., and Jochmus, I. (2001). Therapeutic vaccines for human papillomaviruses. *Intervirology* **44**:167–175.
19. Da Silva, D. M., Eiben, G. L., Fausch, S. C., Wakabayashi, M. T., Rudolf, M. P., Velders, M. P., and Kast, W. M. (2001). Cervical cancer vaccines: Emerging concepts and developments. *J. Cell. Physiol.* **186**:169–182.
20. Hoover, D. R., Carfioli, B., and Moench, E. A. (2000). Attitudes of adolescent/young adult women toward human papillomavirus vaccination and clinical trials. *Health Care Women Int.* **21**:375–391.
21. Dillner, J., Meijer, C. J., von Krogh, G., and Horenblas, S. (2000). Epidemiology of human papillomavirus infection. *Scand. J. Urol. Nephrol.* **205**:194–200.
22. Herrero, R. (2000). Personal communication.
23. van der Burg, S. H., Kwappenberg, K. M., O'Neill, T., Brandt, R. M., Melief, C. J., Hickling, J. K., and Offringa, R. (2001). Pre-clinical safety and efficacy of TA-CIN, a recombinant HPV16 L2E6E7 fusion protein vaccine, in homologous and heterologous prime–boost regimens. *Vaccine* **19**:3652–3660
24. Borysiewicz, L. K., Fiander, A., Nimako, M., Man, S., Wilkinson, G. W., Westmoreland, D., Evans, A. S., Adams, M., Stacey, S. N., Boursnell, M. E., Rutherford, E., Hickling, J. K., and

Inglis, S. C. (1996). A recombinant vaccinia virus encoding human papillomavirus types 16 and 18, E6 and E7 proteins as immunotherapy for cervical cancer. *Lancet* 347:1523–1527.

25. Steller, M. A., Gurski, K. J., Murakami, M., Daniel, R. W., Shah, K. V., Celis, E., Sette, A., Trimble, E. L., Park, R. C., and Marincola, F. M. (1998). Cell-mediated immunological responses in cervical and vaginal cancer patients immunized with a lipidated epitope of human papillomavirus type 16 E7. *Clin. Cancer Res.* 4:2103–2109.

26. Lacey, C. J., Thompson, H. S., Monteiro, E. F., O'Neill, T., Davies, M. L., Holding, F. P., Fallon, R. E., and Roberts, J. S. (1999). Phase IIa safety and immunogenicity of a therapeutic vaccine, TA-GW, in persons with genital warts. *J. Infect. Dis.* 179:612–618.

27. Thompson, H. S., Davies, M. L., Holding, F. P., Fallon, R. E., Mann, A. E., O'Neill, T., and Roberts, J. S. (1999). Phase I safety and antigenicity of TA-GW: a recombinant HPV6 L2E7 vaccine for the treatment of genital warts. *Vaccine* 17:40–49.

28. van Driel, W. J., Ressing, M. E., Kenter, G. G., Brandt, R. M., Krul, E. J., van Rossum, A. B., Schuuring, E., Offringa, R., Bauknecht, T., Tamm-Hermelink, A., van Dam, P. A., Fleuren, G. J., Kast, W. M., Melief, C. J., and Trimbos, J. B. (1999). Vaccination with HPV16 peptides of patients with advanced cervical carcinoma: Clinical evaluation of a phase I–II trial. *Eur. J. Cancer* 35:946–952.

29. Muderspach, L., Wilczynski, S., Roman, L., Bade, L., Felix, J., Small, L. A., Kast, W. M., Fascio, G., Marty, V., and Weber, J. (2000). A phase I trial of a human papillomavirus (HPV) peptide vaccine for women with high-grade cervical and vulvar intraepithelial neoplasia who are HPV 16 positive. *Clin. Cancer Res.* 6:3406–3416.

30. Zhang, L. F., Zhou, J., Chen, S., Cai, L. L., Bao, Q. Y., Zheng, F. Y., Lu, J. Q., Padmanabha, J., Hengst, K., Malcolm, K., and Frazer, I. H. (2000). HPV6b virus like particles are potent immunogens without adjuvant in man. *Vaccine* 18:1051–1058.

31. Evans, T. G., Bonnez, W., Rose, R. C., Koenig, S., Demeter, L., Suzich, J. A., O'Brien, D., Campbell, M., White, W. I., Balsley, J., and Reichman, R. C. (2001). A Phase 1 study of a recombinant viruslike particle vaccine against human papillomavirus type 11 in healthy adult volunteers. *J. Infect. Dis.* 183:1485–1493.

32. Harro, C. D., Pang, Y. Y., Roden, R. B., Hildesheim, A., Wang, Z., Reynolds, M. J., Mast, T. C., Robinson, R., Murphy, B. R., Karron, R. A., Dillner, J., Schiller, J. T., and Lowy, D. R. (2001). Safety and immunogenicity trial in adult volunteers of a human papillomavirus 16 L1 virus-like particle vaccine. *J. Natl. Cancer Inst.* 93:284–292.

33. Adams, M., Borysiewicz, L., Fiander, A., Man, S., Jasani, B., Navabi, H., Lipetz, C., Evans, A. S., and Mason, M. (2001). Clinical studies of human papilloma vaccines in pre-invasive and invasive cancer. *Vaccine* 19:2549-2556.

34. Tindle, R. (1999). "Vaccines for Human Papillomavirus Infection and Anogenital Disease." Austin, TX: RG Landes Company.

35. 18th International Papillomavirus Conference, Barcelona, 2000. Abstracts available at www.ipvsoc.org

36. 19th International Papillomavirus Conference, Florianopolis, 2001. Abstracts available at www.ipvsoc.org

37. Herrero, R., Hildesheim, A., Bratti, C., Sherman, M. E., Hutchinson, M., Morales, J., Balmaceda, I., Greenberg, M. D., Alfaro, M., Burk, R. D., Wacholder, S., Plummer, M., and Schiffman, M. (2000). Population-based study of human papillomavirus infection and cervical neoplasia in rural Costa Rica. *J. Natl. Cancer Inst.* 92:464–474.

38. Schiller, J. T., and Hildesheim, A. (2000). Developing HPV virus-like particle vaccines to prevent cervical cancer: A progress report. *J. Clin. Virol.* 19:67–74.

6 DISEASE STATES AND VACCINES: SELECTED CASES

PART K. *Helicobacter Pylori*

GIUSEPPE DEL GIUDICE AND RINO RAPPUOLI
IRIS Research Center, Chiron SpA, 53100 Siena, Italy

I. BACTERIUM AND INFECTION

The microaerophilic, spiral, gram-negative bacterium *Helicobacter pylori*, first described in 1982, belongs to the long list of new human pathogens that have been discovered during the past 25 years. *H. pylori* is a very specialized pathogen in that it colonizes only the gastric mucosa of humans, a harsh environment where other pathogens would not be able to survive, and consistently remains in the extracellular compartment beneath the mucus layer. Although discovered relatively recently, genetic evidence suggests that *H. pylori* is a very ancient bacterium that has coevolved with human beings [1]. Infection occurs early in life, most probably via the oral–fecal and/or the oral–oral route; the family appears to represent the environment in which transmission more frequently takes place [2]. Once established, the infection persists chronically unless an antibiotic treatment is undertaken. The absence of competition in the gastric environment and the ability to establish a lifelong colonization make *H. pylori* one of the most successful microorganisms. In fact, more than 50% of the human population worldwide are infected with *H. pylori*, and up to 90% in developing countries [3]. Most individuals chronically infected with *H. pylori* will remain symptomless. In about 15–20% of infected subjects, the chronic infection will evolve into peptic ulcer disease, in other individuals to atrophic gastritis, gastric adenocarcinoma, and gastric mucosa-associated lymphoid tissue (MALT) lymphoma [4].

Treatment of *H. pylori* infection is given only to symptomatic patients. It consists of daily administration of a proton pump inhibitor and two antibiotics for 1–2 weeks [5]. Despite its good efficacy (80% or more), drug treatment faces various orders of problems. The high number of tablets to be taken every day can seriously affect the patients' compliance. Side effects are not rare. An increasing number of antibiotic-resistant strains are being reported throughout the world. High reinfection rates have been reported in some areas of high transmission. Finally and importantly, because treatment is only given to symptomatic patients, patients without symptoms would still remain at risk of developing severe complications, such as atrophic gastritis and gastric cancer.

These problems could be overcome by the availability of prophylactic and/or therapeutic vaccines against *H. pylori*, which are under investigation by several groups following different approaches. The knowledge of the entire bacterial genome could be particularly useful in the identification of potential vaccine candidates [6], although the candidates tested so far have been identified during the pregenomic era. Pharmacoeconomic analyses have clearly shown that prophylactic vaccines against *H. pylori* would be highly cost-effective in the prevention of severe complications of the infection, such as peptic ulcer disease and gastric cancer [7, 8].

II. SOME ANTIGENS IMPORTANT IN VIRULENCE AND PATHOGENESIS

Very often antigens are selected as potential vaccine candidates because of some peculiar characteristics they have. For example, vaccine candidates are usually surface-exposed in order to be easily attacked by the immune system; they are abundant in order to be highly immunogenic; they are well-conserved in all bacterial isolates in order to induce a widely protective immune response; and they represent virulence factors involved in the pathogenesis of infection, so that their inactivation may block bacterial colonization and/or survival.

In the case of *H. pylori*, a potential vaccine candidate apparently fulfilling all of these characteristics is urease [9]. This enzyme is well conserved and consists of two moieties, UreA of 27 kDa and UreB of 62 kDa. Urease, which is released by bacteria through autolysis, is very abundant and represents 5–10% of the total protein content. Urease is involved in bacterial colonization: isogenic urease-negative strains are unable to colonize the stomach of gnotobiotic piglets. It also seems to be involved in the activation and adhesion of inflammatory cells at the site of gastric lesions. Despite its abundance, urease does not appear to be particularly immunogenic during natural infection. Anti-urease antibodies are found in approximately 50% of *H. pylori*-infected individuals [10], and very few urease-specific CD4+ T cells are usually isolated from the gastric mucosa of individuals with chronic gastritis or peptic ulcer [11, 12]. Up to now urease has been the *H. pylori* antigen most extensively investigated as a potential vaccine candidate in animals, and it has also been tested in clinical trials in humans (see discussion to follow).

Another well-conserved antigen is the neutrophil-activating protein (NAP). This four-helix bundle protein of 17 kDa assembles to form dodecamers. NAP has an iron-binding capacity, induces chemotaxis and activation of neutrophils and monocytes, and stimulates production of reactive oxigen intermediates. This activity is potentiated by interferon-γ (IFN-γ) and tumor necrosis factor-α

(TNF-α) [13]. Thus, NAP may induce local gastric inflammation through local recruitment and activation of neutrophils and monocytes.

H. pylori isolates have been divided into two large categories. The so-called type I strains are associated with the most severe complications of the infection (peptic ulcer, atrophic gastritis, gastric cancer). In contrast, type II strains are isolated more frequently from individuals with mild forms of gastritis. Type I strains contain a gene (*cagA*) encoding the so-called cytotoxin-associated gene (CagA), which resides in a 40-kb fragment present in the type I, but not type II, strains [1]. This region (*cag*) is a pathogenicity island (PAI) similar to those found in other pathogens, such as enteropathogenic *Escherichia coli, Bordetella pertussis*, etc. The majority (70% or more) of clinical isolates are CagA-positive. CagA is highly immunogenic at both the antibody and cellular level; this immune response has been considered as a marker of severe infection. *H. pylori cag* PAI induces inter leukin (IL-8) production by epithelial cells, activation of NF-κB, remodeling of the cytoskeleton, and formation of cellular pedestals. Importantly, *cag* PAI mediates the translocation of CagA into the eukaryotic cell, where CagA is phosphorylated [14, 15]. This may explain the association that has been found between infection with CagA-positive strains and the development of severe gastric pathology. In agreement with epidemiological observations made in infected humans, CagA-positive strains in Mongolian gerbils have been shown to cause severe gastric pathology including cancer, whereas CagA-negative strains induce only mild gastritis [16]. The active inoculation of CagA into the eukaryotic cell represents a novel mechanism of interaction between a strictly extracellular pathogen and the cell and may also account for the high immunogencity of CagA observed in infected individuals.

H. pylori produces a cytotoxin that induces vacuolation of cells *in vitro* (vacuolating toxin, VacA). The mature protein contains a moiety of 37 kDa essential for the toxic activity and a 58-kDa portion that binds to target cells [15, 17]. VacA monomers oligomerize to form esa- and eptameric flower-like structures. Its sequence is well-conserved among different isolates, except for the midregion of the 58-kDa moiety, which expresses allelic variation. Clinical and experimental data clearly show that both alleles are equally toxic [17]; they could well represent the evolutionary expression at the bacterial level of a genetic polymorphism at the level of the specific host cell receptor for this toxin.

III. AN INEFFICIENT NATURAL IMMUNE RESPONSE TO *H. PYLORI*

Our knowledge of the *H. pylori* infection derives mainly from symptomatic adults chronically infected for a long time. Very little is known of the early stages of infection during the pediatric ages or, thus, of the natural history from the establishment of a successful colonization to the appearance of the overt gastric disease. Infection with *H. pylori* induces an antibody response (detectable in the serum and at the mucosal level) that is utilized for diagnostic purposes. However, the immune response naturally induced by the infection does not appear to confer protective immunity. For example, ferrets successfully treated for a naturally occurring *H. mustelae* infection are still suceptible to reinfection with the same bacterium [18]. Along the same line, high reinfection rates have been reported in individuals living in areas with a high prevalence of *H. pylori* infection [19]. On the contrary, there is a body of evidence suggesting that gastric lesions induced by infection with *H. pylori* could be mediated by the immune response against the bacterium [20].

In any case, the events leading to the establishment of the immune response to *H. pylori* and to the development of gastric lesions are still poorly understood. Natural *H. pylori* infection induces up-regulation of the B7-2 costimulatory molecule and expression of MHC class II antigens on gastric epithelial cells, with an increase in the number of activated mononuclear cells infiltrating the lamina propria. In addition, there is the appearance of CD4+ T lymphocytes, mostly with a phenotype of activated and memory cells, and B lymphocytes. These cells become organized to form lymphoid follicles, which at the most advanced stages become macroscopically evident, and represent the site at which gastric lymphoma may eventually develop in some individuals.

Several lines of evidence have clearly shown that gastric CD4+ T cells from experimentally infected animals and from naturally infected individuals exhibit a Th1-type functional phenotype, with the production of pro-inflammatory cytokines such as IFN-γ, TNF-α, TGF-β, IL-18 [21–23]. Th1 CD4+ T-cell clones are isolated more frequently from the gastric mucosa of patients with peptic ulcer disease, whereas Th0–Th2 clones were found more frequently in patients with milder gastric disease [11, 12]. Remarkably, these clones frequently were specific for CagA, and only few recognized urease and other antigens. A predominant activation of Th1 CD4+ cells has also been observed in murine models of infection with *H. pylori* [24, 25]. The hypothesis that the pathological events observed in the stomach could be mediated by the immune response triggered by *H. pylori* is substantiated by studies carried out in knock-out strains of mice. In fact, the gastric pathology is more severe in IL-10 knock-out mice infected with *H. felis* [26] and in IL-4 [27] or IRF-1 knock-out mice [28] infected with *H. pylori*, suggesting an active role of Th1 cytokines in the induction of the gastric pathology and/or also that Th2-type cytokines may counteract the pro-inflammatory effects of Th1-associated cytokines. Mice naturally infected with the intestinal nematode *Heligmosomoides polygyrus* and, thus, naturally prone to an immune response more polarized toward a Th2-type functional phenotype, exhibit a gastric pathology significantly less severe than that observed in mice not co-infected with the intestinal parasite when they are experimentally challenged with *H. felis* [29].

It can, thus, be concluded that *H. pylori*-induced gastric pathology is associated with a strong Th1-type immune response. In addition, several *H. pylori* antigens (e.g., NAP, *cag* PAI, urease) can actively intervene in potentiating the loop of the pro-inflammatory events mediated by IFN-γ and other cytokines and chemokines, including IL-8. Other antigens, such as NAP, can directly (or synergistically with IFN-γ and TNF-α) enhance the recruitment and activation of inflammatory cells at the colonization sites.

Finally, it should be mentioned that some bacterial epitopes, e.g., of LPS, are shared with the self, such as those of the Lewis blood group, which are expressed by most cells, including gastric epithelial cells. Anti-*H. pylori* (LPS) antibodies cross-reacting with gastric tissues have been demonstrated in naturally infected humans and in experimentally infected animals [30]. The relevance of these cross-reacting antibodies in the induction of chronic gastritis following *H. pylori* infection remains, however, to be formally proven.

IV. FEASIBILITY OF PROPHYLACTIC AND THERAPEUTIC VACCINATIONS IN ANIMAL MODELS

A very large body of evidence exists that protection against *H. pylori* infection can be achieved both prophylactically and therapeutically in animal models. This was first demonstrated in mice infected with *H. felis* (a species that does not infect humans and does not express virulence factors such as VacA and *cag* PAI) immunized orally with bacterial lysates together with cholera toxin (CT) as a mucosal adjuvant. This was then proven with *H. pylori* strains freshly isolated from peptic ulcer patients and adapted to the mice. Prophylactic and therapeutic protection against *H. pylori* has now been demonstrated for a wide variety of native or recombinant antigens, such as urease (and recombinant subunits), heat shock proteins, native and recombinant VacA, CagA, NAP, catalase, and others [31]. More protective antigens will be identified thanks to the information derived from the *H. pylori* genomes. Most of the work on vaccines carried out so far has focused on the oral route of immunization. Work in animals has, however, shown that other routes of immunization, i.e., intranasal, rectal, or parenteral, can be equally effective in conferring protection against *H. pylori* or *H. felis* [31]. However, conflicting results of protection after immunization with urease have been obtained in monkeys.

If prophylactic and therapeutic protection against *H. pylori* has been achieved in different animal models using different antigens and different routes of immunization, the immune mechanisms mediating this protection still remain a conundrum. Specific immunoglobin A (IgA) or IgG antibodies produced locally in the gastric environment have been originally hypothesized. However, protection against *H. pylori* can also be achieved in IgA-deficient mice. In addition, humans with congenital deficiency of IgA antibody do not suffer from a more severe *H. pylori* gastric pathology [32]. Finally, protection against *H. pylori* or *H. felis* can be achieved in μ-MT mice, which are totally unable to produce any antibody [33]. Protection requires the presence of a functional MHC class II-restricted CD4+ T-cell compartment. The exact nature and function of these CD4+ cells remain unclear. It has been proposed that CD4+ T cells bearing the $\alpha 4\beta 7^{hi}$ phenotype are required for protection against *H. felis* after oral immunization [34]. It was shown that homing of these cells in the gastric mucosa was mediated by the mucosal adjuvant used, the CT, but not by *H. felis*; in addition, *in vivo* treatment with a monoclonal antibody against the $\alpha 4\beta 7$ integrin suppressed the protection. It has been shown that a polarized CD4+ Th2-type response is preferentially induced after oral immunization with *H. pylori* antigens and mucosal adjuvants, such as CT [24], LT, or nontoxic LT mutants [31]. In agreement with the observation that mice knocked out at the level of Th2-type cytokines, such as IL-4 and IL-10, have a severe gastric pathology after *Helicobacter* infection [26, 27], it can be

TABLE 1 Clinical Trials in *H. pylori*-Negative and *H. pylori*-Positive Subjects[a]

H. pylori vaccine antigen	Adjuvant or expression system	Dosage and regimen	*H. pylori* status in healthy volunteers	Major findings	Reference
Recombinant urease	Wild-type LT[b]	20, 60, or 180 mg plus 5 μg of LT, 4 times weekly	Positive	Diarrhea in 66% of subjects Serum anti-urease IgA antibodies in subjects receiving 60 or 180 mg Reduction in *H. pylori* load (no correlation with antibody response) No bacterial eradication	37
Urease	*Salmonella typhi* ΔphoP/phoQ	(1–4) × 10^{10} CFU, once Same + boost (day 15) with 60 mg of recombinant urease plus 2.5 μg of wild-type LT	Negative	Gastrointestinal symptoms in 3 out of 6 subjects Antibody response to *Salmonella* antigens No antibody response to urease Diarrhea in 2 out of 3 subjects No antibody response to urease	35
Urease	*Salmonella typhimurium* ΔphoP/phoQ	2 to 8 × 10^7 CFU, once	Negative	Fever in 2 out of 6 subjects Antibody response to *Salmonella* in 5 out of 6 subjects Weak antibody response to urease in 3 out of 6 subjects	36
Whole-cell, formalin-inactivated	LTR192G mutant	2.5 × 10^6 to 2.5 × 10^{10}, 3 times on days 0, 14, 28	Negative and Positive	Diarrhea in 6 out of 41 subjects *H. pylori*-specific mucosal IgA increase at the highest dose No increase of *H. pylori*-specific serum antibodies PBMC proliferative response in 5 out of 10 subjects (*H. pylori*-positive) IFN-γ production in 7 out of 10 subjects (*H. pylori*- positive) No bacterial eradication	38

[a]In all trials carried out to data vaccines were delivered orally.
[b]LT, heat-labile enterotoxin of *Escherichia coli*.

proposed that protective immunizations induce Th2-type cytokines, which intervene by inhibiting the activation of Th1 cells and macrophages for the production of the pro-inflammatory cytokines that are normally produced during *H. pylori* infection. The knowledge of the immunological mechanisms behind the prophylactic and therapeutic protection against *H. pylori* after immunization will be of critical importance for the design of efficacious vaccines to be tested in the clinical trials that will take place in the next years.

V. CLINICAL EXPERIENCE WITH *H. PYLORI* VACCINES: PRESENT AND FUTURE

Up to now, very few vaccination studies have been conducted in humans (Table 1). Two studies using genetically attenuated *Salmonella* spp. expressing urease have shown no or poor immunogenicity [35, 36]. One study has been carried out with urease given orally along with wild-type LT as an adjuvant to *H. pylori*-infected subjects with chronic gastritis [37]. Two-thirds of the subjects developed diarrhea due to the wild-type LT. Urease-specific serum IgA and circulating antibody-secreting cells were shown in subjects receiving the highest doses of antigen (60 or 180 mg given four times weekly). Partial reduction in the number of bacterial colonies was observed, paradoxically, in the group of subjects receiving the lower dose of urease (20 mg). Finally, immunization did not induce amelioration of the preexisting gastric inflammation. The reasons for the very partial results of this trial are not clear. They may be linked to the vaccination regimen employed, to the antigen used, to the mucosal adjuvant utilized, or even to difficulties inherent to the therapeutic approach of vaccination itself that are not understood yet. More recently, the results of a trial using an oral *H. pylori* whole-cell inactivated vaccine adjuvanted with an LT mutant have been reported [38]. The vaccine, given to both *H. pylori*-infected and -uninfected subjects, was immunogenic, but it induced diarrhea (due to the adjuvant used) and did not exhibit therapeutic efficacy in the infected individuals.

Other prophylactic and therapeutic vaccine formulations are currently being tested and will be tested in the near future in humans. The studies will likely answer some of the most crucial questions about the host–microbe interactions. The final hope is that these new vaccines will show the expected efficacy against *H. pylori*-induced diseases and will permit the elimination of this pathogen, which has cohabited with humans for more than 100,000 years.

REFERENCES

1. Covacci, A., Telford, J. L., Del Giudice, G., Parsonnet, J., and Rappuoli, R. (1999) *Helicobacter pylori* virulence and genetic geography. *Science* **284**:1328–1333.
2. Tindberg, Y., Bengtsson, C., Granath, F., Blennow, M., Nyrén, O., and Grandström, M. (2001). *Helicobacter pylori* infection in Swedish school children: Lack of evidence of child-to-child transmission outside the family. *Gastroenterology* **121**:310–316.
3. Parsonnet, J. (1995). The incidence of *Helicobacter pylori* infection. *Aliment. Pharmacol. Ther.* **9**(Suppl. 2):45–51.

4. Uemura, N., Okamoto, S., Yamamoto, S., Matsumura, N., Yamaguchi, S., Yamakido, M., Taniyama, K., Sasaki, N., and Schlemper, R. J. (2001) *Helicobacter pylori* infection and the development of gastric cancer. *New. England J. Medicine* **345**:784–789.

5. Graham, D. Y. (2000). Therapy of *Helicobacter pylori*: Current status and issues. *Gastroenterology* **118**:S2–S8.

6. Alm, R. A., and Trust, T. J. (1999). Analysis of the genetic diversity of *Helicobacter pylori*: The tale of two genomes. *J. Mol. Biol.* **77**:834–846.

7. Rupnow, M. F. T., Owens, D. K., Shachter, R., and Parsonnet, J. (1999). *Helicobacter pylori* vaccine development and use: A cost-effectiveness analysis using the Institute of Medicine methodology. *Helicobacter* **4**:272–280.

8. Stratton, K. R., Durch, J. S., and Lawrence, R. S., Eds. (2000). "Vaccines for the 21st Century. A Tool for Decisionmaking." Washington, DC: National Academy Press.

9. Mobley, H. L. T. (2001). Urease. *In Helicobacter pylori.* Physiology and genetics H. L. T. Mobley, G. L. Mendz, and S. L. Hazell, Eds., pp. 179–191. Washington, DC: ASM Press.

10. Leal-Herrera, Y., Torres, J., Perez-Perez, G., Gomez, A., Monath, T., Tapia-Conyer, R., and Munoz, O. (1999). Serologic IgG response to urease in *Helicobacter pylori*-infected persons from Mexico. *Am. J. Trop. Med. Hyg.* **60**:587–592.

11. D'Elios, M. M., Manghetti, M., De Carli, M., Costa, F., Baldari, C. T., Burroni, D., Telford, J. L., Romagnani, S., and Del Prete, G. (1997). T helper 1 effector cells specific for *Helicobacter pylori* in the gastric antrum of patients with peptic ulcer disease. *J. Immunol.* **158**:962–967.

12. D'Elios, M. M., Manghetti, M., Almerigogna, F., Amedei, A., Costa, F., Burroni, D., Baldari, C. T., Romagnani, S., Telford, J. L., and Del Prete, G. (1997). Different cytokine profile and antigen-specificity repertoire in *Helicobacter pylori*-specific T cell clones from the antrum of chronic gastritis patients with or without peptic ulcer. *Eur. J. Immunol.* **27**:1751–1755.

13. Satin, B., Del Giudice, G., Della Bianca, V., Dusi, S., Laudanna, C., Tonello, F., Kelleher, D., Rappuoli, R., Montecucco, C., and Rossi, F. (2000). The neutrophil-activating protein (HP-NAP) of *Helicobacter pylori* is a protective antigen and a major virulence factor. *J. Exp. Med.* **191**:1567–1576.

14. Covacci, A., and Rappuoli, R. (2000). Tyrosine-phosphorylated bacterial proteins: Trojan horses for the host cells. *J. Exp. Med.* **191**:587–592.

15. Montecucco, C., and Rappuoli, R. (2001). Living dangerously: How *Helicobacter pylori* survives in the human stomach. *Nature Rev. Mol. Cell Biol.* **2**:457–466.

16. Ogura, K., Maeda, S., Nakao, M., Watanabe, T., Tada, M., Kyutoku, T., Yoshida, H., Shiratori, Y., and Omata, M. (2000). Virulence factors of *Helicobacter pylori* responsible for gastric diseases in Mongolian gerbils. *J. Exp. Med.* **192**:1601–1610.

17. Reyrat, J. M., Pelicic, V., Papini, E., Montecucco, C., Rappuoli, R., and Telford, J.L. (1999). Towards deciphering the *Helicobacter pylori* cytotoxin. *Mol. Microbiol.* **34**:197–204.

18. Batchelder, M., Fox, J., Hayward, A., Yan, L., Shames, B., Murphy, J. C., and Palley, L. (1996). Natural and experimental *Helicobacter mustelae* reinfection following successful antimicrobial eradication in ferrets. *Helicobacter* **1**:34–42.

19. Ramirez-Ramos, A., Gilman, R. H., Leon-Barua, R., Recavarren-Arce, S., Watanabe, J., Salazar, G., Checkley, W., McDonald, J., Valdez, Y., Cordero, L., and Carrazco, J. (1997). Rapid recurrence of *Helicobacter pylori* infection in Peruvian patients after successful eradication. Gastrointestinal Physiology Working Group of the Universidad Peruana Cayetano Heredia and the Johns Hopkins University. *Clin. Infect. Dis.* **25**:1027–1031.

20. Ernst, P. B., and Gold, B. D. (2000). The disease spectrum of *Helicobacter pylori*: The immunopathogenesis of gastroduodenal ulcer and gastric cancer. *Annu. Rev. Microbiol.* **54**:615–640.

21. Bamford, K. B., Fan, X., Crowe, S. E., Leary, J. F., Gourley, W. K., Luthra, G. K., Brooks, E. G., Graham, D. Y., Reyes, V. E., and Ernst, P. B. (1998). Lymphocytes in the human gastric mucosa during *Helicobacter pylori* infection have a T helper cell 1 phenotype. *Gastroenterology* **114**:482–492.

22. Lindholm, C., Quiding-Jarbrink, M., Lonroth, H., Hamlet, A., and Svennerholm, A. M. (1998). Local cytokine response in *Helicobacter pylori* infected subjects. *Infect. Immun.* **66**:5964–5971.

23. Tomita, T., Jackson, A. M., Hida, N., Hayat, M., Dixon, M. F., Shimoyama, T., Axon, A. T. R., Robinson, P. A., and Crabtree, J. E. (2001). Expression of IL-18, a Th1 cytokine, in human gastric mucosa is increased in *Helicobacter pylori* infection. *J. Infect. Dis.* **183**:620–627.

24. Saldinger, P. F., Porta, N., Launois, P., Louis, J. A., Wanders, G. A., Bouzourene, H., Michetti, P., Blum, A. L., and Corthesy-Theulaz, I. E. (1998). Immunization of BALB/c mice with *Helicobacter* urease B induces a T helper 2 response absent in *Helicobacter* infection. *Gastroenterology* **115**:891–897.

25. Eaton, K. A., Mefford, M., and Thenenot, T. (2001). The role of T cell subsets and cytokines in the pathogenesis of *Helicobacter pylori* gastritis in mice. *J. Immunol.* **166**:7456–7461.

26. Berg, D. J., Lynch, N. A., Lynch, R. G., Mauricella, D. M. (1998). Rapid development of severe hyperplastic gastritis with gastric epithelial dedifferentiation in *Helicobacter felis*-infected IL-10-/- mice. *Am. J. Pathol.* **152**:1377–1386.

27. Smythies, L. E., Waites, K. B., Lindsey, J. R., Harris, P. R., Ghiara, P., and Smith, P. D. (2000). *Helicobacter pylori*-induced mucosal inflammation is Th1 mediated and exacerbated in IL-4, but not IFN-γ, gene-deficient mice. *J. Immunol.* **165**:1022–1029.

28. Sommer, F., Faller, G., Roellinghoff, M., Kirchner, T., Mak, T. W., and Lohoff, M. (2001). Lack of gastritis and of an adaptive immune response in interferon regulatory factor-1-deficient mice infected with *Helicobacter pylori*. *Eur. J. Immunol.* **31**:396–402.

29. Fox, J. G., Beck, P., Dangler, C. A., Whary, M. T., Wang, T. C., Shi, H. N., and Nagler-Anderson, C. (2000). Concurrent enteric helminth infection modulates inflammation and gastric immune responses and reduces helicobacter-induced gastric atrophy. *Nature Med.* **6**: 536–542.

30. Moran, A. P., and Prendergast, M. M. (2001). Molecular mimicry in *Campylobacter jejuni* and *Helicobacter pylori* lipopolysaccharides: Contribution of gastrointestinal infections to autoimmunity. *J. Autoimmunity* **16**:241–256.

31. Del Giudice, G., Covacci, A., Telford, J. L., Montecucco, C., Rappuoli, R. (2001). The design of vaccines against *Helicobacter pylori* and their development. *Annu. Rev. Immunol.* **19**: 523–563.

32. Bogstedt, A. K., Nava, S., Wadstrom, T., and Hammarstrom, L. (1996). *Helicobacter pylori* infection in IgA deficiency: Lack of role for the secretory immune system. *Clin. Exp. Immunol.* **105**:202–204.

33. Ermak, T. H., Giannasca, P. J., Nichols, R., Myers, G. A., Nedrud, J., Weltzin, R., Lee, C. K., Kleanthous, H., and Monath, T. P. (1998). Immunization of mice with urease vaccine affords protection against *Helicobacter pylori* infection in the absence of antibodies and is mediated by MHC class II-restricted responses. *J. Exp. Med.* **188**:2277–2288.

34. Michetti, M., Kelly, C. P., Kraehenbuhl, J. P., Bouzourene, H., and Michetti, P. (2000). Gastric mucosal α4β7-integrin-positive CD4 T lymphocytes and immune protection against *Helicobacter* infection in mice. *Gastroenterology* **119**:109–118.

35. Di Petrillo, M. D., Tibbetts, T., Kleanthous, H., Killeen, K. P., and Hohmann, E. L. (1999). Safety and immunogenicity of *phoP/phoQ*-deleted *Salmonella typhi* expressing *Helicobacter pylori* urease in adult volunteers. *Vaccine* **18**:449–459.

36. Angelakopoulos, H., and Hohmann, E. L. (2000). Pilot study of *phoP/phoQ*-deleted *Salmonella enterica* serovar Typhimurium expressing *Helicobacter pylori* urease in adult volunteers. *Infect. Immun.* **68**:2135–2141.

37. Michetti, P., Kreiss, C., Kotloff, K., Porta, N., Blanco, J. L., Bachmann, D., Herranz, M., Saldinger, P. F., Corthesy-Theulaz, I., Losonsky, G., Nichols, R., Simon, J., Stolte, M., Ackerman, S., Monath, T. P., and Blum, A. L. (1999). Oral immunization with urease and *Escherichia coli* heat-labile enterotoxin is safe and immunogenic in *Helicobacter pylori*-infected adults. *Gastroenterology* **116**:804–812.

38. Kotloff, K., Sztein, M. B., Wasserman, S. S., Losonsky, G., Di Lorenzo, S. C., and Walker, R. I. (2001). Safety and immunogenicity of oral inactivated whole-cell *Helicobacter pylori* vaccine with adjuvant among volunteers with or without subclinical infection. *Infect. Immun.* **69**:3581–3590.

6 DISEASE STATES AND VACCINES: SELECTED CASES

PART L. Genome-Based Approach for a Vaccine against *Neisseria meningitidis*

VEGA MASIGNANI, MARIAGRAZIA PIZZA, AND RINO RAPPUOLI

IRIS Research Center, Chiron SpA, 53100 Siena, Italy

I. *NEISSERIA meningitidis*

A. Bacterium and Disease

One disease that has struck alarm in many western and developed countries and constitutes an ordinary plague in sub-Saharan Africa and Arabia is represented by meningococcal meningitis and septicemia. The etiological cause of this disease is *Neisseria meningitidis* (meningococcus), a gram-negative, capsulated bacterium, which belongs to the phylogenetic order of the β-subgroup of Proteobacteria. On the basis of the organism's capsular polysaccharide, 12 serogroups of N. *meningitidis* have been identified. Among these, only five have been associated with the disease and are thus considered as pathogenic serogroups: A, B, C, Y, and W135 [1, 2].

When meningococcus causes invasive disease in a susceptible individual, the process involves invasion of the respiratory tract epithelia and the underlying endothelia of the microvascular system, followed by systemic dissemination via the bloodstream [3]. The outcome is life-threatening septicemia and metastatic spread to the meninges and cerebrospinal fluid (meningitis). The reported annual incidence of meningococcal disease varies from 0.5 to 10 per 100,000 persons [4]. However, during epidemics the incidence can rise above 400 per 100,000. The case–fatality rate ranges from 5 to 15%, and up to 25% of survivors are left with neurological sequelae, which can range from deafness to serious mental retardation [5].

During the 1980s and early 1990s, the other major causes of meningitis were *Haemophilus influenzae* type b and *Streptococcus pneumoniae* (Fig. 1A).

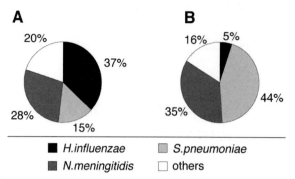

A

20%
37%
28%
15%

B

16% 5%
44%
35%

■ *H.influenzae*   ▦ *S.pneumoniae*
■ *N.meningitidis*   □ others

FIGURE 1   Epidemiological distribution of causative agents of meningitis before (A) and after (B) the introduction of the glycoconjugate vaccine against *H. influenzae.*

In addition to these three bacterial species, which were responsible for the bulk of bacterial meningitis, a number of other species were also found to cause meningitis. Among them, *Escherichia coli* K1 and *Streptococcus agalactiae* are particularly associated with neonatal meningitis [6]. Following the success of the conjugate vaccine against *H. influenzae*, which was introduced into clinical practice in 1988, the incidence of *H. influenzae*-associated meningitis decreased to 5% of total cases [7]. *N. meningitidis* became the major cause of meningitis for all ages in the United States in 1995 and the second leading cause worldwide (Fig. 2B) [4].

Among the various virulence factors known for this pathogen [8] two in particular play a critical role: the polysaccharide capsule, which confers an important survival advantage to meningococcus through the inhibition of macrophage and neutrophil phagocytosis in the bloodstream [9], and the pili, long filamentous structures that protrude from the bacterial surface and are essential for adhesion to human epithelial and endothelial cells [10]. As a consequence, acapsulated bacteria are inactive *in vivo* because they cannot persist in the blood, and pilus-deficient meningococci can still cause damage but to a significantly lesser extent than piliated bacteria.

In a similar manner, most of the known immunodominant proteins of this pathogen are associated with the surface of the bacterium. The more abundant proteins are the two major porins, PorA and PorB [11] and the opacity proteins (Opas) [12], which globally represent more than 40% of the total outer membrane repertoire of the bacterium (Fig. 2).

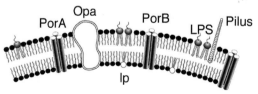

Opa
PorA   PorB   LPS   Pilus
lp

FIGURE 2   Schematic representation of the outer surface of *Neisseria meningitidis* along with the major proteins that compose it. Abbreviations: PorA, porin, class 1 OMP; PorB, porin, class 2/3 OMP; Opa, adhesion, opacity protein, class 5 OMP; LPS, lipopolysaccharide; lp, lipoproteins.

B.   Prevention

A number of different formulations are available today for the treatment of at least some serogroups of N. meningitidis; others are in the phase of clinical testing. The meningococcal vaccine currently in use is a tetravalent polysaccharide vaccine composed of serogroups A, C, Y, and W135 [1, 2]. Although efficacious in adolescents and adults, it induces a poor immune response and a short duration of protection, and it cannot be used in infants. This is because polysaccharides are T-cell-independent antigens that induce a weak immune response, which cannot be boosted by repeated immunization. Following the success of the vaccination against H. influenzae, conjugate vaccines against serogroups A and C have been developed and are in the final stage of clinical testing [13–15].

However, the more critical target for vaccination is meningococcus B (MenB), because this serogroup is responsible for 32% of all cases of meningococcal disease in the United States and for 45–80% or more of the cases in Europe [16]. Unfortunately, a polysaccharide-based vaccine approach cannot be used for this serogroup because the MenB capsular polysaccharide is a polymer of $\alpha(2–8)$-linked N-acetylneuraminic acid, which is also present in the tissues of mammals, who should be tolerant to it and, therefore, should not mount an immune response upon vaccination [17, 18].

An alternative approach to MenB vaccine development is based on the use of surface-exposed proteins contained in outer membrane preparations (outer membrane vesicles, OMV). These vaccines have been shown both to elicit serum bactericidal antibody responses and to protect against developing meningococcal disease in clinical trials [19]. However, the applicability of these vaccines is limited by the sequence variability of major protein antigens, which therefore are able to elicit an immune response only against homologous strains [6]. In a similar way, other major surface structures of N. meningitidis undergo phase variation and are not expressed in some strains (OpA, OpC, PilC) [20, 21]. For all of these reasons, both the capsular-based and the OMV-based approaches seem far from the final goal of providing a multivalent vaccine against meningitis.

II.   REVERSE VACCINOLOGY: THE MenB EXPERIENCE

A.   Biocomputing and Whole-Genome Sequencing

The introduction of methods for automated DNA sequence analysis nearly a decade ago, together with the more recent advances in the field of bioinformatics, has revolutionized biology and medicine and has guided research into the new era of genomic science [22, 23]. Computational biology can help researchers circumvent conventional time-intensive and laborious experimental techniques and proceed directly from gene sequencing to protein functional assignments and, ultimately, from genome sequence to functional genomics.

The latest step in this ever-growing field of biocomputing science has been the challenge of whole-genome sequencing. The first sequence of a free-living organism (H. influenzae) was accomplished at The Institute for Genomic Research (TIGR) and published in 1995 [24]. After that, during the period

1995–1997 12 complete microbial genomes were determined, including four Archaea and the first Eukaryotae, yeast [25]. Today, the microbial database maintained at TIGR lists 55 completed bacterial plus four eukaryotic genomes, and more than 100 other microorganisms are being sequenced in various laboratories around the world (www.tigr.org/tdb).

Genomics has actually changed the aspect and quality of modern biology: rapid progress can now be made in elucidating a variety of biosynthetic pathways that were previously difficult to exploit and also in examining the complexity of transcriptional signals, which regulate the communication between the pathogen and its environment. Obviously, the principal goal of sequencing pathogenic bacteria is aimed at the understanding of the disease process; along with such understanding should come the ability to develop molecular diagnostic probes and to define new drug targets and vaccines to treat infections caused by these organisms. However, despite the increasing amount of information available today on the web, genomic studies of bacterial pathogens have not yet led to new advances in therapeutic or preventive measures.

B. The Reverse Vaccinology Approach

Biochemical, immunological, and microbiological methods have been used for decades to identify microbial components important for an understanding of pathogenesis and useful for vaccine development. This kind of approach has been successful in many cases; however, the antigens that are generally identified are the most abundant and often variable in sequence. Once the suitable antigens are identified, they need to be purified in large amounts for testing in the animal model, and this can be achieved by starting from the microorganism itself or by cloning the gene and hyperexpressing the antigen. A number of antigens have been identified using this conventional approach; however, in most cases decades have been required to characterize and propose these proteins as suitable vaccine components. Furthermore, it has been totally unsuccessful for those pathogens whose protective antigens are difficult to identify or are expressed only during infection or under particular growth conditions.

The advent of Genomics has been a revolution in the field of vaccinology. By taking advantage of the complete sequence of a pathogen and with the application of sophisticated computer programs, it becomes possible to choose potentially surface-exposed proteins in a reverse manner, starting from the genome rather than from the microorganism. We have named this novel approach "reverse vaccinology" [26, 27].

C. Genome Analysis and Antigen Selection in *Neisseria meningitidis B*

The "shotgun sequencing" approach was used to obtain the complete genomic sequence of MenB (Fig. 3) [28]. Of the 2158 ORFs identified by the annotation, only 1158 could be assigned a putative biological role on the basis of sequence homology criteria, whereas the remaining 1000 are considered as proteins of unknown function.

The reverse vaccinology approach takes advantage of the complete genome sequence generated at TIGR and applies computer programs in order

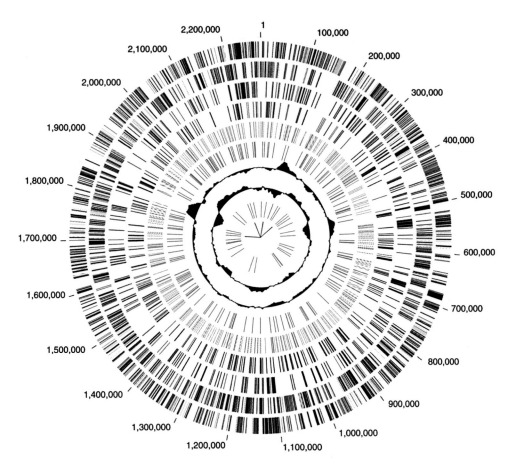

FIGURE 3 Circular representation of the *N. meningitidis* strain MC58 genome. First and second circles, predicted coding regions on the plus and minus strands color-coded by role categories. Third and fourth circles, predicted coding regions on the plus and minus strands color-coded by function in virulence. Fifth and sixth circles, *Neisseria* DNA uptake sequences (USSs). Seventh circle, atypical nucleotide composition curve. Eighth circle, %G+C curve. Ninth circle, tRNAs. Tenth circle, rRNAs. Reprinted with permission from Tettelin *et al. Science* **287**:1809–1815. Copyright 2000 American Association for the Advancement of Science.

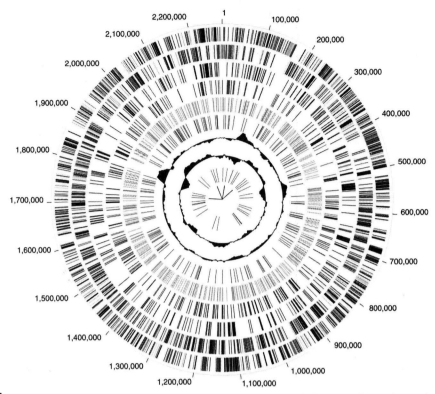

FIGURE 3   Circular representation of the *N. meningitidis* strain MC58 genome. First and second circles, predicted coding regions on the plus and minus strands color-coded by role categories. Third and fourth circles, predicted coding regions on the plus and minus strands color-coded by function in virulence. Fifth and sixth circles, *Neisseria* DNA uptake sequences (USSs). Seventh circle, atypical nucleotide composition curve. Eighth circle, %G+C curve. Ninth circle, tRNAs. Tenth circle, rRNAs. (See color insert). Reprinted with permission from Tettelin *et al. Science* **287**:1809–1815. Copyright 2000, American Association for the Advancement of Science.

to predict *in silico* all those genes that could express potentially surface-associated proteins (Fig. 4), which are more likely to induce an immunogenic response [29]. For this purpose, currently available computational methods, which range from homology searches, to detection of functional motifs, to prediction of signal sequences for secretion, have been applied to finally yield 570 novel ORFs, which could therefore represent new potential vaccine candidates. These have been classified with respect to their topological features (Fig. 5). The histogram in Fig. 5 shows that the higher proportion is represented by integral membrane proteins and transporters, followed by periplasmic-predicted proteins and lipoproteins. The smallest group includes predicted outer membrane and secreted proteins and proteins that have been selected exclusively on the basis of sequence homology criteria.

D. Experimental Strategy

By means of the polymerase chain reaction (PCR), the DNA sequences of the 570 genes detected by computer analysis were amplified from meningococcus

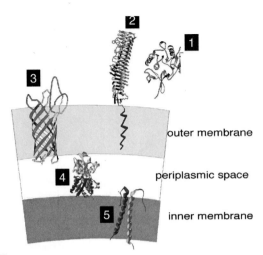

FIGURE 4   Schematic representation of the classes of proteins that have been selected as potential antigens by the reverse vaccinology approach: (1) secreted proteins, (2) outer membrane proteins and lipoproteins, (3) porin-like structures, (4) periplasmic proteins, and (5) integral membrane proteins.

and cloned in *E. coli* to express each polypeptide as His-tagged or glutathione-S-transferase (GST) fusion proteins. The recombinant proteins were purified and used to immunize mice. Immune sera were then tested in ELISA, fluorescence-activated cell sorter (FACS) analyses, and Western blotting on outer membrane preparations to verify whether those proteins were present on the surface of MenB. Finally, all of the immune sera were tested for bactericidal activity (BCA) [30], as this assay has been reported to correlate with protection in humans [31].

Of the 85 novel proteins found to be strongly positive in at least one of the preceding assays (FACS, ELISA, or BCA), seven representative proteins (genome-derived *Neisseria* Antigens) that were positive in all three assays were selected for further study (Fig. 6). Interestingly, some of the antigens were able to confer bactericidal titers that were similar in magnitude to those induced by OMV, which are known to confer protection in humans against homologous strains [6].

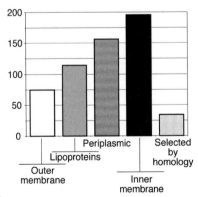

FIGURE 5   Distribution of the 570 novel MenB vaccine candidates according to their topological features.

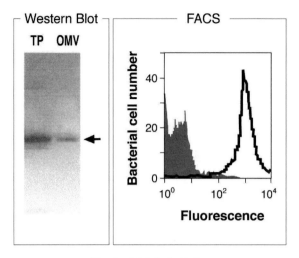

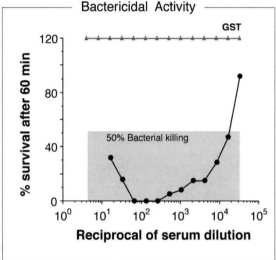

FIGURE 6 Serological analyses obtained for one of the novel MenB antigens. Western blot analysis shows a reactive band on both MenB total protein extract (TP) and outer membrane vesicles (OMVs). The FACS analyses show that the antigen is surface-exposed in MenB (gray and black profiles show binding of preimmune and immune sera, respectively). The bactericidal assay shows that the antibodies are able to induce 50% bacterial killing at a dilution higher than 1:32,000.

To test the suitability of these novel proteins as candidate antigens for conferring protection against different MenB strains, and not just against the homologous strain, it was important to investigate whether the new candidate molecules were expressed and conserved in sequence among a number of different strains. For this analysis, a collection of 31 *N. meningitidis* strains from all serogroups, isolated worldwide, and over many years was used. Each of the seven candidates was amplified from the 31 selected strains of *N. meningitidis* and directly sequenced. PorA was included in the analysis because it is a well-characterized membrane-associated protein that displays sufficient sequence diversity [13, 32] to render it unsuitable as an effective vaccine.

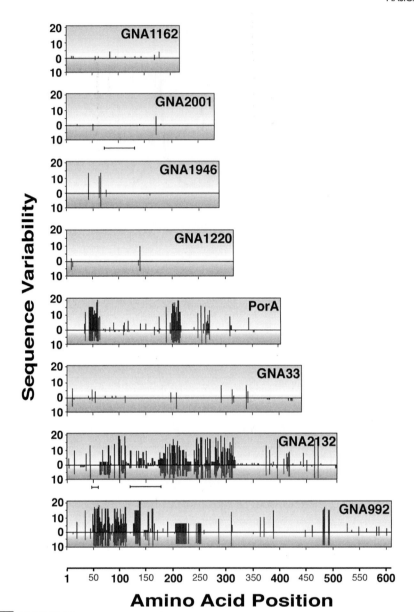

FIGURE 7 Schematic representation of amino acid sequence variability within *N. meningitidis* of the seven novel antigens and PorA. Abscissa reports amino acid position, and ordinate reports the number of strains analyzed. Line 0 represents the sequence of the MC58 reference strain. Amino acid differences from the sequence of MC58 within the 22 strains of MenB are indicated by black lines above the 0. Amino acid differences within the nine *N. meningitidis* strains from serogroups A, C, Y, X, Z, and W135 are indicated by gray lines below the 0. The height of the lines represents the number of strains with amino acid changes. Highly variable domains are represented by black and gray peaks, respectively. Reprinted with permission from Pizza *et al. Science* **287**:1816–1820. Copyright 2000, American Association for the Advancement of Science.

The results of the multiple sequence alignment are shown in a diagram (Fig. 7) in which every mutation is reported as a peak whose length is proportional to the number of strains that are affected by that mutation with respect to the reference strain MC58. With the exception of antigen GNA2132 and the N-terminal portion of GNA992, the level of sequence conservation of the selected candidates is striking, and this is even more surprising given the fact that all of the candidates under investigation are surface-exposed molecules. In fact, in three decades of studies, with only one exception [33], only antigenically variable outer membrane proteins have been described for *N. meningitidis* [34].

III. CONCLUSIONS

Compared with conventional microbiological approaches, the genome analysis of meningococcus B has allowed the identification of a far higher number of novel surface-exposed proteins, which are highly conserved among distantly related strains and serotypes and are also able to induce bactericidal antibodies. These novel antigens hopefully will be the basis for the clinical development of a vaccine not only against group B *N. meningitidis* but also against other serogroups and species of pathogenic *Neisseria*.

In conclusion, the meningococcus experience represents the first success of reverse vaccinology and shows the potential of the use of genome analysis for vaccine development, suggesting that such an approach could be applied to microorganisms for which vaccines are not yet available. In fact, following the success of the MenB project, this strategy has been recently used to identify potential vaccine candidates for *Streptococcus pneumoniae* [35] and *Porphyromonas gingivalis* [36].

REFERENCES

1. Gotschlich, E. C., Liu, T. Y., and Artenstein, M. S. (1969). Human immunity to the meningococcus. 3. Preparation and immunochemical properties of the group A, group B, and group C meningococcal polysaccharides. *J. Exp. Med.* **129**(6):1349–1365.
2. Gotschlich, E. C., Goldschneider, I., and Artenstein, M. S. (1969). Human immunity to the meningococcus. V. The effect of immunization with meningococcal group C polysaccharide on the carrier state. *J. Exp. Med.* **129**(6):1385–1395.
3. van Deuren, M., Brandtzaeg, P., and van der Meer, J. W. (2000). Update on meningococcal disease with emphasis on pathogenesis and clinical management. *Clin. Microbiol. Rev.* **13**(1):144–166.
4. Schuchat *et al.* (1997). Bacterial meningitis in the United States in 1995. *New England J. Medicine* **337**:970–976.
5. Naess, A., Halstensen, A., Nyland, H., Pedersen, S. H., Moller, P., Borgmann, R., Larsen, J. L., and Haga, E. (1994). Sequelae one year after meningococcal disease. *Acta Neurol. Scand.* **89**(2):139–142.
6. Poolman, J. T. (1995). Development of a meningococcal vaccine. *Infect. Agents Dis.* **1**:13–28.
7. Lindberg, A. A. (1999). Glycoprotein conjugate vaccines. *Vaccine* **17**(Suppl. 2):S28–36.
8. Virji, M. (1996). Meningococcal disease: Epidemiology and pathogenesis. *Trends Microbiol.* **4**(12):466–469.
9. Cartwright, K., Ed. (1995). "Meningococcal Disease." John Wiley & Sons, Chichester, United Kingdom.

10. Virji, M., Kayhty, H., Ferguson, D. J., Alexandrescu, C., Heckels, J. E., and Moxon, E. R. (1991). The role of pili in the interactions of pathogenic *Neisseria* with cultured human endothelial cells. *Mol. Microbiol.* 5(8):1831–1841.

11. Minetti, C., Song, J., Colombini, M., and Blake, M. S. (1998). Meningococcal PorA class 1 proteins exist in nature as a heterotrimeric porin with PorB proteins. *In* "Eleventh International Pathogenic *Neisseria* Conference," X. Nassif, Ed. Nice, France.

12. Malorny, B., Morelli, G., Kusecek, B., Kolberg, J., and Achtman, M. (1998). Sequence diversity, predicted two-dimensional protein structure, and epitope mapping of neisserial Opa proteins. *J. Bacteriol.* 180(5):1323–1330.

13. Zollinger, W. D. (1997). New and improved vaccines against meningococcal disease. *In "New Generation Vaccines,"* 2nd ed., M. M. Levine, G. C. Woodrow, J. B. Kaper, and G. S. Cobon, Eds., pp. 469–488. New York: Marcel Dekker.

14. Lieberman *et al.* (1996). Safety and immunogenicity of a serogroups A/C *Neisseria meningitidis* oligosaccharide–protein conjugate vaccine in young children. A randomized controlled trial. *J. Am. Med. Assoc.* 275: 1499–1503.

15. Costantino, P. *et al.* (1992). Development and phase I clinical testing of a conjugate vaccine against meningococcus A and C. *Vaccine* 10: 691–698.

16. Scholten, R. J., Bijlmer, H. A., Poolman, J. T., Kuipers, B., Caugant, D. A., Van Alphen, L., Dankert, J., and Valkenburg, H. A. (1993). Meningococcal disease in The Netherlands, 1958–1990: A steady increase in the incidence since 1982 partially caused by new serotypes and subtypes of *Neisseria meningitidis*. *Clin. Infect. Dis.* 16(2):237–246.

17. Finne, J., Bitter-Suermann, D., Goridis, C., and Finne, U. (1987). An IgG monoclonal antibody to group B meningococci cross-reacts with developmentally regulated polysialic acid units of glycoproteins in neural and extraneural tissues. *J. Immunol.* 138(12):4402–4407.

18. Hayrinen, J., Jennings, H., Raff, H. V., Rougon, G., Hanai, N., Gerardy-Schahn, R., and Finne, J. (1995). Antibodies to polysialic acid and its N-propyl derivative: Binding properties and interaction with human embryonal brain glycopeptides. *J. Infect. Dis.* 171(6):1481–1490.

19. Tappero, J. W., Lagos, R., Ballesteros, A. M., Plikaytis, B., Williams, D., Dykes, J., Gheesling, L. L., Carlone, G. M., Hoiby, E. A., Holst, J., Nokleby, H., Rosenqvist, E., Sierra, G., Campa, C., Sotolongo, F., Vega, J., Garcia, J., Herrera, P., Poolman, J. T., and Perkins, B. A. (1999). Immunogenicity of 2 serogroup B outer-membrane protein meningococcal vaccines: A randomized controlled trial in Chile. *J. Am. Med. Assoc.* 281: 1520–1527.

20. Stern, A., and Meyer, T. F. (1987). Common mechanism controlling phase and antigenic variation in pathogenic neisseriae. *Mol. Microbiol.* 1: 5–12.

21. Jonsson, A. B., Nyberg, G., and Normark, S. (1991). Phase variation of gonococcal pili by frameshift mutation in pilC, a novel gene for pilus assembly. *EMBO J.* 10:477–488.

22. Andrade, M. A., and Sander, C. (1997). Bioinformatics: From genome data to biological knowledge. *Curr. Opin. Biotechnol.* 8:675–683.

23. Brutlag, D. L.(1998). Genomics and computational molecular biology. *Curr. Opin. Microbiol.* 1:340–345.

24. Fleischmann, R. D., Adams, M. D., White, O., Clayton, R. A., Kirkness, E. F., Kerlavage, A. R., Bult, C. J., Tomb, J. F., Dougherty, B. A., Merrick, J. M. *et al.* (1995). Whole-genome random sequencing and assembly of *Haemophilus influenzae* Rd. *Science* 269: 496–512.

25. Goffeau, A., Barrell, B. G., Bussey, H., Davis, R. W., Dujon, B., Feldmann, H., Galibert, F., Hoheisel, J. D., Jacq, C., Johnston, M., Louis, E. J., Mewes, H. W., Murakami, Y., Philippsen, P., Tettelin, H., and Oliver, S. G. (1996). Life with 6000 genes. *Science* 274: 546, 563–567.

26. Rappuoli, R. (2000). Reverse vaccinology. *Curr. Opin. Microbiol.* 3(5):445–450.

27. Rappuoli, R. (2001). Reverse vaccinology, a genome-based approach to vaccine development. *Vaccine* 19(17–19):2688–2691.

28. Tettelin, H., Saunders, N. J., Heidelberg, J., Jeffries, A. C., Nelson, K. E., Eisen, J. A., Ketchum, K. A., Hood, D. W., Peden, J. F., Dodson, R. J., Nelson, W. C., Gwinn, M. L., DeBoy, R., Peterson, J. D., Hickey, E. K., Haft, D. H., Salzberg, S. L., White, O., Fleischmann, R. D., Dougherty, B. A., Mason, T., Ciecko, A., Parksey, D. S., Blair, E., Cittone, H., Clark, E. B., Cotton, M. D., Utterback, T. R., Khouri, H., Qin, H., Vamathevan, J., Gill, J., Scarlato, V., Masignani, V., Pizza, M., Grandi, G., Sun, L., Smith, H. O., Fraser, C. M., Moxon, E. R., Rappuoli, R., and Venter, J. C. (2000). Complete genome sequence of *Neisseria meningitidis* serogroup B strain MC58. *Science* 287:1809–1815.

29. Pizza, M., Scarlato, V., Masignani, V., Giuliani, M. M., Arico, B., Comanducci, M., Jennings, G. T., Baldi, L., Bartolini, E., Capecchi, B., Galeotti, C. L., Luzzi, E., Manetti, R., Marchetti, E., Mora, M., Nuti, S., Ratti, G., Santini, L., Savino, S., Scarselli, M., Storni, E., Zuo, P., Broeker, M., Hundt, E., Knapp, B., Blair, E., Mason, T., Tettelin, H., Hood, D. W., Jeffries, A. C., Saunders, N. J., Granoff, D. M., Venter, J. C., Moxon, E. R., Grandi, G., and Rappuoli, R. (2000). Identification of vaccine candidates against serogroup B meningococcus by whole-genome sequencing. *Science* **287**:1816–1820.

30. Mandrell, R. E., Azmi, F. H., Granoff, D. M. (1995). Complement-mediated bactericidal activity of human antibodies to poly α (2→8) N-acetylneuraminic acid, the capsular polysaccharide of *Neisseria meningitidis* serogroup B. *J. Infect. Dis.* **172**: 1279–1289.

31. Goldschneider, I., Gotschlich, E. C., and Artenstein, M. S. (1969). Human immunity to the meningococcus. II. Development of natural immunity. *J. Exp. Med.* **129**(6):1327–1348.

32. Peltola, H. (1998). Meningococcal vaccines. Current status and future possibilities. *Drugs* **55**:347–366.

33. Martin, D., Cadieux, N., Hamel, J., and Brodeur, B. R. (1997). Highly conserved *Neisseria meningitidis* surface protein confers protection against experimental infection. *J. Exp. Med.* **185**(7):1173–1183.

34. Maskell, D., Frankel, G., and Dougan, G. (1993). Phase and antigenic variation—the impact on strategies for bacterial vaccine design. *Trends Biotechnol.* **11**(12):506–510.

35. Wizemann, T. M., Heinrichs, J. H., Adamou, J. E., Erwin, A. L., Kunsch, C., Choi, G. H., Barash, S. C., Rosen, C. A., Masure, H. R., Tuomanen, E., Gayle, A., Brewah, Y. A., Walsh, W., Barren, P., Lathigra, R., Hanson, M., Langermann, S., Johnson, S., and Koenig, S. (2001). Use of a whole genome approach to identify vaccine molecules affording protection against *Streptococcus pneumoniae* infection. *Infect. Immun.* **69**(3):1593–1598.

36. Ross, B. C., Czajkowski, L., Hocking, D., Margetts, M., Webb, E., Rothel, L., Patterson, M., Agius, C., Camuglia, S., Reynolds, E., Littlejohn, T., Gaeta, B., Ng, A., Kuczek, E. S., Mattick, J. S., Gearing, D., and Barr, I. G. (2001). Identification of vaccine candidate antigens from a genomic analysis of *Porphyromonas gingivalis*. *Vaccine* **19**(30):4135–4142.

7 ECONOMICS OF VACCINES: FROM VACCINE CANDIDATE TO COMMERCIALIZED PRODUCT

AMIE BATSON,* SARAH GLASS,* AND ERICA SEIGUER†

*Health, Nutrition, and Population Unit, Human Development Network, The World Bank, Washington, DC 20433

†Global Health Program, Bill & Melinda Gates Foundation, Seattle, Washington 98102

I. INTRODUCTION: WHY ECONOMICS MATTERS

Historically, the vaccine world has relied on private investment to finance most of the costs of vaccine product development and commercialization. However, increasingly, scientists and public health experts are recognizing that they must understand the economics behind such decisions to ensure that priority vaccines move through development, manufacturing, regulatory, and production phases as quickly and efficiently as possible. Ultimately, optimization of the use of limited resources, or economics, is an important factor in all decisions whether it is which candidates to develop, which vaccine to scale up, or which vaccine to purchase for use in national immunization programs.

The private sector's[1] investment in the development and manufacture of vaccines benefiting developing countries is not as large as the public sector[2] believes it should be. The increasing divergence in products tailored for and used by industrialized and developing countries is further exacerbating this investment gap. In 1992, only 4% of the $55.8 billion[3] spent on global research and development went toward disorders that accounted for most of the disease burden in low- and middle-income countries [1]. The situation had not changed much by 2000, when a study by the Global Forum for Health Research found that only 10% of global research funds were dedicated to the 90% of disease burden that affects the world's poorest people [2]. More

---

[1]Private sector is defined as commercial vaccine manufacturers.

[2]Public sector is defined as governments, academia, foundations, UN agencies, development banks, and other nongovernmental organizations.

[3]Monetary amounts are given in U.S. dollars.

specifically, research and development of vaccines for global priority diseases such as HIV, malaria, and tuberculosis remain critically underfunded. Global investment in the development of preventive AIDS vaccines has been limited as well [3]. In 1999, only $350 million were spent by the public and private sectors on such work, and nearly two-thirds of this amount was disbursed by the U.S. National Institutes of Health [4].

Investment in the development of priority products is not determined simply by disease burden or scientific know-how. One need only consider that industry has invested hundreds of millions of dollars to develop vaccines against lyme disease and drugs against baldness, but almost nothing on vaccines against dengue fever, meningococcal A, or malaria, which cause great morbidity and mortality in developing countries, to illustrate this point.

In fact, the transition from idea to marketable product is marked by a series of economic decisions in addition to scientific ones. Each successful step in a product's development results in new investment decisions—decisions that are based largely on an economic evaluation of the costs and risks of investment and the expected return on future sales.

An understanding of the economic factors encouraging or inhibiting investment in priority products allows public partners to work with the vaccine industry in new ways, such as public–private partnerships designed to share risk and returns. These partnerships can open the door to public sector partners playing a more active role in shaping the economics governing critical investment decisions.

In this chapter, we discuss the role that economics plays in vaccine development. First, we describe the characteristics of the vaccine market and how they influence investment in vaccine research and development. With this background, we explore the economic bottlenecks inhibiting vaccine development, illustrating the issues with a case study on meningococcal A conjugate vaccine. Finally, we discuss a series of options for addressing the economic challenges, categorized as "push" and "pull" approaches, and the types of public–private partnerships needed to implement them.

II. THE MARKET: VACCINE DEMAND, SUPPLY, AND PRICING

When faced with the decision to invest resources in a new product, a manufacturer makes comparisons across all products in its portfolio. Products promising high potential profits are more attractive investments than those products with more limited expected earnings. Industry leaders look to market demand to assess potential earnings from a candidate product. Low or uncertain demand for a product discourages industry from investing and, if there is an option, may result in resources being redirected toward products with strong markets.

Although the developing country market has tremendous needs, it is still considered a weak market because governments and international partners historically have been slow to finance and introduce new vaccines. As a consequence of this market uncertainty, near-term and priority vaccines needed in developing countries are lower priority for commercial manufacturers and therefore are often underfinanced.

A. Need Does Not Equal Demand

The concept of the market and how it is defined are critical as companies and public health organizations often use the same term to describe very different items. The *potential* market for vaccines in developing countries is very large. Annually, 64 million infants are born in low-income countries, and another 48 million are born in middle-income countries [5]. However, the *actual* market is the portion of this need that translates into demand or purchased doses at a given price. Thus, the market for a manufacturer is the revenue generated by sales of the product. Although the *potential* need for a vaccine may be large given the number of individuals who would benefit from it, the *actual* demand is much smaller because those who need the vaccine are often least able to pay for it. Despite 85% of the global population living in low- and middle-income countries [5], the vaccine market in the developing world is estimated to be only 10–15% of the global vaccine market of $5 billion and less than 0.2% of the global pharmaceutical market, which is estimated to approach $340 billion annually [6].

B. Who Uses and Who Buys Vaccines?

Children are the main recipients of vaccines, receiving between 8 and 12 different vaccines in the first year of life. These vaccines protect children from deadly and debilitating diseases, such as measles, hepatitis, and diphtheria. In most countries, however, children and their families do not purchase the vaccines. As a preventive intervention that protects not only the individual child but all who come in contact with him or her, vaccines are "public goods" and are primarily financed by governments and public health partners, including bilateral donors, multilateral institutions, development banks and foundations. Figure 1, showing immunization financing for four countries, highlights, the fact that each country has a different mix of funding sources; however, in

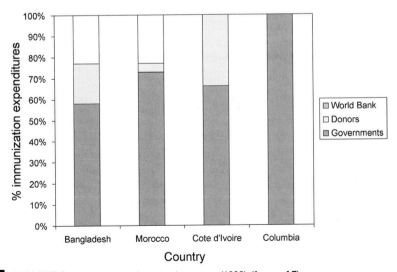

FIGURE 1 Immunization financing by source (1998) (from ref 7).

most cases, the government finances the majority of the costs of the national immunization program [7].

The decision by governments and partners to purchase vaccines depends on the absolute resources available and how these resources are allocated given other needs. Disease burden, the potential impact of the vaccine, the vaccine's price, and its cost of delivery are all important considerations in deciding whether scarce resources should be allocated to introduce a vaccine [8]. Historically, the public health community's estimates of vaccine need (confusingly called demand) have been far larger than the actual national and international willingness to pay for the vaccine. The resulting gap between expected demand based on public health need and real demand based on willingness and capacity to pay has increased the risk of serving the developing country vaccine market. In addition, the gap has undermined the credibility of public sector partners urgently calling for investment in new vaccines "needed" to combat diseases such as diarrhea or malaria.

C.   The Market

The vaccine market encompasses distinct product and target customer groups. The international vaccine market comprises vaccines sold by manufacturers licensed in multiple countries and so does not include vaccines produced by government facilities for the national vaccine program. This market can be divided into four primary categories, as shown in Fig. 2: developing country procurement agents (UNICEF–PAHO), the adult vaccine market, the industrialized country pediatric market, and new proprietary products, including the most recently introduced product, Prevnar, a pneumococcal conjugate vaccine

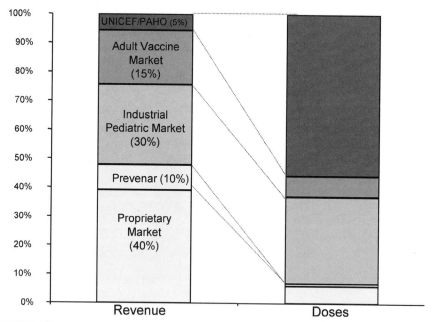

FIGURE 2   International vaccine market by revenue and volume (with permission from Ref. 35).

manufacturered by Wyeth-Lederle. Vaccine revenues are increasingly dominated by new proprietary vaccines, which are tailored to industrialized country markets. In 2000, the year of its introduction, sales of Prevnar reached nearly $500 million and were predicted to top $1 billion in 2001 [9].

Approximately half of all vaccine doses purchased on the international market are procured by UNICEF and PAHO on behalf of developing countries. However, whereas procurement by UNICEF and PAHO accounts for a tremendous number of doses, this bulk purchase constitutes only 5% of global revenues (e.g., 5% of the market).

D. Vaccine Supply

The vaccine industry is dominated by only a handful of companies, as shown in Fig. 3: American Home Products (Wyeth-Lederle Vaccines), Aventis Pasteur, GlaxoSmithKline, and Merck have seen their share of the vaccine market rise from approximately 50% in 1988 to about 80% today. Small- to medium-sized companies, notably Chiron and Baxter, as well as emerging companies in Korea, India, Indonesia, and other sites, vie for the remaining revenues [6]. Although the number of manufacturers of essential vaccines has grown in the past 20 years, virtually all of these new firms are smaller manufacturers based in middle- and low-income countries.

During the 1980s and early 1990s, declining projections of vaccine market growth were paralled by a dramatic decline in the number of industrialized country manufacturers supplying vaccines. Many manufacturers merged, sold, or shut down their vaccine business so that, of the 20 internationally active companies in the 1960s, less than 12 remained by the early 1990s [10]. However, the vaccine market is now much stronger. Even before the interest in vaccines stimulated by bioterrorism concerns, new product opportunities created by biotechnology advances had been rejuvenating the market, with growth predictions of as high as 11% over the coming years [6].

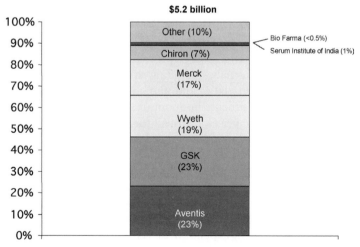

FIGURE 3 The vaccine players in vaccine market (based on 2000 market data) (with permission from Ref. 35).

Although not captured in an analysis of the international vaccine market, a tremendous number of doses used in the world are supplied by domestic producers in China, Brazil, India, and other countries. Most of these vaccines are licensed and used only within the home market. The vaccine manufacturers are often wholly or partially government owned. A series of studies by the World Health Organization (WHO) on behalf of the Children's Vaccine Initiative (CVI) evaluating domestic vaccine production in 10 countries was conducted from 1993 to 1995 [11–18]. These studies found that vaccine prices were often set by the government and in many cases did not cover the full costs of the manufacturer, making it difficult for these entities to maintain the facilities, train staff, or invest in new vaccine research and development or production. Domestic production is a very important component of global vaccine supply, but it is a relatively small part of the international vaccine market.

E. Vaccine Pricing

The traditional vaccines [measles, bacilli Calmette–Guérin (BCG), diphtheria–pertussis–tetanus (DPT), Tetanus toxoid (TT), oral polio vaccine (OPV)] are "mature," off-patent products, having been on the market for decades. These vaccines are widely used around the world and reach, on average, 50–80% of infants each year. Due in part to increasingly efficient production resulting from both learning over time and economies of scale (i.e., costs per dose decline as volume increases because of more efficient use of fixed costs), as well as the impact of competition, these vaccines are sold at very low prices and provide low marginal rates of return to producers [19].

Newer vaccines that have been available in industrialized countries for many years, such as hepatitis B and *Haemophilus influenzae* type B (Hib), have faced very slow introductions into developing countries. Although these vaccines were provided to developing countries at prices far below those charged in industrialized markets, the prices were still far higher than the cost of traditional products, which range from $0.05–0.15/dose. For example, as demand has shifted toward DTP combination products containing hepatitis B and/or Hib antigens, the prices being offered to countries through UNICEF and PAHO range from $0.90 to $3.50/dose [20]. In addition to "sticker shock," various other barriers have prevented the rapid introduction of cost-effective, life-saving vaccines, including weak delivery systems, inadequate national disease burden data, and the unwillingness of governments and donors to increase investments in immunization [8].

The extremely low prices charged for vaccines in developing countries relative to prices charged in industrialized markets are the result of differential pricing (also called tiered pricing, equity pricing, and price discrimination), whereby different prices are charged for the same products in different regions. By using differential pricing, manufacturers can charge a low price in the poorest developing country market, allowing these countries access to the product while still charging higher prices in the industrialized markets to recoup their research and development expenditures. As shown in Fig. 4, oral polio vaccine offers one compelling example of tiered pricing: in a given year, the highest price at which polio vaccine was offered was 250 times the lowest price [21].

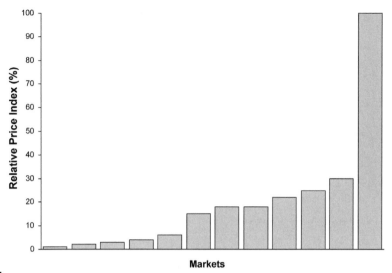

FIGURE 4 Differential pricing for one vaccine (OPV) (with permission from Ref. 35).

Figure 5 illustrates the distortion in manufacturer's product portfolios created by differential pricing; a few products have very high unit prices and very small volumes, whereas others have enormous volumes at low prices. Vaccine manufacturers must manage how this price–volume imbalance is perceived within their companies. Parent pharmaceutical firms often question the value of supplying the high volume–low price market, especially as the price differentials between old and new vaccines grow. As shown in Fig. 6, the new vaccines are significantly more expensive than traditional vaccines. Prices have reached $46/dose (for Prevnar in the United States) and are routinely above $15/dose. From industry's perspective, the introduction of higher margin products, like

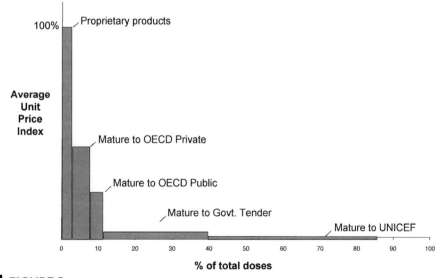

FIGURE 5 Typical vaccine market profile for a supplier (from UNICEF, Mercer analysis).

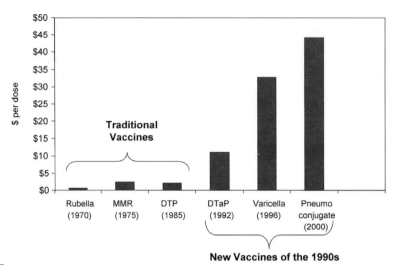

FIGURE 6 Price of traditional and new vaccines (with permission from CDC contract prices, 1970–2000) (with permission from Ref. 35).

Prevnar, makes investment in low-margin vaccines for the developing world even less attractive. In fact, the combined volume of total vaccine doses purchased by UNICEF and PAHO amounted to over 100 times the number of doses of Prevnar used in the United States, but less than half the sales revenue.

Pooled procurement has helped facilitate differential pricing. UNICEF and PAHO purchase large quantities of vaccines at very low prices. The ability of the international procurement agencies to demand vaccine at low prices and the willingness of various suppliers to produce vaccine at these prices have been important factors in the success of national immunization programs. However, it has also had some less desirable consequences. In comparison to other products and markets, the UNICEF and PAHO market generates little revenue and even less profit for manufacturers—two key factors driving a company's decision to maintain production lines and invest in new research and development and new capacity. Companies must internally justify their continued involvement in this "marginal" market. The result is low levels of investment in vaccine research and development and production capacity for the products needed by the poorest countries. As the vaccines used by the industrialized and developing markets diverge, this gap between the markets is becoming increasingly problematic [22].

F. Diverging Products

An analysis by WHO found that products are increasingly diverging between industrialized and developing countries [23]. The divergence of products is due to increased focus on both vaccines serving regional rather than "global" needs and second–generation vaccines against diseases already widely controlled by vaccines. There is increasing availability of vaccines to address regional diseases such as meningococcal C conjugate vaccine, which is only relevant to a few markets.

In the second trend, different vaccines for the same disease are now available for different markets. In the past, a disease was typically controlled by

Baseline	Tailored to the Developing Market	Tailored to the Industrial Market
Measles	Measles	MMR
DTP	DTwP	DTaP
OPV	OPV	IPV
TT	TT (Td in some areas)	Td
Hep B	Monovalent, DTwP-Hep B	Monovalent, DTaP-Hep B, DTaP-Hep B-IPV-Hib, Hep B-Hib, Hep A-Hep B
Hib	Monovalent, DTwP-Hib, DTwP-Hep B-Hib	Monovalent, DTaP-HepB-IPV-Hib, Hep B-Hib
Mening A/C polysaccharide	Meningitis A/C conjug (wanted)	Meningitis C conjugate, (Meningitis BC conjugate)
Pneumo polysaccharide	11-valent conjugate-wanted	7-valent conjugate; 11-valent under development
Product presentations	Multidose, thimerosal	Single dose, no thimerosal

FIGURE 7 Diverging products used in developing and industrialized markets (from ref 23).

one vaccine used in both industrialized and developing countries. However, today the same diseases are being controlled with different vaccines, depending on geography and cost. For example, in the 1990s some industrialized countries adopted DTaP (containing a purified acellular pertussis) as the vaccine of choice over DTwP (containing whole cell pertussis) [23]. Although more expensive, DTaP vaccine has roughly the same efficacy and fewer nonserious adverse events associated with it than DTwP, and so has enjoyed greater acceptance among parents in industrialized countries [24]. The divergence in the number and type of vaccines used in developing and industrialized markets is illustrated in Fig. 7.

III. BRINGING A VACCINE TO MARKET: COSTS, RISKS, AND CONSTRAINTS

A company with a portfolio of research and development projects must assess both an individual project's costs and likelihood of success as well as the project's relative cost and benefit compared to other research and development options. In addition, a company must ensure that product revenues cover all the research and development costs from both successful and failed projects.

Assessment and finance of risks are a major part of the business of pharmaceutical firms. The risks encompass not only a vaccine's development but span all the steps in the product's life cycle. It is a combination of known costs plus estimated risks and their possible cost implications that determines a company's decision to invest in a product. A manufacturer must believe that all risks and costs can be justified (e.g., completely paid for) by the product's future sales. Although a great deal of basic research is funded by the public sector, the traditional model of product development relied on private manufacturers financing most of the scientific, technical, political, and market-based risks associated with product development, production scale-up, and sales. Manufacturers are less and less willing to accept all of the product risk, particularly when the product has only a small market and the risk is one over which they have no control. These risks may be influenced by other partners including WHO, UNICEF, and bilaterals who, for example, can help shape the timing and scope of new vaccine introduction into developing countries.

A. Baseline Costs

For each vaccine, there is a set of defined costs that constitutes the investment needed to develop and scale up the product in the absence of any failures, complications, or delays. These generic costs include clinical trials, production scale-up, licensing fees, on-going production, and quality control. The costs vary based on the particular characteristics of the disease, vaccine, and production technology. Figure 8 shows an estimate of the cost of developing an HIV–AIDS vaccine for one target population.

From industry's perspective, the baseline costs associated with product development correspond to four distinct stages: (1) basic, preclinical research; (2) identifying a vaccine candidate through studies and tests of safety and immune response in humans; (3) developing and testing a candidate for a single market; and (4) scaling up manufacturing capacity for that market. Based on an analysis of the cost to develop an HIV–AIDS vaccine [25], Figure 9 shows that these investments can be quite significant. Each investment becomes a decision gate initiating a reevaluation of the product's risks and potential return.

To assure a vaccine for global use would require two additional stages of investment: (5) "relevance," adapting and testing a vaccine to ensure safety

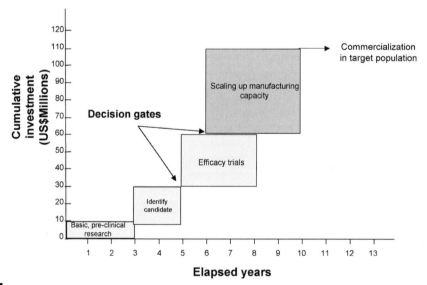

FIGURE 8 Baseline investments and decision gates in developing an HIV–AIDS vaccine for one target population (from ref 25).

and efficacy in additional populations, and (6) "supply," ensuring adequate manufacturing capacity and funding–pricing structures to enable broad developing country access. To make a new vaccine broadly available at the earliest possible technical and regulatory opportunity, a manufacturer must make development and capital investments explicitly to support supply to the developing world markets. It is unlikely that these incremental investments will be entirely recouped by sales to the industrialized market, so they must be covered by the developing world market or other special financing interventions.

B.   Risks

In addition to these baseline costs, every product faces a range of predictable risks. Should they become real, any one of these risks can necessitate tremendous additional investment or—if the product fails entirely—result in wasted investment. Figure 9 illustrates many of the risks that exist at each stage of the product's life from earliest research to final sales. The following section describes many of these risks in more detail.

1.   Research and Development

Faced with finite human and financial resources to support research and development projects in the pipeline, the decision to invest in one candidate will absorb resources that could have been used for other products. The opportunity cost of supporting one project versus another has very real costs to the firm,

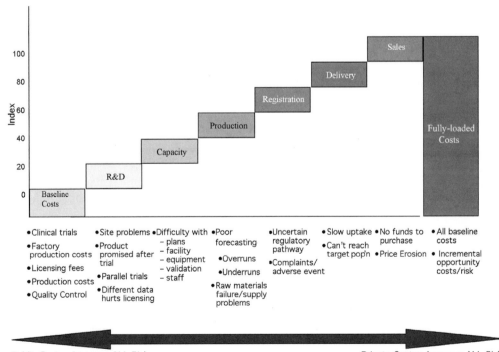

FIGURE 9   Risk at each stage of the vaccine life cycle (with permission from Ref. 35).

especially if one product is likely to result in large revenues whereas another promises only a small market.

Phase 3 trials are relatively expensive. Each additional trial can add between $10 and 30 million to the product's development cost. If there are not good animal models or correlates of immunity, for example, as with HIV–AIDS, vaccine candidates tested in phase 3 trials face a lower likelihood of success than products for which experiments in suitable animal models have shown immunogenicity and efficacy.

In addition, clinical trials sited in developing countries with different epidemiological conditions may result in data that compromises licensing, and thus revenues, in industrialized country markets.

Finally, as the number of antigens in a combination increases, so does the risk that at least one of the antigens will have reduced immunogenicity or efficacy.

2. Capacity and Production

To assure timely licensure and availability of new products, companies invest in production capacity before the data from the phase 3 trial are released and before the product is licensed for use in a given market. If the product fails at a late stage, most of this investment is lost. As companies are facing growing internal competition for production capacity, the opportunity cost of building or using a facility for one product versus another must be considered.

In addition, sizing the facility appropriately given uncertainties about future demand is very difficult. On one hand, a larger facility requires greater investment, but because of economies of scale it can result in more efficient, lower cost production in the long run if large volumes of the product are demanded. However, if demand is lower than expected, the facility will be underutilized, a costly waste of resources. On the other hand, a smaller facility requires less total investment, but the firm is less able to take advantage of economies of scale and so the long-term cost per dose may be higher. If demand is greater than expected, the company not only loses the market to a competitor but also must invest in resizing its facility, a costly mistake.

3. Regulatory and Licensing Issues

Regulatory requirements are becoming increasingly stringent, and the preparation of applications for marketing authorization is hampered by the differing requirements across countries. To obtain separate authorizations for each market in which a manufacturer proposes to sell a vaccine is a long, costly, and risky process.

The increasing difficulty in finding appropriate regulatory pathways for developing market vaccines produced in the United States and Europe also increases the risk of licensure. Most new vaccines are produced in either the United States or European countries. The FDA and EMEA (European Agency for the Evaluation of Medicinal Products) each have a responsibility to regulate products for their home markets. However, these agencies are also called upon to regulate products intended for developing countries but that are developed and produced by manufacturers situated within their borders. If these agencies refuse to apply their regulatory expertise for products that would not be marketed within their borders, the risks and costs to the companies will increase significantly.

Finally, unexpected adverse events picked up through postmarketing surveillance may result in a product losing its license, as in the case of Wyeth Lederle's Rotashield, a rotavirus vaccine. The millions of dollars invested to develop and scale up production for this product were lost once the product was removed from the market.

4. Production

Vaccine production is very tightly regulated. Every batch is tested before being released. Often, seemingly inconsequential changes to the production materials or process, such as a change in the brand of cleaning fluid used to wash out a fermenter, can cause a batch to fail.

There is also a long "lead time" before the initial investment in vaccine production results in vaccine doses for sale. Unexpected changes in demand can be costly to manufacturers as they result in excess inventory, inefficient changes to the production plan, or market shortages.

5. Delivery

The immunization delivery systems within countries determine whether a new vaccine will even be able to reach target populations. Although public–private partnerships such as the Global Alliance for Vaccines and Immunization (GAVI) provide support to strengthen immunization infrastructure, the challenge is great [26].

6. Sales

Demand and potential profit should balance all of the costs and risks that have been outlined. However, for the developing country market there is a great deal of uncertainty about not only which vaccines will be demanded and financed by governments and partners but when these vaccines will be demanded and introduced. The slow and uncertain uptake of vaccines to date has increased uncertainties about the market.

This uncertainty extends to intellectual property rights. Protection of intellectual property rights (IPR) helps ensure that manufacturers will have a market in which they can recoup their investment. Weak IPR protection increases the risk of multiple suppliers and so further weakens the developing country market, thus reducing the incentive for companies to invest in products for the developing world.

C. Each Vaccine Is Different

The relative importance of these different risks varies tremendously from vaccine to vaccine. Each vaccine faces a unique mix of obstacles that may be best overcome by different types of interventions, as shown in Fig. 10, which highlights two key factors, the potential market revenue and the degree of scientific certainty. For instance, a vaccine against a disease like malaria, which faces a number of scientific hurdles [27] as well as low revenue potential, faces high development risk, high opportunity costs, and high demand risk. By contrast, a meningococcal A conjugate vaccine for Africa has high scientific

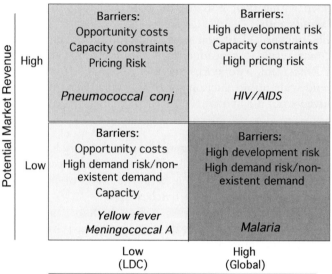

FIGURE 10 Barriers given degree of scientific certainty and potential market revenue (with permission from Ref. 35).

and technical certainty but almost no revenue opportunity. In this case, the primary barriers are the opportunity costs, constraints on capacity, and high demand risk. HIV vaccines, for which the revenue opportunity is high but the scientific and technical hurdles are also high, represent a tantalizing, often frustrating, opportunity for manufacturers. Acceleration of its development may require incremental investments to tailor a vaccine to the needs of developing countries once a successful approach is discovered and innovative ways to ensure differential pricing.

IV. THE CASE OF MENINGOCOCCAL A CONJUGATE VACCINE

The meningococcal A–C conjugate vaccine is a clear illustration of the importance of market incentives. Although the technology for this vaccine is well-known and nonproprietary, there has been limited investment in developing a vaccine tailored for the African market. The low level of research and development investment in this vaccine resulted from doubts about the ability to deliver the vaccine in some of the poorest countries with weak immunization infrastructure, the future prioritization of introducing meningococcal vaccines versus other disease control efforts, and the future willingness to actually purchase a vaccine at a "reasonable" price. Without public–private partnership aimed at both sharing and lowering risks, it is unlikely that this vaccine will be developed and distributed in a timely fashion and at an affordable price.

A. The Missed Opportunity

Although the technology to develop and produce meningococcal conjugate vaccines against A and C is known, there is no appropriate vaccine against the serogroup that causes the majority of epidemics in Africa, meningococcal A. Three meningococcal C conjugate vaccines were developed for and licensed in the United Kingdom in 1999 after a specific request by the UK Government to industry. For epidemiological and political reasons, the UK Government specified that it would only use a meningococcal C product. In response, industry halted efforts to develop a meningococcal A–C conjugate and focused solely on the C component. Reports document that the vaccines used in the United Kingdom are safe and highly effective in decreasing the incidence of serogroup C meningococcal disease in the United Kingdom [28].

B. Pushing and Pulling a Meningococcal A Vaccine

In late 1999, concerned that all vaccine manufacturers had stopped development of meningococcal A vaccine candidates, the World Health Organization (WHO), Program for Appropriate Technology in Health (PATH), and the Centers for Disease Control and Prevention (CDC) launched a project to spur the vaccine's development. A broad call for expressions of interest was issued to all manufacturers followed by visits to vaccine manufacturers willing to explore a public–private partnership towards the development of a meningococcal A vaccine for Africa.

C. Identifying the Hurdles

A few key hurdles were identified for the development and scale-up of a meningococcal A conjugate vaccine as part of the larger effort to map the relative importance of different incremental risks and costs for different vaccines. Figure 11 shows a generic cost–risk framework tailored to a meningococcal conjugate vaccine and highlights the most significant bottlenecks and risks, allowing public and private partners to tailor investments and solutions to share the risks [29]. The size of each box provides an indication of the relative importance of different costs and risks, helping to pinpoint the barriers or decision points that might be influenced by public sector intervention. For instance, the bottleneck identified for ongoing manufacturing results from the growing internal competition for capacity combined with the limited production flexibility from increasing regulatory pressures. The clinical trials and research and development risks represent other points at which public sector investment could be helpful. Clinical trials often represent a "go–no go" decision point, particularly for trials in developing countries. Not surprisingly, a final critical issue was the concern that the product would not be purchased because of low national demand, weak delivery systems, and the absence of any guarantees of purchase.

As the project evolved and manufacturers expressed interest in participating in a public–private partnership, another partner, the Bill & Melinda Gates Foundation, joined the effort with a grant of $70 million [30]. This grant will support the public–private activities to develop, scale up, create demand for, and introduce a meningococcal A conjugate vaccine in Africa. As this book goes to press, the global community is following the progress of this potential public–private partnership with interest as it will provide practical information and insight.

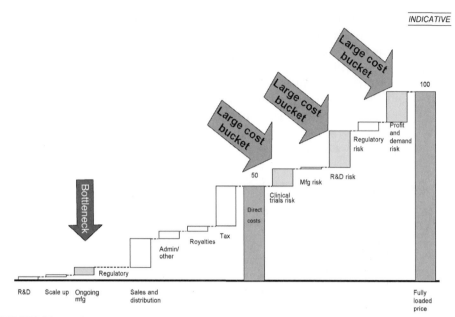

FIGURE 11   Identifying the most significant risks and costs for public–private risk sharing (from McKinsey & Co. analysis).

V.   PUSH AND PULL MECHANISMS: CHANGING THE ECONOMIC EQUATION

A.   Push and Pull Mechanisms

Two basic approaches to accelerate vaccine investment, development, and scale-up of priority products such as a meningococcal A conjugate vaccine have been identified as "push" and "pull" mechanisms. Though different shades of definitions have been offered, the simplest defines "push" mechanisms as those that reduce the risks and costs of investments and "pull" options as those that assure a future return in the event that a product is produced. Although the merits of push and pull interventions have been debated extensively [31–33], much work remains to translate the two concepts of financing and market creation into action. Historically, the public sector has invested substantially in basic research "push" mechanisms, particularly basic research and some early product development. There are fewer examples of pull mechanisms being implemented.

To ensure the early development of and access to priority vaccines will require new ways to assess and share risk. To accelerate the development, production scale-up, and introduction of priority vaccines into developing countries, significant investments must be made early in the product's life cycle. However, the private sector will not make these investments if they promise nothing but losses, nor will the public sector channel resources into the product's development or purchase if they are not assured of a product at a reasonable price. Creative public–private partnerships that share specific risks and costs and protect the interests of all partners are essential for global vaccine access.

It has become increasingly clear that there is no "silver bullet." No one mechanism, either push or pull, will single-handedly solve the current vaccine problems, in part because although the end goals of the public and private sectors may be similar, the incentives of the two sectors are diametrically opposed. For the public sector, the ideal scenario would be to avoid all risk by only buying a product once it is developed, the demand is present, and the vaccine is available at a very reasonable price. For the private sector, however, the ideal situation would be one in which industry's risk is reduced through some agreement that "locks" the public sector into purchasing vaccine (probably at a specific price and quantity) before investment is made in the product's development and production capacity.

It is unlikely that the public sector will ever have the resources either to entirely finance all of the costs of vaccine development and production or to purchase product at prices equivalent to those paid in the United States or Europe. The public sector must leverage its resources by targeting its actions and its funds as directly as possible to the particular obstacles inhibiting a vaccine's progress. If the risks are linked to uncertainty about the science, as in the case of an HIV–AIDS vaccine, then push mechanisms may prove more valuable than pull mechanisms, which are too far in the future and too low probability [25]. If the risks are linked to the market with little uncertainty about the science, as in the case of the meningococcal A conjugate vaccine, then pull mechanisms become most important. As shown in Fig. 12, different mechanisms can be tailored to best address risks occurring at different stages of the life cycle.

B. Push Mechanisms

Push options offer a number of benefits, the most important of which is that they reduce risk and thereby encourage investment where it might not otherwise occur (e.g., by funding a clinical trial in a geography that a manufacturer

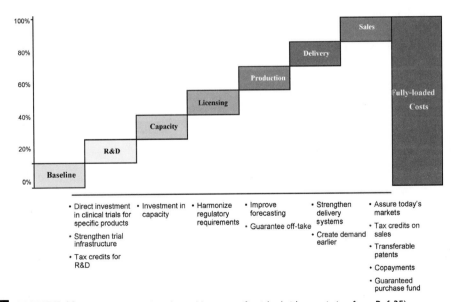

FIGURE 12 Approaches tailored to address specific risks (with permission from Ref. 35).

has little incentive to choose). Another important benefit is the familiarity and proven track record of certain push interventions that have been used successfully in the past. When structured to support broad objectives such as the development of animal models or field studies of the natural exposure and resistance to disease, this mechanism can also help to advance the development of all products targeted against a given disease.

However, certain types of push mechanisms, such as financing a clinical trial or funding research and development on a specific strain of a vaccine, necessitate specifying or picking one candidate or project. It is possible that better options may emerge at a later date or that manufacturers will drop promising candidates. Product "failures" must also be expected, as picking one project early in the process increases the risks. Finally, push options do not promise a successful outcome. When projects or products do fail, the money will already have been spent and, in most cases, will be unrecoverable.

1. Direct Financing

Direct financing is used to fund or directly implement activities necessary to develop a vaccine. The main goal of this type of intervention is to increase the number of candidate vaccines and generate data needed to assess a vaccine and recommend its use. Public sector funds may go toward financing either cross-cutting activities or product-specific activities. In the category of cross-cutting activities, the public sector might fund research on geographical disease burden, diagnostics, impact of different vaccine presentations, correlates of immunity, animal models, or reagents. The public sector could also play a role in helping the private sector to navigate regulatory requirements and establish clinical trial infrastructure.

For specific products, the public sector can support clinical trials, particularly the expensive phase 3 efficacy trials. Although the private sector expects to fund trials relevant for licensing the vaccine in industrialized countries, it has fewer incentives to finance trials in developing countries. Not only are developing country trials primarily of value to introduce the vaccine into this "uncertain, weak" market, but on occasion they can generate data or negative publicity that can threaten licensing in the industrial market.

The public sector could also help to finance production scale-up. For some vaccines, the incremental costs of scaling up production to meet developing country demand may be too costly given the expected market return. Without public sector help, the private sector may be unwilling to incur these costs.

2. Facilitating Research

Aside from providing money, there is an important role for the public sector to play in two other areas of clinical trial work: facilitation of the trials and adherence to ethics. Potential roles for the public sector include working with local officials to inform and educate country participants of the need for and importance of clinical trials and assisting in managing any problems that arise during the trial. The first role, education, is extremely important, particularly for diseases like HIV–AIDS and malaria. Enrollees must understand that participating in the trial will not necessarily protect them against the disease and, equally importantly, that if they do contract the disease, their participation in the trial is not necessarily the cause. The public sector could also help manufacturers to clarify expectations and mechanisms for posttrial access to the vaccine. It is not uncommon for clinical trial participants to assume that their community will have guaranteed and continued access to a vaccine being tested. However, as the time between a clinical trial and licensure of a vaccine is often lengthy, and the introduction of a vaccine into developing countries is even longer, this expectation needs to be addressed. In cases in which the private sector is already ambivalent about conducting a trial in a developing country, demands for posttrial vaccine access may further reduce interest.

3. Harmonizing Regulatory Requirements

The development and ultimate use of biological products like vaccines are carefully regulated to ensure their safety, immunogenicity, and efficacy. This regulatory function is carried out by the National Regulatory Authority (NRA) responsible for licensing and controlling the quality of all biological products used within the country. Currently, many NRAs have their own unique licensing requirements, compelling manufacturers to prepare and submit separate licensing applications, an expensive, time-consuming, and many argue inefficient process. Efforts are underway to harmonize requirements in order to ensure product safety but minimize the time and cost of making a vaccine available to populations [10].

4. Tax Credits for Vaccine Research

The structure of tax credits to encourage vaccine research and development is another push intervention under consideration. Like direct financing, tax credits have the benefit of being a familiar policy tool with a proven track record. The tax code of the United States currently contains an allowance for a 20% credit on research and development expenditures, and though it must be renewed on an annual basis, it is noncontroversial and nearly unanimously supported by members of Congress and industry. There remain two issues with using tax credits for vaccine research and development as a push mechanism. First, like other push mechanisms, the existence of the credit does not guarantee results. Second, tax credits can be difficult to implement and monitor.

C. Pull Mechanisms

At the other end of the spectrum, pull mechanisms can also help to reduce the risks of investing in priority products. For the public sector, pull mechanisms may be less risky as they can be structured to commit money only if a desirable and affordable vaccine becomes available. However, if financial commitments to specific products must be made early, it will be very difficult to determine the desired outcomes and prices far in advance of having an actual product. If a "winner" is chosen and a price is agreed, the public sector runs the risk of being "locked in" to an outcome that may not be best. Finally, the fact that these mechanisms have not been tested increases the risk, for example, that they will be subject to political "changes of heart." For manufacturers who must invest early and heavily, "changes of heart" have serious financial implications.

1. Accelerating the Uptake of Existing Vaccines

Increasing the uptake of existing vaccines sends a clear, credible signal to industry that the public sector is committed to purchasing and introducing new vaccines. The immediate allocation of hard cash is the best way to strengthen the market. However, although strengthening the market is the most credible way to assure industry of public support for vaccines in the long term, it is likely to be inadequate to overcome the immediate financial risks and costs inhibiting development in the short term.

2. Prizes and Tournaments

Prizes are generally offered to whomever achieves a prespecified goal or product, whereas tournaments generally offer a reward to whomever has progressed farthest toward the target by a given date. The size of the prize generally corresponds to the level of investment required to win. Although structured slightly differently, prizes and tournaments have similar advantages and risks. Both mechanisms theoretically encourage competition and, therefore, could increase the number of potential vaccine candidates. They also would be relatively easy for the public sector to implement. For the private sector, good public relations may be a valuable side effect of winning a prize or a tournament. However, prizes have a number of serious deficiencies, most notably industry's lack of enthusiasm for them. Industry representatives have commented that they are "in the business of making vaccines, not winning prizes" [25]. Another serious drawback to prizes and tournaments is that they do not guarantee that a viable, affordable vaccine will be developed or produced, however, the prize or tournament prize must be awarded to the winner regardless.

3. Extending and Transferring Patents

Transferable patents would give a manufacturer the right to extend the patent of any product in its portfolio in several large industrial markets. To allow a firm that successfully develops a priority product to extend any patent in its portfolio is a powerful mechanism because the firm could extend the revenue stream of blockbuster products. This mechanism captures the attention and interest of large pharmaceutical players as well as small firms because it would trigger a "trickle-down" effect. Pharmaceutical and venture capital companies would likely increase investment flows to smaller firms in hopes of gaining the right to extend a different patent in their portfolio. This trickle-down effect in some ways mitigates a disadvantage of transferable patents, which is that they tend to favor large pharmaceutical companies that are more likely to have blockbuster drugs than smaller start-ups. There are two significant issues with this mechanism. First, the willingness of consumers to essentially subsidize vaccines by accepting higher prices on their patented drugs is unknown. Second, the political and legal implications of implementing this mechanism present a serious roadblock. In order for a patent to be transferred or extended, major exceptions to patent laws must be made in each country in which the patent would be affected. These exceptions are not likely to be politically popular.

4. Copayments

A copayment—probably some amount not equal to the full price of the vaccine—is guaranteed to manufacturers in lieu of trying to lock in to a particular price for a vaccine. Copayments are attractive to industry because they are viewed as a commitment today, which is more credible than a promise for future payment. They are attractive to the public sector because they delay negotiations on the full price for a vaccine. However, it is unclear how the "correct" copay amount should be derived and if the copayment amount, which could end up being a fraction of the vaccine price, would offer enough of an incentive to industry.

5. Differential (Tiered) Pricing

As noted earlier in the chapter, differential pricing refers to the practice of selling vaccines at different prices in different markets. In order for differential pricing to work, the vaccine market must be segmented by purchasing power, have minimal risk of parallel imports, and a willingness to accept and enforce differential prices, particularly in the markets with the higher prices. Differential pricing is currently the only mechanism that allows affordable access to a vaccine in developing and industrialized countries, while still allowing industry to recoup its investment on a vaccine. In many cases, the time delay before the product is fully "mature" and has a significantly lower tiered price is 15–20 years. However, to allow industry to charge "what the market will bear" in industrialized countries provides a significant incentive for investment, an incentive that would not exist if the price was set by the purchasing ability of sub-Saharan Africa. Finally, differential pricing is a familiar mechanism, as it has been used in nearly ever other sector from toothpaste to batteries and is often integral to widespread use of products. Differential pricing, however, is difficult to implement. Safeguards must be instituted to prevent lower tier prices from becoming the index down to which wealthy countries negotiate. The risk of products flowing back into wealthy markets either between countries or in the same country, though small to date, must also be managed. Finally, for differential pricing to be successful, government procurement agencies, politicians, and the public in industrialized nations must be willing to pay more for vaccines than the poorest developing countries, an idea that is often at odds with political agendas.

6. Market Assurances

Market assurances are a type of "guaranteed" off-take. The assurance may be structured as a commitment to an individual manufacturer or as a commitment to purchase a product from any manufacturer. Three variables may be part of a guarantee between the public to buy and the private sector to supply a given vaccine: (1) the specific product, (2) the number of doses, and/or (3) the time frame. The contract or agreement may be backed by either "soft" or "hard" money guarantees, depending on the trade-off between opportunity cost and credibility: money put aside now is the most credible to industry, but it carries the highest opportunity cost. Although the credibility of market assurances theoretically can be increased through legally binding commitments, in reality it is difficult to imagine how they would be enforced against public institutions like WHO, UNICEF, or the World Bank. Issues of credibility aside, industry may still deem the commercial return to be too distant and uncertain to be worthwhile given the immediate, high-risk investments under consideration.

7. Intellectual Property Right (IPR) Enforcement–Protection

The public sector could potentially make a concerted commitment to enforce IPRs. IPRs provide a company with a period of product exclusivity during which to recover the costs of its investment. IPRs are the foundation upon which the profit of pharmaceutical companies is based. When IPRs are violated, the incentives for industry to invest in vaccine research and development are greatly diminished. The public sector has already taken steps in this direction [e.g., setting up the WTO to enforce the Agreement on Trade Related Aspects of Intellectual Property Rights (TRIPS)], but improvements remain to be made. However, simply put, IPRs provide continuing incentives for industry to invest in vaccine research, development, and distribution. IPRs, however, are difficult to implement, difficult to enforce, and politically very unpopular with those who contend that cheap drugs need to be made available to bridge the gap between the "haves" and the "have nots." Questions remain about whether it will be possible to enforce TRIPs and whether this mechanism provides enough of an incentive to ensure adequate investments.

8. Tax Credits for Vaccine Sales

Like tax credits for vaccine research and development, tax credits for vaccine sales are perceived as a credible and familiar policy tool. From the standpoint of the public sector, tax credits are attractive because money is only spent once a product has been developed and sold. The down sides, however, are many. First, industry has not expressed much interest in this mechanism possibly because tax credits on vaccine sales are too far in the future to justify investment that may not even result in a successful vaccine. In the event that a tax credit is enacted, it may be difficult to determine the appropriate level of credit. Finally, it is unclear how priorities will be set in the event that a number of these vaccines become available simultaneously, all competing within the same authorized cap.

D. Implementing Push–Pull Mechanisms: Public–Private Partnerships

The push and pull mechanisms discussed in the previous sections are potential solutions to many of the economic challenges inhibiting rapid development and scale-up of priority vaccines. However, many of these ideas have yet to translate into concrete actions by either public and private sector partners.

Ultimately, public–private partnerships allow the sectors to share the risks and costs of developing and introducing priority vaccines in novel ways. However, both sectors have concerns about entering into partnerships. A history of low uptake of vaccines and changing public priorities reduces industry's incentive to tightly partner with the public sector. Conversely, concerns about industry's motives as well as the perceptions of the popular press are disincentives for the public sector to engage in partnerships.

Despite these fears, partners on the GAVI board (see box on The Role of GAVI and the Vaccine Fund) believe that by harnessing the strengths of both private industry and public partners they can achieve the ambitious goal of developing and making accessible priority vaccines for developing countries. To move forward with credible partnerships, two basic conditions must be met. First, both partners must understand the costs, risks, and benefits driving the partnership. With this knowledge, partners can identify those costs and risks that are important and sensitive to public sector support—those with the highest "leverage." Second, both partners must be confident that the mechanisms or agreements that define the partnership protect each of their interests. For the public sector, this means ensuring that public investments result in more rapid development, expanded capacity, and/or lower prices. For the private sector, it means public "promises" translate into real financial commitments that cover investments.

The Role of GAVI and the Vaccine Fund

The Global Alliance for Vaccines and Immunization (GAVI) represents a historic alliance of public and private sector partners assembled into a worldwide network. These partners include developing country governments, the Bill & Melinda Gates Foundation, public health and research institutions, UNICEF, WHO, the World Bank, the International Federation of Pharmaceutical Manufacturers Association (IFPMA), bilateral governments, the Developing Country Vaccine Manufacturers Network, and others. GAVI's Board of Directors consists of top officials from its members and is currently chaired by Ms. Carol Bellamy, Executive Director of UNICEF, for a 2-year term.

To fulfill its mission of protecting children of all nations and of all socioeconomic levels against vaccine-preventable diseases, GAVI has established six strategic objectives [34]:

- Improve access to sustainable immunization services
- Expand the use of all existing, safe, and cost-effective vaccines where they address a public health problem
- Accelerate the development and introduction of new vaccines and technologies
- Accelerate research and development efforts for vaccines needed primarily in developing countries
- Make immunization coverage a centerpiece in international development efforts
- Support national and international accelerated disease control targets for vaccine-preventable diseases

GAVI partners believe they can best achieve their objective to accelerate the development of and access to priority new vaccines for the developing world by harnessing the complementary product expertise of public organizations and private firms. The GAVI Board has mandated that partners move the public–private sector discussion from rhetoric to action, identifying and addressing the specific scientific, financial, logistical, and policy hurdles currently blocking the rapid development and use of three priority vaccines.

Selecting Priority Vaccines

The GAVI research and development task force targeted three priority vaccines for special GAVI support: meningococcal A conjugate to control meningitis epidemics in Africa, pneumococcal conjugate vaccine to help prevent respiratory disease, and rotavirus vaccines to help prevent severe diarrhea. These vaccines were selected on the basis of the following criteria: disease burden in the developing world, scientific feasibility of the vaccine, projected time until a vaccine would be available for introduction, alternative public health measures available to control the disease, and public perception of the disease.

Current and Proposed Implementation

Through the four GAVI task forces (research and development, advocacy, financing–economics, country coordination), GAVI partners are working to identify and address the barriers to rapid development, scale-up, introduction, and use. As part of the comprehensive work plan, an analysis of the economic obstacles inhibiting development and the optimal vaccine strategies to accelerate access will be conducted.

About the Vaccine Fund

The Vaccine Fund works in parallel with GAVI to support the immunization objectives. The Vaccine Fund was created with a generous initial grant of $750 million from the Bill & Melinda Gates Foundation. Since that time, the governments of Canada, Denmark, the Netherlands, Norway, the United Kingdom, and the United States have all contributed to support the Vaccine Fund, pushing its total resources to above $1 billion for 2001–2005. More countries are expected to contribute to the Vaccine Fund, and contributions from corporations and foundations will also be pursued. The Vaccine Fund is designed to be catalytic and focused, complementing more sustainable, comprehensive program funding from governments, development banks, and bilaterals.

As this book goes to press, the Vaccine Fund has opened all three of its possible funding windows: (1) provision of new and underused vaccines, with corresponding safe immunization equipment, (2) funding to help governments strengthen their basic immunization services, and (3) funding to support activities accelerating the development of priority vaccines and technologies.

VI. NEXT STEPS AND NEW APPROACHES

As daunting as the science is, the risks and uncertainties of vaccine development are amplified by the economic implications of each investment decision. By understanding the current vaccine market, its suppliers and consumers, and the investment decisions occurring along the pathway from candidate to commercialized product, the reader can understand the role played by economics in shaping the vaccine world. Public sector partners are increasingly using this growing knowledge to explore possible solutions—push and pull mechanisms—that might change the economics of high-priority products for the developing world.

New mechanisms and partnerships will provide concrete evidence about both the critical obstacles inhibiting progress on specific vaccines and the value and feasibility of particular push and pull mechanisms. This evidence will increase the immunization community's understanding of the economic forces influencing the development of and access to the highest priority vaccines.

The successful development and introduction of vaccines against the great plagues of our time will be the ultimate test of the immunization community's scientific knowledge and economic savvy. By working together now, public and private partners not only will identify the most feasible, high-impact solutions but more importantly will learn how to work together to jointly tackle the biggest challenges facing the world—vaccines that can prevent HIV–AIDS, malaria, and tuberculosis.

REFERENCES

1. Kettler, H. E. (2000). "Narrowing the Gap between Provision and Need for Medicines in Developing Countries." London: Office of Health Economics.
2. Global Health Forum (2000). "Creating Global Markets for Neglected Drugs and Vaccines: A Challenge for Public–Private Partnership." Global Health Forum I: Consensus Statement. San Francisco.
3. International AIDS Vaccine Initiative (2000). Scientific Blueprint 2000, "Accelerating Global Efforts in AIDS Vaccine Development." New York: International AIDS Vaccine Initiative.
4. Ainsworth, M., and Batson, A. (2000). "Accelerating an AIDS vaccine for developing countries: Recommendations for the World Bank." (http://iaen.org/vacc/accelerateb.pdf).
5. World Bank (1994). "World Bank Population Projections 1994–1995." Baltimore, MD: Johns Hopkins University Press.
6. Siwolop, S. (2001). Big Steps for the Vaccine Industry. *The New York Times*, 25 July 2001, Section C1, C17.
7. Kaddar, M., Levin, A., Doughterty, L., and Maceira D. (2000). "Costs and Financing of Immunization Programs: Findings of Four Case Studies." Special Initiatives Report 26. Bethesda, MD: Partnerships for Health Reform Project, Abt Associates, Inc.
8. Shann, F., and Steinhoff, M. C. (1999). Vaccines for children in rich and poor countries. *The Lancet* 354(Suppl. II).
9. American Home Products Corporation (2000). Annual Report.
10. CVI Forum (1996). Special Vaccine Industry Issue, Number 11, Geneva, Swizerland.
11. CVI Mission to the Philippines (1993). Task Force on Situation Analysis Report, Geneva, Switzerland.
12. CVI Mission to Brazil (1994). Task Force on Situation Analysis Report, Geneva, Switzerland.
13. CVI Mission to Iran (1993). Task Force on Situation Analysis Report, Geneva, Switzerland.
14. CVI Mission to Bangladesh (1992). Task Force on Situation Analysis Report, Geneva, Switzerland.

15. CVI Mission to Nigeria (1996). Task Force on Situation Analysis Report, Geneva, Switzerland.
16. CVI Mission to Pakistan (1993). Task Force on Situation Analysis Report, Geneva, Switzerland.
17. CVI Mission to Egypt (1993). Task Force on Situation Analysis Report, Geneva, Switzerland.
18. CVI Mission to the Republic of South Africa (1993). Task Force on Situation Analysis Report, Geneva, Switzerland.
19. Batson, A. (1998). Win-win interactions between the public and private sectors. *Nature Medicine Vaccine Suppl.* 4(5):487-491.
20. www.supply.unicef.dk/insideSD/immunization.htm, January 24, 2002.
21. Mercer Management Consulting (1997). Presentation on the Economic Framework for Global Vaccine Supply: Optimal Methods to Meet Global Demand. Presented at the CVI/Rockefeller Conference on Global Supply of New Vaccines, Bellagio, Italy, February 3–7, 1997.
22. Whitehead, P. (1999). Global Alliance Board Discussion Paper: Public Sector Vaccine Procurement Approaches, October 28, 1999, New York.
23. Milstien, J. *et al.* (2001). Divergence of Products for Public Sector Immunization Programs. Presentation to the WHO Scientific Advisory Group of Experts (SAGE).
24. CVI Forum (1998). Combination Vaccines—Juggling with the Options, Number 16, July 1998, Geneva, Switzerland.
25. Batson, A., and Ainsworth, M. (2001). Obstacles and Solutions: Understanding Private Investment in HIV/AIDS Vaccine Development. *WHO Bull.* 79(8):721–728.
26. GAVI (2001). Revised Guidelines and Forms for Countries to Use in Submitting Proposals to GAVI and the Vaccine Fund.
27. "Vaccine Challenges: Why is Developing a Malaria Vaccine Difficult?" found at www.malariavaccine.org/mal-vac2-challenge.htm, January 25, 2002.
28. Miller, E., Salisbury, D., and Ramsay, M. (2002). Planning, Registration, and Implementation of an Immunisation Campaign Against Meningococcal Serogroup C Disease in the U.K.: A Success Story. *Vaccine* 20:S58–S67.
29. Batson, A., and Bekier, M. (2001). Vaccines where they're needed. *The McKinsey Quarterly*, Special Edition on Emerging Markets, No. 4, 103–112.
30. http://www.who.int/vaccines/en/meningACproject.shtml, January 25, 2002.
31. Kremer, M. (2000). "Creating Markets for New Vaccines. Part II: Design Issues." Cambridge: National Bureau of Economic Research, May 12, 2000.
32. Madrid, Y. (2001). "A New Access Paradigm: Public Sector Actions to Assure Swift, Global Access to AIDS Vaccines." New York: International AIDS Vaccine Initiative.
33. Performance Innovation Unit (2001). "Tackling the Diseases of Poverty." A Report by the Performance Innovation Unit (PIU), UK Cabinet, May 2001. http://www.cabinet-office.gov.uk/innovation/healthreport/default.htm on January 16, 2001.
34. www.vaccinealliance.org, January 25,2002.
35. Batson, A. (2002). Cost and economics of modern vaccine development. (F. Brown and I. Gust, Eds.: Orphan Vaccines). *Dev. Biologicals* Vol: 110, pp. 15–24, Karger, Basel, Switzerland.

8 VACCINE SAFETY: REAL AND PERCEIVED ISSUES

NEAL A. HALSEY
Institute for Vaccine Safety, Johns Hopkins University, Baltimore, Maryland 21205

I. INTRODUCTION

Vaccines are the most effective tools available for the prevention and control of infectious diseases. Widespread use of vaccines has prevented millions of premature deaths, paralysis, blindness, and neurological damage. Nevertheless, since the smallpox vaccine was developed more than 200 years ago, vaccines have been controversial because of concerns about safety. Vaccines, which are administered to healthy people, are held to a higher safety standard than medications to treat people who are already ill, and vaccines are often given universally to infants and children so that even a very low risk of serious side effects can result in a substantial population-attributable risk.

The tolerance for adverse events associated with vaccines varies because of real and perceived differences in the risks and severity of the illness prevented. Increased public awareness of product safety has led to decreased tolerance for vaccine-associated adverse events. As with air and highway travel, food products, and toys, advocacy groups argue for increased attention to the safety of medications and vaccines. When infections such as measles, diphtheria, and polio were common in our society, it was easier to accept occasional rare serious side effects from vaccines as a necessary risk. As immunization programs have become more successful and the risk of contracting preventable diseases has diminished, the acceptance of side effects from immunizations has also decreased.

Smallpox vaccine caused several serious adverse events, including eczema vaccinatum, encephalitis, and progressive debilitating infections in patients with immunodeficiency disorders [1]. When the risk of smallpox diminished to near zero in the United States, the acceptance of these serious adverse events decreased, and routine immunization against smallpox ended in 1972, 6 years prior to the interruption of transmission of smallpox in Africa [2]. Communication about the benefits and the risks from vaccination has become much more complicated due to the increased number of vaccines available, the declining incidence of some vaccine-preventable diseases, and the fact that some new vaccines protect against diseases that do not normally cause serious complications.

II. HOW VACCINES ARE EVALUATED FOR CAUSAL ASSOCIATIONS WITH ADVERSE EVENTS

Causal associations can usually be determined by isolating a live vaccine agent in affected tissue or through epidemiologic studies demonstrating an increased risk of the disorder in vaccine recipients compared to appropriate controls. A more detailed discussion of this process can be found in a recent publication [4].

A. Identification of Vaccine Agents in Affected Tissues

Individual case reports usually provide insufficient evidence to establish causal associations. However, if a vaccine virus or bacterium is isolated from affected tissue, the organism is not found in controls, there is no evidence of contamination, and there is no other explanation for the illness, the evidence is strongly suggestive of a causal association. For example, bacilli Calmette–Guérin (BCG) vaccine occasionally causes osteomyelitis or joint infections, as evidenced by isolation of the organism from bone or joint tissue in affected patients [4]. Similarly, measles vaccine virus has been identified in the lung tissue of children with leukemia and in one patient with HIV infection [5, 6]. If the vaccine organism routinely infects the affected tissue and the identification is made during the window of time when the organism would be found, other possible explanations need to be ruled out before accepting the evidence that the agent caused the disorder. If, however, the vaccine agent is usually present in blood or body tissues for only a short window of time following vaccination and subsequently is not detectable in normal individuals, identification of the vaccine agent in persons who had been vaccinated much earlier provides suggestive evidence that the persistence of the agent may contribute to a causal relationship. One must be cautious, however, when interpreting the findings because it is possible that some infectious agents persist in lymph nodes, brain, or other tissues [3, 7, 8]. Numerous false assumptions about causative agents for multiple sclerosis have been made on the basis of laboratory tests that were later found to be false positives, or the agent was found in normal tissue as well as in persons affected by the disease [7, 9]. Contamination of specimens at the time of collection, during processing, or during laboratory analyses has resulted in false assumptions that the agent was present in affected tissue. The use of molecular techniques to identify infectious agents, including immunohistochemical staining or PCR, has resulted in a proliferation of investigations of infectious agents as possible causes of chronic disorders. Unfortunately, these techniques are often associated with false positive results [8]. Therefore, most experts wait for confirmation by several investigators using specimens collected and processed separately before accepting evidence of the presence of the organism in affected tissue.

B. Epidemiologic Studies

Before vaccines are licensed by regulatory authorities, controlled trials are performed to compare individuals who receive vaccine with individuals who receive placebo or a control vaccine. These controlled studies provide the most powerful evidence for establishing causal associations between vaccines and adverse events. For example, when the first live attenuated measles vaccines were developed, comparison of children who received measles vaccine with children who did not receive the vaccine revealed increased rates of fever from 5 to 10 days after vaccination as compared with children who received only immune serum globulin (Fig. 1) [10]. The rates of fever were higher than the rates (5–15%) noted after the further attenuated vaccines that are in use today [11]. A small proportion of children who did not receive vaccine developed febrile illnesses during the study due to intercurrent infections. Therefore, it can be difficult with any individual child to determine whether a fever (or other adverse event) occurring during the window of time when an increased risk occurs is due to a reaction to the vaccine or to some other illness; however, these studies can establish whether there is an increased risk of a disorder during a specified period of time following vaccination. Controlled trials are very useful for identifying common adverse events that occur within a relatively short time after vaccination. Prelicensure, prospective randomized studies usually are not designed to detect delayed-onset adverse events, and studies usually are limited to 5,000 vaccinees and an equal number of controls. These studies can detect a doubling of the adverse events that occur in the control population at a rate of 1 in 100 or higher, but these studies have insufficient power to detect rare adverse events or adverse events that might occur

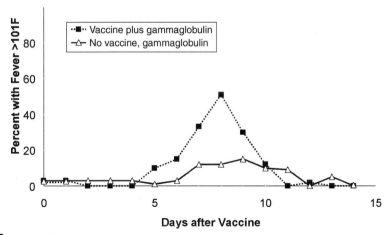

FIGURE 1   Percent of children with fever following Edmonston B measles vaccine (1963). Adapted with permission from ref 10.

months or years after vaccination [12, 13]. In order to increase the ability of studies to detect these rare events, expanded trials involving 10,000–50,000 individuals are needed. For vaccines that are likely to be given to all children, such studies can be justified, but the cost of performing such studies is high.

After licensure, monitoring of adverse events following vaccination involves healthcare providers who observe and report adverse events, vaccine manufacturers, and regulatory authorities. In the United States, reports of adverse events are submitted to the Vaccine Adverse Events Reporting System (VAERS), which is maintained jointly by the Centers for Disease Control and Prevention (CDC) and the Food and Drug Administration (FDA) [14]. The purpose of this system is to monitor reports of adverse events that might signal the need for further study. For example, prelicensure studies identified only 5 children who developed intussusception among the 10,000 who participated in clinical trials of rhesus rotavirus vaccine, and there was no consistent pattern of timing or dose of vaccine in the children who developed the intussusception [15]. After several hundred thousand children had been vaccinated, 9 reports to VAERS of intussusception occurring within 15 days after vaccination triggered case–control and cohort studies conducted by CDC that revealed a causal association [16, 17]. The risk of intussusception is now estimated to be approximately 1 in every 5,000–10,000 children vaccinated, a rate too low to be detected in the prelicensure studies.

Black [18] has shown that the generation of controlled data from large numbers (10,000–40,000) of individuals postlicensure is possible and practical in large health maintenance organizations. Such studies are being conducted with the recently licensed pneumococcal conjugate vaccine. Chen [19] has pointed out the potential limitations of such studies, including the potential for less healthy children to not receive vaccines, "confounding by contradiction." Efforts are made in the analyses of data generated from these studies to evaluate and adjust for other measures of healthcare-seeking behavior.

CDC has implemented a program to collect controlled data on large numbers of individuals by linking immunization records with all health outcomes in the Vaccine Safety Datalink (VSD) [14, 19]. This program includes approximately 2.5% of the entire United States birth cohort and provides the opportunity to compare vaccinated and unvaccinated individuals in the same geographic area for adverse events and to adjust for factors that might contribute to these events.

C. Misunderstanding Causality Assessment

Unfortunately, there have been misunderstandings as to how case reports and the VAERS system can be used [14, 20]. Limitations of this program include incomplete reporting, which precludes the verification of diagnoses, the absence of denominator information regarding the number of individuals vaccinated, and the absence of rate data in persons who did not receive the vaccine.

Too often, affected individuals and their physicians incorrectly assume that a vaccine administered prior to the onset of a disorder provides conclusive evidence that the vaccine caused the disorder [21–25]. Reports of temporal associations do not constitute evidence for causal associations, but these reports can provide clues to indicate the need for additional studies to determine whether there is a causal association with the vaccine. In the absence of a specific laboratory test, as discussed earlier, a temporal association can only be used to generate hypotheses. The number of reported events alone is not evidence of a causal association, but if the number of events exceeds the number expected due to chance, this can signal the need for more formal controlled studies. In large countries such as the United States, it is possible to collect many hundreds of cases who have developed specific disorders after vaccination, even if the disorders are relatively rare. For example, false assumptions were made that silicone breast implants were responsible for causing autoimmune disorders based upon collections of reports of women who had breast implants and subsequently developed autoimmune disorders [26–28]. The use of silicone breast implants was stopped and the manufacturer filed for bankruptcy because of the large number of lawsuits. Several years later, the scientific evidence from controlled observations indicated that there was no increased risk of autoimmune disorders associated with silicone breast implants [26–28]. Increased efforts need to be made to introduce better science into the legal process and to avoid similar mistakes with regard to adverse events following vaccination [29].

III. TYPES OF ADVERSE EVENTS CAUSED BY VACCINES

Vaccines are known to cause adverse events by several different mechanisms (Table 1).

A. Injection Process

1. Pain

Most vaccines are given by injections, which cause pain at the injection site. Pain, the most common adverse event associated with immunizations, is usually mild to moderate in severity and short-lived. Pain can be reduced by distraction methods and techniques that stimulate alternative stimulation of other areas, such as pressure or rubbing, or feeding the individual with sugar just prior to injections [30–32].

Table 1　Mechanisms Involved in Adverse Events Caused by Vaccines

1.	Injection process
2.	Incomplete inactivation of vaccine agent
3.	Replication of a live vaccine agent
4.	Inadvertent contamination of vaccine with other live agent
5.	Direct effect of vaccine component (pyrogens, adjuvants, preservatives, etc.)
6.	Host immune response to vaccine component (normal or aberrant)

2. Fainting

Rarely, serious adverse events have occurred due to fainting after receiving vaccines, including head injuries that have resulted in skull fracture, cerebral bleeding, or cerebral contusions [33]. The majority (63%) of fainting episodes occurred within 15 min after vaccination, and a disproportionate number of episodes occurred in adolescents. In order to minimize injuries from such events, expert committees often advise that individuals should be observed for 15 min after immunization to minimize the potential adverse events associated with fainting while walking down stairways or other places more prone to cause injury than sitting in a chair [34].

Other serious events associated with the injection process are shown in Table 2.

3. Tissue Injury

The most common injury associated with needle sticks has been damage to nerves from the needle stick. Sciatic nerve damage now occurs less frequently since the World Health Organization, CDC, and American Academy of Pediatrics discouraged use of the buttocks for vaccine administration [34–36]. Because alternative sites are almost always available for administering vaccines, there is little justification to administer vaccine in the buttock region. Moreover, the large fat pad in this region can result in subcutaneous injections and decreased immunogenicity, as occurred with hepatitis B vaccines [37].

4. Provocation Polio

When children incubating wild-type poliovirus infections receive injections, there is an increased likelihood of residual paralytic diseases in the extremity injected [38]. The damage to small nerve endings probably provides entrance to the nervous system for polioviruses circulating in the bloodstream, which subsequently travel to the spinal cord and damage the motor neuron. A study in Romania revealed that multiple injections were associated with residual paralysis from oral poliovirus vaccines (discussed later) [38].

5. Errors in Reconstitution

Vaccines that have been lyophilized require reconstitution in diluent (usually water) provided by the manufacturer. Occasionally, healthcare workers have mistakenly used similar-appearing vials of medications to reconstitute vaccines, resulting in overdoses and unintended effects [39]. Commonly recognized mistakes have included the administration of agents such as

TABLE 2   Injection-Related Serious Adverse Events

1.	Pain
2.	Fainting and associated injuries
3.	Tissue injury
4.	Provocation polio
5.	Errors in reconstitution
6.	Contamination of multidose vials

succinylcholine or pavulon (pancuronium bromide), resulting in temporary paralysis or respiratory arrest. These problems can be avoided by storing vaccines separately from other medications, packaging vaccines with the diluents, and training healthcare workers to carefully read the vials of all diluents prior to administration.

6. Contamination of Multidose Vials

Vaccines that come in multidose vials should be used within a few hours of opening if they do not contain a preservative. In at least three countries, multidose vials of measles vaccines that were inappropriately kept overnight became contaminated with *Staphylococcus aureus*, which multiplied and caused septic shock or toxic shock syndrome [39]. Multidose vials of DTP contain preservatives to minimize the potential for bacterial contamination. However, thimerosal in whole cell DTP was insufficient to prevent the growth of *Streptococcus pyogenes* (group A streptococcus), and several clusters of cellulitis, sepsis, and abscesses have been reported [39, 40]. These problems could be prevented by using more effective preservatives or single-dose vials without preservatives.

B. Incomplete Inactivation of Vaccine Agent

Historically, serious adverse events have been caused by the inadvertent administration of wild-type infectious agents instead of inactivated agents. In 1955, several companies produced inactivated poliovirus vaccines following procedures modified from methods used for vaccines produced for experimental field trials [41–43]. One manufacturer's product was associated with paralytic disease due to incomplete inactivation of the wild-type polioviruses. When the manufacturer scaled up production from 50- to 500-ml vials, sediment formed at the bottom of the vials, allowing for protection of the wild-type virus from the action of formaldehyde [41–43]. This event resulted in the establishment of the Division of Biological Standards (currently the Center for Biologics Evaluation and Research of the Food and Drug Administration), which monitors the safety of all vaccines and related biological products [44]. Current good manufacturing practices should prevent recurrences of this type of problem, as all lots of inactivated vaccines must be shown to have complete inactivation of vaccine agents. In addition, rigorous safety testing and annual reviews are performed for all steps in the manufacture of vaccines [45].

C. Replication of Live Vaccine Agent

For live attenuated vaccines, replication of the vaccine agent in the body produces a mild infection, which results in fever, malaise, myalgia and other adverse events triggered by the release of cytokines. For example, increased rates of fever occurred in children who received the original attenuated measles vaccine in the 5–10 days after vaccination as compared to children who received immune globulin only (Fig. 1) [10].

Increased rates of rash occurred during a similar time window. Similarly, about 15% of children who receive varicella vaccine develop a mild fever, and

3–4% develop a mild varicella-like rash 10–42 days after vaccination [46]. Some live attenuated vaccine agents can cause diseases similar to the wild-type agent. For example, BCG can cause bone or joint infections [47]. In normal hosts, these infections usually are self-limited and mild.

Vaccine-associated paralytic poliomyelitis (VAPP) is a rare complication of live oral poliovirus vaccination, occurring in approximately 1 in 760,000 first vaccinations[48]. Approximately one-fourth of affected individuals are found to have a definable immunodeficiency disorder, but the majority of cases of VAPP occur in otherwise normal hosts. Future technological developments such as the use of genetic arrays may provide further insight into why some people develop these complications in the absence of other definable immunodeficiency states.

D. Inadvertent Contamination of Vaccines with Other Live Agents

In 1962, simian virus 40 (SV40) virus was discovered to be a contaminant of the monkey kidney cells used to produce oral and inactivated polio vaccines [49]. This infectious agent had not been previously identified because the virus does not cause cytopathic effects in the cell lines that had been used for safety testing. SV40 infection was shown to be associated with selected tumors in animals and several investigators have identified SV40 virus in mesotheliomas and other tumors [50]. SV40 virus has been identified in persons who never received vaccines that might have contained the virus, and other investigators have not found evidence of these viruses in tumors. A causal relationship between SV40 viruses and any human disease has not been demonstrated.

In 1942, an outbreak of hepatitis occurred involving 25,585 United States military recruits who had received yellow fever vaccine. The source of the infection was immediately suspected to be the human serum used as a stabilizer in the vaccine, which was replaced with bovine serum in 1942. In 1987, epidemiological studies of individuals who had received the contaminated vaccines and controls demonstrated that the human serum had been contaminated with hepatitis B virus [51]. In 1966, avian leukosis virus was also found to be a contaminant in 17D yellow fever vaccines [52]. All vaccines produced since the early 1970s are free of this virus.

Current manufacturing practices include intensive testing of all vaccine additives to assure the absence of detectable infectious agents. Although questions have been raised with regard to the use of bovine serum due to theoretical concerns about the possibility of transmission of bovine spongiform encephalopathy, experts agree that this risk is extremely unlikely [53]. Nevertheless, regulatory authorities now require that any bovine products used in vaccine production must come from countries that are free of bovine spongiform encephalopathy [54] (http://www.who.int/vaccines-diseases/safety/hottop/bse.shtml, http://www.fda.gov/cber/bse/risk.htm).

Some vaccines under consideration for human testing will require new cell lines for production, including continuous cell lines that have been transformed by molecular techniques. These considerations have raised theoretical concerns about potential infectious agents, including oncogenic viruses in cell lines that might be used for vaccine production. A conference on this topic

summarized the concerns and the steps that can be taken to test for these effects [55]. Regulatory authorities must depend upon applying the best scientific methods available at any point in time to assure the safest possible vaccine production. As new information and tools become available, testing methods need to be updated.

E. Direct Effect of Vaccine Component

Vaccines, especially whole bacterial vaccines, often contain pyrogens that cause fever by the release of cytokines from macrophages. For example, whole cell pertussis vaccines induce fever in 30–50% of vaccine recipients [56]. Adjuvants enhance the antibody response to vaccines, but aluminum hydroxide and aluminum phosphate often induce local reactions, such as induration and swelling by stimulating or enhancing an inflammatory response [57]. Other vaccine components may have undesirable effects. The preservative thimerosal has been used for many years in a variety of vaccine products. Thimerosal can induce hypersensitivity reactions, which are usually localized [58]. One of the breakdown products of thimerosal is ethylmercury, which can cause neurologic damage when administered in large doses [58]. An FDA analysis revealed that the use of multiple thimerosal-containing vaccines in infants could result in cumulative exposures that exceed some federal guidelines for methylmercury [58]. In 1999, the United States Public Health Service, Academy of Pediatrics, and European EMA issued statements encouraging the reduction or elimination of the use of thimerosal in vaccines administered to infants as soon as possible [59]. A review by the Institute of Medicine concluded that current evidence was insufficient to determine whether harmful effects were caused by thimerosal exposures in vaccines [60]. Ongoing follow-up studies of children who had high and low exposures should provide further information on whether there is any evidence of neurodevelopmental effects from these exposures in the United States. The amount of exposure to thimerosal was much less in most other countries of the world, as many European authorities had been phasing out this preservative and other countries had not added as many new vaccines that contained thimerosal as a preservative to the routine infant schedule [61, 62].

F. Host Immune Response to Vaccine Component

Hypersensitivity reactions, including hives, anaphylaxis, and Stevens Johnson syndrome, have been observed following the administration of many different vaccines [63, 64]. Although these reactions are usually very rare, they can be life-threatening. Hypersensitivity can be generated to vaccine agents, preservatives, stabilizers, adjuvants, or residual antimicrobial agents. For many years, there was concern about immediate hypersensitivity reactions to measles vaccines produced in chick embryo tissue culture in children with egg allergies. Careful studies using sensitive techniques have not detected residual egg protein in measles and other vaccines produced in chick embryo tissue culture [11]. Children who have had hypersensitivity reactions following measles-containing vaccines have been shown to react to the gelatin stabilizer and not to egg

protein [65, 66]. In addition, children with true hypersensitivity reactions to egg protein can be safely administered measles–mumps–rubella (MMR) vaccine [34, 67, 68]. Administration of vaccines produced in eggs, such as influenza and yellow fever vaccines, is contraindicated in persons with immediate hypersensitivity reactions to eggs, as some residual egg protein is present in these vaccines [69].

Most hypersensitivity reactions to neomycin, which is commonly used during vaccine production, are mild local reactions, and adverse reactions to the small amounts in vaccines have not been documented [36]. Other antibiotics that are commonly used for treating infections, such as penicillin and cephalosporins, are not used in vaccine production.

An increased risk of Gullain–Barre syndrome, an autoimmune disorder, was observed in persons who received the swine influenza vaccine developed in 1976 [70]. The attributable risk was approximately 1 in every 110,000 persons vaccinated. Subsequent studies demonstrated no increased risk associated with influenza vaccines administered in the 1980s, but a small increased risk of approximately 1 per 1,000,000 vaccinees was noted following influenza vaccines administered in the United States during 1992–1994 [70,71]. No other autoimmune disorders have been shown to have been caused by any vaccine (see later discussion).

IV. IMMUNE DEFICIENCY DISORDERS

Although persons with underlying immune deficiency disorders may not benefit, they are not at increased risk of complications from inactivated and subunit vaccines. Because these vaccines may provide partial or complete protection, most expert groups recommend the administration of these vaccines to all immunodeficient patients if the vaccines are otherwise indicated [34, 72].

There are many different disorders of the immune system; most are mild and do not alter the risk of adverse events from vaccines. Persons with disorders of macrophage function, such as chronic granulomatous disease, are not at increased risk of complications from viral infections. Therefore, there is no reason to expect increased complications from live viral vaccines. Many affected individuals have been vaccinated with live oral poliovirus vaccine and measles, mumps, and rubella vaccines without serious adverse events before they were diagnosed with immune deficiency disorders. However, persons with macrophage disorders are at increased risk from BCG vaccine (Table 3).

In patients with T-cell immune deficiency disorders, unchecked replication of live vaccine agents can result in severe infections and death. A progressive fatal pneumonitis developed when measles vaccine was administered to children with leukaemia [5]. Children with leukemia in prolonged remission and patients who are 2 or more years post successful bone marrow transplants can receive live viral vaccines if they are not on severe immunosuppressive therapy [72]. In general, patients with underlying T-cell immunodeficiency disorders should not receive live viral vaccines, with the exception of patients with HIV infection as noted next. Whenever there is doubt about specific disorders, consultation with an immunologist or infectious disease specialist is indicated.

TABLE 3 Vaccines Contraindicated in Patients with Underlying Immune Deficiency Disorders (United States Guidelines)

Immune deficiency	Contraindicated vaccine
B cell	OPV and live bacterial (*S. typhi* 21a and BCG; "consider" giving measles and varicella vaccines)
T cell	All live vaccines
Phagocyte	Live bacterial (BCG, Ty 21a)
HIV	OPV, BCG (measles and varicella, see Table 4)
Suppressive therapy	All live, depending on status

A. HIV Infection

Infection with the human immunodeficiency virus (HIV) induces a progressive immune deficiency state and increased risk of complications from numerous infectious agents. Severe complications from BCG vaccine have occurred in HIV-infected children and adults [73–75]. One HIV-infected adult with severe immune deficiency developed a progressive fatal pneumonitis after receiving measles vaccine, and one instance of pneumonia caused by varicella vaccine has been reported [6, 76]. However, HIV-infected persons with no or minimal evidence of immune suppression can be safely immunized with these vaccines [11]. In the United States where resources are available for routine testing, advisory groups have recommended the administration of these vaccines to HIV-infected persons with CD4 counts above cutoff points depending upon age (Table 4). There have been two persons who developed VAPP following OPV, but many hundreds of thousands of HIV-infected people have been immunized, and there is no convincing evidence that the risk of VAPP is increased in HIV-infected children [73]. In developing countries in which routine HIV testing is not performed, the advantages of routine administration with OPV and measles vaccines far outweigh the theoretical risks of complications from these vaccines. In addition, vaccination early in life often results in an adequate immune response before HIV-induced immunosuppression develops. Therefore, the World Health Organization and individual countries recommend routine universal immunization with these vaccines [73].

TABLE 4 Advisory Groups' Recommendations for Vaccine Administration to HIV-Infected Persons with CD4 Counts above Age-Dependent Cutoff Points

Vaccine	CD4 (%)	CD4 count		
		<1 year	1–5 years	6–12 years
Measles	≥15%	≥750	≥500	≥200
Varicella	≥25%	≥1500	≥1000	≥500

V.   MISUNDERSTANDINGS AND FALSE ACCUSATIONS REGARDING THE SAFETY OF COMMONLY USED VACCINES

During the past few years, there have been several false concerns raised about vaccines possibly causing serious diseases. For most of the diseases in question, medical science has generated an incomplete understanding of the etiology or pathogenesis of the disease, which provides an opportunity for speculation about the role of vaccines. In addition, in several instances, individual investigators have made observations and speculated beyond their data to imply causal relationships between vaccines and the disorders.

A.   Hepatitis B, Multiple Sclerosis, and Other Demyelinating Diseases

Multiple sclerosis is an autoimmune disorder. Epidemiologic evidence from many countries indicates that environmental factors as well as genetic predisposition contribute to the risk of developing multiple sclerosis. Infectious agents have been suggested as possible priming or triggering factors [7, 9]. When individuals developed the first episodes of multiple sclerosis after receiving a vaccine, they and some of their physicians hypothesized that the immune response to the hepatitis B vaccine (or other vaccines) could have contributed to the development of the disease. In 1994, the Institute of Medicine vaccine safety committee reviewed the available evidence regarding multiple sclerosis and hepatitis B vaccine [63]. The committee concluded that the case reports available at that time provided insufficient evidence to establish a causal relationship. Unfortunately, the committee also concluded that there was biologic plausibility for a possible association between hepatitis B vaccine and multiple sclerosis based on a study in rabbits. Investigators had found a short amino acid sequence in the myelin basic protein of rabbits that was identical to a sequence in the hepatitis B virus [77]. When these investigators immunized rabbits with an experimental protein based upon the sequence with Freund's complete adjuvant, some of the rabbits developed an autoimmune encephalomyelitis. The protein used is not present in hepatitis B surface antigen vaccines, and the genetic sequence in question is not present in human myelin basic protein. Some individuals interpreted the Institute of Medicine conclusion that "the evidence was insufficient to accept or reject a possible causal relationship" as implying that there was evidence supporting the relationship. In France, large-scale programs had been implemented in 1997 and 1998 to immunize adolescents and young adults against hepatitis B, including individuals in the age group when multiple sclerosis usually presents (20–40 years of age). Some individuals developed the onset of multiple sclerosis symptoms within 2 months after receiving hepatitis B vaccine. Although a quickly conducted case–control study did not reveal a significantly increased odds ratio for multiple sclerosis patients having received hepatitis B vaccine as compared to persons without multiple sclerosis, the minister of health decided on October 1, 1998, to terminate the hepatitis B vaccine program for adolescents and adults pending further investigations. This action was interpreted by some reports in the popular press as evidence that the government of France had evidence that hepatitis B vaccine caused multiple sclerosis. Subsequently, carefully conducted cohort and case–control studies have documented that there is no increased risk of multiple sclerosis or other demyelinating diseases following hepatitis B immunization, and there is no evidence of any vaccines triggering relapses of multiple sclerosis [78].

B. Type 1 Diabetes Mellitus and Vaccines

Type 1 diabetes mellitus is an autoimmune disease. Based on individual case reports of temporal associations and population-based increases in the incidence of type 1 diabetes, Classen [79] believed that the introduction of *Haemophilus influenzae* type b vaccines caused the disorder. In addition, mice and rats genetically predisposed to develop diabetes could be prevented from developing the disease by early immunization with BCG or other vaccines [79]. The investigator misunderstood how ecological data showing population-based changes in incidence can be used for assessing causal relationships. The incidence of diabetes is increasing in multiple age groups in countries throughout the world [81]. Noting an increased incidence after the introduction of a vaccine (or any other change in the population) does not provide evidence to support a causal relationship. Careful studies in Finland demonstrated a continuous increasing incidence of type 1 diabetes before and after the introduction of *Haemophilus influenzae* type b vaccines and no evidence of any significant difference in risk of diabetes for children who received multiple doses of this vaccine in infancy as compared to children who received only a single dose at 18 months of age [81]. Two separate expert panels reviewed the preceding data and numerous other studies and concluded that there is no evidence to suggest a causal relationship between vaccines and increased risk of diabetes [82, 84].

C.   MMR and Autism

In 1998, a gastroenterologist published a brief article implying that MMR vaccine contributed to the development of autism [84]. He had been studying the possible role of measles in inflammatory bowel disease [85]. Twelve patients who had autism had been referred to him for evaluation of gastrointestinal disorders. When he asked the parents of the twelve children (whose average age was 6 years) if the onset of disease had occurred within 2 weeks after receiving MMR, eight of the twelve parents said yes. In statements to the popular press, the investigator went beyond his data and stated that he had evidence that MMR was the cause of these children's disease. Dr Wakefield also believed that the incidence (or prevalence) of autism increased in the United States and the United Kingdom after the introduction of MMR. He believed that the simultaneous administration of measles, mumps and rubella vaccines constituted an "atypical" exposure to measles, which predisposed the vaccinee to persistent measles vaccine virus infection of the intestine and an associated inflammatory disorder that resulted in the absorption of toxins from the gastrointestinal tract, causing neurologic damage. Because the cause of autism was largely unknown, many parents of affected children were seeking explanations for their children's disease, and they were willing to accept the belief of someone who appeared to be an expert. In-depth reviews of these hypotheses were undertaken by the Academy of Pediatrics [3] and the Institute of Medicine [86]. Both groups concluded that the available data did not support the conclusions promoted by the investigator. Simultaneous administration of measles, mumps, and rubella vaccines results in immunologic responses to each of the vaccines that are similar to these of the vaccines administered separately, and there is no increased risk of adverse events with the combined vaccines, especially gastrointestinal disorders. The evidence for possible persistence of measles viruses and other paramyxoviruses in the intestinal tract and other body tissues is inconclusive, but several investigators in respected institutions around the world were unable to find evidence of measles virus in tissue biopsies from children and adults with inflammatory bowel disease [87]. There is no evidence that inflammation of the gastrointestinal tract contributes to the abnormal absorption of toxins and neurologic damage. Several epidemiologic studies have shown no association between the timing of the introduction of MMR vaccine and apparent increases in the prevalence of autism in several countries [88–90]. Dr. Wakefield had misinterpreted ecological data, which have limited value for the assessment of causal relationships, and he assumed that temporal associations implied causal associations. Expert groups in many countries strongly endorse the simultaneous administration of measles, mumps, and rubella vaccines in MMR as the most effective way to prevent these diseases [11, 34].

Other false assumptions and concerns about vaccines and diseases including asthma and neurologic damage following whole cell DTP have been reviewed by McPhillips and Marcuse [91].

VI. CONCLUSION

All vaccines and related products carry some risk of adverse events. Fortunately, most adverse events caused by vaccines are mild, and serious adverse events caused by vaccines are usually rare. When compared to the risk of serious complications from the diseases that the vaccines protect against, the risks of serious consequences are usually 1000-fold or more greater with the natural disease than the vaccine (Table 5).

There will be future vaccine safety issues. Up-to-date information can be found at a variety of reputable websites (see Table 6). Efforts need to be made to increase understanding by the general public and healthcare practitioners of

Table 5   Relative Incidence of Severe Complications from Diseases and the Vaccines Used to Protect against These Diseases

Disease–Vaccine	Complication	Disease	Vaccine	Ratio
OPV	Paralysis	1–5/1000	1/1,000,000	1,000+
Measles	Encephalitis	1/1000	1/1,000,000	1,000
Varicella	Cerebellar ataxia	4/1000	1/1,000,000	4,000+
Tetanus	Death	1/10	< 1/1,000,000	100,000+

◼◼◼ **TABLE 6** **Websites for Vaccine Information**

(AAP)	American Academy of Pediatrics	www.aap.org
(CDC/NIP)	Centers for Disease Control and Prevention	www.cdc.gov/nip
(IAC)	Immunization Action Coalition	www.immunize.org
(IOM/Immunization Safety Review Committee)		www.iom.edu/IOM/IOMHome.nsf/ pages/immunization + safety + review
(IVS)	Institute for Vaccine Safety	www.vaccinesafety.edu
(NNii)	National Network Immunization Information	www.immunizationinfo.org
(FDA/VAERS)	Food and Drug Administration Vaccine Adverse Event Reporting System	www.fda.gov/cber/vaers/vaers.htm

how vaccine safety issues are properly investigated and what constitutes evidence for causal relationships in order to minimize the risks for future large-scale misunderstandings. Efforts to closely monitor vaccines for safety before and after licensure must continue and expand as needed in order to assure the public that the vaccines used to protect against disease are as safe as possible.

REFERENCES

1. Lane, J. M., Ruben, F., Neff, J. M., and Millar, J. D. (1970). Complications of smallpox vaccination, 1968: Results of ten statewide surveys. *J. Infect. Dis.* **122**:303–309.
2. Centers for Disease Control and Prevention. (2001). Recommendations of the Advisory Committee on Immunization Practices (ACIP) of the Centers for Disease Control and Prevention. Vaccinia (Smallpox) Vaccine. *Morbidity and Mortality Weekly Report* 50(No. RR-10):1–26.
3. Halsey, N. A., and Hyman, S. L. (2001). Measles–mumps–rubella vaccine and autistic spectrum disorder: Report from the New Challenges in Childhood Immunizations Conference convened in Oak Brook, IL, June 12–13, 2000. *Pediatrics* **107**(5):E84.
4. Kroger, L., Korppi, M., Brander, E. *et al.* (1995). Osteitis caused by bacillus Calmette–Guerin vaccination: A retrospective analysis of 222 cases. *J. Infect. Dis.* **172**:574–576.
5. Siegel, M. M., Walter, T. K., and Ablin, A. R. (1997). Measles pneumonia in childhood leukemia. *Pediatrics* **60**:38–40.
6. Centers for Disease Control and Prevention. (1996). Measles pneumonitis following measles–mumps–rubella vaccination of a patient with HIV infection, 1993. *MMWR* **45**(28):603–606.
7. Cermelli, C., and Jacobson, S. (2000). Viruses and Multiple Sclerosis [Review]. *Viral Immunol.* **13**(3):255–267.
8. Fredricks, D. N., and Relman, D. A. (1999). Application of polymerase chain reaction to the diagnosis of infectious diseases. *Clin. Infect. Dis.* **29**:475–486, 487–488.
9. Johnson, R. T. (1998). Inflammatory and demyelinating diseases. *In* "Viral Infections of the Nervous System," 2nd ed., pp. 227–264. Philadelphia, PA: Lippincott-Raven Publishers.
10. Martin, C. M. (1963). *Am. J. Dis. Children* **106**:270.
11. Watson, J. C., Hadler, S. C., Dykewicz, C. A., Reef, S., and Phillips, L. (1998). Measles, mumps, and rubella—Vaccine use and strategies for elimination of measles, rubella, and congenital rubella syndrome and control of mumps. Recommendations of the Advisory Committee on Immunization Practices (ACIP). *Morbidity and Mortality Weekly Report* **47**(RR-8):1–57.
12. Ellenberg, S. S. (2000). Evaluating the Safety of Combination Vaccines. *CID* **33** (Suppl. 4):S319–322.
13. Silvers, L. E., Ellenberg, S. S., Wise, R. P., Varricchio, F. E., Mootrey, G. T., and Salive, M. E. (2001). The epidemiology of fatalities reported to the vaccine adverse event reporting system 1990–1997. *Pharmacoepidemiol. Drug Safety* **10**(4):279–285.

14. Chen, R. T., Rastogi, S. C., Mullen, J. R., Hayes, S. W., Cochi, S. L., Donlon, J. A. *et al.* (1994). The Vaccine Adverse Event Reporting System (VAERS). *Vaccine* 12:542–550.

15. Rennels, M. B., Parashar, U. D., Holman, R. C. *et al.* (1998). Lack of an apparent association between intussusception and wild or vaccine rotavirus infection. *Pediatr. Infect. Dis. J.* 17:924–925.

16. Niu, M. T., Erwin, D. E., and Braun, M. M. (2001). Data mining in the US Vaccine Adverse Event Reporting System (VAERS): Early detection of intussusception and other events after rotavirus vaccination. *Vaccine* 19(32):4627–4634.

17. Murphy, T. V., Gargiullo, P. M., Massoudi, M. S., Nelson, D. B., Jumaan, A. O., Okoro, C. A., Zanardi, L. R., Setia, S., Fair, E., LeBaron, C. W., Wharton, M., and Livingood, J. R. (2001). Intussusception among infants given an oral rotavirus vaccine. *New England J. Medicine* 344(8):564–572.

18. Black, S. (2001). Perspectives on the Design and Analysis of Prelicensure Trials: Bridging the Gap to Postlicensure Studies. *Clin. Infect. Dis.* 33(Suppl. 4):S323–326.

19. Chen, R. T., DeStefano, F., Davis, R. L., Jackson, L. A., Thompson, R. S., Mullooly, J. P., *et al.* (2000). The Vaccine Safety Datalink: Immunization research in health maintenance organizations in the USA. *Bull. WHO* 78:186–194.

20. Chen, R. T., Pool, V., Takahashi, H., Weniger, B. G., and Patel, B. (2001). Combination Vaccines: Postlicensure Safety Evaluation. *CID* 33(Suppl. 4):S327–333.

21. McPhillips, H. A., Davis, R. L., Marcuse, E. K., and Taylor, J. A. (2001). The rotavirus vaccine's withdrawal and physicians' trust in vaccine safety mechanisms. *Arch. Pediatr. Adolesc. Med.* 155(9):1051–1056.

22. Nadler, J. P. (1993). Multiple Sclerosis and Hepatitis B Vaccination. *CID* 17:928–929.

23. Pirmohamed, M., and Winstanley, P. (1997). Hepatitis B Vaccine and neurotoxicity. *Postgrad. Med. J.* 73(861):462–463.

24. Tourbah, A., Gout, O., Liblau, R., Lyon-Caen, O., Bougniot, C., Iba-Zizen, M. T., and Cabanis, E. A. (1999). Encephalitis after hepatitis B vaccination: Recurrent disseminated encephalitis or MS? *Neurology* 53:396–401.

25. Kaplanski, G., Retornaz, F., Durand, J. M., and Soubeyrand, J. (1995). Central Nervous System Demyelination After Vaccination Against Hepatitis B and HLA Haplotype [letter]. *J. Neurol. Neurosurg. Psychiatry* 58(6):758–759.

26. Angell, M. (1996). Evaluating the health risks of breast implants: The interplay of medical science, the law, and public opinion. *New England J. Medicine* 334:1513–1518.

27. Hulka, B. S., Kerkvliet, N. L., and Tugwell, P. (2000). Sounding Board: Experience of a Scientific Panel Formed to Advise the Federal Judiciary on Silicone Breast Implants [Editorial]. *New England J. Medicine* 342(11):812–815.

28. Janowski, E. C., Kupper, L. L., and Hulka, B. S. (2000). Meta-analyses of the relation between silicone breast implants and the risk of connective-tissue diseases. *New England J. Med.* 342(11):781–790.

29. Alper, P. R., Borzelleca, J. F., Botts, M. K. *et al.* (1997,1998). Silicone Breast Implants: Why Has Science Been Ignored? American Council on Science and Health, Brochure available at URL: http://www.acsh.org/publications/booklets/implants.html.

30. Reis, E. C., Jacobson, R. M., Tarbell, S., and Weniger, B. G. (1998). Taking the sting out of shots: Control of vaccination-associated pain and adverse reactions. *Pediatr. Ann.* 27(6):375–386.

31. Barnhill, B. J., Holbert, M. D., Jackson, N. M., and Erickson, R. S. (1996). Using pressure to decrease the pain of intramuscular injections. *J. Pain Symptom Manage.* 12(1):52–58.

32. French, G. M., Painter, E. C., and Coury, D. L. (1994). Blowing away shot pain: A technique for pain management during immunization. *Pediatrics* 93(3):384–388.

33. Braun, M. M., Patriarca, P. A., and Ellenberg, S. S. (1997). Syncope after immunization. *Arch. Pediatr. Adolesc. Med.* 151(3):255–259.

34. American Academy of Pediatrics (2000). *In* "2000 Red Book: Report of the Committee on Infectious Diseases," 25th ed., L. K. Pickering, Ed. Elk Grove Village, IL: American Academy of Pediatrics.

35. WHO Guidelines on vaccine administration sites.

36. Centers for Disease Control and Prevention (1994). General Recommendations on Immunization. Recommendations of the Advisory Committee on Immunization Practices (ACIP). *MMWR* 43(No. RR-01):1–38.

37. Centers for Disease Control and Prevention (1991). Hepatitis B Virus: A Comprehensive Strategy for Eliminating Transmission in the United States Through Universal Childhood Vaccination. Recommendations of the Advisory Committee on Immunization Practices (ACIP) of the Centers for Disease Control and Prevention. *Morbidity and Mortality Weekly Report* **48**(No. RR-13):1–25.

38. Strebel, P. M., Ion-Nedelcu, N., Baughman, A. L., Sutter, R. W., and Cochi, S. L. (1995). Intramuscular Injections Within 30 Days of Immunization with Oral Poliovirus Vaccine—A Risk Factor for Vaccine-Associated Paralytic Poliomyelitis. *New England J. Med.* **332**(8): 500–530.

39. World Health Organization (1996). Vaccine supply and quality: Surveillance of adverse events following immunization. *Weekly Epidemiol. Record* **71**(No. 32):237–244.

40. Bernier, R. H., Frank, J. A., Jr, and Nolan, T.F., Jr. (1981). Abscesses complicating DTP vaccination. *Am. J. Dis. Child.* **135**(9):826–828.

41. Nathanson, N., and Langmuir, A. (1963). The Cutter Incident: Poliomyelitis following formaldehyde-inactivated poliovirus vaccination in the United States during the spring of 1955. I Background. *Am. J. Hyg.* **78**:16–28.

42. Nathanson, N., and Langmuir, A. (1963). The Cutter Incident: Poliomyelitis following formaldehyde-inactivated poliovirus vaccination in the United States during the spring of 1955. II. Relationship of poliomyelitis to Cutter vaccine. *Am. J. Hyg.* **78**:29–60.

43. Nathanson, N., and Langmuir, A. (1963). The Cutter Incident: Poliomyelitis following formaldehyde-inactivated poliovirus vaccination in the United States during the spring of 1955. III. Comparison of the clinical character of vaccinated and contact cases occurring after use of high rate lots of Cutter vaccine. *Am. J. Hyg.* **78**:61–81.

44. Parkman, P. D., and Hardegree, M. C. (1999). Regulation and Testing of Vaccines. *In* "Vaccines," 3rd ed., S. A. Plotkin, W. A. Orenstein, Eds., p. 1131. Philadelphia: W.B. Saunders Company.

45. Ebbert, G. B., Mascolo, E. D., and Six, H. R. (1999). Overview of Vaccine Manufacturing and Quality Assurance. *In* "Vaccines," 3rd ed., S. A. Plotkin, W. A. Orenstein, Eds., p. 40. Philadelphia: W.B. Saunders Company.

46. Centers for Disease Control and Prevention (1996). Prevention of Varicella. Recommendations of the Advisory Committee on Immunization Practices (ACIP) of the Centers for Disease Control and Prevention. *Morbidity and Mortality Weekly Report* **45**(No. RR-11):1–36.

47. Smith, K. C., and Starke, J. R. (1999). Bacille Calmette–Guerin Vaccine. *In* "Vaccines," 3rd ed., S. A. Plotkin, W. A. Orenstein, Eds., p. 111. Philadelphia: W.B. Saunders Company.

48. Strebel, P. M., Sutter, R. W., Cochi, S. L., Biellik, R. J., Brink, E. W., Kew, O. M., *et al.* (1992). Epidemiology of poliomyelitis in the United States one decade after the last reported case of indigenous wild virus-associated disease. *Clin. Infect. Dis.* **14**:568–579.

49. Mortimer, E. A., Jr, Lepow, M. L., Gold, E., Robbins, F. C., Burton, G. J., and Fraumeni, J. F., Jr. (1981). Long-term follow-up of persons inadvertently inoculated with SV40 as neonates. *New England J. Med.* **305**(25):1517–1518.

50. Shah, K. V. (2000). Does SV40 infection contribute to the development of human cancers? *Rev. Med. Virol.* **10**(1):31–43.

51. Williams, G. (1959). Hepatitis and Hell to Pay. *In* "Virus Hunters," 1st ed. New York: Alfred A. Knopf.

52. Monath, T. P. (1999). Yellow Fever. *In* "Vaccines," 3rd ed., S. A. Plotkin, and W. A. Orenstein, Eds., p. 815. Philadelphia: W.B. Saunders Company.

53. Marwick, C. (2000). FDA calls bovine-based vaccines currently safe. *J. Am. Med. Assoc.* **284**(10):1231–1232.

54. Public Health Service (2000). Notice to Readers: Public Health Service Recommendations for the use of Vaccines Manufactured with Bovine-Derived Materials. *Morbidity and Mortality Weekly Report* **49**(50): 1137–1138.

55. Brown, F., Lewis, Jr., A. M, Peden, K., Krause, P., Eds. (2001). Evolving Scientific and Regulatory Perspectives on Cell Substrates for Vaccine Development. Proceedings of an International Association of Biologicals Symposium, September 7–10, 1999, Rockville, MD. Basel, Switzerland: Karger.

56. Steinhoff, M. C., Reed, G. F., Decker, M. D., Edwards, K. M., Englund, J. A., Pichichero, M. E., Rennels, M. B., Anderson, E. L., Deloria, M. A., and Meade, B. D. (1995). A randomized

comparison of reactogenicity and immunogenicity of two whole cell pertussis vaccines. *Pediatrics* 96:567–570.

57. Edelman, R. (1997). Adjuvants for the Future. *In* "New Generation Vaccines," 2nd ed., M. M. Levine, G. C. Woodrow, J. B. Kaper, and G. S. Cobon, Eds., pp. 137–192. New York: Marcel Dekker.

58. Ball, L. K., Ball, R., and Pratt, R. D. (2001). An Assessment of Thimerosal Use in Childhood Vaccines. *Pediatrics* 107:1147–1154.

59. AAP and PHS (1999). Joint Statement of the American Academy of Pediatrics (AAP) and the United States Public Health Service (PHS). *Pediatrics* 104:568–569.

60. Stratton, K., Gable, A., and McCormick, M. C., Eds. "Immunization Safety Review: Thimerosal-Containing Vaccines and Neurodevelopmental Disorders." Washington, DC: National Academy Press.

61. Freed, G. L. (2001). History of Thimerosal Concern and Comparative Policy Actions. Presented at Thimerosal-Containing Vaccines and Neurodevelopmental Outcomes (sponsored by IOM), July 16, 2001. Cambridge, MA.

62. Halsey, N. A., and Goldman, L. (2001). Balancing risks and benefits: Primum non nocere is too simplistic. *Pediatrics* 108(2):466–467.

63. Institute of Medicine (1994). Adverse Events Associated with Childhood Vaccines: Evidence Bearing on Causality. Washington, DC: National Academy Press.

64. Mortimer, E. A., Jr, Ball, R., Ball, L. K., Wise, R. P., Braun, M. M., Beeler, J. A., and Salive, M. E. (2001). Stevens-Johnson Syndrome after Vaccination. *Pediatr. Infect. Dis. J.* 20(8):818–819.

65. Kelso, J. M., Jones, R. T., and Yunginger, J. W. (1993). Anaphylaxis to measles, mumps, and rubella vaccine mediated by IgE to gelatin. *Allergy Clin. Immunol.* 91(4):867–872.

66. Sakaguchi, M., Nakayama, T., and Inouye, S. (1996). Food allergy to gelatin in children with systemic immediate-type reactions, including anaphylaxis, to vaccines. *J. Allergy Clin. Immunol.* 98(6 Pt. 1):1058–1061.

67. James, J. M., Burks, A. W., Roberson, P. K., and Sampson, H. A. (1995). Safe administration of the measles vaccine to children allergic to eggs. *New England J. Med.* 332(19):1262–1266.

68. Greenberg, M. A., and Birx, D. L. (1988). Safe administration of mumps–measles–rubella vaccine in egg-allergic children. *J. Pediatr.* 113(3):504–506.

69. Centers for Disease Control and Prevention. (2001). Prevention and Control of Influenza. Recommendations of the Advisory Committee on Immunization Practices (ACIP) of the Centers for Disease Control and Prevention. *Morbidity and Mortality Weekly Report* 50(No. RR-04):1–36.

70. Hurwitz, E. S., Schonberger, L. B., Nelson, D. B., and Holman, R. C. (1981). Guillain–Barre syndrome and the 1978–1979 influenza vaccine. *New England J. Med.* 304(26):1557–1561.

71. Lasky, T., Terracciano, G. J., Magder, L., Koski, C. L., Ballesteros, M., Nash, D., Clark, S., Haber, P., Stolley, P. D., Schonberger, L. B., and Chen, R. T. (1998). The Guillain–Barre syndrome and the 1992–1993 and 1993–1994 influenza vaccines. New *England J. Med.* 339(25):1797–1802.

72. Centers for Disease Control and Prevention (1996). Measles pneumonitis following measles–mumps–rubella vaccination of a patient with HIV infection, 1993. *MMWR* 45(28):603–606.

73. Moss, W. J., and Halsey, N. A. (2001). Immunization of HIV-infected Children. *Bull. WHO*, in press.

74. Msellati, P., *et al.* (1991). BCG vaccination and pediatric HIV infection—Rwanda, 1988–1990. *Morbidity and Mortality Weekly Report* 40(48):833–836.

75. Bhat, G. J., Diwan, V. K., Chintu, C., Kabika, M., and Masona, J. (1993). HIV, BCG and TB in children: A case control study in Lusaka, Zambia. *J. Trop. Pediatr.* 39(4):219–223.

76. Sharrar, R. G., LaRussa, P., Galea, S. A., Steinberg, S. P., Sweet, A. R., Keatley, R. M., Wells, M. E., Stephenson, W. P., and Gershon, A. A. (2000). The postmarketing safety profile of varicella vaccine. *Vaccine* 19(7–8):916–923.

77. Fujinami, R. S., and Oldstone, M. B. (1985). Amino acid homology between the encephalitogenic site of myelin basic protein and virus: Mechanism for autoimmunity. *Science* 230(4729):1043–1045.

78. Confavreux, C., Suissa, S., Saddier, P., Bourdes, V., and Vukusic, S. (2001). Vaccinations and the risk of relapse in multiple sclerosis. Vaccines in Multiple Sclerosis Study Group. *New England J. Med.* 344(5):319–326.

79. Classen, J. B. (1996). The timing of immunization affects the development of diabetes in rodents. *Autoimmunity* **24**(3):137–145.

80. LaPorte, R. E., Matusushima, M., and Change, Y. F. (1997). Prevalence and incidence of insulin-dependent diabetes. *In* "Diabetes in America," 2nd ed., NIH Publication 95–1468 M. I. Harris, C. C. Cowie, M. P. Stern, E. J. Loyko, G.E. Reiber, and P. H. Bennett, Eds., pp. 37–46. Bethesda, MD: National Diabetes Data Group of the National Institute of Diabetes and Digestive and Kidney Diseases, National Institutes of Health.

81. Karvonen, M., Cepaitis, Z., and Tuomilehto, J. (1999). Association between type 1 diabetes and *Haemophilus influenzae* type b vaccination: Birth cohort study. *Br. Med. J.* **318**:1169–1172.

82. Institutes for Vaccine Safety (1999). Childhood immunizations and type 1 diabetes: summary of an Institute for Vaccine Safety Workshop. *Pediatr. Infect. Dis. J.* **18**:217–222.

83. (1998). Childhood Immunization Schedule and Diabetes: An Unfounded Hypothesis. *Clin. Infect. Dis.* **27**(2): hot page.

84. Wakefield, A. J., Murch, S. H., Anthony, A., Linnell, J., Casson, D. M., Malik, M., Berelowitz, M., Dhillon, A. P., Thomson, M. A., Harvey, P., Valentine, A., Davies, S. E., and Walker-Smith, J. A. (1998). Ileal–lymphoid–nodular hyperplasia, non-specific colitis, and pervasive developmental disorder in children. *Lancet* **351**(9103):637–641.

85. Wakefield, A. J., Montgomery, S. M., and Pounder, R. E. (1999). Crohn's disease: The case for measles virus. *Ital. J. Gastroenterol. Hepatol.* **31**:247–254.

86. Institute of Medicine (2001). "Immunization Safety Review: Thimerosal-Containing Vaccines and Neurodevelopmental Disorders," K. Stratton, A. Gable, and M. C. McCormick, Eds. Washington DC: National Academy Press.

87. Afzal, M. A., Minor, P. D., Ghosh, S., and Jin, L. (2001). Measles virus persistence in specimens of inflammatory bowel disease and autism cases. *Dig. Dis. Sci.* **46**(3):658–660.

88. Taylor, B., Miller, E., Farrington, C. P. *et al.* (1999). Autism and measles, mumps and rubella vaccine: No epidemiological evidence for a causal association. *Lancet.* **353**:2026–2029.

89. Kaye, J. A., del mar Melero-Montes, M., and Jick, H. (2001). Mumps, measles, and rubella vaccine and the incidence of autism recorded by general practitioners: A time trend analysis. *Br. Med. J.* **322**:460–463.

90. Dales, L., Hammer, S. J., and Smith, N. J. (2001). Time Trends in Autism and in MMR Immunization Coverage in California. *J. Am. Med. Assoc.* **285**:1183–1185.

91. McPhillips, H., and Marcuse, E. K. (2001). Vaccine safety. *Curr. Prob. Pediatr.* **31**(4):91–121.

9 INTRODUCTION OF NEW VACCINES IN THE HEALTHCARE SYSTEM

EDWARD KIM MULHOLLAND*,# AND BJARNE BJORVATN†,#

*Department of Paediatrics, University of Melbourne, Royal Children's Hospital, Melbourne, Australia

†Centre for International Health and Department of Microbiology and Immunology, University of Bergen, 5020 Bergen, Norway

#Both authors contributed equally to the writing of this chapter

During the course of the twentieth century, vaccination developed from a largely experimental procedure to prevent or treat disease to become the focus of national and ultimately global programs to control and even eradicate disease. All countries in the world now have national vaccination programs, mainly directed at children. During the 1990s, a number of important trends emerged in the field of vaccination. The number of new vaccines available increased markedly. The resulting complexity of immunization programs led to an increasing demand for combination vaccines to simplify program implementation and minimize trauma for children. These developments were accompanied by an alarming increase in vaccine price. Traditionally, vaccination has been regarded as one of the most cost-effective public health tools. Advocates have been able to use cost-effectiveness analyses to demonstrate the great advantages that vaccination presents over other public health strategies. Now with new vaccines that could have a major public health impact carrying price tags close to $200[1] per child, detailed cost-effectiveness analyses are being used to reject the introduction of new vaccines or to help manufacturers establish the highest price that may be borne by the community. For example, a cost-effectiveness study of pneumococcal conjugate vaccine introduction in the United States [1] concluded that pneumococcal conjugate vaccine would be cost-saving for society at a price of $46 per dose, a price remarkably similar to the eventual price. Importantly, more than half of the cost savings in the U.S. study were attributable to indirect costs. The vaccine was approved for universal use in the United States soon after. In Australia, a similar economic analysis

---

[1]All monetary amounts are given in U.S. dollars.

resulted in the rejection of universal use of the same vaccine, largely because indirect costs were not included in the analysis.

Experience has shown that arguments used to introduce new vaccines are complex, multifactorial, and not always rational. This chapter will review the factors that must be considered when making such decisions, highlighting important areas where problems have occurred. Examples of new vaccines that are under consideration for introduction in many countries will be presented, and case studies will outline some of the experience of countries that have introduced new vaccines. There are great differences between countries with respect to attitudes to vaccination and its risks, vaccination schedules used, and the nature of the political and scientific decision-making process. Thus, this chapter cannot be comprehensive, but instead it will focus on common problems and lessons learned.

I. THE DECISION PROCESS

A. Organizational Features

Although there is wide variety of organizational models for the decision-making process, most countries have a committee of experts, usually appointed by the Ministry of Health, with the mandate to assess proposals for the inclusion of new vaccines into the national immunization program. In some countries, several committees are involved, some looking into technical and others into programmatic or cost-effectiveness aspects of such proposals. Suggestions for new vaccines may originate from within the medical and scientific community, from professional societies both national and international, from other groups such as nongovernmental organizations, from multilateral agencies such as WHO and UNICEF, or from the pharmaceutical industry itself. Occasionally major outbreaks may trigger public demand for large-scale vaccination or even the development of a new vaccine. The latter occurred when meningococcal B epidemics struck Scandinavia and Latin America, prompting the development of targeted vaccines in Norway and Cuba. This situation has more recently been repeated in New Zealand. In some countries, the role of the vaccine manufacturing industry in the process is perceived as relatively strong, due in part to the growing number of industry-sponsored conferences and symposia attended by key personnel. In general, the decision-making process is both scientific (evidence-based) and political, the latter often reflecting public perceptions about the disease or the vaccine. Since 1998, the WHO policy on the use of a number of internationally licensed novel vaccines has been published in the *Weekly Epidemiological Record* (http://www.who.int/wer).

B. Scientific Decision-Making Process: Assessment of Disease Burden and Cost Effectiveness

Where a vaccine is known to be safe, inexpensive, and effective, it is sufficient to know that the disease in question is a significant public health problem and therefore worth the effort involved in delivering the vaccine. Even so, this can require a significant amount of work. No fewer than 315 lameness surveys were undertaken between 1977 and 1991 to support the introduction of polio vaccination into developing countries [2]. The arrival of hepatitis B vaccine, which in the early years was extremely expensive, brought the need for cost-effectiveness analyses to support its introduction. As a result, countries considering introduction of the vaccine sought information on the burden of disease to assist with the decision-making process. Carriage rates for hepatitis B can be measured with simple surveys, and many such studies have been undertaken, often with support from the pharmaceutical industry. However, the true burden of disease associated with hepatitis B is manifested as chronic liver disease and liver cancer in lateral life. Data from some countries, particularly Gambia and Taiwan [3], have provided reasonable estimates to link carriage rates with later disease. However, with hepatitis B the long period between the prevention of infection and the improved health outcome still makes cost-effectiveness studies of this intervention difficult to conduct and interpret. Furthermore, the notion of vaccinating an infant to prevent an adult disease proved difficult for some agencies, such as UNICEF, to accept.

Vaccines against such a dramatic and well-defined condition as bacterial meningitis should present a fairly simple task when it comes to assessment of disease burden. Surprisingly, this has not always proved to be the case. Although the burden of disease due to *Haemophilus infuenzae* type b (Hib) extends beyond meningitis, the measurable burden of Hib meningitis has been sufficient to justify the introduction of Hib vaccine into the United States, United Kingdom, and many other western countries. In some regions, particularly Asia, this has proved to be extremely difficult to measure due to overuse of antibiotics and reluctance to perform lumbar punctures, and this now represents a major obstacle to the introduction of Hib vaccine. For developing countries, it is important to consider the numerically greater burden of Hib pneumonia, although measurement of this presents significant difficulties (discussed later). Estimation of the burden of disease becomes much more complex when considering diseases for which risk persists through life, but the disease burden is age-dependent. Examples of this include meningococcal disease, typhoid, human papilloma virus (HPV), and the difficult problem of penumococcal disease that is discussed in the box.

Economic analyses of new vaccine introduction endeavor to evaluate the associated cost savings by calculating, in economic terms, the costs attributable to the disease. Then the proportion that can be prevented by the vaccine is estimated on the basis of known vaccine efficacy, expected coverage, and issues that might be specific to particular vaccines, such as herd immunity or serotype coverage. The cost of vaccine introduction is then calculated, taking into account issues like transportation and storage requirements as well as vaccine price. Increasingly, vaccine price tends to overwhelm other costs. As this is variable, and indeed negotiable, the price is usually introduced as a variable parameter in a sensitivity analysis. It then becomes a simple task to calculate the vaccine price at which the total cost of vaccine introduction equals the costs saved, referred to as the "break-even price." Too often this becomes the focus of the analysis. This is a misleading approach. Although it is true that the argument in favor of introducing a "cost-saving" vaccine is strong, to imply that this should be a requirement for vaccine introduction is to completely disregard the human cost associated with the prevention of disease, disability, and death.

The science of the economic analysis of the impact of new vaccine introduction is at present in its infancy. The arrival of new, expensive vaccines places great pressure on health economists to improve to their techniques of analysis

---

Disease Burden and Cost Effectiveness—The Example of Pneumococcal Conjugate Vaccines

One of the most complex examples of estimating the vaccine-preventable burden of disease is presented by the burden of disease attributable to *Streptococcus pneumoniae* (pneumococcus). For most developing countries, the etiological diagnosis of pneumococcal meningitis is a difficult enough task, but meningitis represent only a small fraction of the burden of pneumococcal disease. Much more morbidity and mortality are due to the other pneumococcal manifestations, particularly pneumonia. Globally, pneumonia is the most important cause of severe illness and death in children, with current estimates of the total child mortality due to pneumonia being in the region of 2 million deaths per year [4]. At the level of the individual case, it is usually impossible to determine the etiology of pneumonia, although when appropriate bacteriological investigations are performed, pneumococcus is almost always found to be the leading bacterial cause. Its contribution is greatest in the most severe forms of pneumonia, and it is quite possible that pneumococcus may cause more than half of all very severe or fatal episodes of pneumonia in children. When one considers milder episodes of pneumonia, the usual clinical manifestion in developed countries, the proportion that can be shown to bacterial (or pneumococcal) is smaller, perhaps leading to underestimation of the importance of pneumococcus by suggesting that the organism is less important as a cause of less severe disease.

The most useful measure of disease burden is the burden of pneumonia that can be prevented by a pneumococcal vaccine. It is now recognized that this can only be estimated in the context of a randomized, controlled trial. An initial analysis of the results of the first pneumococcal conjugate vaccine trial to address the prevention of pneumonia (S. Black, personal communication) suggested that an unexpectedly large proportion of the "mild" pneumonia episodes in Californian children was prevented by this vaccine. This points to a larger role for pneumococcus as a cause of less severe pneumonia than was previously thought. The results of ongoing pneumococcal conjugate vaccine trials in South Africa, the United States (Navajo people), Gambia, and the Philippines should resolve this issue.

However, pneumococcal burden is far more than pneumonia and meningitis, although these manifestations account for most of the deaths. Acute otitis media is the most common reason for U.S. infants to receive antibiotics, and pneumococcus is known to be an important cause. For this reason, many were disappointed when the U.S. and Finnish trials of the seven-valent Pnc-CRM pneumococcal conjugate vaccine showed reductions in overall acute otitis media of no more than 7% and 6% [5, 6]. In the Finnish study, the vaccine efficacy against acute otitis media due to vaccine serotypes was only 57%. The final efficacy was then a product of the true vaccine efficacy, the proportion of episodes that were due to pneumococcus, and the proportion of pneumococci that were of the types included in the vaccine. The outcome was further offset by an apparent increase in the incidence of disease due to nonvaccine serotypes, a phenomenon known as "serotype replacement," which could also impact the efficacy of the vaccine against pneumonia in high-carriage areas. This example serves to emphasize the fact that the measure of disease burden that is most relevant to the prediction of vaccine impact is the burden that can be prevented by an effective vaccine. For conditions like pneumonia or otitis media, this can only measured in the context of a randomized, controlled trial. Because of knowledge that pneumococcal vaccines will be expensive and that their use is likely to require supportive evidence of their cost effectiveness and overall impact, there is now pressure on those conducting field trials of pneumococcal conjugate vaccines to ensure that the vaccine-preventable disease burden is measured accurately.

In the case of a condition like pneumonia, this extends to qualitative aspects as well. As those pneumonia cases that are prevented by an effective pneumococcal vaccine will probably be the more severe cases, a measure of the fractional reduction in pneumonia cases may miss a much larger and more important reduction in more severe cases requiring, for example, prolonged oxygen therapy or chest drainage. At the other end of the disease spectrum, there is already evidence that pneumococcal conjugate vaccines prevent nonspecific febrile episodes that often require medical attention and antibiotics. All of these factors need to be included in the economic analysis that now becomes increasingly complex and heavily reliant on the results of vaccine trials. A useful analysis of the cost effectiveness of pneumococcal vaccine introduction should estimate the cost savings in terms of all of the manifestations of pneumococcal disease, taking into account qualitative differences in pneumonia cases prevented. Thus, a realistic evaluation of the direct cost savings is very difficult and heavily reliant on extrapolation from the results of vaccine trials. The indirect cost savings are also a substantial component of the cost savings associated with vaccine introduction, and these should not be ignored.

Curiously, we have few data on the long-term effects of pneumococcal pneumonia. It may be that severe pneumonia results in lung damage that is not apparent until later in life. As there are no data on long-term morbidity, this aspect is not considered. Thus, for a number of reasons, current estimates of cost savings associated with the prevention of pneumococcal disease are likely to underestimate the true savings.

and to demand better quality disease burden and vaccine trial data, as these are the essence of such analyses. In the future we will almost certainly see examples of proposed new vaccines that are cost-effective or even cost-saving in economic analyses, but that are not introduced as the money cannot be found. Inevitably considerations of cost effectiveness are overruled by considerations of cost alone.

C.   Assessment of Individual Vaccines: Efficacy and Effectiveness

When discussing the introduction of a new vaccine, parameters such as efficacy, effectiveness, duration of protection, adverse effects, and vaccine price are usually considered. However, a number of vaccine-specific features, such as formulation, route, number of administrations, and adaptability to the existing immunization program, deserve consideration. These may prove critical as they affect vaccine wastage and vaccination staff workload, as well as professional and public acceptance of the vaccine. All of these parameters impact vaccine effectiveness, directly or indirectly. In addition, the priority of the new vaccine relative to other health needs in the population should be carefully considered and discussed [7]. The community perception of a particular vaccine-preventable disease should not be underestimated. In the United Kingdom, this factor was a major impetus in the introduction of meningococcal C vaccine, whereas in sub-Saharan Africa the prevention of measles remains a major concern for communities.

Randomized, double-blind, phase 3 clinical trials serve as the gold standard for studies of vaccine efficacy. Efficacy may be measured directly as clinical protection rate or inferred from serological responses. Unfortunately, both clinical protection rates and serological responses are very sensitive to trial design, characteristic of the involved target groups, and environmental factors. Data from trials conducted in highly sophisticated settings may not be directly applicable to poor areas with insufficient infrastructure and potential differences in the prevailing microbial strains. The experience with rotavirus and Hib vaccines presents clear examples of vaccines that were effective in developed countries (where infection pressure is relatively low), but that failed in developing countries or developing-country-like settings [8, 9]. Authorities responsible for the introduction of new vaccines may face apparently conflicting results of previous efficacy studies and conclude that a national trial will have to be conducted. However, in most countries large-scale clinical trials are unnecessary and impractical. Careful analysis of the outcome of trials in comparable settings should provide sufficient information for rational decision making.

Most efficacious vaccines induce over 80% protection in vaccinated children. This may indicate that over 80% of children are fully protected ("all or nothing" protection) or that all children have over 80% protection (incomplete protection). Where there is high vaccination coverage, such vaccines may control or even eradicate the disease in question. On the other hand, vaccine efficacy of less than 60% is unlikely to control a highly communicable disease, even when vaccine coverage is high. In some settings, vaccines of low efficacy may still be useful as supplements to other control measures or when realistic alternatives for reducing the spread of disease are missing. Thus, in spite of its shortcomings, the baccille calmette–Guérin (BCG) vaccine is still the most widely used vaccine in the world, and future vaccines against other global health problems such as HIV and malaria may be considered valuable, even with efficacy rates below 50%. The decision as to whether to initiate large-scale immunization in situations in which the vaccine is only modestly efficacious will depend on a number of factors, some determined by the vaccine and others by epidemiological or environmental characteristics.

Whereas efficacy is a measure of the protective impact of a vaccine under the relatively ideal conditions of a clinical trial, vaccine effectiveness is a measure of the protection achieved in the real-life setting of a large-scale immunization program. Effectiveness takes into account imperfections in vaccine storage, transportation, and handling at the clinic level, as well as the erratic presentation of children for vaccination. It also takes into account changes in microbial population biology, such as serotype replacement, that may cause a vaccine's effectiveness to wane with time. On the positive side, effectiveness takes into account the issue of herd immunity, which has proven to be particularly important for some vaccines (e.g., Hib conjugate vaccines). Therefore, the effectiveness of a vaccine may be lower than its efficacy, although in well-functioning national programs the difference should be marginal, or it may be higher than its efficacy if herd immunity is marked. For immunization policy-makers, the results of effectiveness studies of new vaccines from comparable settings may be valuable when considering the introduction of such vaccines into their own programs. Most efficacy studies conducted in developing countries have some features of effectiveness studies, and these should be taken into account when considering such studies. Some have argued that effectiveness studies should be specifically set up to support the introduction of new vaccines [10]. This may be the case for some vaccines, but new studies should always be set up to address specific scientific questions. When these questions relate to effectiveness issues not addressed in existing trials, new studies may be warranted.

The protective efficacy of a vaccine is normally measured within a few years following completion of the course of immunization. The long-term duration of protection is less well-understood or even unknown in the case of newer vaccines. Traditional immunogens such as polysaccharide vaccine do not elicit a T-cell response and are known to provide short-lived protection [11]. Although protection following immunization with live, attenuated viral vaccines or polysaccharide–protein conjugates may last for decades, there are concerns that ultimately protection will fade and necessitate large-scale revaccination of adults and possibly teenagers. This could be an important concern in future years [12]. As the duration of protection is essential for the planning of any immunization program, uncertainty in this regard complicates the decision process.

There are some diseases for which partial vaccination of a community may result in a paradoxical increase in disease or in more severe disease. For these conditions, the local epidemiology and likely vaccine coverage should be considered carefully before embarking on a vaccination program. The best example of this program is rubella (see the following box).

Paradoxical Effects of Vaccine Introduction—The Example of Rubella Vaccine

Large-scale vaccination against rubella is motivated by prevention of the congenital rubella syndrome (CRS) that may affect the babies of mothers who contract the infection during early pregnancy. Unfortunately, in many countries epidemiological surveillance for rubella and the congenital rubella syndrome (CRS) is unreliable or nonexistent. In such countries, sero-epidemiological studies of representative sample of women of childbearing age may be the most efficient way of assessing the risk of CRS in the population. Where most of the young females are naturally immune and the allocation of resources to the immunization program is limited, childhood vaccination against rubella should receive low priority [13].

On the other hand, childhood rubella immunization should be considered where there is high measles vaccination coverage, and where the country can afford the extra cost of using combined measles and rubella (MR) or measles, mumps, and rubella (MMR) vaccines rather than measles vaccine alone in its vaccination programs. The MR and MMR combination vaccines are safe, efficacious, and widely used in developed countries. Where rubella vaccination of children can be sustained at coverage levels of 80% or more, CRS will be gradually eliminated. Childhood immunization coverage below this level may reduce the circulation of rubella virus in the population, but will not stop it. This could ultimately result in increased numbers of susceptible young adults and a paradoxical increase in the rate of CRS [14]. The relatively widespread rubella vaccination of young children in the private sector of some countries could have a similar epidemiological impact. To avoid the potential increase in CRS as a consequence of recently initiated or inadequately implemented childhood rubella immunization, equally high priority should be given to rubella vaccination of all adolescent girls and/or women of childbearing age. Large-scale immunization of adolescents and adults alone is unlikely to affect significantly rubella transmission in the community at large.

This is the clearest example of a vaccine that, when partially introduced, can have a paradoxical negative effect on disease rates. Similar concerns were raised about childhood polio vaccination in the early years of its introduction in developing countries, but these concerns proved to be unfounded. So far, it is not known whether these questions also apply to vaccines against varicella and hepatitis A, diseases that also tend to run a more serious course in adolescents and adults than in children.

D. Antimicrobial Resistance

We are accustomed to considering vaccination as a strategy that prevents individuals from acquiring a disease, but its effects on the epidemiology of the disease in question can be more subtle. Herd immunity is the best known example of unexpected beneficial effects of vaccination. Another important example is the potential of vaccines to reduce the spread of drug-resistant microorganisms. The last decades of the twentieth century saw a dramatic increase in the incidence of antimicrobial resistance. This global phenomenon is a direct result of the widespread use, and misuse, of antimicrobial agents. Drug resistance has resulted in spiraling costs for those countries that can afford the newer, alternative therapies, and treatment failures and deaths for those countries that cannot. In this bleak field, bacterial vaccines offer the best hope of a solution. The example of pneumococcal vaccine is given here, but similar arguments apply for other bacterial vaccines.

Pneumococcal vaccination may affect resistance at different levels. At the simplest level, effective vaccines will reduce the number of episodes of disease requiring antibiotics and thereby antibiotic consumption, but other factors may be more important. Most antibiotic-resistant pneumococcal strains are serotypes included in the vaccines. As the vaccines are known to affect carriage, it was not surprising to find that pneumococcal vaccination had a profound effect on the circulation of resistant pneumococci. In South Africa, carriage of penicillin-resistant pneumococci was reduced from 41 to 21% in recipients of nine-valent Pnc-CRM vaccine [15], so that it can be assumed that large-scale use of the vaccine in the community will reduce resistance rates, at least temporarily. Furthermore, these vaccines have the potential to reduce and simplify empiric antibiotic therapy. The argument to treat a child with fever or with fever and cough with antibiotics is greatly weakened if the child has received effective pneumococcal and Hib vaccines, whereas the case for adding vancomycin to the initial therapy of suspected meningitis is also weaker, as the probability of penicillin-resistant pneumococci being responsible is very small. Thus, these new vaccines present the pediatric community with an unprecedented opportunity to promote and implement the judicious use of antibiotics, and this more than anything could curb the spread of resistance. The same arguments will apply to other vaccines against conditions that usually lead to empiric antibiotic therapy, such as shigella dysentery and typhoid. Up to now, no work has been undertaken to estimate the potential savings attributable to this effect.

E. Vaccine Safety

Acceptability of new vaccines among medical professionals as well as the general public depends on many factors, including the force and skill with which the vaccine is promoted by the responsible authorities. A key issue in terms of acceptability is the perceived risk of adverse effects. The tolerance of a population for adverse vaccination effects is related to the perceived health risk associated with the disease in question. However, most people will accept considerably higher risks of complications as a consequence of natural disease as compared to risks associated with prophylactic measures. It is a common experience that at least mild and moderate adverse effects of a vaccine are well-tolerated during period of high disease incidence, whereas public concern about such effects rises when the disease is under control. Thus, vaccines such as BCG and whole cell pertussis vaccines are well-accepted in regions still experiencing a substantial disease burden, whereas their large-scale use has been discontinued in countries in which tuberculosis and pertussis have been largely eliminated. This predictable shift in public acceptance results in reduced compliance, which is a growing problem in many industrialized countries [16].

In more recent years, an increasing number of individuals and groups who for philosophical or other reasons have been actively working against the concept of large-scale immunization. Some of these people claim that the concept of immunization is in conflict with nature, whereas others claim that calls for obligatory childhood immunization interfere with personal freedom [17]. In affluent communities, some do not accept the idea that prevention of diseases through childhood vaccination is a human right. Efficient control of vaccine-preventable diseases can only be achieved through active community participation in national immunization programs [18]. In some areas the antivaccine movement has significantly reduced compliance with vaccination programs [19]. Those responsible for the national programs should be acquainted with the antivaccine arguments and prepare for the introduction of new vaccines with carefully designed information campaigns, for both medical professionals and the public. Those responsible for the evaluation of new vaccines should also ensure that adequate attention is paid to the assessment of long-term safety as well as the efficacy of the vaccines.

Vaccine Safety Issues—The Introduction of Acellular Pertussis Vaccine

For many years, inactivated whole cell pertussis vaccines (wP), usually in combination with vaccines against diptheria and tetanus (DTP), have been part of national childhood immunization programs all over the world, finally reaching about 80% of the world's children. Although they are frequently associated with minor adverse reactions such as fever and local swelling, wP vaccines are inexpensive and best the wP vaccines show protection rates of 80% or more. These vaccines are produced in many parts of the world, including several developing countries. In the 1970s, safety issues, in particular fear of encephalopathy, led to reduced public acceptance of wP vaccines. In some industrialized countries pertussis vaccination was practically abandoned, which resulted in a dramatic increase in the number of pertussis cases [19].

A new generation of pertussis vaccines based on selected bacterial components rather than the whole bacterial cell became available in the 1980s, and first combination of acellular pertussis (aP), diptheria, and tetanus vaccines were licenced in the early 1990s. In comparing the best vaccines in each group, acelluar pertussis vaccines (aP) are equally efficacious, but less reactogenic than the wP vaccines and therefore more acceptable. In some countries, such as Sweden and Germany where wP vaccine was temporally abandoned from the national immunization programs, availability of the aP vaccine was seen as a prerequisite for the reintroduction of immunization against this disease. The aP vaccine is now preferred in most industrialized countries [20].

The reactogenicity of wP vaccines is unlikely to vary between developing and industrialized countries. However, as long as the seriousness of pertussis in the early life is apparent for medical professionals as well as for parents, the acceptability of wP in developing countries will probably remain high. Besides, the current high price of aP renders this vaccine unavailable for large-scale immunization in most poor countries. On the other hand, in countries in which successful vaccination and improved socioeconomic conditions result in the falling pertussis disease rates, vaccine reactogenicity is likely to become an important issue. Therefore, those responsible for the national immunization programs may need to weigh the advantage of the low price of wP vaccine against reduced popular acceptance of the vaccine. This dilemma may be particularly difficult to handle in countries with well-established national production of wP vaccine.

Vaccine Safety Issues—The RotaShield Story

Since its discovery, rotavirus has been recognised as a major cause of morbidity in all countries and mortality in developing countries. Efforts to develop an effective vaccine have been underway for many years, and in August 1998, the U.S. FDA approved the tetravalent rhesus–human reassortant rotavirus vaccine (RRV-TV; Rotashield®, Wyeth Lederle Vaccines). Trials in the United States, Finland, and Venzuala had shown the vaccine to be efficacious. Despite urging from WHO, no trials had been conducted in developing countries in which mortality from diarrhea is high. At the time of licensure, five episodes of acute intussusception had been described after RRV-TV vaccination. As this was not statistically significantly different from the incidence of intussusception in controls, it did not stop licensure, although a warning of intussusception was included in the package insert [21]. In May 1999, reports of intussusception following rotavirus vaccination were received by the U.S. Vaccine Adverse Event Reporting System, prompting the withdrawal of the product by the company and suspension of rotavirus vaccination. Since then, a detailed analysis has been undertaken which has confirmed the relationship between RRV-TV vaccination and intussusception 3–14 days following vaccination. The effect is mainly seen after the first or second dose and effects 1 in 5,000–10,000 infants vaccinated [22]. At the time when this effect was first described, several studies of RRV-TV that were ready to start in developing countries were abandoned. No further studies with vaccine have been undertaken, although the development of live rotavirus vaccines based on bovine and human strains is continuing. The manufacturer of the bovine vaccine (Merck & Co.) has already started trials in infants and plans to use a sequential analysis model to evaluate whether any episode of intussusception that occur are due to the vaccine. They estimate that the study will enrol 50,000–70,000 infants.

Critics of the withdrawal RRV-TV have pointed to WHO estimates of rotavirus–associated mortality, which indicate that 20% of diarrhea deaths may be due to rotavirus on the basis of extrapolation from hospital-based studies. In the settings in which infant mortality rates exceed 100 in 1000 live births, 0.5–1% of the children may die of rotavirus disease. It seems absurd to abandon a vaccine because of a side effect that occurs in less than 0.02% of recipients. However, the situation is more complex. Reports from some Asian countries indicate much higher rates of natural intussusception than are seen in western countries [23]. Furthermore, death from diarrhea costs only a few cents to prevent (with oral rehydration solution), so that children who continue to die of rotavirus disease are those who lack the most basic form of healthcare and are least likely to have access to new and expensive vaccine. An important lesson to be learned from this episode is the importance of evaluating new vaccines, both for safety and for efficacy, in the settings in which they are to be used to greatest effect, usually developing countries. Under our present system of vaccine development, the motivation for the development of new vaccines is commercial profit. As a result, the process is driven by the needs of the most lucrative market, i.e., the United States, rather than those of developing countries in which disease prevention needs are greatest. Whether the benefits associated with RRV-TV use in the developing countries would have justified its use even in the face of small risk of intussusception will now never be known.

F. Adaptability to National Immunization Schedules

Efficacy data for new vaccines are, in principle, representative only of the schedules under which they were developed. Vaccines intended for use in infants are usually harmonized with timing of DTP schedule. However, very few countries have identical national childhood immunization programs. Even within the European Community schedules differ considerably [24]. Before incorporation of vaccine for infants into such a program, evidence should be provided that the vaccine will work satisfactorily in the local setting. This does not necessarily require the duplication of large efficacy studies. "Bridging" immunogenicity studies will usually suffice. Thus, the effectiveness of a new vaccine in Sweden and Norway, administered with DTP at 3, 5, and 10–12 months, should be similar to its effectiveness in the United Kingdom and Germany where DTP is given at 2, 3, and 4 months, provided comparable immunogenicity data are available [24].

Coverage rates of other vaccines have been shown to be good predictors of the uptake of new vaccines into national immunization programs [25]. On the other hand, new vaccine introduction may have a negative impact on coverage. Thus, in some countries with acceptable national programs, there is a risk that the introduction of new vaccines may impair the delivery of traditional vaccines such as BCG, DTP, measles, and polio. This risk is likely to increase when new vaccines are provided free or at attractive prices for a limited period of time. For this reason, major donors such as the Global Alliance for Vaccines and Immunization (GAVI) are paying increasing attention to improving the infrastructure and coverage of existing national immunization programs before supporting the introduction of new vaccines.

In the developing countries, high rates of vaccine wastage are common. These usually reflect the provision of inexpensive vaccines in multidose vials to peripheral facilities where the number of children attending on any given day may be very small. Health facility staff often have the choice of opening a new vial, knowing that most will be wasted, or asking the children to return on a specified day, knowing that many will not return. Thus, the replacement of inexpensive DTP vaccine with an expensive combination vaccine inevitably will lead to moves to curb vaccine wastage, which in turn may lead to reduced coverage. This is one of the less obvious consequences of new vaccine introduction.

In most developing countries, large-scale immunization is limited to programs for infants and pregnant women, with little capacity for the vaccination of older children. For this reason, vaccines against important diseases such as cholera, typhoid, and meningococcal meningitis (A and C)—all of which currently are formulated for use only in children >1 year of age—may be unavailable for important target groups in some highly endemic regions. In this respect, rigid adherence to the EPI schedule has been a significant barrier to the use of some valuable vaccines (see the following box).

Vaccines with Different Schedule Requirements—Introduction of Vaccines against Typhoid Fever

Typhoid fever is a serious public health problem in many countries in Asia, Africa, and South America. Each year approximately 16 million cases occur, 600,000 of which end fatally. Transmission occurs mainly through fecal contamination of water or food by chronic intestinal carriers of *Salmonella typhi*. Traditionally schoolchildren and young adults are most affected, although studies from India have shown that, at least in some highly endemic regions, peak incidence may occur in children 1–5 years old [26]. Drug-resistant *S. typhi* is becoming a problem all over the world, and large-scale vaccination is considered the only realistic tool to control this disease, at least in the medium term.

The traditional inactivated whole cell vaccine is still available in some countries. However, this vaccine is only moderately efficacious and is quite reactogenic. During the past 15 years, two new typhoid vaccines have been licensed internationally: the Vi polysaccharide vaccine for parenteral use and the Ty21a vaccine for oral use. Both vaccines are efficacious and well-tolerated. In adults and children >5 years of age, protective efficacy of 50–70% or more appears to be sustained for several years. Unfortunately, none of these vaccines are currently recommended for use in children under 2 years of age.

Ideally, vaccines against childhood diseases should be easily adapted to schedules and timing of national childhood immunization programs. The current formulation of Vi polysaccharide vaccine requires one subcutaneous or intramuscular injection and revaccination every 3 years to maintain protection. The Ty21a vaccine is administrated orally as enteric coated capsules or as a liquid formulation, the latter meant for children as young as 2 years of age. Following the three initial doses, which are given at intervals of 2 days, a booster dose is recommended every 3 years.

The current typhoid vaccines are used mainly by adult travelers to endemic areas. However, as the greatest burden of disease is in preschool and school-aged children, effective typhoid vaccination programs need to either introduce the vaccines for preschool and school-aged children, a group not currently targeted by routine immunization programs, or develop a new vaccine that can be used in infants. In endemic regions, it may be difficult to reach preschool children with existing vaccines. Although vaccination of schoolchildren may have a considerable effect on disease incidence in that age group, the disease is unlikely to be controlled unless transmission can be stopped in children of preschool age. Thus, a vaccine is required that can be administered in infancy and will provide long-lasting protection. Protein–polysaccharide conjugate vaccine development is a logical approach. A newly developed Vi conjugate vaccine has been evaluated in Vietnam and found to provide 92% protection in 2- to 5-year-old children [27]. If shown to be immunogenic in children under 2 years of age, such a vaccine could be given at the time of the traditional EPI vaccines.

G. Vaccine Price

The price is the single most important barrier to the introduction of new vaccines in developing countries. Whether a new vaccine is considered affordable depends on the perceived public health burden of the disease and the country's economic situation, as well as on political will. It is a global phenomenon that the public will accept high costs for therapeutics, but is reluctant to pay for important prophylactic tools, particularly vaccines.

National authorities may purchase new vaccines directly from the manufacturer or via international agencies such as UNICEF or PAHO. In more recent years, WHO and UNICEF have launched a tiered pricing system, whereby essential vaccines for developing countries may be purchased at prices that are GDP-dependent [28]. This system allows the poorest countries to obtain essential vaccines at the lowest price. Hopefully tiered prices will also provide access in developing countries for the new Hib and pneumococcal conjugate vaccines [29], but with current prices there is a long way to go before these vaccines become truly available for the developing world. The development of new vaccines is an expensive process, and whereas public sector input usually has been considerable, substantial private sector investment is required to drive a new vaccine through the development process toward international licensure. Pharmaceutical companies expect significant return on their investment, and increasingly the result is high prices for new vaccines.

A large number of national and international organizations sponsor both the development and the purchase of vaccines for developing countries. An important development in this field was the creation of the Global Alliance for Vaccines and Immunization (GAVI) in 1999. GAVI was established as a network of major private and public sector partners with the objectives of improving access to safe and cost-effective vaccines and sustainable vaccine services, accelerating the development of vaccines needed primarily in developing countries, and making immunization coverage a key issue in development aid. The GAVI Vaccine Fund provides funding for the purchase of underutilized vaccines and for the strengthening of national immunization services. By December 2001 the Vaccine Fund had approved applications for support from 52 developing countries, representing a 5-year commitment of well over $600 million (http://www.vaccinealliance.org/reference/vaccinefund.html). This is a major achievement for the countries concerned, but it will increase the extent to which those countries are dependent on donors. This in turn makes immunization programs vulnerable to donor fatigue or shifts in foreign policy. In addition, as an externally applied program, decisions are more likely to come from outside the country, thereby diluting national responsibility. At present, 36 African and 5 Asian countries are completely dependent upon donors for the purchase of their routine vaccines [30]. The strengthening and development of vaccination programs, including both traditional and new vaccines, require a long-term view. Sustainability should be the central theme of all new initiatives in this area.

H. The Role of the Private and Nongovernment Sectors

The private sector may influence the introduction of new vaccines into national immunization programs in various ways. Successful introduction into the private sector could facilitate public acceptance of important vaccines, but on the other hand, the promotion of vaccines of doubtful public health importance could unduly influence decision makers. Support from nongovernmental organizations may be vital for the operation of national programs, but they may also have their own political priorities and direct funding accordingly.

With regard to the role of the vaccine manufacturing industry, new vaccines are usually available in the private health sector long before they may become part of the national immunization program. Although presence on the private market may stimulate professional and public interest in a particular product, limited private sector immunization may not be sufficient to drive public acceptance. It is difficult to assess the role of the vaccine manufacturing industry in the decision process leading to the acceptance or rejection of a vaccine for use in national immunization programs. However, once an official decision has been made to adopt immunization against a particular disease, the efforts made by the vaccine manufacturing industry to promote their products may become quite aggressive.

All over the world, numerous nongovernment organizations (NGOs) are involved in the promotion of immunization. For many years, large, international associations such as Save the Children Fund, Médicins Sans Frontières, World Vision International, the Red Cross and Red Crescent Societies, Rotary International, PATH, and Epicentre have contributed to vaccine funding, delivery, and research as supportive partners and advocates. In many countries, local NGOs working directly at the household or community level may also have a remarkable impact on vaccine delivery and acceptance, particularly among underprivileged sections of the population. Due to their independent political position and operational expertise, NGOs may sometimes openly criticize the performance of national immunization programs and provide essential information on disease burden and vaccination coverage in areas in which public surveillance is inadequate. In this way, their influence on national immunization policy may be considerable. Governments have increasingly acknowledged the role of NGOs as providers of complementary skills and resources. Through international networking and pooling of resources, their influence on national as well as global vaccine policy is increasing. The entry of GAVI into this scene is likely to accelerate this development significantly.

II. CASE STUDIES

The events that lead to the introduction of a new vaccine are different and reflect the economy of the country, social and political environments, and local epidemiology, as well as knowledge about the vaccine and its performance elsewhere. The following are two examples of the process.

A. Case Study 1: Introduction of Hib Vaccine into Australia

Details are courtesy of R. Hall (Western Pacific Regional Office, WHO). Prior to the arrival of Hib vaccines, the burden of Hib disease was well-understood in Australia. The concentration of the population into a number of large cities meant that most cases of invasive Hib disease were managed at large tertiary hospitals. In the case of epiglottitis, which was relatively common in Australia, a sophisticated emergency transport system brought cases occurring in rural centers into the tertiary centers, so that estimates of disease incidence were quite accurate. Following the licensure of Hib conjugate vaccines for use in infants in the United States in December 1990, pressure began to mount from among the pediatric community for the vaccine to be used in Australia. However, the Australian immunization system was not well-equipped to deal with a new and expensive vaccine. The system was quite fragmented and vaccine purchases were made by state governments, with significant differences often emerging in the prices paid for vaccines. Approval by the Therapeutic Goods

Administration (TGA, the Australian equivalent of the FDA) was required for a new vaccine to be marketed in the country, and marketing was required before a product could be included in the standard vaccination schedule. As a reactive organization, the TGA could only respond to submissions from manufacturers and, although some provisions existed to accelerate the submission of desirable products, these were not enacted. In addition, the process of TGA review of dossiers was relatively slow at the time. (Performance standards have since been introduced). The first Hib conjugate was marketed in Australia in May 1992 (PRP-D for use with children 18 months and older), and within 12 months the other three vaccines suitable for infants were approved (PRP-T, HbOC, and PRP-OMP). The Communicable Diseases Subcommittee of the National Health and Medical Research Council (NHMRC) was asked to make recommendations on the use of Hib vaccine. In early 1993, the group approved the use of Hib conjugate vaccines, but this was not supported by funding. The ministry had proposed a plan for hepatitis B immunization the year before and been rejected, but the proposal for Hib vaccination in 1993 was viewed more sympathetically by the Minister of Health. A document was then prepared that was submitted to the Budget Committee of Cabinet, where the proposal was approved and allocated $7 million Australian per year (later reduced to $6 million). While the logistics of how this was to be implemented were being worked out, Australia was in the midst of an election campaign. The incumbent Minister of Health (who was reelected) promised to provide the vaccine on a claim-back basis so that parents could purchase the vaccine in the private sector and be reimbursed later. Meanwhile, the government negotiated an excellent price from one manufacturer, and by late 1993 universal infant Hib immunization was in place and disease rates began to fall sharply. Invasive Hib disease, once the scourge of Australian pediatrics, is now largely of historical interest, as the younger generation of Australian pediatricians have rarely seen Hib meningitis or epiglottitis.

B. Case Study 2: Introduction of Hepatitis B Vaccine into Vietnam

Details are courtesy of David Hipgrave, Program for Appropriate Technology in Health (PATH), and government of Vietnam. Dang Duc Trach, Nguyen Thu Van, Nguyen Van Cuong, Beverley Biggs, and James Maynard. Having established a successful EPI program with high coverage and a marked decrease in the incidence of vaccine-preventable diseases, Vietnam committed itself to the introduction of hepatitis B vaccine (HepB) in October 1997. This was decided by the Vietnam Ministry of Health, following a recommendation and request for support by the national EPI. By that time, Vietnam already had a substantial local vaccine production capacity, producing its own DTP, oral polio, BCG, and a number of other non-EPI vaccines. National researchers had developed a plasma-derived HepB with the support of WHO and technology transfer from the Kitasato Institute in Japan. The development program commenced in 1990, and, following a number of field trials, the locally produced HepB was licensed in 1997 and has been supplied to the EPI since that time.

Vietnam's HepB program started slowly, initially limited to the main urban centers of Hanoi and Ho Chi Minh City. By mid-2001, the government was purchasing around 1 million doses of the local vaccine per year, enabling coverage of around 20% of the nation's annual infant cohort of 1.6 million. The vaccine is provided free of charge as part of the EPI in predominantly urban areas in many of Vietnam's 61 provinces. Expanded introduction to include infants born in rural areas has been prevented by a shortage of funds to pay for the vaccine and difficulties with supply of the raw material used in its production. Until the advent of GAVI, no funding source had been identified to support broader access of Vietnamese infants to vaccination against hepatitis B, although there are plans to develop a locally produced, recombinant HepB with the support of a loan for technology transfer from the government of Korea. Production of this vaccine is due to commence in 2002. The seroprevalence of hepatitis B in Vietnam is high, with one study showing carriage rates of 12.5 % among infants aged 9–18 months, reaching 18% by 4–6 years. The "e" antigen, a marker of high infectivity, was present in around one-third of adult carriers, suggesting that between 3 and 5% of all infants in the study population were infected at birth. If these carriage rates are representative, HepB infection may be responsible for up to 5% of all deaths in Vietnam. Despite the high risk of perinatal infection, Vietnam's HepB vaccination program has not so far included a birth dose. HepB is currently given as a separate injection in three doses, along with the three doses of DTP. At present, no vaccines are given until the first contact with the EPI, usually at or after the age of 1 month. The need to protect infants against perinatally acquired HepB infection is now being addressed.

Vietnam successfully applied for support from GAVI to provide sufficient HepB to vaccinate the balance of its birth cohort not currently receiving the local vaccine. A phased expansion reaching all newborns is planned over the next 2 years and will exploit the heat stability of HepB by including a birth dose using single-dose Uniject devices labeled with vaccine vial monitors and stored outside cold chain for up to 1 month in areas without refrigeration. During the 5-year period of support from GAVI, the production capacity of the local recombinant HepB is anticipated to reach a level sufficient for national needs, but funding to purchase this vaccine from the local manufacturer in the future is not yet assured. Vaccination of newborns is likely to be met with some resistance in the community. Some support for health worker education and community mobilization that will be needed for this initiative has been identified from GAVI partners. The effectiveness of Vietnam's expanded introduction of HepB will be monitored carefully over the next few years.

III. CONCLUDING REMARKS

Attitudes toward the introduction of new vaccines vary with epidemiological, sociological, economic, and cultural settings. In industrialized countries, people tend to reject vaccines against diseases that are now rare, largely because of vaccination. Parents' reluctance to immunize their children is often related to concerns about adverse effects. Antivaccine movements are fueling these concerns and may significantly reduce popular acceptance of new vaccines. In developing countries, barriers to the introduction of new vaccines include lack of information on local disease burden, uncertainty about vaccine efficacy, poor capacity of national immunization programs, and lack of funding. Developments in vaccinology are characterized not only by the development of important new vaccines but also by constructive collaboration between major players in this field, making new vaccines available in some of the poorest countries. Although these developments are very exciting, it should be emphasized that international sponsors of new vaccine introduction assume considerable responsibility. New vaccines must not be introduced at the cost of existing programs. Thus, countries introducing new vaccines may require financial and technical support for national immunization programs in order to avoid overloading their programs. Furthermore, if there is no realistic chance that the new vaccine may be sustained over time, the ethics of new vaccine introduction should be seriously questioned. In this context, the unique role of the WHO network in international surveillance, counseling, and policy making is vital. More important than anything else, however, is a stronger local commitment toward national immunization programs, particularly in highly endemic areas. It is a sad fact that, in the poorest countries with the greatest need for effective vaccination, spending on health is inadequate in both absolute and relative terms. In those countries, and indeed throughout the world, there is a need for national authorities to appreciate the true value of vaccination with both existing and new vaccines, so that support for vaccination programs assumes a higher profile in national planning.

REFERENCES

1. Black, S., Lieu, T. A., Ray, G. T., *et al.* (2001). Assessing the costs and cost effectiveness of pneumococcal disease and vaccination within Kaiser Permanente. *Vaccine* **19**(Suppl.):S83–86.
2. World Health Organization Expanded Programme on Immunization (1996). EPI Information System: Global Summary. *Document WHO/EPI/CEIS/96.07*, 196.
3. Ryder, R. W., Whittle, H. C., Sanneh, A. B., *et al.* (1992). Persistent hepatitis b infection and hepatoma in the Gambia, Africa. A case–control study of 140 adults and their 603 family contacts. *Am. J. Epidemiol.* **136**:1122–1131.
4. Mulholland, E. K. (1999). Magnitude of the problem of childhood pneumonia in developing countries. *Lancet* **354**:590–592.
5. Black, S. B., Shinefield, H., Firemnan, B., *et al.* (2000). Efficacy, Safety and Immunogenicity of heptavalent pneumococcal conjugate vaccine in children. *Pediatr. Infect. Dis. J.* **19**:187–195.
6. Eskola, J., Kilpi, T., Palmu, A., *et al.* (2001). Efficacy of a pneumococcal conjugate vaccine against acute otitis media. *New England J. Med.* **344**(6):403–409.
7. Levine, M. M., and Levine O. S. (1997). Influence of disease burden, public perception, and other factors on new vaccine development, implementation and continued use. *Lancet* **350**:1386–1392.
8. Jafari, H. S., Adams, W. G., Robinson, K. A., Plikaytis, B. D., and Wenger, J. D. (1999). Efficacy of *Haemophilus influenzae* type b conjugate vaccines and persistence of disease in disadvantaged populations. The *Haemophilus influenzae* Study Group. *Am. J. Public Health* **89**:364–368.
9. Joensuu, J., Kosenniemi, E., Pang, X. –L., and Vesikari, T. (1997). Randomised placebo-controlled trial of rhesus–human reassortant rotavirus vaccine for prevention of severe rotavirus gastroenteritis. *Lancet* **350**:1205–1209.
10. Clemens, J., Brenner, R., Rao, M., Tafari, N., and Lowe, C. (1996). Evaluating new vaccines for developing countries. *J. Am. Med. Assoc.* **275**:390–397.
11. Reingold, A. L., Broome, C. V., Hightower, A. W., Ajello, G. W., Bolan, G. A., Adamsbaum, C., Jones, E. E., Phillips, C., Tiendrebeogo, H., and Yada, A. (1985). Age-specific differences in

duration of clinical protection after vaccination with meningococcal polysaccharide A vaccine. *Lancet* **2**(8447):114–118.

12. Mulholland, E. K. (1995). Measles and pertussis in developing countries with good vaccine coverage. *Lancet* **345**:305–307.

13. Global Programme for Vaccines and Immunization (2000). The WHO position paper on rubella vaccines. *Weekly Epidemiol. Record* **75**:161–169.

14. Robertson, S. E., Cutts, F. T., Samuel, R., and Diaz-Ortega, J.-L. (1997). Control of rubella and congenital rubella syndrome (CRS) in developing countries, part 2: Vaccination against rubella. *Bull. WHO* **75**:69–80.

15. Mbelle, N., Huebner, R. E., Wasas, A., *et al.* (1999). Immunogenicity and impact on nasopharyngeal carriage of a nonavalent pneumococcal conjugate vaccine. *J. Infect. Dis.* **180**:1171–1176.

16. Feudtner, C., and Marcuse, E. K. (2001). Ethics and immunization policy: Promoting dialogue to sustain consensus. *Pediatrics* **107**:1158–1164.

17. Poland, G. A., and Jacobson, R. M. (2001). Understanding those who do not understand: A brief review of the anti-vaccine movement. *Vaccine* **19**:2440–2445.

18. King, S. (1999). Vaccination policies: Individual rights vs community health. We can't afford to be half hearted about vaccination programmes. *Br. Med. J.* **319**:1448–1449.

19. Gangarosa, E. J., Galazka, A. M., Wolfe, C. R., *et al.* (1998). Impact of anti-vaccine movements on pertussis control: The untold story. *Lancet* **351**:1019–1021.

20. Global Programme for Vaccines and Immunization (1999). The WHO position paper on pertussis vaccines. *Weekly Epidemiol. Record* **74**:137–143.

21. Rennels, M. B., Parashar, U. D., Holman, R. C., *et al.* (1998). Lack of an apparent association between intussusception and wild or vaccine rotavirus infection. *Pediatr. Infect. Dis. J.* **17**:924–925.

22. Murphy, T. V., Gargiullo, P. M., Massoudi, M. S., *et al.* (2001). Intussusception among infants given an oral rotavirus vaccine. *New England J. Med.* **344**:564–572.

23. Guo, J. Z., Ma, X. Y., Zhou, Q. H. (1986). Results of air pressure enema reduction of intussusception: 6,396 cases in 13 years. *J. Pediatr. Surg.* **21**:1201–1203.

24. European Commission COST/STD Initiative (1996). Report of the Expert Panel V: Harmonization of European Vaccination Programmes. *Vaccine* **14**: 611–623.

25. Miller, M. A., and Flanders, W. D. (2000). A model to estimate the probability of hepatitis B- and *Haemophilus influenzae* type b-vaccine uptake into national vaccination programs. *Vaccine* **18**:2223–2230.

26. Sinha, A., Sazawal, S., Kumar, R., Sood, S., Reddaiah, V. P., Singh, B., Rao, M., Naficy, A., Clemens, J. D., and Bhan, M. K. (1999). Typhoid fever in children aged less than 5 years. *Lancet* **354**:734–737.

27. Lin, F. Y. C., Ho, V. A., Kheim, H. B., *et al.* (2001). The efficacy of a *Salmonella typhi* Vi conjugate vaccine in two-to-five-year-old children. *New England J. Med.* **344**:1263–1269.

28. Batson, A. (1998). Sustainable introduction of affordable new vaccines: The targeting strategy. *Vaccine* **16**(Suppl.):S93–98.

29. Mahoney, R. T., Ramachandran, S., and Xu, Z. (2000). The introduction of new vaccines into developing countries II. Vaccine financing. *Vaccine* **24**:2625–2635.

30. Hausdorff, W. P. (1996). Prospects for the use of new vaccines in developing countries: Cost is not the only impediment. *Vaccine* **14**:1179–1186.

10 FUTURE CHALLENGES FOR VACCINES AND IMMUNIZATION

PAUL-HENRI LAMBERT* AND BARRY R. BLOOM†

*Centre of Vaccinology, Department of Pathology, University of Geneva, 1290 Geneva 4, Switzerland

†Harvard School of Public Health Boston, Massachusetts 02115

In the short time in human history since vaccines were introduced on a global scale, they have changed the world. Yet despite the staggering increase in knowledge about both the pathogenesis of infectious diseases and immunological responses, we do not understand fully the mechanisms by which most of the existing vaccines provide protection. As the previous chapters have indicated, vaccines have eradicated smallpox, one of the great epidemic scourges of humankind, protected most of the world against polio, prevented millions of children from dying or suffering tragic disabilities from childhood infectious diseases, and contributed to an increase in life expectancy in most countries of the world. Yet most current vaccines have only had to protect children against common infectious diseases that occur within the first few years of life. In contrast, many of the new infectious disease challenges, such as AIDS, tuberculosis, and sexually transmitted diseases, are not primarily diseases of early childhood. Thus, new models need to be developed for immunizing teenagers, adolescents, and adults. Similarly, as a consequence of sparing the lives of children, in all countries we can expect a rising "epidemic of chronic diseases," including cardiovascular disease, cancer, neurological disease, and diabetes. These represent vaccine challenges for the future. However, even if the scientific community develops an understanding in molecular terms of how to engage the immune response to provide protection against these diseases and comes forth with new and effective vaccines, how will they be used, how will these new populations be reached, and how will they be accepted by the public? Will new vaccines be so valued that people and their governments will be willing to pay for them and give them high priority in their public policies? These are realms that we see as future challenges to vaccines and immunization.

I. CROSS-CUTTING ISSUES IN VACCINES AGAINST INFECTIOUS DISEASES

CHALLENGE #1   Dealing with antigenic diversity and antigenic variation. Can we circumvent microbial escape?

Naturally occurring mutations and microbial selection under immunological pressure can influence the epidemiology of microorganisms targeted by vaccination strategies. This is certainly the case for influenza viruses. It is only through international monitoring of viral isolates, organized at a global level under World Health Organization (WHO) auspices, that vaccine-based control can be planned on an annual and sustained basis. However, the world has to be ready for a suddenly emerging, major antigenic change in influenza virus, which would require the rapid development and production of appropriate vaccines. Antigenic variation can be fast, and even though most infections initially are generally clonal, variation can occur even within the infected individual, as seen with HIV, hepatitis C, malaria, and trypanosomiasis. This is often a major impediment to the selection of optimal vaccine antigens and has evolved to favor microbial escape mechanisms.

The multiplicity of antigenic variants may also open the way to "replacement" strains. For example, new pneumococcal vaccines, which only prevent a limited number of dominant serotypes, can be effective in decreasing the infant carrier rate for the serotypes included in the vaccine. However, this may lead to colonization by pneumococci of different serotypes not included in the vaccine formulation. Thus far, the overall impact of pneumococcal vaccines remains beneficial, because the newly colonizing strains are much more sensitive to antibiotics than those included in the vaccine. However, such replacement phenomena will necessitate the continual monitoring of pneumococcal strains and, if needed, regular changes in the vaccine formulation.

Some questions have been raised regarding the potential epidemiologic changes that might occur as a result of the emergence of antigenic variants of hepatitis B and *Bordetella pertussis*, but there is no evidence yet of any related clinical significance.

CHALLENGE #2   To ensure enduring protection. Generating immunological memory and long-lasting antibody responses.

For many of the current childhood vaccines, the period of greatest vulnerability to infection and disease is within the first few years. Long-term memory may not be so important because, after a basic level of immunization, reexposure to the antigens or reinfection can serve to "boost" the level of response. It must be noted, however, that particularly in the cases of tetanus and measles, immunity wanes over time and boosters are required after 10–15 years. Memory B-cell responses are not always sufficient: there is a need for ensuring long-lasting antibody production and persistence of protective antibody levels to prevent toxin-mediated pathology (e.g., tetanus) or rapidly invasive infections (e.g., group C meningococcal disease).

In the case of HIV–AIDS, tuberculosis, and malaria, the risk of infection is great for adults throughout much of their lifetime. For most existing vaccines given in early life, it is unclear how long protection can be counted on, and for these diseases especially, long-term memory will be required. Even in animal models, the necessary conditions for generating long-term T helper cell memory for either antibody production or cell-mediated immune responses are not clear. Currently, with the exception of MF59 used for only one vaccine, there is essentially a single adjuvant licensed for use in humans, aluminum hydroxide, which was developed almost a century ago. Can we expect a single immunization in childhood to protect against tuberculosis reactivation 60 years after primary infection or, in the case of HIV–AIDS, even for 20 years? How can immunological memory be generated and sustained? Can subunit vaccines generate it, or will immunization with live, attenuated vectors or vaccines be necessary with the attendant greater risks of adverse effects?

CHALLENGE #3   To define relevant immunological indicators of protection. Clinical trials made easier to perform.

Vaccine trials are a long, complex, and expensive process. Most existing vaccines are thought to provide protection against infectious diseases primarily by the production of protective or neutralizing antibodies. This humoral response can be easily assessed. But the major vaccine-preventable diseases for which new vaccines are desperately needed, HIV–AIDS, tuberculosis, and malaria, are likely to have multiple and complex mechanisms of protection, including antibodies and cell-mediated immunity.

Tuberculosis infection carries with it a lifetime risk of reactivation, so that immunization must be long-lasting. In all of these cases, it may be more feasible to prevent disease than to prevent infection. Consequently, we must recognize that it may be unreasonable to expect the degree of protection with new vaccines that we have taken for granted in the current childhood vaccines. Furthermore, the science of vaccines must develop end points for vaccines against these diseases that do not require the assessment of life and death—the world does not have time to wait for that.

This means that we need to understand the mechanisms of immunity—those responses that are necessary and sufficient to engender protection. That knowledge begins with identifying "correlates" of protection, i.e., immunological measures or tests that appear to parallel protection. Ultimately, such correlates that are established will be assessed in clinical trials as "surrogates" for immunity, i.e., laboratory tests that predict protection with high accuracy. In HIV–AIDS, it has been very difficult for candidate vaccines to engender neutralizing antibodies. The antibodies are not comparable to those produced against childhood diseases, which seem to correlate best with protection, although in most cases the infection of host cells generally is not prevented. It may be possible to develop protection against disease, if not infection, by means of cell-mediated immunity. Yet it is not known which cellular responses are necessary or sufficient: T helper cell activity, T-cell-produced lymphokine and cytokine production, cytotoxic T-lymphocyte (CTL) activity, or a combination. In malaria, different immune responses may be required for different forms of the parasite, for example, antibodies that prevent the mosquito-derived sporozoites from infecting liver cells, CTLs that kill infected liver cells, antibodies that kill the blood forms, and still other antibodies that neutralize the gametocytes and hence block transmission. They could all be assayed *in vitro*, but which of these will correlate best with protection remains to be established in clinical trials.

In the case of tuberculosis, all of the evidence suggests that protection is mediated by cellular immune mechanisms, but here again it is unclear which type of cellular immune responses and which T-cell functions are critical correlates of protection. All evidence indicates that antibodies generated by infection are not protective, but it is conceivable that new vaccines could engender antibodies different from those arising in natural infection that might prevent the pathogen from reaching the lungs or from spreading systemically from the lungs. One of the challenges of vaccines is to improve on nature and, frequently, to do better than natural infection.

A major thrust of vaccine research must be to employ the best methodology and technology available in clinical trials and in animal models to analyze those immune responses that correlate best with protection, so that they can be useful end points. Such correlates or surrogate end points obviously will accelerate the period of clinical testing, reduce costs, and allow more candidates to be compared.

Finally, we may have to set our sights not only on preventive vaccines, that is, those that prevent infection or progression to disease, but also on therapeutic vaccines, for which there has been little past success. Perhaps research on vaccines for chronic diseases will provide insights that can be applied to the therapy of already existing infectious diseases, particularly those causing chronic infections such as papillomavirus associated with cervical and anal cancer, *Helicobacter pylori* associated with stomach cancer, and chlamydial infections possibly involved in atherosclerosis.

II.   VACCINES FOR NONINFECTIOUS DISEASES: A NEW FRONTIER FOR VACCINE RESEARCH

The development of vaccine-based preventive or therapeutic approaches for chronic diseases such as cancer, autoimmune diseases, allergies, and Alzheimer's disease represents new and attractive applications of vaccine research (Table 1). These vaccine strategies differ from those aiming at the prevention of infectious diseases.

TABLE 1   Vaccines for Noninfectious Diseases: Diverging Objectives

Cancer, Alzheimer's	Autoimmune Diseases	Allergic Diseases
Induce "autoimmune" responses against overexpressed normal or mutated host antigens on tumor cells	Restore immunological tolerance to host antigens (functional tolerance)	Redirect immune responses toward nonpathogenic responses

The objective is to elicit responses against host antigens (e.g., cancer antigens, amyloid peptide) or, conversely, to suppress undesirable responses. This can be done through redirecting antigen-specific autoimmune or allergic responses toward nonpathogenic responses (e.g., from Th1 to Th2 in type 1 diabetes, from IgE to IgG in honeybee anaphylaxis) or through restoring some form of functional tolerance (e.g., induction of CD4$^+$ CD25$^+$ regulatory T cells). Considerable efforts are now invested in this direction in relation to the huge potential economic impact of the expected products.

CHALLENGE #4   Cancer vaccines.

Cancer vaccines in development aim at inducing *tumor antigen-specific* responses that lead to the destruction or growth inhibition of primary tumors or metastases. This differs from immunopotentiating approaches, which have previously aimed at an overall stimulation of the immune system with an associated enhancement of spontaneous antitumor responses. In the past few years, increasing information about the molecular basis of tumor–host interactions has been gathered. Many different intracellular proteins are known to represent human cancer antigens. The numerous genomic alterations that occur in a cancer cell may lead to overexpression of individual proteins or expression of mutated proteins, which provide opportunities for the emergence of cancer antigens. The main categories of antigens are: (1) normal nonmutated differentiation antigens that are expressed exclusively on specific cells (e.g., melanoma), (2) cancer antigens widely expressed on a variety of epithelial tumors as well as on testis and placental tissue, (3) normal proteins that contain mutations or translocations that give rise to unique epitopes, and (4) nonmutated shared antigens that are overexpressed on cancers (e.g., carcinoembryonic antigen, p53, and Her-2/neu). Apart from approaches targeting antigens acting as receptors for tumor growth factors (e.g., Her-2/neu), most passive or active immunization approaches are based on recognition by T cells, particularly HLA-A class I restricted CD8$^+$ lymphocytes (see the review in ref 1). Obviously, many of the viruses associated with oncogenesis can also present proteins on the induced cancers that can serve as targets for immune attack.

The identification of tumor-specific antigens recognized on human tumors by autologous cytolytic T cells has led to the use of defined antigens for the therapeutic vaccination of cancer patients [2]. For example, several clinical trials are based on the use of antigens encoded by genes of the MAGE family as tumor-specific targets that are expressed on many tumors of various histological types. In a trial involving three monthly vaccinations with a MAGE-3 peptide, which is expressed in about three out of four metastatic melanoma patients [3], but is not expressed in normal cells, with the exception of the male germline cells, significant tumor regressions were observed in 7 out of 25 melanoma patients [4]. In one of these responding patients, a monoclonal CTL response against the MAGE-3 antigen has been actually demonstrated [5]. Gene transfer strategies are also developed to actively immunize against tumor-associated antigens through a wide variety of gene transfer technologies, including gene delivery into dendritic cells (DCs).

One novel approach being investigated for therapeutic cancer vaccines derives from the finding that certain heat-shock proteins can bind small antigenic peptides from cancers in animals and present them in an immunogenic way to the host. Although the nature of the tumor-specific peptides is not determined, immunization with the heat-shock proteins from tumor cells can generate therapeutic responses against experimental tumors, and they are currently undergoing clinical trials in humans [6, 7].

Probably the main remaining challenge for cancer vaccines is to overcome limiting factors, which allow tumor cells to escape immunological control. In many tumors, T cells can be found that react against tumor antigens, but they may be present at insufficient levels or may not have sufficient efficacy to mediate tumor destruction. A variety of active mechanisms may also limit the effectiveness of immunization, including a relative nonresponsiveness of T cells resulting from the lack of expression of appropriate costimulatory molecules on the tumor or from the down-regulation of T-cell-receptor signal transduction. The tumor cell may be an active participant in specifically depressing the immune response, protecting itself from immune destruction. We hope the present efforts to understand the mechanisms used by the tumor to escape destruction and to optimize immunization will lead to improvements in anticancer vaccination strategies.

CHALLENGE #5   Vaccines for Alzheimer's diseases.

It is estimated that approximately 12,000,000 people in the United States and European Union Member States are affected by Alzheimer's disease (AD). The major neuropathological hallmarks of AD are extracellular deposits of "senile" amyloid plaques, intraneuronal neurofibrillary tangles, synapse loss, and the death of neurons. The present target for a potential vaccine is the amyloid plaque. Plaques are complex extracellular deposits that contain β-amyloid, a peptide that is 39–43 amino acids long and is produced in normal cells by proteolytic cleavage of the β-amyloid precursor protein (APP). There are two major identified pathways of APP cleavage, and one of them can generate either a 40 (Aβ-40) or a 42 (Aβ-42) amino acid peptide. The shorter form, Aβ-40, is more soluble and aggregates slowly. The Aβ-42 fibrils are insoluble and interact to form β-pleated sheets, which are the key component of the amyloid plaques.

Exciting studies have shown that simple immunization of transgenic mice (PDAPP) that overexpress a mutant APP with the more amyloidogenic Aβ-42 peptide can reduce the deposition of amyloid onto existing plaques and partly clear established senile plaques that are present in the brain (Fig. 1) [8]. Other studies performed with passively administered anti-Aβ monoclonal antibodies suggest that this effect may be due to an antibody-mediated removal of Aβ deposits from the brain through an efflux of brain Aβ to the cerebrospinal fluid (CSF) and into circulation [9]. Antibodies that cross the blood–brain barrier (BBB) may also bind to amyloid plaques, activate microglial cells, and locally induce the clearance of preexisting amyloid [10]. In one study in mice, the antibodies that reduced plaques even were able to reverse some of the cognitive defects.

Human phase 1 safety studies of the first candidate Aβ vaccine (AN-1792) have now been completed. AN-1792 has been administered to 100 patients with mild to moderate Alzheimer's disease in a variety of dosage regimens. preliminary results from Elan Corporation indicated that the vaccine was well-tolerated and that a portion of patients developed a sufficient immunological response to warrant initiation of additional clinical studies. Such results are promising, but it is obvious that conclusions regarding the relative potential of the vaccine approach will only be possible when human efficacy trials have been completed. In fact, a phase 2A study was temporarily suspended in

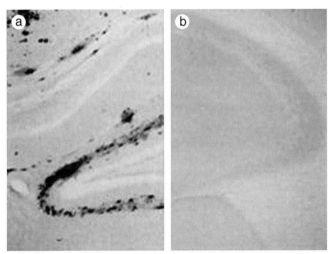

FIGURE 1 Hippocampal Aβ deposition, neuritic plaque formation, and cortical astrocytosis in PBS- and Aβ-42-injected mice (13-month-old mice). Hippocampal amyloid plaques in PBS- (a) and Aβ-42-injected (b) mice. Abundant amyloid deposition is seen in the outer molecular layer of the hippocampal dentate gyrus in a PBS-treated animal (a), whereas no detectable amyloid is seen in this region in an Aβ-42-immunized (b) mouse. Reprinted with permission from schenk et al. Nature **400**:173–177. Copyright (1999), Macmillan Megazines Limited.

January 2002 after 4/97 patients receiving the vaccine in France were reported to have clinical signs consistent with inflammation in the central nervous system.

CHALLENGE #6   Vaccines for autoimmune diseases.

Autoimmune diseases result from a dysfunction of the immune system in which the body attacks its own organs. They comprise more than 70 clinically distinct autoimmune diseases, with an immune-mediated injury localized to a single organ [e.g., type 1 diabetes, multiple sclerosis (MS)] or affecting several or many different organs (e.g., systemic lupus erythematosus, rheumatoid arthritis). Whereas some are rare, it is estimated that over 30 million people are affected by one or another autoimmune disease in the United States and the European Union countries. The social and financial burden of these chronic diseases is immense, and this is amplified by the fact that some of them (e.g., type 1 diabetes) are now showing a significant increase in prevalence.

Disease-specific vaccine-based therapeutic approaches are under development. They aim at stopping the autoimmune process before the occurrence of tissue injury or at reducing the level of ongoing autoimmune pathological processes. Efforts to induce immunological tolerance initially focused on the oral administration of antigens. Encouraging responses were obtained in animal studies in which the oral administration of high or low doses of antigen was shown to block the response to subsequent systemic administration of antigen. This approach has been evaluated in several human diseases (e.g., oral insulin in type 1 diabetes, DPT-1 trial). However, the results obtained thus far do not appear to meet the level of expectation. A number of other tolerogenic approaches are now being pursued in autoimmune diseases, including parenteral administration of antigen in the form of small peptides, concomitant with immunomodulatory agents (e.g., blocking of CD40-dependent costimulation of the immune system, cytokine modulation). For example, trials are now underway using low-dose parenteral insulin peptides in newly diagnosed individuals with type 1 diabetes and in at-risk nondiabetic relatives of individuals with type 1 diabetes.

Other approaches aim not at the induction of immunological tolerance but at the interruption or prevention of interactions between antigen-presenting cells and T cells, which produce inflammatory molecules responsible for tissue damage. For example, an immunodominant T-cell epitope known to be recognized by the patient's cells can be used in an altered form to block subsequent antigen presentation to disease-causing auto-reactive T cells. Preliminary results of clinical trials in patients with recent-onset type 1 diabetes, using a heat-shock protein (HSP-60) derived peptide (positions 437–460, "Diapep 277") shown to be recognized by diabetogenic T cells, are promising (Cohen *et al.*, personal communication) [11]. The interim results from phase 2 randomized double-blind placebo-controlled studies in 60 newly diagnosed type 1 diabetes patients (children and adults), after 10 months of treatment, are now available.

The requirement for insulin stabilizes in patients treated with Diapep277 when compared with patients receiving a placebo, whose insulin requirement increases with the progression of the disease (communicated on http://www.peptor.co.il/2.htm). This suggests that, even in a disease involving multiple antigens, the immune response to an appropriately selected single antigen can have a major influence on disease progression. Immune deviation with altered peptide ligands is now being assessed in patients with multiple sclerosis.

Vaccines based on the use of DNA plasmids may also have a future for early therapy in autoimmune diseases. It was shown that a DNA vaccine encoding for the GAD-65 antigen of pancreatic islet cells could prevent the development of diabetes in the NOD mouse model [12].

CHALLENGE #7   Vaccines for allergic diseases.

In the industrialized world, the prevalence of asthma and other allergic diseases has increased over the past two decades, especially in young children (Fig. 2). This trend in the prevalence of allergic diseases is often explained by a declining incidence of many infectious diseases and delayed infant colonization with commensal microorganisms. This would favor an increase in the Th2 bias of immune responses toward environmental allergens (hygiene hypothesis). Allergic diseases affect as many as 30% of individuals in some countries, and in the United States the CDC reports that more than 17 million people currently suffer from asthma, about 7% of the population. In addition, 20% of the population suffer from allergic rhinitis. Anaphylaxis from insect stings (e.g., honeybees, yellow jackets) is estimated to occur in 0.5–5% of the general population, with 40–100 deaths per year [13]. Overall, the economic cost of allergic diseases is substantial: in the United States, allergy-associated health costs probably exceeded $14 billion in the year 2000 [14, 15].

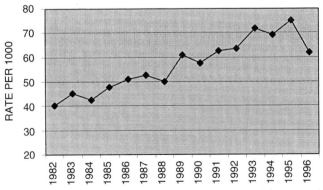

FIGURE 2   Evolution of asthma prevalence in the United States population under 18 years of age, from 1982 to 1996 (from Ref. 15a).

Protein-based allergen immunotherapy has proven somewhat effective in the prevention of the immediate and late-phase hypersensitivity responses. Unfortunately, although over 90% of patients with anaphylactic hypersensitivity to stinging insects develop clinical tolerance, only 30–50% of allergic rhinitis patients and fewer asthmatics respond. New formulations that would be more immunogenic and less allergenic would be attractive alternatives. Such "vaccines" have now become feasible because many clinically important allergens have been identified, purified, cloned, epitope-mapped, and produced as biologically active recombinant proteins. Approaches that combine non-antigen-specific methods (e.g., second signal blockade and cytokine modulation) and antigen-specific tolerance appear very promising in animal models. Several of these are now entering or will shortly enter phase 1 and/or phase 2 clinical studies in humans, e.g., peptide vaccines representing T-cell epitopes of allergen, immunostimulatory oligonucleotides that elicit Th1 immune responses when coadministered with protein allergens, or plasmid DNA encoding recombinant allergens (see the review in ref 16).

III. REQUIREMENTS RELATED TO NEW VACCINATION STRATEGIES

CHALLENGE #8   New methods to deliver vaccines.

Despite its relative rusticity, the oldest human vaccine, vaccinia, has been one of the easiest to use. It does not require injections because it is given into the skin, it is rather stable, and it is efficacious after only a single administration. Such properties represent a remarkable advantage for a health tool to be used in the most difficult conditions, and they have probably contributed to the success of smallpox eradication. Although it requires a logistically demanding cold chain and repeated administration, the second easiest vaccine to deliver happens to be the oral polio vaccine, and again this is facilitating a move toward massive global use, which should eventually lead to elimination of the disease.

Vaccine delivery systems are important determinants of the effectiveness of vaccination strategies. In developing countries, there is a considerable comparative advantage for vaccines that would not require injections, and thus avoid the risk of contamination. It is estimated that, in the year 2000, close to one-third of all injections carried out in Africa were not done within the desirable safety standards. Thus, the risk of bacterial contamination or blood-borne disease transmission could not be fully excluded. Syringes that cannot be reused and vaccines that can be given orally, such as OPV, completely avoid the injection-associated risk. They are particularly well-accepted and well-suited for vaccination campaigns. Vaccines that are relatively heat-stable, e.g., DTP, and do not require a strict cold chain system also bear an intrinsic advantage.

It is a real challenge to ensure that the delivery issue be considered for vaccines under development. Some of the new emerging products are of special interest, such as nasal vaccines that use live vectors or subunit formulations with appropriate adjuvants, as well as DNA vaccines that can be administered using the gene gun technology.

CHALLENGE #9 Appropriate vaccines for the elderly, newborns, and adolescents. Vaccination in the context of an increasing prevalence of immunodeficiency.

We have learned a great deal about how to deliver vaccines to young children to protect them against childhood infectious diseases. This has required responding to extraordinary logistical challenges in every country, developing ways of integrating immunization programs into public health and primary care programs, and maintaining a global cold chain to ensure the immunologic integrity, quality, and safety of vaccines. The new challenges will be how to develop comparably responsive immunization systems to reach new populations. One is the elderly, who present immunological challenges at two levels: how to immunize individuals whose immune system is often wearing out, and how to reach them with vaccines. In the United States from 1995 to 1998, it is estimated that 26,000 people over 60 years of age died annually from two vaccine-preventable diseases: influenza and pneumococcal pneumonia [17, 18]. The death or disability of anyone who could have been protected by an existing vaccine is a tragedy. Economic analyses indicate that these vaccines are highly cost-effective, yet even in the United States there is no organized system or program for adult immunization. It will be a challenge to set up immunization for adults as a national priority in most countries. The public health leaders will have to inform and motivate the public and give incentives to healthcare professionals, hospitals, nursing homes, and public health systems to immunize the elderly against the infectious diseases that put them most at risk.

At the other end of the spectrum is the immunological challenge of engendering strong and protective immune responses in newborns. That is a time in which the immune system is not yet fully developed, such that many carbohydrates and even some protein antigens fail to engender adequate protective immunity. We have learned by conjugating carbohydrates to common protein antigens, as in the case of Hib and pneumococcal vaccines, that T cells can be engaged to recognize the protein epitopes so as to provide helper activity that augments the quantity and quality of B-cell responses to carbohydrates. Here is a case in which new "carrier" proteins and new kinds of immunological adjuvants need to be developed. They could augment immune responses in newborns, decrease the number of booster shots, and decrease the time required to achieve protective levels of immune responses. This will be relevant to any new vaccine in which carbohydrate or lipid antigens are essential targets of immune responses as, for example, in vaccines against streptococcal, staphylococcal, and salmonella infections.

Finally, the HIV–AIDS epidemic, which is increasing dramatically in many developing countries and is the major cause of death among infectious diseases, is paralleled by coepidemics of tuberculosis and hepatitis C. They affect young adults primarily, and these and other sexually transmitted infections require new vaccines and an entirely new vaccine delivery strategy. For adolescents in schools, there is the opportunity to set up school immunization programs. For example, the increase in measles and atypical measles in previously vaccinated young people has led to a revaccination and "booster" program for young teenagers in schools throughout the United States. Yet in developing countries, it is difficult to add further responsibilities to overburdened school systems where teachers are overworked. In many resource-poor countries, a large percentage of young people are not in schools and therefore are difficult to access. New social institutions with the ability to provide reproductive counseling and immunization of adolescents and young people will need to be created. Because of the urgency of the threat of HIV–AIDS, it is essential now to initiate some imaginative experiments on reaching young people in this age group.

Immunization of immunodeficient individuals poses special problems. First is the risk of adverse events, particularly if live attenuated vaccines in immunocompetent individuals are unable to be controlled and cause disease. We know that children suffering from serious immunodeficiency diseases frequently suffer fatal infections from OPV and measles, yet the experience in adults is less clear. Because of the long latency in HIV before the symptoms of AIDS develop, there is the possibility of providing significant levels of protection prior to major immunodeficiency. In several studies of BCG immunization, the incidence of adverse effects in HIV-seropositive children was no greater than in healthy children. Thus, at the first level, the risks from immunization, even from live attenuated vaccines, are not entirely clear.

Second, to ensure safety, it will be advantageous to immunize with either effective subunit vaccines or live attenuated vaccines that have been sufficiently genetically modified to ensure that they cannot cause disease even in immunodeficient individuals. At a third level, even if one can safely vaccinate immunodeficient individuals, there is the question of what the implications would be if the level of immunity falls short of that seen in the general population. How will that affect the transmission and persistence of infections? These are important questions for which we need more information from studies in animal models, from epidemiological analyses of experiences in individuals who were later found to develop immunodeficiency, and from studies in highly endemic populations.

Ultimately, there may be ethical issues raised similar to those raised by the rotavirus vaccine. In this case, although the vaccine provided a major reduction in death from the infection, which kills 800,000 young children worldwide annually, it had a serious adverse effect, intussusception or strangulation of the intestine, in a small number of children. The safety concern required that the vaccine be removed from the market in the United States. The public health argument would be that many more lives could have been saved than the small fraction adversely affected had the vaccine been used in high endemic countries, although the vaccine would have caused some deaths. It is conceivable that there will be effective vaccines that will raise similar ethical concerns due to adverse effects in immunodeficient individuals.

IV. PUBLIC ACCEPTANCE OF VACCINATION

CHALLENGE #10 How far can we go with new vaccines and new vaccination strategies? Support of the public for vaccines?

Many new vaccines are now in development, and we can expect that some of them will soon reach a final stage, be produced commercially, and put on the market. Key questions that will have to be addressed are as follows: Will these vaccines be used? Will they be integrated into appropriate vaccination strategies?

In several industrialized countries with a relatively high average income, public resistance to new vaccines being added to the existing pediatric schedule is now emerging. This is often reflected in parental resentment of what they consider a medical "aggression" for their healthy child. It is also directly related to the number of injections required for completing the infant vaccination plan. In several European countries, this feeling is further reinforced by particular medical groups, which are proponents of new "natural" preventive or therapeutic attitudes. These negative reactions partly reflect a decreased perception of the risk of infectious disease in countries in which efficient disease control or elimination strategies have been successful.

The combination of several vaccines should reduce the fear of multiple injections, and such products are now appearing for key pediatric vaccines (e.g., hexavalent DTP–IPV–HBV–Hib vaccine). Surprisingly, without any scientific evidence, such products are sometimes attacked for being an "excessive aggression" for the maturing infant immune system, which may result in some kind of immunodeficiency. An overall decrease in public confidence in vaccine safety is also the result of unsubstantiated allegations of vaccine-induced diseases (e.g., autism or bowel disease after MMR or multiple sclerosis after hepatitis B vaccination) (see Chapter 8). Therefore, in the absence of legal requirements, future recommendations issued by public health authorities regarding the addition of new vaccines to the existing pediatric schedule will not be *ipso facto* accepted by all parents nor even by the whole medical community. Another challenge is economic in nature. It is usually well-appreciated that economic obstacles still limit the use of existing vaccines in many developing countries in which no solution would be expected without international aid and relevant financing plans (see Chapter 7). However, at a time of rapidly expanding health costs in every country of the world, economic issues more and more will influence the use of newly developed vaccines everywhere, including most industrialized countries. For example, public health authorities in some of the wealthiest European countries do not recommend universal use in infants of the pneumococcal conjugates, as a result of perceived insufficient cost-effectiveness potential.

Unfortunately, other economic issues are likely to slow down the investment of major vaccine manufacturers, under shareholder pressure, in the area of prophylactic vaccines. Indeed, these vaccines appear more and more as low-profit products, with an increasing litigation risk, as compared to some of the new high-return blockbusting drugs, often produced by another section of the same giant pharmaceutical company. Such considerations may favor the development of therapeutic vaccines for which one can expect a higher acceptance of mild side effects, a less stringent requirement for clinical efficacy, the use of multiple dose schedules, and the emergence of pricing policies common to the drug market.

It is our hope that the real value of preventive vaccination will be taken into proper consideration at the political and general public levels in both developing and industrialized countries. We see already that disease eradication campaigns may bear sufficient political value to attract open support from some major national leaders, and this is helping to achieve goals often considered as utopian. Such reconsideration and greater recognition of the social value of prophylactic vaccination will be essential in order to allow humankind to benefit from the considerable scientific and economic efforts that support the development of new vaccines. Vaccines have saved millions of lives, and newer vaccines under development have the potential to save even more, but we must work to maintain and strengthen the global commitment to immunization and vaccines. As Oscar Wilde wrote, "A map of the world without utopia on it is not worth glancing at."

REFERENCES

1. Rosenberg, S. A. (2001). Progress in human tumour immunology and immunotherapy. *Nature* **411**:380–384.
2. van der Bruggen, P., Traversari, C., Chomez, P., Lurquin, C., De Plaen, E., Van den Eynde, B., Knuth, A., and Boon, T. (1991). A gene encoding an antigen recognized by cytolytic T lymphocytes on a human melanoma. *Science* **254**:1643–1647.
3. Brasseur, F., Rimoldi, D., Lienard, D., Lethe, B., Carrel, S., Arienti, F., Suter, L., Vanwijck, R., Bourlond, A., Humblet, Y., *et al.* (1995). Expression of MAGE genes in primary and metastatic cutaneous melanoma. *Int. J. Cancer* **63**:375–380.
4. Marchand, M., van Baren, N., Weynants, P., Brichard, V., Dreno, B., Tessier, M. H., Rankin, E., Parmiani, G., Arienti, F., Humblet, Y., Bourlond, A., Vanwijck, R., Lienard, D., Beauduin, M., Dietrich, P. Y., Russo, V., Kerger, J., Masucci, G., Jager, E., De Greve, J., Atzpodien, J., Brasseur, F., Coulie, P. G., van der Bruggen, P., and Boon, T. (1999). Tumor regressions observed in patients with metastatic melanoma treated with an antigenic peptide encoded by gene MAGE-3 and presented by HLA-A1. *Int. J. Cancer* **80**:219–230.
5. Coulie, P. G., Karanikas, V., Colau, D., Lurquin, C., Landry, C., Marchand, M., Dorval, T., Brichard, V., and Boon, T. (2001). A monoclonal cytolytic T-lymphocyte response observed in a melanoma patient vaccinated with a tumor-specific antigenic peptide encoded by gene MAGE-3. *Proc. Natl. Acad. Sci. USA* **98**:10290–10295.
6. Srivastava, P. K. (2000). Immunotherapy of human cancer: Lessons from mice. *Nat. Immunol.* **1**:363–366.
7. Srivastava, P. K., and Amato, R. J. (2001). Heat shock proteins: The "Swiss Army Knife" vaccines against cancers and infectious agents. *Vaccine* **19**:2590–2597.

8. Schenk, D., Barbour, R., Dunn, W., Gordon, G., Grajeda, H., Guido, T., Hu, K., Huang, J., Johnson-Wood, K., Khan, K., Kholodenko, D., Lee, M., Liao, Z., Lieberburg, I., Motter, R., Mutter, L., Soriano, F., Shopp, G., Vasquez, N., Vandevert, C., Walker, S., Wogulis, M., Yednock, T., Games, D., and Seubert, P. (1999). Immunization with amyloid-β attenuates Alzheimer-disease-like pathology in the PDAPP mouse. *Nature* **400:**173–177.

9. DeMattos, R. B., Bales, K. R., Cummins, D. J., Dodart, J. C., Paul, S. M., and Holtzman, D. M. (2001). Peripheral anti-Aβ antibody alters CNS and plasma Aβ clearance and decreases brain Aβ burden in a mouse model of Alzheimer's disease. *Proc. Natl. Acad. Sci. USA* **98:**8850–8855.

10. Bard, F., Cannon, C., Barbour, R., Burke, R. L., Games, D., Grajeda, H., Guido, T., Hu, K., Huang, J., Johnson-Wood, K., Khan, K., Kholodenko, D., Lee, M., Lieberburg, I., Motter, R., Nguyen, M., Soriano, F., Vasquez, N., Weiss, K., Welch, B., Seubert, P., Schenk, D., and Yednock, T. (2000). Peripherally administered antibodies against amyloid β-peptide enter the central nervous system and reduce pathology in a mouse model of Alzheimer disease. *Nat. Med.* **6:**916–919.

11. Regner, M., and Lambert, P. H. (2001). Autoimmunity through infection or immunization? *Nat. Immunol.* **2:**185–188.

12. Balasa, B., Boehm, B. O., Fortnagel, A., Karges, W., Van Gunst, K., Jung, N., Camacho, S. A., Webb, S. R., and Sarvetnick, N. (2001). Vaccination with glutamic acid decarboxylase plasmid DNA protects mice from spontaneous autoimmune diabetes and B7/CD28 costimulation circumvents that protection. *Clin. Immunol.* **99:**241–252.

13. Neugut, A. I., Ghatak, A. T., and Miller, R. L. (2001). Anaphylaxis in the United States: An investigation into its epidemiology. *Arch. Intern. Med.* **161:**15–21.

14. Centers for Disease Control and Prevention (1998). CDC Surveillance Summaries. *Morbidity Mortality Weekly Rep.* **47.**

15. Centers for Disease Control and Prevention. (1998). Forcasted State-Specific Estimates of Self-Reported Asthma Prevalence—United States. *Morbidity Mortality Weekly Rep.* **47:**1022–1025.

15a. Centers for Disease Control and Prevention (2000). Measuring childhood asthma prevalence before and after the 1997 redesign of the National Health Interview Survey—United States. Morbidity Mortality Weekly Rep. 49:908–911.

16. Horner, A. A., Van Uden, J. H., Zubeldia, J. M., Broide, D., and Raz, E. (2001). DNA-based immunotherapeutics for the treatment of allergic disease. *Immunol. Rev.* **179:**102–118.

17. Centers for Disease Control and Prevention (1999). Reasons reported by medicare beneficiaries for not receiving influenza and pneumococcal vaccinations—United States, 1996. *Morbidity Mortality Weekly Rep.* **48:**886–890.

18. Robinson, K. A., Baughman, W., Rothrock, G., Barrett, N. L., Pass, M., Lexau C., Damaske, B., Stefonek, K., Barnes, B., Patterson, J., Zell, E. R., Schuchat, A., and Whitney, C. G. (2001). Epidemiology of invasive *Streptococcus pneumoniae* infections in the United States, 1995–1998: Opportunities for prevention in the conjugate vaccine era. *J. Am. Med. Assoc.* **285:**1729–1735.

■ INDEX